Open Issues in Core Collapse Supernova Theory

PROCEEDINGS FROM THE INSTITUTE FOR NUCLEAR THEORY

Series Editors: Wick C. Haxton *(Univ. of Washington)*
Ernest M. Henley *(Univ. of Washington)*

Published

National Institute for Nuclear Theory,
University of Washington, Seattle
22–24 June 2004

Proceedings from the Institute for Nuclear Theory – Vol.14

Open Issues in Core Collapse Supernova Theory

editors

Anthony Mezzacappa
Oak Ridge National Laboratory, USA

George M. Fuller
University of California, San Diego, USA

World Scientific

NEW JERSEY • LONDON • SINGAPORE • BEIJING • SHANGHAI • HONG KONG • TAIPEI • CHENNAI

Published by

World Scientific Publishing Co. Pte. Ltd.

5 Toh Tuck Link, Singapore 596224

USA office: 27 Warren Street, Suite 401-402, Hackensack, NJ 07601

UK office: 57 Shelton Street, Covent Garden, London WC2H 9HE

British Library Cataloguing-in-Publication Data
A catalogue record for this book is available from the British Library.

Acknowledgment for Cover:
Volume rendering by Ross Toedte (ORNL) of the development of the "stationary accretion shock instability (SASI)" in a three-dimensional simulation performed by John Blondin (NCSU). The simulation was initiated from conditions before the onset of a core collapse supernova and was performed under the auspices of the DOE SciDAC TeraScale Supernova Initiative led by Tony Mezzacappa (ORNL). Entropy is rendered in this image, with cyan, yellow, and magenta correspond to increasing entropy, respectively. The gross asphericity of the postshock flow and the distortion of the supernova shock wave (the outer "surface" in the rendering), both induced by the SASI, are evident. In addition, strong, SASI-induced rotational flows can be inferred as well.

OPEN ISSUES IN CORE COLLAPSE SUPERNOVA THEORY
Proceedings from the Institute for Nuclear Theory — Vol. 14

ISBN 981-256-314-8

Printed in Singapore by World Scientific Printers (S) Pte Ltd

SERIES PREFACE

The National Institute for Nuclear Theory Series

The national Institute for Nuclear Theory (INT) was established by the US Department of Energy in March, 1990. The goals of the INT include:

1. Creating a productive research environment where visiting scientists can focus their energies and exchange ideas on key issues facing the field of nuclear physics, including those crucial to the success of existing and future experimental facilities;
2. Encouraging interdisciplinary research at the intersections of nuclear physics with related subfields, including particle physics, astrophysics, atomic physics, and condensed matter;
3. Furthering the development and advancement of physicists with recent Ph.D.s;
4. Contributing to scientific education through graduate student research, INT summer schools, undergraduate summer research programs, and graduate student participation in INT workshops and programs;
5. Strengthening international cooperation in physics research through exchanges and other interactions.

While the INT strives to achieve these goals in a variety of ways, its most important efforts are the three-month programs, workshops, and schools it sponsors. These typically attract 300 visitors to the INT each year. In order to make selected INT workshops and summer schools available to a wider audience, the INT and World Scientific established the series of books to which this volume belongs.

In January 2004 the INT and several partners, Argonne National Laboratory, Michigan State University, and the Joint Institute for Nuclear Astrophysics, began a new workshop series to explore physics questions connected with the proposed Rare Isotope Accelerator (RIA). This volume summarizes the inaugural workshop, which focused on the r-process, the mechanism by which many of the heavy elements were synthesized under explosive conditions typical of core-collapse supernovae and neutron star collisions. Organized by Yong-Zhong Qian, Ernst Rehm, Hendrik Schatz, and Friedrich-Karl Thielemann, the workshop sought to connect the astrophysics of this process with the new information on short-lived, neutron-rich isotopes that will become available with RIA.

This volume is the 13th in the INT series. Earlier series volumes include the proceedings of the 1991 and 1993 Uehling summer schools on Nucleon Resonances and Nucleon Structure and on Phenomenology and Lattice QCD; the 1994 INT workshop on Solar Modeling; the 1997 Jefferson Lab/INT workshop on Nucleon Resonance Physics; the tutorials of the spring 1997 INT program on Tunneling in Complex Systems; the 1998 and 1999 Caltech/INT workshops on Nuclear Physics with Effective Field Theory; the proceedings of the 1998 RHIC Winter Workshop on Quarkonium Production in Relativistic Nuclear Collisions; and the proceeding of Confinement III, of Exclusive and Semiexclusive Reactions at High Momentum, of Chiral Dynamics 2000, and of the Phenomenology of Large-N QCD, all collaborative efforts with Jefferson Laboratory. We intend to continue publishing those proceedings of INT workshops and schools which we judge to be of broad interest to the physics community.

Wick C. Haxton and Ernest Henley
Seattle, Washington, August, 2004

PREFACE

Efforts to uncover the explosion mechanism of core collapse supernovae and to understand all of their associated phenomena have been ongoing for nearly four decades. This is a problem of the utmost importance in astrophysics, generally, and nuclear and particle astrophysics, in particular. Core collapse supernovae are the dominant source of the elements between oxygen and iron and believed to be responsible for producing half the elements heavier than iron. As a result, they are arguably the single most important source of elements in the Universe. In addition, they are among the most energetic events in the Universe, exhibiting extremes in density, temperature, and composition that make them ideal laboratories (given detailed models that will be developed) for fundamental physics that is inaccessible in terrestrial experiment, they give birth to neutron stars and pulsars and stellar mass black holes, and play a significant role in the evolution of galaxies.

During the past four decades, observations have continually "raised the bar," uncovering an extraordinary array of core collapse supernovae, from "ordinary" core collapse supernovae to "hypernovae," originating from a range of progenitors, exhibiting a broad spectrum of explosion energies, and producing a variety of nucleosynthetic yields. We have, for example, SN1998bw, the first of the observed hyper-energetic core collapse supernovae, or "hypernovae," characterized by remarkably smooth spectra (the spectral features are smeared by the large kinetic energies in the explosion). SN1998bw originated from a 40 $M_\odot$ progenitor, had an explosion energy of 30 foe (1 foe = 1×10^{51} erg), and produced 0.5 $M_\odot$ of nickel. SN2002ap, on the other hand, another hypernova, originated from a $20 - 25$ $M_\odot$ progenitor, had an explosion energy of $4 - 10$ foe, and produced 0.07 $M_\odot$ of nickel, the amount produced in the "ordinary" core collapse supernova SN1987A. Moreover, the gamut of observed core collapse supernovae include "ordinary" core collapse supernovae with energies only a fraction of a foe originating from massive progenitors in the $30 - 40$ $M_\odot$ range. In addition, we now have confirmation of an association between core collapse supernovae and "long, soft" gamma ray bursts. Thus, we are at the moment in an observationally dominated field, and the observations have laid a complex mosaic for which detailed multidimensional models must now follow.

The neutrinos predicted from a core collapse supernova event were for the first time detected, in SN1987A, confirming the basic picture of collapse to nuclear density at the heart of all core collapse supernova models. The detection of these neutrinos represented a significant victory for the core collapse supernova modeling community. So, it is now observationally established that a tremendous sea of neutrinos of all flavors, comprising some 10% of the cold neutron star remnant's rest mass, is an inevitable result of the core collapse process. Discerning precisely what role the neutrinos play in the explosion mechanism has become more complicated now that we know from the laboratory that neutrinos have mass and can change their flavors. There may well be new physics from the neutrino/weak interaction sector that has not yet shown up, and perhaps never will show up, in the laboratory, but could nevertheless radically alter our picture of core collapse. Sterile neutrinos are one example of this, new neutrino couplings are another. Developing our understanding of core collapse supernovae to the point where it becomes a viable and unique "laboratory" for probing/constraining this kind of weak interaction physics is a lofty goal, but one that may be attainable.

In general, our theoretical understanding of core collapse supernovae remains rather limited. Two- and three-dimensional modeling of these events is in its infancy. Most of the modeling efforts over the past four decades have by necessity been constrained to spherical symmetry, with the first two-dimensional, albeit simplified, models appearing only during the last decade. Simulations to understand the complex interplay between the turbulent stellar core fluid flow, its magnetic fields, the neutrinos produced in and emanating from the proto-neutron star, the stellar core rotation, and the strong gravitational fields have yet to be performed. Only subsets of these fundamental ingredients have been included in the models thus far, with gross approximations sometimes invoked.

In addition, recent advances in nuclear structure theory and in the theory of neutrino weak interactions have better clarified the significant challenges that must be addressed to describe, among other things, heavy nuclei in the stellar core during collapse and their interactions with the neutrinos produced and the equally complex many-body system of nucleons, and potentially other particles, underpinning the equation of state during stellar core collapse, particularly at super-nuclear densities.

Fortunately, we have seen computing power increase steadily and dramatically, with "terascale" computing now a reality and "petascale" computing on the near horizon. And we have seen concomitant advances in the computational science infrastructure built around supercomputing platforms. In addition, the core collapse supernova problem is now being addressed by multidisciplinary teams that have been assembled, including astrophysicists, nuclear physicists, and particle physicists, among others, a necessity for a multiscale, multiphysics application such as ours. In light of these developments, the core collapse supernova problem can now be addressed in earnest, in all of its complexity, in a systematic and steadfast way.

Thus, the purpose of this workshop was to bring together a multidisciplinary group of scientists to identify the outstanding issues that remain in order to come to a complete understanding of these important astrophysical events. While the workshop included many active core collapse supernova modelers, experts in fundamental areas such as computational astrophysical hydrodynamics, magnetohydrodynamics, radiation transport, and radiation hydrodynamics, in numerical relativity, and in nuclear structure, hadron, and equation of state theory were included so that we could identify fundamental issues that need to be addressed in important sectors of a core collapse supernova model before reliable models can be developed.

The collection of papers presented here is the result of this workshop. With its focus on "open issues," our hope is it will serve as a guide for both advanced researchers and students for years to come.

In closing, we would like to thank the administrative staff at the Institute for Nuclear Theory, especially Nancy Tate and Linda Vilett, for their tireless efforts on our behalf, as part of the Supernovae and Gamma Ray Bursts Program (INT-04-2), the Open Issues in Understanding Core Collapse Supernovae Workshop, and this proceedings. We also would like to thank Wick Haxton and the Institute for Nuclear Theory for their hospitality and, generally, for making the program, workshop, and this volume possible.

Anthony Mezzacappa
Oak Ridge, August 2005

George M. Fuller
San Diego, August 2005

CONTENTS

Section 6: The Equation of State

Section 7: Nucleosynthesis and Light Curves

Section 1
Overview

OPEN ISSUES IN CORE COLLAPSE SUPERNOVA THEORY: AN OVERVIEW

A. MEZZACAPPA

Physics Division, Oak Ridge National Laboratory
Bldg. 6025, MS 6354, P.O. Box 2008
Oak Ridge, TN 37831-6354, USA
E-mail: mezzacappaa@ornl.gov

More than four decades have elapsed since modeling of the core collapse supernova mechanism began in earnest. To date, the mechanism remains illusive, at
least in detail, although significant progress has been made in understanding these
multi-scale, multi-physics events. One-, two-, and three-dimensional simulations
of or relevant to core collapse supernovae have shown that (1) neutrino transport,
(2) fluid instabilities, (3) rotation, and (4) magnetic fields, together with proper
treatments of (5) the sub- and super- nuclear density stellar core equation of state,
(6) the neutrino interactions, and (7) gravity are all important, quantitatively and
qualitatively. The importance of these "ingredients" applies to both the explosion mechanism and to phenomena directly associated with the mechanism, such
as neutron star kicks, neutrino and gravitational wave emission, and spectropo-
larization. Not surprisingly, current two- and three-dimensional models have yet
to include (1)–(4) with sufficient realism. One-dimensional spherically symmetric models have achieved a significant level of sophistication but, by definition,
cannot incorporate (2)–(4), except phenomenologically. Fully general relativistic
spherically symmetric simulations with Boltzmann neutrino transport do not yield
explosions, demonstrating that some combination of (2), (3), and (4) is required
to achieve this. Systematic layering of the dimensionality and the physics will
be needed to achieve a complete understanding of the supernova mechanism and
phenomenology. The past modeling efforts alluded to above have illuminated that
core collapse supernovae may be neutrino driven, MHD driven, or both, but uncertainties in the current models prevent us from being able to answer even this
most basic question. And it may be that more than one possibility is realized in
Nature.

1. The Core Collapse Supernova Paradigm

Stars more massive than ~ 10 $M_\odot$ evolve to an onion-like configuration,
with an iron core surrounded by successive layers of silicon, oxygen, carbon,
helium, and finally hydrogen. In addition to iron group nuclei, the core is
composed of electrons, positrons, photons, and a small fraction of protons

and neutrons. The pressure in the core, which supports it against the inward pull of gravity, is dominated at this stage by the electrons, and the balance between the electron pressure and gravity is only marginally stable. As a result of electron capture on the free protons and nuclei in the core and as a result of nuclear dissociation under the extreme densities and temperatures, electron and thermal pressure support are reduced, and the core becomes unstable and collapses.

The velocity of infalling matter in the core increases linearly with radius, characteristic of the "homologous" collapse expected of a fluid whose pressure is dominated by relativistic, degenerate electron pressure. The sound speed, on the other hand, decreases with density and, therefore, radius. Thus, with increasing radius, the infall velocity eventually exceeds the local sound speed; i.e., the infall becomes supersonic. Consequently, during infall the core splits into an inner homologously and subsonically infalling core and an outer supersonically infalling outer core.

Beginning with central densities $\sim 10^{9-10}$ g/cm^3, the collapse proceeds through nuclear matter densities ($\sim 1 - 3 \times 10^{14}$ g/cm^3). At this point the inner core undergoes a phase transition from a two-phase system of nucleons and nuclei to a one-phase system of bulk nuclear matter. One may view the inner core at this point as one giant nucleus. The pressure in the inner core increases dramatically as the result of Fermi effects and the repulsive nature of the nucleon–nucleon interaction potential at short distances. As a result of this dramatic increase in pressure, the inner core becomes incompressible and rebounds.

Any information about the rebounding inner core would be conveyed to the outer core via pressure waves that propagate radially outward at the speed of sound. When these waves reach the point at which the infall is supersonic—i.e., the "sonic point"—they are swept in as fast as they attempt to propagate outward. The net result: No information about the rebounding inner core reaches the infalling outer core, which in turn sets up a density, pressure, and velocity discontinuity in the flow—i.e., a shock wave. This shock wave will ultimately be responsible for propagating outward through the star, disrupting the star in a core collapse supernova. Schematically, the shock wave is launched and energized by the rebounding inner core piston.

If the shock were to propagate outward without stalling, we would have what has been called a "prompt explosion." All of the realistic models completed to date suggest that this does not occur. Because the shock loses energy in dissociating the iron nuclei that pass through it as it propagates outward, the shock is enervated. The shock loses additional energy in the form of electron neutrinos. The copious production of electron neutrinos occurs when the core electrons capture on the newly dissociation-liberated protons. These neutrinos are initially trapped but escape when the shock moves out beyond the electron neutrinosphere. This gives rise to the "electron neutrino burst" in a core collapse supernova, which is the first of three major phases of the three-flavor neutrino emission during these events. As a result of these two enervating mechanisms, the shock stalls in the iron core. *How the shock is reenergized in a "delayed shock mechanism" is currently the central question in core collapse supernova theory.*

At the time the shock stalls, the core configuration is composed of a central radiating object: the proto-neutron star, which will go on to form a neutron star or black hole. The proto-neutron star has a relatively cold inner part, composed of unshocked bulk nuclear matter, together with a hot "mantle" of nuclear matter that has been shocked but not expelled. The ultimate source of energy in a core collapse supernova is the $\sim 10^{53}$ erg of gravitational binding energy associated with the formation of the neutron star. This gravitational binding energy is released after core bounce over ~ 10 seconds in the form of a three-flavor neutrino "pulse." This marks the second phase of the neutrino emission from a core collapse supernova. Electron neutrinos are produced during stellar core collapse by electron capture on protons and nuclei, but after bounce, in the hot proto-neutron star mantle, all three flavors of neutrinos and their antineutrinos are produced and are emitted as the mantle cools and contracts during its "Kelvin-Helmholtz" cooling phase. The neutrinos are emitted from their respective neutrinospheres. The neutrinospheres are defined in a way similar to the way the photosphere of the Sun is defined. They are the surfaces of last scattering for each neutrino energy and flavor. Equivalently, they are at radii at which the respective neutrino depths are 2/3. The neutrino luminosities during this phase are maintained at their average values $\sim 10^{52}$ erg/s by mass accretion onto the proto-neutron star (the kinetic energy of infall is converted into thermal energy when the material hits the proto-neutron star surface). After explosion is initiated, the accretion luminosity decreases dramatically, and the neutrino pulse enters its third and final stage: the exponential decay of the neutrino luminosities characteristic of neutron star formation and cooling.

2. Neutrino Transport

The stalled supernova shock is thought to be revived, at least in part, by the charged-current absorption of electron neutrinos and antineutrinos that emerge from the proto-neutron star, a fraction of which are absorbed by protons and neutrons behind the shock. This is known as the "delayed shock" or "neutrino-heating" mechanism, originally proposed by Wilson and Bethe[1,2]. While the total energy emitted in neutrinos is two orders of magnitude greater than what is required for the generation of an $\sim 10^{51}$ erg explosion, deciphering the precise role of this neutrino heating in the supernova mechanism is, as we will discuss, difficult.

Between the neutrinosphere and the shock, the material both heats and cools by electron neutrino and antineutrino emission and absorption. The neutrino heating and cooling have different radial profiles; consequently, this region splits into a net cooling region and a net heating region, separated by a "gain" radius at which heating and cooling balance. We refer to the region between the gain radius and the shock as the "gain region."

The neutrino heating in the gain region can be written as

$$\dot{\epsilon} = \frac{X_n}{\lambda_0^a} \frac{L_{\nu_e}}{4\pi r^2} < E_{\nu_e}^2 >< \frac{1}{F} > + \frac{X_p}{\lambda_0^a} \frac{L_{\bar{\nu}_e}}{4\pi r^2} < E_{\bar{\nu}_e}^2 >< \frac{1}{\bar{F}} > . \qquad (1)$$

The first (second) term corresponds to the absorption of electron neutrinos (antineutrinos). It depends linearly on the neutrino luminosity and "inverse flux factor," which is a measure of the isotropy of the neutrino distribution, and quadratically on the neutrino spectrum.

In addition to the dependence on the three key neutrino quantities in the heating rate above, the revival of the stalled supernova shock depends on a complex interplay of neutrino heating, mass accretion through the shock, and mass accretion through the gain radius[3]. It is mass accretion through the shock and gain radii that determines the amount of mass in the gain region, the former being a source of mass in the gain region and the latter being a sink. Moreover, the mass accretion through the gain radius serves to both sustain the neutrino luminosities and to undermine the pressure in the gain region, simultaneously—i.e., it serves a supporting and a detrimental role.

All three quantities in the neutrino heating rate must be computed accurately[3-8], which requires that we solve the neutrino Boltzmann neutrino transport equations, but given the quadratic dependence on the neutrino spectrum it is imperative that the spectrum be computed accurately. This requires the use of multi-neutrino energy (a.k.a. multi-frequency or "multi-group") transport.

The most compelling argument for the use of multifrequency neutrino transport can be made, of course, by simply taking stock of the results from all past core collapse supernova simulations. *With the exception of Wilson's spherically symmetric models that invoke the doubly diffusive neutron finger instability in the proto-neutron star to boost the neutrino luminosities, no simulation to date performed with multifrequency neutrino transport has yielded an explosion. This is a sobering fact*[7,9–15]. [a] And this list now includes both one- and two-dimensional simulations. Moreover, without neutron fingers, whose existence is a matter of current debate, Wilson does not obtain explosions[17].

Generally, we anticipate that the magnetic fields in the collapsed stellar core may be amplified through a variety of mechanisms, as we will discuss, to become important in the supernova mechanism. In light of the above mentioned difficulties associated with generating core collapse supernovae via neutrino heating when realistic multifrequency neutrino transport is used and in light of the expectation that magnetic fields may play a role in the supernova mechanism, a number of investigations have been performed and have concluded that, *if the magnetic fields are in fact sufficiently amplified, they may aid the neutrinos in powering the explosion or even replace them as the central vehicle whereby energy in the collapse is converted into outflow kinetic energy*[18–21]. In the case where explosion is powered by neutrinos, the neutrinos serve as the conduit between gravitational binding energy and internal energy behind the shock. In the case where explosion is driven by magnetic fields, the magnetic fields serve as the conduit between rotational energy in the collapsing core and the kinetic energy of outflow. The latter possibility would correspond to a "paradigm shift." Hence, even the basic neutrino-heating-mediated delayed-shock paradigm has been called into question. We might already anticipate that both the neutrinos and the magnetic fields will act in concert to produce these explosions. But an answer to these basic questions will require three-dimensional neutrino radiation magnetohydrodynamics simulations, which have not yet been performed. Even guidance on this issue from two-dimensional simulations would be welcome.

After four decades of core collapse supernova research, detailed spher-

---

[a] A recent marginal exception to this trend was found by[16], where a *weak* explosion of a less-massive 11 $M_\odot$ progenitor was reported in the context of a two-dimensional model.

ically symmetric simulations that now include state of the art neutrino interactions, an industry standard equation of state, and multiangle, multifrequency, Boltzmann neutrino transport in full general relativity have finally been performed[22]. It is important to remember that we are working in phase space. Therefore, simulations that assume spherical symmetry and require a solution of the neutrino Boltzmann equation are actually simulations in three dimensions (radius or mass, direction cosine, and energy), not one dimension (radius). When cast in this light, it is not surprising that such three-dimensional multiphysics simulations have taken four decades to complete. This is also an omen of what will be required to perform such simulations in three spatial dimensions (six phase space dimensions). Despite this seemingly overwhelming requirement, we must proceed in a systematic and steadfast manner. Efforts by several groups are now underway to develop Boltzmann neutrino transport for the two- and three-dimensional cases[23–26].

As for the physics "neglected" in these spherically symmetric models, there are two possibilities: (1) Physics included in the models could have been treated more accurately. (2) Physics essential to the explosion mechanism was not included in any approximation. Ongoing efforts to improve the neutrino opacities and high-density equation of state fall under category (1) and are examples of efforts to develop supernova models quantitatively. We will discuss this work in more detail later, but it is unlikely that advances in this area will change the outcome of current supernova models *qualitatively*–that is, lead to explosion in models that currently do not explode. Nonetheless, our ultimate goal is to produce explosions and to predict *quantitatively* all of the supernova-associated observables. It has been demonstrated that changes in the "microphysics" (weak interaction rates, sub- and super-nuclear density equation of state) lead to changes in the models that will affect these observables. Ongoing efforts, which we now discuss, to include fluid instabilities, rotation, and magnetic fields in core collapse supernova models fall under category (2). We will see how they have no doubt fundamentally altered supernova theory, but we still do not have in hand the long-sought-after mechanism by which explosions are "guaranteed."

Before proceeding, it is important to make the following point: At the moment, supernova modelers have identified seven major components of a detailed supernova model: neutrino transport, fluid instabilities, rotation, magnetic fields, gravity, the neutrino weak interactions, and the sub- and super-nuclear equation of state. *Two- and three-dimensional models developed to date have included only a subset of these known important components.* Thus, multidimensional supernova modeling is in its infancy. Over the next five years, multidimensional supernova models will undergo a dramatic change in realism. We will be provided with far more realistic two- and three-dimensional models that will give us far better guidance on the roles of fluid instabilities, rotation, and magnetic fields, and their coupling, in generating and defining supernova explosions. Of course, quantitatively accurate models will require fully general relativistic simulations. These will not be completed in the near term. They will require a decade to develop. Nonetheless, efforts to include corrections for general relativity in multidimensional supernova models are underway[27].

3. Fluid Instabilities

The potential role of fluid instabilities in the post-stellar-core-bounce dynamics was first articulated decades ago and explored phenomenologically for some time in the context of spherically symmetric models. Fortunately, two- and three-dimensional models that have emerged over the past decade, albeit incomplete, have fundamentally changed our ability to model these instabilities and to assess their role in the supernova mechanism.

It is important to classify the regions in which the various instabilities may occur, and their fundamental nature. Fluid instabilities in core collapse supernovae can be put in one of three fundamental categories: (1) convection, (2) doubly diffusive instabilities, and (3) shock wave instabilities. Convection includes both Ledoux convection, which results from both entropy and composition gradients (in this case lepton fraction gradients) in the stellar core, and Schwarzchild convection, which results from entropy gradients alone. Doubly diffusive instabilities result from a competition between the transport of entropy and leptons in the stellar core and occur in regions in which there are crossed or competing gradients in entropy and composition, with one being stabilizing and the other destabilizing.

Instabilities in the stellar core, whether we are discussing convection or doubly diffusive instabilities, occur throughout the region below the supernova shock wave. If they occur beneath the neutrinosphere, we refer to them as proto-neutron star (PNS) instabilities. PNS instabilities may boost the neutrino luminosities emerging from the PNS and thereby boost the neutrino energy deposition beneath the stalled shock. Essentially, these instabilities may lead to the more efficient transport of neutrinos by advection, rather than diffusion, outward in radius to the neutrinosphere.

Directly beneath the stalled shock, between the gain radius and the shock, where the material is undergoing net neutrino heating, an instability develops as the result of the heating-associated entropy gradient. This instability is referred to as neutrino-driven convection. As has been pointed out by a number of authors[15,28–30], *neutrino-driven convection fundamentally alters the scenario under which neutrino shock reheating occurs.* Neutrino-driven convection allows for both an explosion, in which the shock wave and material behind it begin to move radially outward, and continued accretion, which maintains the neutrino luminosities sufficiently high so as to sustain the heating. *In spherical symmetry, explosion and accretion are mutually exclusive.* Neutrino-driven convection also acts hydrodynamically on the shock, pushing it out to larger radii and shallower points in the gravitational potential, thereby facilitating explosion, and can boost the neutrino heating efficiency by moving heated matter upward towards the shock—in so doing, the heated matter expands, its temperature drops, and its neutrino cooling losses are reduced, thereby keeping more of the neutrino-deposited energy in the gain region.

Whereas the vigor and extent of PNS instabilities and, consequently, their impact on the supernova mechanism is currently a matter of debate, a debate that will ultimately require two- and three-dimensional radiation magnetohydrodynamics simulations to resolve, the development of neutrino-driven convection is a characteristic feature of all two- and three-dimensional supernova models that have been performed to date, although it does not *guarantee* explosions.

A fundamentally new hydrodynamic instability has been discovered recently[31] that may play a significant role in generating core collapse supernovae and some of their key observables (e.g., neutron star kicks, spectropolarimetry, pulsar spin up). The postbounce stellar core flow is best characterized as an accretion flow through a quasi-stationary shock. It has been shown in two- and three-dimensional hydrodynamics studies constructed to reflect the conditions during the postbounce shock reheating epoch that *nonspherical perturbations of the accretion shock lead to the development of a "stationary accretion shock instability (SASI)."* Recent studies[32] confirm the existence of the SASI instability in two-dimensional models that include radial-ray neutrino transport. The potential ramifications of the SASI for the supernova mechanism and phenomenology were first elaborated by Blondin, Mezzacappa, and DeMarino[31]. (1) The SASI may aid in generating the supernova explosion itself. Simulations indicate that, much like the neutrinos, the SASI may act as a conduit between gravitational binding energy and the kinetic energy of outflow. In addition to aiding the explosion *directly*, the SASI may aid the explosion *indirectly* by aiding the neutrino heating[32]. (2) The SASI may also define the gross asymmetry of the explosion. Two-dimensional simulations of the SASI lead to bipolar explosions and to a self-similarity in the flow at late times with an aspect ratio consistent with the supernova spectropolarimetry data[33]. Three-dimensional simulations yield a far more complex outcome[34]. Moreover, these simulations offer a proof of principle that *the SASI is also capable of imparting significant angular momentum to the proto-neutron star, even beginning with spherically symmetric initial conditions*[34].

The discovery of the SASI must fundamentally alter the way we think about core collapse supernovae. We must certainly retire many of our past prejudices regarding (1) the role of rotation in defining the nature of the explosion and (2) the evolution of the angular momentum distribution in the stellar core during collapse and after bounce given the initial angular momentum distribution of the progenitor model. For example, although somewhat restricted, two-dimensional simulations indicate core rotation may not be required to generate bipolar-like outflows[31]. And in three dimensions, the SASI has been shown to be capable of spinning up the proto-neutron star significantly even when we begin with spherically symmetric initial conditions. The discovery of the SASI must also force us to reexamine how we interpret core collapse supernova observations. For example, the connection one might make between supernova spectropolarimetry data and stellar core rotation and magnetic fields may be complicated by the SASI. Generally speaking, the connections between supernova observables and theory are likely far more complex than we had ever assumed in the past.

4. Rotation

Neutrino transport in core collapse supernovae has been studied extensively over the last four decades, culminating recently in fully general relativistic simulations with Boltzmann neutrino transport that have essentially closed the book on spherically symmetric models (at least in the absence of neutrino mixing)[13]. And a number of two-dimensional simulations have been performed over the past decade to explore the dynamics of fluid instabilities in the stellar core after bounce and their impact on the supernova mechanism, culminating in simulations that include sophisticated radial-ray neutrino transport that captures a significant amount of realism in the models[15]. In contrast, very few simulations have been performed to date that include rotation, and no contemporary, sufficiently realistic simulation has been performed that includes magnetic fields.

The work of several groups [15,21,30,35] has shown that rotation can significantly influence stellar core collapse and the postbounce dynamics in a variety of ways. (1) Centrifugal forces will slow collapse along the equator and, if sufficiently large, can lead to a low-density bounce. (2) Gravitational binding energy will be channeled differently during core collapse, relative to the spherically symmetric case, partitioned between the internal energy of the matter and neutrinos and the rotational energy of the core. (3) The neutrinospheres will be distorted, and the neutrino luminosities and rms energies may be noticeably changed. (4) The preshock accretion ram pressure along the rotation axis and the equator may differ significantly. And rotation (5) will alter the development of fluid instabilities below the shock, (6) may provide through viscous dissipation a new source of internal energy in the postshock flow that may augment the energy supplied by neutrino heating, and (7) may have a dramatic impact on the growth of magnetic fields in the stellar core after bounce.

5. Magnetic Fields

Again, the work of several groups [e.g., see [19,21,35–37]] has shown that, much like rotation, magnetic fields may influence stellar core collapse and the supernova mechanism in a number of ways: (1) The development of significant magnetic pressure may have an impact on stellar core collapse and the postbounce flow. Magnetic fields may (2) alter the development of fluid instabilities in the stellar core after bounce, (3) provide additional channels for the generation of internal energy through viscous dissipation, (4) alter the weak interactions in the stellar core, and (5) provide a conduit through which rotational energy in the collapsed core may be channeled into the outflows of explosion.

Arguably, the fundamental question is whether the magnetic fields will organize into large-scale configurations that will help drive and collimate outflows from the stellar core (point 5 above). The pioneering simulations of LeBlanc and Wilson [18] and Symbalisty [19] were the first to explore the evolution of stellar core magnetic fields during core collapse and their impact on the explosion mechanism. These simulations exhibited the development of a magnetic bubble deep in the core owing to the dramatic increase in magnetic pressure close to the rotation axis as core field lines are dragged inward and compressed. This magnetic bubble led to buoyant, bipolar outflows that culminated in bipolar explosions (the LeBlanc–Wilson (LW) jet).

More recently, this idea has been extended[20]. Owing to stellar core differential rotation, an initially poloidal field threading the stellar core could be wound up into a potentially significant toroidal field. Moreover, the field may be wound up quickly before it has a chance to expand vertically, later expanding in a spring-like fashion along the rotation axis. In so doing, the field would evolve into an open helix. In this way, material in the core could be driven outward along the rotation axis in a "spring and fling" manner, the latter arising because the material is accelerated along the rotating field lines by centrifugal forces. The so called "hoop stresses" would serve to collimate the flow.

The most advanced simulations of stellar core collapse to include magnetic fields were performed more than twenty years ago. Symbalisty[19] concluded that inordinately large rotation and magnetic fields strengths were required for an explosion to develop through the original Leblanc–Wilson mechanism. The magnetic fields in the stellar core can be amplified in the way suggested by Leblanc and Wilson or through wrapping, as described above, but the growth of field strength through other mechanisms must be considered.

Magnetic fields during stellar core collapse may be amplified in one of four basic ways: (1) Through collapse, as in the Leblanc–Wilson scenario. (2) By wrapping. (3) Through a "dynamo" effect combining the action of fluid instabilities, such as convection, and rotation. (4) Through shear (the magnetorotational instability discussed below). In particular, at the time of the early simulations by Leblanc, Wilson, and Symbalisty, the magnetorotational instability (MRI) was not yet discovered. The MRI was first discovered in the context of the differential rotation in accretion disks[38] and is believed to be the mechanism whereby angular momentum is transported in such disks. But it may also operate in a differentially rotating core after stellar core bounce. Field amplification through the MRI is fundamentally different than through compression and wrapping. In the case of the MRI, the field amplification occurs through the *stretching* of the field lines in a strongly differentially rotating core.

Akiyama et al.[21] were the first to propose that the MRI could be important in the core collapse supernova context. They argued that, with sufficient differential rotation, the MRI could amplify the magnetic field strengths in the stellar core *exponentially quickly*, rather than linearly, and lead to magnetohydrodynamic "luminosities" $\sim 10^{52}$ erg/s, rivaling the

14

neutrino luminosities. It is important to note, in this scenario, the magnetic fields act primarily as a conduit, channeling rotational energy into outflows[20]. They do not power the outflows directly. In short, magnetic fields may play an important role in supernova dynamics. Even initially small magnetic fields in the stellar core may be amplified quickly after core bounce, through a variety of mechanisms, to participate in the supernova dynamics. However, determining their precise role will require realistic two- and three-dimensional simulations, which have not yet been performed.

6. Neutrino Interactions

Improvements in modeling the macrophysics of stellar core radiation magnetohydrodynamics must be matched by improvements in modeling the neutrino weak interactions in the stellar core and the stellar core equation of state.

First among the weak interactions of concern is electron capture on nuclei. (For an extensive discussion of the physics of electron capture on nuclei in this context, the reader is referred to[39–44].) Electron capture during stellar core collapse determines the extent to which the core is "deleptonized" during collapse, which in turn sets the mean electron fraction in the core and the size of the inner homologous core at bounce[45]. The size of the inner core determines the mass at which the shock forms and the energy imparted to it—in short, the initial conditions for the postbounce core evolution, shock revival, and consequent supernova explosion.

Electron capture on nuclei is dominated by Gamow–Teller transitions. Until recently, the "standard" weak interactions used in supernova simulations included electron capture on nuclei as described by an "independent particle model"[40,41,46]. In this model, nucleons in the nucleus are assumed to be independent, and electron capture is Fermi blocked for nuclei with $N > 40$[41]. The net result: In this approximation the electron capture during core collapse is dominated by capture on free protons, not nuclei. In reality, the opposite is the case.

(1) The nucleus is an interacting many-body system. Correlations owing to the residual interaction between nucleons excite nucleons in the nucleus and unblock these transitions (this is known as "configuration mixing")[47]. (Nucleons are assumed to move in a mean field. The residual interaction corrects the mean field to better represent the complete interaction experienced by each nucleon.) Moreover, as the core becomes "neutronized" during stellar core collapse via electron capture on protons and nuclei, the

nuclei increase in mass (the nuclear size results from a competition between Coulomb and surface effects in the nucleus, the latter of which favors larger nuclei[48]). In fact, electron capture rates will be needed for nuclei well above mass 100[49]. This will present a significant technical challenge. Nuclear structure theorists will have to perform these rate calculations, taking into account configuration mixing, in large Hilbert spaces. (2) Current nuclear structure calculations are typically performed at zero temperature, focused on ground state properties, not at the finite temperatures found in the stellar core during collapse, which are sufficient to excite nuclei and unblock the Gamow–Teller transitions[41,50]. As we will describe below, "hybrid" models have been constructed to contend with both finite temperatures and configuration mixing. (3) Once the neutrinos become trapped in the core during stellar core collapse, the inverse reaction of electron-neutrino capture on nuclei becomes significant, and, eventually, electron capture and its inverse reaction reach an equilibrium, at which point the electron fraction in the core is set by the nuclear partition functions, which must therefore be determined self-consistently (e.g., see[40,44]).

To advance the state of the art in nuclear structure theory to satisfy all of the above requirements will take considerable effort, and advances must be made through the implementation of models of intermediate sophistication.

Another compelling recent development in the theory of weak interactions in core collapse supernovae was the discovery that nucleon–nucleon bremsstrahlung and neutral-current neutrino–antineutrino annihilation are in fact the most important source of muon and tau neutrinos and antineutrinos in the proto-neutron star mantle after core bounce[51,52]. Prior to this discovery, the sole mechanism whereby neutrinos and antineutrinos of these two flavors were produced in the models was through the annihilation of electrons and positrons[53]. The latter has now been shown to be insignificant in the production of these neutrinos relative to the contributions from bremsstrahlung and neutral-current neutrino–antineutrino annihilation.

In addition to the realization that new neutrino interaction channels may be important in core collapse supernovae, such as in the cases mentioned above, significant improvements have been made recently in the computation of the weak interaction rates traditionally used in supernova models. The inclusion of nucleon recoil, Fermion blocking (a form of correlation), and degeneracy in the charged- and neutral-current interactions of neutrinos on free nucleons is one example[54–57]. These improvements led to significant changes in the neutrino emissivities and opacities (factors of 2– 3) and the realization that energy exchange in neutrino–nucleon scattering in the pre- and post-bounce core cannot be ignored[58].

Generally, the naive picture of neutrinos interacting on free nucleons and nuclei is now being replaced by the more accurate picture of neutrinos interacting with correlated nuclei and nucleons at sub- and super-nuclear density. Calculations of neutrino interactions on nuclei during stellar core collapse that include correlations among nuclei[59], on the extended correlated structures present in the nuclear "pasta" phase during the transition from nuclei to nuclear matter[60], and on the strongly interacting correlated nucleons in the proto-neutron star[54-57] have either been completed or are well underway (for a review, the reader is referred to[61,62]). Nonetheless, more remains to be done, and the calculations will be increasingly challenging.

However, it is important to keep in mind that, although the overall neutrino emissivities and opacities can be dramatically altered when either new channels are included or improved rates are calculated, *the impact of new and/or improved neutrino emissivities and opacities on the proto-neutron star evolution and supernova mechanism can only be determined when they are included in the postbounce neutrino radiation hydrodynamics models with all of the concommitant feedbacks.* We cannot know *a priori* whether a large change in the weak interaction rates will lead to a large or small change in the models. Both scenarios are possible.

Finally, it is important to note that any of the above weak interaction rates that involve a nuclear force model—e.g., neutrino scattering on interacting nucleons at high densities in the stellar core before and after bounce—must be computed self-consistently with the nuclear equation of state[55], in which a nuclear force model is assumed.

7. Equation of State

Foundational to any determination of the equation of state of matter in the stellar core during core collapse and after core bounce, at both sub- and super-nuclear densities, is the determination of the energy per particle (for more detail on what we are about to discuss, the reader is referred to[63,64]). Given this quantity, the equation of state is readily derived from standard thermodynamics. However, our ability to compute this quantity in what is generally a strongly interacting environment is, of course, limited. Nonetheless, at present there are several fundamentally different approaches that have been adopted: (1) semi-empirical compressible liquid drop models, (2) phenomenological models, and (3) realistic models. Moreover, there are both nonrelativistic and relativistic versions for models in categories (2) and (3), in which a nuclear force model is assumed in the determination of V_{NN}, the nucleon–nucleon interaction potential.

In the compressible liquid drop model, the energy per particle is assumed to be expressible as a polynomial in the symmetry (neutron to proton ratio) and compressibility (density relative to the saturation density of symmetric nuclear matter) of the matter. The starting point are the empirically determined values of the symmetry energy and compressibility for symmetric nuclear matter at saturation. The compressible liquid drop model is, of course, appropriate only for sub-nuclear densities.

In the "phenomenological models, a parameterized (e.g., Skyrme) nucleon–nucleon interaction potential is assumed, where the parameters are fit to reproduce observables such as ground state properties of nuclei, symmetric nuclear matter at saturation, neutron star masses and radii, etc. There are a total of 10–15 adjustable parameters in such a parameterization, fit to hundreds of data points. However, there are, for example, 90 different Skyrme parameterizations of the nucleon–nucleon interaction potential, although only about one third of them are able to reproduce neutron star properties. Phenomenological models are capable of describing both sub- and super-nuclear-density matter.

In the "realistic" models, a nucleon–nucleon interaction potential with 40–60 adjustable parameters fit to several thousand data points from free nucleon–nucleon scattering and properties of the deuteron is assumed. However, when applied to dense, interacting matter the potentials must obviously be "renormalized" using Brueckner–Hartree–Fock and Dirac–Brueckner–Hartree–Fock techniques in the nonrelativistic and relativistic cases, respectively. Moreover, realistic potentials are computed only for symmetric nuclear matter and pure neutron matter. Interpolation is then required to describe the matter present in the cores of supernovae and in neutron stars, which is neither symmetric nor purely neutronic. While "realistic" models can be used to describe matter at super-nuclear densities, they cannot yet be used at sub-nuclear densities and, therefore, in a self-consistent description of matter in the stellar core at both sub- and super-nuclear densities during stellar core collapse.

Despite the uncertainties in our knowledge of the equation of state of matter in stellar cores at both sub- and super-nuclear densities, the availability of such a diverse set of models allows us to explore the sensitivity of the supernova mechanism and quantitative predictions of supernova models to variations in this physics.

Two additional points should be made: (1) At 2–3 times the saturation density of symmetric nuclear matter, heavy baryons (e.g., hyperons) and meson condensates (e.g., pion and kaon condensates) and possibly deconfined quarks may be present. Our discussions above, on the other hand, focus soley on different approaches at calculating the nucleon–nucleon interaction potential. The interactions between nucleons and the heavy baryons, and between the heavy baryons themselves, are not as well known[65], an uncertainty which also casts uncertainty on the densities at which we might expect these more exotic constituents to be present in our stellar cores. For a path forward, the reader is referred to Barnes' contribution in this volume [66]. (2) During stellar core collapse at both sub- and super-nuclear densities, the equation of state must be developed for temperatures that reach tens of MeV. To date, only a few equations of state have been developed for finite temperature[67–70].

In all simulations of stellar core collapse thus far performed, a mean-nucleus approximation has been used in the determination of the sub-nuclear density equation of state. In this approximation, the thermodynamic state of the core is computed assuming the core is composed of a single representative nucleus, as opposed to the ensemble of nuclei actually present. While the thermodynamic state at sub-nuclear densities is well approximated in such a mean nucleus approach, weak interactions on the nuclei actually present in the core will vary significantly from nucleus to nucleus and will, consequently, not be well approximated by computing the weak interactions on a mean nucleus. Therefore, hybrid computations using an NSE mean-nucleus equation of state in combination with an NSE network must be performed. The progress detailed above in the implementation of improved electron capture rates on nuclei during stellar core collapse involved such a hybrid implementation[71,72].

In addition to the use of a mean nucleus rather than an ensemble of nuclei, *we also have to consider what restrictions we have placed in the past on the nature of these nuclei.* In particular, in all of the industry standard equations of state used in supernova models performed to date (and discussed above), the nuclei are assumed to be spherical in shape and organized in a b.c.c. ("body-centered, cubic") lattice during stellar core collapse. In the critical transition region between inhomogeneous matter composed of nuclei and nucleons and homogeneous nuclear matter—the so called "pasta phase"—an ensemble of complex shapes are anticipated (spheres, tubes, sheets, ...)[73]. Clearly, the assumption that nuclei are spherical is not compatible with the fact that we expect such complex shapes.

Three-dimensional models that abandon the restriction to spherical nuclei have been developed for $T = 0$[74]. Efforts are now underway to develop a three-dimensional equation of state at finite temperature[75], for both homogeneous (super-nuclear-density) and inhomogeneous (sub-nuclear-density) matter. Of course, this development must be accompanied by efforts to calculate the interactions of neutrinos on the complex structures that will be admitted by such an equation of state[60].

8. Gravity

One should expect the gravitational fields around proto-neutron stars to deviate significantly from their Newtonian values and lead to significant changes in the collapse and postbounce hydrodynamics and neutrino transport. Detailed comparisons of Newtonian and fully general relativistic collapse have been performed in spherical symmetry[12,13] and confirm this expectation.

General relativistic effects can be expected to substantially modify the *hydrodynamics* of the core at high densities. An extreme example of this, of course, is the possibility that in the general relativistic limit one can have continued collapse and the formation of an event horizon. The *neutrino transport* will also be modified by general relativity, directly through gravitational redshift and aberration, and indirectly through its strong coupling to the general relativistic modified hydrodynamics.

It is important to note the complex feedbacks active during stellar core collapse and the postbounce evolution that are made evident through these detailed comparisons. It would be simplistic to assume that general relativity, in increasing the gravitational potential and, hence, the compactness of the postbounce stellar core configuration and in leading to the gravitational redshift of neutrinos, would lead to a postbounce state far less conducive to explosion. While the compactness is increased by including general relativistic gravity in the models, so too are the neutrino luminosities and rms energies. It remains to be seen, in the case of a complete model with general relativity and explosion, what final role general relativity will play[13].

9. Neutrino Mass and Mixing

It is now an experimental fact that neutrinos have mass and, therefore, mix in flavor. Observations of Solar and atmospheric neutrinos, and experiments at LSND, indicate there may be as many as three independent values of the difference in the square of the neutrino masses (δm^2), and four mixing angles, which would require three active and at least one sterile neutrino. Although we await confirmation of the LSND findings, the data already strongly suggest that neutrino mixing should be included in core collapse supernova models. Neutrino mixing may significantly affect one or more of the following: the supernova mechanism, supernova nucleosynthesis, and terrestrial supernova neutrino detection.

As we discussed earlier, all three active neutrino flavors are involved in core collapse supernova dynamics. Electron, muon, and tau neutrinos and their antineutrinos are produced primarily through thermal emission, nucleon–nucleon bremsstrahlung, and neutral-current neutrino–neutrino annihilation in the hot mantle of the proto-neutron star after core bounce. Owing to the lack of charged-current interactions among the muon and tau neutrinos and antineutrinos, electron neutrino and antineutrinos decouple at lower densities given their larger total interaction cross sections. Decoupling at lower densities and, consequently, lower temperatures results in softer relative spectra for the electron flavor neutrinos[76]. Herein lies the essential relevance of neutrino mixing to the supernova mechanism: If flavor conversion between electron flavor and muon/tau flavor neutrinos were to occur below the supernova shock wave in the neutrino heating epoch after stellar core bounce, the neutrino heating behind the shock, which is mediated predominantly by the charged-current absorption of electron neutrinos and antineutrinos, could be significantly increased[77,78]. The softer electron neutrino flavor spectra would be replaced by the harder muon/tau neutrino flavor spectra in this region.

The argument to consider neutrino mixing in the core collapse supernova context, particularly with an eye toward the explosion mechanism, has been made even more compelling recently with the discovery that *neutrino mixing may occur deep in the stellar core after bounce at small values of δm^2*. This mixing may arise as a result of the neutrino background in and above the proto-neutron star, which would be fundamentally different than MSW mixing induced by the matter background, or vacuum mixing [79,80]. Neutral-current neutrino–neutrino forward scattering increases the neutrino effective mass, much as charged-current electron-neutrino scattering increases the electron-neutrino effective mass in the MSW case. The net result may be near maximal mixing of neutrino flavors in the environment of the proto-neutron star after bounce[79,80], with obvious ramifications for the supernova mechanism, nucleosynthesis, and neutrino signatures.

While the experimental evidence for neutrino mixing is now clear, and while a number of past exploratory studies have elucidated some of the possible ramifications neutrino mixing may have for core collapse supernova dynamics, the precise impact of such flavor transformation remains to be determined. Neutrino mixing is a coherent, quantum mechanical phenomenon, unlike the incoherent collisional phenomena included in the Boltzmann kinetic equations discussed earlier. A more complete (quantum kinetic) treatment of neutrino transport in stellar cores beyond (classical) Boltzmann transport will be needed if we are to accurately and fully explore the impact of neutrino mixing on core collapse supernova dynamics.

10. The Future

Three-dimensional, general relativistic, radiation magnetohydrodynamics models of core collapse supernovae must be developed before we are confident we truly understand the supernova mechanism and are able to accurately predict all of the associated supernova observables. This will take a decade of systematic research. We currently do not have models of this level of sophistication even in two spatial dimensions, let alone three, and three-dimensional models with multi-angle, multi-frequency neutrino transport will likely require "petascale" supercomputers, which will not be available for at least several years.

Nonetheless, there is much work to be done on the road forward. Until now, we had no two-dimensional models that implemented two-dimensional multifrequency neutrino transport and all of the relevant terms in the transport equations—i.e., realistic two-dimensional transport. There still are no simulations, two- or three-dimensional, that include magnetic fields and realistic neutrino transport. And all of the two- and three-dimensional models performed to date are largely confined to the Newtonian limit. Moreover, the precise implications of neutrino mixing are essentially unexplored in any supernova model.

Turning to the input nuclear and weak interaction physics, nuclear structure theory and the theory of nuclear matter in supernovae have certainly been challenged by the requirements of realistic supernova models, as we have discussed. Responding to these challenges will require a concomitant long-term effort on the part of the nuclear physics community to advance nuclear structure models for heavy nuclei and to pin down the high-density equation of state with sufficient accuracy, where what constitutes "sufficient" is yet to be fully determined.

Despite the long road ahead, this is a time for optimism, not pessimism—much progress has been made, as the following chapters will show.

Acknowledgments

This work was supported at the Oak Ridge National Laboratory, which is managed by UT-Battelle, LLC for the DOE under contract DE-AC05-00OR22725 and in part by a DOE Scientific Discovery through Advanced Computing grant. I would like to acknowledge many invaluable discussions with my colleagues, especially John Blondin, Steve Bruenn, Christian Cardall, George Fuller, Raph Hix, Matthias Liebendörfer, Bronson Messer, Jirina Stone, and Michael Strayer, and the hospitality of the Institute for Nuclear Theory where the Open Issues workshop took place. I would especially like to thank my co-editor, George Fuller, for his part in assembling the Open Issues Workshop and for his encouragement, guidance, expertise, and friendship over the years.

References

1. J. R. Wilson, in *Numerical Astrophysics*, edited by J. M. Centrella, J. M. LeBlanc, and R. L. Bowers (Jones and Bartlett, Boston, 1985), pp. 422–434.
2. H. A. Bethe and J. R. Wilson, Astrophysical Journal **295**, 14 (1985).
3. H.-T. Janka, Astronomy and Astrophysics **368**, 527 (2001).
4. A. Burrows and J. Goshy, Astrophysical Journal Letters **416**, L75 (1993).
5. H.-T. Janka and E. Müller, Astronomy and Astrophysics **306**, 167 (1996).
6. O. E. B. Messer, A. Mezzacappa, S. W. Bruenn, and M. W. Guidry, Astrophysical Journal **507**, 353 (1998).
7. A. Mezzacappa, A. C. Calder, S. W. Bruenn, J. M. Blondin, M. W. Guidry, M. R. Strayer, and A. S. Umar, Astrophysical Journal **495**, 911 (1998).
8. A. Mezzacappa, M. Liebendörfer, O. E. B. Messer, W. R. Hix, F.-K. Thielemann, and S. W. Bruenn, Physical Review Letters, **86**, 1935 (2001).
9. J. R. Wilson and R. W. Mayle, Physics Reports **227**, 97 (1993).
10. F. Swesty and J. Lattimer, Astrophysical Journal **425**, 195 (1994).
11. M. Rampp and H.-T. Janka, Astrophysical Journal **539**, L33 (2000).
12. S. W. Bruenn, K. R. DeNisco, and A. Mezzacappa, Astrophysical Journal **560**, 326 (2001).
13. M. Liebendörfer, A. Mezzacappa, F. Thielemann, O. E. Messer, W. R. Hix, and S. W. Bruenn, Physical Review D **63**, 103004 (2001).
14. T. A. Thompson, A. Burrows, and P. A. Pinto, Astrophysical Journal **592**, 434 (2003).
15. R. Buras, M. Rampp, H.-T. Janka, and K. Kifonidis, Physical Review Letters **90**, 241101 (2003).
16. H.-T. Janka, R. Buras, K. Kifonidis, A. Marek, and M. Rampp, in *Supernovae*, edited by J. Marcaide and K. Weiler (Springer Verlag, 2004).

17. J. Wilson, private communication (2004).

18. J. M. LeBlanc and J. R. Wilson, Astrophysical Journal **161**, 541 (1970).

19. E. M. D. Symbalisty, Astrophysical Journal **285**, 729 (1984).

20. J. C. Wheeler, D. L. Meier, and J. R. Wilson, Astrophysical Journal **568**, 807 (2002).

21. S. Akiyama, J. C. Wheeler, D. L. Meier, and I. Lichtenstadt, Astrophysical Journal **584**, 954 (2003).

22. M. Liebendörfer, A. Mezzacappa, F. Thielemann, O. E. Messer, W. R. Hix, and S. W. Bruenn, Physical Review D **63**, 103004 (2001).

23. C. Cardall and A. Mezzacappa, Physical Review D **68**, 023006 (2003).

24. C. Cardall, in *Numerical Methods for Multidimensional Radiative Transfer Problems*, edited by R. Rannacher and R. Wehrse (Springer Publishing Company, 2004).

25. E. Livne, A. Burrows, R. Walder, I. Lichtenstadt, and T. Thompson, Astrophysical Journal **609**, 277 (2004).

26. C. Cardall, E. Lentz, and A. Mezzacappa, Physical Review D, in press (2005).

27. A. Marek, H. Dimmelmeier, H.-T. Janka, E. Müller, and R. Buras, astro-ph/0502161 (2005).

28. M. Herant, W. Benz, W. R. Hix, C. L. Fryer, and S. A. Colgate, Astrophysical Journal **435**, 339 (1994).

29. A. Burrows, J. Hayes, and B. A. Fryxell, Astrophysical Journal **450**, 830 (1995).

30. C. L. Fryer and M. S. Warren, Astrophysical Journal **601**, 391 (2004).

31. J. M. Blondin, A. Mezzacappa, and C. DeMarino, Astrophysical Journal **584**, 971 (2003).

32. H.-T. Janka, R. Buras, F. Kitaura Joyanes, A. Marek, M. Rampp, and S. L., Nucl. Phys. A, in press (2004).

33. L. Wang, D. A. Howell, P. Höflich, and J. C. Wheeler, Astrophysical Journal **550**, 1030 (2001).

34. J. Blondin and A. Mezzacappa, Nature, submitted (2005).

35. T. A. Thompson, E. Quataert, and A. Burrows, Astrophysical Journal **620**, 861 (2005).

36. J. A. Miralles, J. A. Pons, and V. A. Urpin, Astrophysical Journal **574**, 356 (2002).

37. H. Duan and Y. Qian, Physical Review D **69**, 123004 (2004).

38. S. Balbus and J. Hawley, Astrophysical Journal **376**, 214 (1991).

39. G. M. Fuller, W. A. Fowler, and M. J. Newman, Astrophysical Journal Supplement **42**, 447 (1980).

40. G. M. Fuller, W. A. Fowler, and M. J. Newman, Astrophysical Journal **252**, 715 (1982).

41. G. M. Fuller, Astrophysical Journal **252**, 741 (1982).

42. G. M. Fuller, W. A. Fowler, and M. J. Newman, Astrophysical Journal Supplement **48**, 279 (1982).

43. G. M. Fuller, W. A. Fowler, and M. J. Newman, Astrophysical Journal **293**, 1 (1985).

44. J. Pruet and G. M. Fuller, Astrophysical Journal Supplement **149**, 189 (2003).

45. A. Yahil and J. M. Lattimer, in *NATO ASIC Proc. 90: Supernovae: A Survey of Current Research* (1982), pp. 53–70.

46. H. Bethe, G. Brown, J. Applegate, and J. Lattimer, Nucl. Phys. A **324**, 487 (1979).

47. K. Langanke, E. Kolbe, and D. J. Dean, Physical Review C **63**, 032801 (2001).

48. J. Cooperstein and E. Baron, in *Supernovae*, edited by A. Petschek (New York: Springer-Verlag, 1990), pp. 213–266.

49. O. Messer, W. Hix, M. Liebendörfer, and A. Mezzacappa, Astrophysical Journal, submitted (2005).

50. J. Cooperstein and J. Wambach, Nucl. Phys. A **420**, 591 (1984).

51. S. Hannestad and G. Raffelt, Astrophysical Journal **507**, 339 (1998).

52. R. Buras, H. Janka, M. T. Keil, G. G. Raffelt, and M. Rampp, Astrophysical Journal **587**, 320 (2003).

53. S. W. Bruenn, Astrophysical Journal Supplement **58**, 771 (1985).

54. S. Reddy, M. Prakash, and J. M. Lattimer, Physical Review D **58**, 013009 (1998).

55. S. Reddy, M. Prakash, J. M. Lattimer, and J. A. Pons, Physical Review **C59**, 2888 (1999).

56. A. Burrows and R. F. Sawyer, Physical Review **C58**, 554 (1998).

57. A. Burrows and R. F. Sawyer, Physical Review C **59**, 510 (1999).

58. T. A. Thompson, A. Burrows, and J. E. Horvath, Physical Review C **62**, 035802 (2000).

59. C. J. Horowitz, Physical Review D **55**, 4577 (1997).

60. C. J. Horowitz, M. A. Pérez-García, and J. Piekarewicz, Physical Review C **69**, 045804 (2004).

61. A. Burrows and T. Thompson, in *Stellar Collapse*, edited by C. Fryer (Dordrecht: Kluwer Academic Publishers, 2004), pp. 133–174.

62. A. Burrows, S. Reddy, and T. Thompson, Nucl. Phys. A, in press (2004).

63. J. R. Rikovska Stone, J. C. Miller, R. Koncewicz, P. D. Stevenson, and M. R. Strayer, Physical Review C **68**, 034324 (2003).

64. J. Stone, in *Open Issues in Core Collapse Supernova Theory*, edited by A. Mezzacappa and G. Fuller (World Scientific, Singapore, 2005), in press.

65. H. Heiselberg and V. Pandharipande, Annual Reviews of Nuclear and Particle Science **50**, 481 (2000).

66. F. Barnes, in *Open Issues in Core Collapse Supernova Theory*, edited by A. Mezzacappa and G. Fuller (World Scientific, Singapore, 2005), in press.

67. W. Hillebrandt and R. G. Wolff, in *Nucleosynthesis : Challenges and New Developments*, edited by D. Arnett and J. Truran (University of Chicago Press, Chicago, 1985), pp. 131–+.

68. J. Lattimer and F. D. Swesty, Nuclear Physics **A535**, 331 (1991).

69. M. Onsi, H. Przysiezniak, and J. M. Pearson, Physical Review C **55**, 3139 (1997).

70. H. Shen, H. Toki, K. Oyamatsu, and K. Sumiyoshi, Nuclear Physics **A637**, 435 (1998).

71. K. Langanke, G. Martínez-Pinedo, J. M. Sampaio, D. J. Dean, W. R. Hix, O. E. Messer, A. Mezzacappa, M. Liebendörfer, H.-T. Janka, and M. Rampp,

Physical Review Letters **90**, 241102 (2003).

72. W. R. Hix, O. E. Messer, A. Mezzacappa, M. Liebendörfer, J. Sampaio, K. Langanke, D. J. Dean, and G. Martínez-Pinedo, Physical Review Letters **91**, 201102 (2003).

73. D. G. Ravenhall, C. J. Pethick, and J. R. Wilson, Physical Review Letters **50**, 2066 (1983).

74. P. Magierski and P.-H. Heenen, Physical Review C **65**, 045804 (2002).

75. W. Newton, Ph.D. Thesis, Oxford University, in preparation (2005).

76. A. Burrows and T. Thompson, in *Stellar Collapse*, edited by C. Fryer (Dordrecht: Kluwer Academic Publishers, 2004), pp. 133–174.

77. G. M. Fuller, R. Mayle, J. R. Wilson, and D. Schramm, Astrophysical Journal **322**, 795 (1987).

78. G. M. Fuller, R. Mayle, B. S. Meyer, and J. R. Wilson, Astrophysical Journal **389**, 517 (1992).

79. Y.-Z. Qian and G. Fuller, Physical Review D **52**, 656 (1995).

80. G. Fuller and Y.-Z. Qian, Physical Review D, submitted (2005).

Section 2
Fundamental Issues in Radiation Magnetohydrodynamics

RADIATION DIFFUSION: AN OVERVIEW OF PHYSICAL AND NUMERICAL CONCEPTS[1]

FRANK GRAZIANI

Lawrence Livermore National Laboratory
Livermore, CA, 94550, USA

An overview of the physical and mathematical foundations of radiation transport is given. Emphasis is placed on how the diffusion approximation and its transport corrections arise. An overview of the numerical handling of radiation diffusion coupled to matter is also given. Discussions center on partial temperature and grey methods with comments concerning fully implicit methods. In addition finite difference, finite element and Pert representations of the div-grad operator is also discussed

1. The "So What?" Question: Why Radiation Transport Matters

Photons, be they in the radio, optical. X-ray or gamma-ray portion of the radiation spectrum leave their mark across the fabric of the universe in a multitude of ways. At the "smallest" astronomical scales, radiation transport is crucial to understanding the atmospheres of planets. At the largest scales the universe is bathed in an afterglow of its birth called the cosmic background radiation. As is well known, whole fields in astronomy are devoted to the study of different portions of the electromagnetic spectrum.

Photons can act as a signature of some astronomical event. In addition, because of the density and temperatures encountered in many astrophysical applications, photons can effect the movement of a gas or fluid and the movement of the gas or fluid can in turn affect the behavior of the photons. Radiation pressure, changes is spectral shape due to moving fluids, and PdV work done on the radiation field are all important examples of the interaction of matter and photons.

With the advent of large scale computing, the complex system of equations involving radiation transport and fluid dynamics could be solved. Currently, with the introduction of parallel computing it is now possible to model 3D astrophysical phenomena such as supernovae with unprecedented accuracy and with the inclusion of complex physics. In all of these simulations radiation transport remains an exciting but challenging obstacle. In multi-physics codes it tends to dominate CPU time. This is easy to see when one considers that in 3D dynamic applications the solution of the photon transport problem involves solving a seven dimensional Boltzmann equation[1]. This equation is in general highly non-linear and non-local. In addition, its coupling to the fluid modifies both the fluid equations and the usual hydrodynamic equations. This interaction between radiation and fluids defines the field of radiation hydrodynamics. It is a vast field with excellent references by Mihalas and Mihalas[2], Pomraning[3], Bowers and Wilson[4], and Castor[5]. Besides being very good guides to radiation hydrodynamics they are excellent sources for the field of radiation transport in general.

The challenges of solving the transport problem have led researchers to solving a simpler problem. In many applications the physics allows one to solve the diffusion approximation to the full transport problem. By going to the diffusion limit of the transport equation, the numbers of degrees of freedom in 3D are reduced from seven to five for multi-group and seven to four for Planckian. This approximation is by far the most used approximation to the transport equation. In fact, it makes the radiation problem so tractable that the diffusion approximation is used in regimes where only a transport description is valid. The use of transport corrected diffusion such as flux limiters helps extend the applicability of diffusion. The benefit is of course the cheaper cost of diffusion over transport but at the price of reduced accuracy.

This paper is devoted to discussing the general framework of radiation transport and in particular how diffusion arises from it. In addition a review of the numerical treatments of the diffusion operator and how the coupled radiation material equations are handled is given. Due to the lack of space, the subject of radiation hydrodynamics is not given. Interested readers are urged to read the above listed resources. However, the work presented here should always be thought of in the larger context of a multi-physics code. In addition, subjects not covered here include opacities and scattering. Detailed discussions of these topics can also be found in the above listed references.

2. Review of Radiation Transport Concepts

2.1. *Classical and Quantum Properties of the Radiation Field*

The classical manifestation of the radiation field is based on the wave properties of light. A classical description of radiation is consistent with the properties of polarization, diffraction, and refraction. Unfortunately, the fact that a classical radiator predicts a Rayleigh-Jeans Law for the emission spectra, in violation of the experimental data, means that the wave description of light is not complete. As is well known, the quantum mechanical description of radiation describes a wealth of observed phenomena. The photoelectric effect, where photons which are impinging on a metal surface release electrons above a certain threshold frequency is a well known example that earned Einstein the Nobel Prize. Compton scattering, where the frequency of hard X-ray photons is shifted downwards due to the incoming photons scattering off of stationary electrons and transferring some of their energy and momentum to the electrons, is another example. The observed emission spectra from atoms are a classic example where the quantum mechanical description of radiation explained observations that were previously unexplained. The quantum mechanical treatment of photons and the description of the blackbody spectrum is a famous and singular success that heralded the beginning of a new age. Finally, the beauty behind the quantum mechanical description of radiation culminated with the unification, by Dirac, of the particle and wave descriptions.

2.1.1. *The Boltzmann Description of Radiative Transfer*

The standard description of radiative transfer rests on describing the radiation field as a photon gas moving with the speed of light and interacting with a medium via absorption, emission, and scattering. For simplicity, the effects of material motion are ignored here as are the effects of refraction, diffraction, and dispersion. The radiation field is assumed to consist of point particles (photons). Associated with each photon is a frequency ν, energy $h\nu$, and momentum $h\nu/c$. At any time t, six variables in 3D are required to specify the position of the photon in phase space. There are three position variables and three momentum variables. The three momentum variables are written in terms of the speed of light c and the photon direction Ω. Using the gas analogy, a photon distribution function $f_\nu(r,\Omega,t)$ is defined such that

$$dn = f_\nu(r,\Omega,t)d^3r d^2\Omega \tag{1}$$

is the number of photons at time t and position r, contained in the differential element d^3r, with frequency ν, traveling in the direction Ω subtending a solid angle element $d^2\Omega$. In the literature, the photon distribution function is

32

rarely used. Instead, the specific intensity or angular flux is the quantity most used and the object referred to in this paper when the radiation field is mentioned. It is defined by

$$I_\nu(r,\Omega,t) = ch\nu f_\nu(r,\Omega,t) \tag{2}$$

In general, $I_\nu(r,\Omega,t)$ or $f_\nu(r,\Omega,t)$ should have an additional index describing the particular polarization state. There are in general four components of the specific intensity, called Stokes parameters, necessary to describe a polarized beam of radiation. One of these components is the radiation specific intensity defined above. The other three components come from the plane of polarization and the ellipticity of the beam. In this paper, it is assumed that the radiation is either unpolarized or the polarization states have been averaged over. For details on the radiative transfer equation for polarized radiation please consult either Chandrasekhar[6] or Pomraning[3].

Using the specific intensity as the fundamental quantity of interest, a number of physically relevant objects can be defined.

$$u = \int_0^\infty d\nu \int d^2\Omega (h\nu) f_\nu(r,\Omega,t) = \frac{1}{c}\int_0^\infty d\nu \int d^2\Omega I_\nu(r,\Omega,t)$$

$$F = \int_0^\infty d\nu \int d^2\Omega (c\Omega)(h\nu) f_\nu(r,\Omega,t) = \int_0^\infty d\nu \int d^2\Omega \Omega I_\nu(r,\Omega,t)$$

$$M = \int_0^\infty d\nu \int d^2\Omega (\frac{h\nu\Omega}{c}) f_\nu(r,\Omega,t) = \frac{1}{c^2}\int_0^\infty d\nu \int d^2\Omega \Omega I_\nu(r,\Omega,t) \tag{3}$$

$$P = \int_0^\infty d\nu \int d^2\Omega (c\Omega)(\frac{h\nu\Omega}{c}) f_\nu(r,\Omega,t) = \frac{1}{c}\int_0^\infty d\nu \int d^2\Omega \Omega\Omega I_\nu(r,\Omega,t)$$

The quantities u, F, M, and P are respectively the energy density, radiative flux, momentum density, and pressure tensor of the radiation field.

In writing down the equation of radiative transfer, a pedagogic approach is usually taken. By using the analogy of a classical gas, a semi-classical equation can be written down based solely on conservation of photons. Terms due to sink and source terms and scattering are included in a semi-phenomenological fashion.

$$\frac{1}{c}\frac{\partial I_v(r,\Omega,t)}{\partial t} + \Omega \bullet \nabla I_v(r,\Omega,t) = j_v - \sigma_v I_v(r,\Omega,t)$$

$$+ \int dv' \int d\Omega \frac{v}{v'}\sigma_s(v \to v, \Omega \bullet \Omega') I_v(r,\Omega,t)\left[1 + \frac{c^2 I_v(r,\Omega,t)}{2hv^3}\right] \quad (4)$$

$$- \int dv' \int d\Omega \sigma_s(v \to v, \Omega \bullet \Omega') I_v(r,\Omega,t)\left[1 + \frac{c^2 I_v(r,\Omega,t)}{2hv'^3}\right]$$

The right hand side of equation [4], represents the interaction of the radiation field with matter through three basic mechanisms; (1) emission (2) absorption (3) scattering. The quantity j_v represents the emission of photons from the material and is called the emissivity. The quantity σ_v is the absorption coefficient and has dimensions 1/cm. It is related to the opacity κ_v and the density ρ by the simple relation $\sigma_v = \kappa_v \rho$. The quantity σ_s is the scattering kernel. Typically, it will represent the process of Compton scattering[3]. For simplicity, In the rest of this paper scattering is ignored.

The radiative transfer equation is semi-classical in nature. In the kinetic equation, the photons are treated like any other gas. The quantum mechanical effects come through the absorption, emission and scattering terms. Each of these three processes describes at a micro-physical level the quantum mechanical interaction of matter and radiation. In the next section, a simple example of the micro-physical origins of the emission and absorption mechanisms will be given along with consequences of the matter field being in thermodynamic equilibrium.

2.1.2. *The Einstein Coefficients and the Planck Distribution*

In this section, a derivation of the Planck distribution will be given. The derivation presented here (based on the book by Pomraning[3]) elucidates the nature of local thermodynamic equilibrium (LTE) and non-LTE (NLTE) and yields insight into the sink and source terms of the radiative transfer equation. Consider an atom with a number of bound states. Consider two levels with energies E_n and E_m and statistical weights g_n and g_m . The probability per time that an atom in state m, exposed to radiation of frequency v, will absorb a photon from the radiation field is given by.

$$P_{mn} = B_{mn}I(v_{nm})d^2\Omega \quad (5)$$

$$h \nu_{nm} = E_n - E_m \tag{6}$$

B_{mn} is one of the Einstein coefficients. It is a constant of proportionality representing the transition rate at which the presence of the radiation field induces an upward transition in energy. $I(\nu_{nm})$ is the specific intensity of the radiation field which we will define in a moment and $d^2\Omega$ is the solid angle subtended by the photon. Besides being absorbed, the atom can also emit a photon. The probability per time that an atom in state n will emit a photon of frequency ν is given by.

$$P_{nm} = \left[A_{nm} + B_{nm} I(\nu_{nm}) \right] d^2\Omega \tag{7}$$

The term proportional to A_{nm} is the last Einstein coefficient. It represents the transition rate at which the atom undergoes spontaneous emission. That is, there is a finite probability that an atom in a state n will emit a photon and undergo a downward transition in energy in the absence of a radiation field. The second term proportional to B_{nm} represents the effect of stimulated emission. It is the transition rate at which the radiation field induces an atom to undergo a downward transition in energy. That is, the presence of a radiation field itself will enhance the emission process. This stimulated emission is a consequence of the quantum statistics obeyed by bosons. The A and B coefficients are related. Given the Hamiltonian of an atom in a radiation field, time dependent perturbation theory can be used to compute the transition rates for absorption and emission. The fundamental relationship can be derived,

$$A_{nm} / B_{nm} = h \left(\nu_{nm} \right)^3 / c^2 \tag{8}$$

At this point in the discussion, nothing has been assumed about the absorption and emission processes. The arguments are completely general and are applicable to both systems in and out of thermodynamic equilibrium. If however, thermodynamic equilibrium is assumed, then additional results regarding the Einstein coefficients and the associated radiation field can be derived.

In complete thermodynamic equilibrium, the principle of detailed balance holds. That is, there exists a detailed balance between all absorption and emission processes. Mathematically, this means the probability for emission exactly equals the probability for absorption.

$$N_n\left[A_{nm} + B_{nm}I(\nu_{nm})\right] = N_m B_{mn}I(\nu_{nm}) \qquad (9)$$

N_n refers to the number density of atoms in state n. Solving for the radiation field yields

$$I(\nu_{nm}) = \frac{A_{nm}/B_{nm}}{\left[(N_m B_{mn}/N_n B_{nm})-1\right]} \qquad (10)$$

Since complete thermodynamic equilibrium has been assumed, the populations are distributed according to the Boltzmann distribution

$$N_n\Big/N_m = \frac{g_n}{g_m}\,e^{-h\nu_{nm}/kT} \qquad (11)$$

Therefore, the Planck distribution is obtained

$$I(\nu_{nm}) = \frac{2h(\nu_{nm})^3/c^2}{\left[(B_{mn}/B_{nm})e^{h\nu_{nm}/kT} - 1\right]} \qquad (12)$$

2.1.3. *Local Thermodynamic Equilibrium (LTE) and Kirchoff's law*

The simple example above proves the point that solving the radiative transfer equation when the matter emits, absorbs, and scatters radiation is complex. This is because, in general, detailed knowledge of the atomic populations and ionization states making up the material must be known if the absorptivities and emissivities are to be calculated. The concept of LTE is a simplifying assumption of the matter that greatly reduces the complexity of trying to solve the radiative transfer problem when radiation-matter interactions are important.

The essential point in establishing LTE in any given material is that the properties of the matter are dominated by atomic collisions which establish thermodynamic equilibrium locally at a space-time point (r,t) and the radiation field does not destroy this equilibrium. Therefore, the main difference between complete thermodynamic equilibrium and LTE is that LTE does not require the radiation field to be Planckian. The implications are that at a given space-time point, only the atomic composition and two thermodynamic quantities (density and temperature) need be specified. The LTE assumption of course assumes very specific features of the states of the atoms and molecules making up the material. The first is kinetic equilibrium, that is, the electron and ion distributions obey a Maxwellian. The second feature is excitation equilibrium, that is, the population density of the excited states of every species must obey a Boltzmann distribution. Third, ionization equilibrium, that is, the particle densities for neutrals, electrons, and ions obey a Boltzmann like distribution involving ionization potentials. This is the so-called Saha equation[8]. Fourth is the Kirchoff-Planck relation. This is an amazing relation in that it reduces the emissivity to a product of the absorption coefficient and a Planck function whose temperature is characteristic of the local material temperature. Simply put, $j_v = \sigma_v B_v(T)$. Therefore, the material-radiation interaction reduces to a study of the absorption mechanisms in a plasma where kinetic, excitation, and ionization equilibrium holds.

When does LTE hold? In order for atomic collisions to dominate over radiative processes, it is clear the plasma must be dense. Griem[7], has constructed a criterion based on the ratio of radiative to collisional rates in a hydrogen-like atom where the plasma is optically thin enough that photons once emitted are not re-absorbed. The condition for LTE of atomic level n is

$$1.23 \times 10^{-5} \left(\frac{n}{n-1} \right)^2 \left(\frac{Z^{*7}}{n^{17/2}} \right) \left(\frac{T(eV)}{Z^{*2}(13.6eV)} \right)^{1/2} \left(\frac{1}{\rho} \right) << 1 \qquad (13)$$

The effective ionization is denoted by Z^*. Griem's criterion clearly shows that for high density and/or hot plasmas LTE is a good approximation. In this paper, LTE will always be assumed unless otherwise noted.

So far the discussion has focused on the radiation field. In most applications with optically thick matter, the radiation is absorbed, remitted and the material temperature changes. Using the assumption of LTE, every piece of matter acts like a blackbody radiator emitting photons with a Planckian spectrum characteristic of the temperature of the material. Therefore, ignoring scattering, using Kirchoff's law for the emissivity and writing an energy balance relation for the material we have,

$$\frac{1}{c} \frac{\partial I_v(r,\Omega,t)}{\partial t} + \Omega \bullet \nabla I_v(r,\Omega,t) = \sigma_v B_v(T) - \sigma_v I_v(r,\Omega,t) \qquad (14)$$

$$\frac{\partial [\rho C_v T(r,t)]}{\partial t} = -\int dv \sigma_v \left[B_v(T) - \int d^2\Omega I_v(r,\Omega,t) \right] \qquad (15)$$

In writing down the latter equation it is assumed that the opacities are independent of angle and all conduction effects are negligible.

2.1.4. *The Equilibrium Radiation Field*

The isotropic and homogeneous distribution obeyed by a photon gas in complete thermodynamic equilibrium at temperature T, is the Planck function

$$I_\nu(r,\Omega,t) = B_\nu(T) = \frac{2h\nu^3}{c^2} \frac{1}{e^{h\nu/kT} - 1} \tag{16}$$

This should not be confused with LTE. The Planck function can be derived in a variety of ways. We have seen one such method in section 2.1.2. Physically, it is a consequence of the photons undergoing three basic processes (1) absorption (2) stimulated emission (3) spontaneous emission and the fact that the photons are bosons. The isotropic nature of the equilibrium radiation field means that the energy density, radiative flux, momentum flux, and pressure tensor quantities can be easily evaluated. Substituting equation into equation yields

$$u = \frac{8\pi^5 k^4}{15h^3 c^3} T^4 \equiv aT^4$$

$$F = M = 0 \tag{17}$$

$$P = \frac{aT^4}{3} \begin{pmatrix} 1 & 0 & 0 \\ 0 & 1 & 0 \\ 0 & 0 & 1 \end{pmatrix} = 45.7 Mbar(T[keV])^4 \begin{pmatrix} 1 & 0 & 0 \\ 0 & 1 & 0 \\ 0 & 0 & 1 \end{pmatrix}$$

As expected for an isotropic gas, the radiative flux and momentum density are zero and the pressure tensor is diagonal. The concept of a temperature for the photon gas is uniquely defined here. That is, for an equilibrium distribution of photons, the temperature is defined by the Planck distribution. One could think of the relationship between energy density and temperature as a defining relation. However, for non-Planckian distributions this is not true. Several definitions of temperature are encountered in multi-group diffusion and transport[8]. The above relations for energy density and pressure give a strong indication of the power of the radiation field to move matter. At T=1 keV, the

38

pressure exerted by the photons on matter is 45.7 Mbar. In addition, this pressure rapidly rises with temperature. This fact means that for high temperatures, the radiation field can make a substantial or dominant contribution to the overall pressure in a fluid.

2.1.5. *Assumptions of a Kinetic Theory and the Micro-physical Foundations of Radiative Transfer*

The specific form of the radiative transfer equation, whose numerical solution is the focus of this paper, rests on a number of assumptions. The first set of these assumptions is not inherent in the kinetic theory of radiative transfer but rather a simplification. As mentioned previously, polarization effects are ignored. In addition, refraction and dispersion effects are also ignored. A radiative transfer equation incorporating these effects can be readily constructed (Pomraning[3]) but for simplicity they are ignored in this paper. The second class of assumptions is inherent in the semi-classical kinetic approach adopted here. Since photons are treated classically, their wave behavior is ignored. Therefore, the possibility of interference between different photons is ignored. Hence the spread of the photon wave packet is assumed small on the resolution we are interested in. Since the photons are treated like a classical gas, other quantum phenomena such as photon number fluctuation are ignored. Finally, all collision, absorption, and emission processes occur instanteously.

The fundamental description of photons and their interaction with matter is based on quantum electrodynamics (QED). Is it possible to arrive at a fully self-consistent description of radiative transfer from a fundamental description? The answer is yes. Although a detailed description of the derivation would take us to far a field, suffice it to say that a number of authors have derived kinetic equations for the photons starting with the Hamiltonian of QED (Gelinas and Ott[9], Cannon[10], Degl'Innocenti[11], Graziani[12]). An example of the procedure involves using either the density matrix formalism or the Heisenberg equations of motion to construct dynamic equations for the number operators of the plasma and photon degrees of freedom. A quantum mechanical distribution function, similar to the Wigner distribution can be constructed from products of photon operators. The Wigner distribution is an operator that is a quantum mechanical generalization of the specific intensity. The quantum radiative transfer equation comes from the dynamic evolution of this Wigner function. The plasma degrees of freedom enter naturally into the quantum radiative transfer equation because of the interaction terms present in the QED Hamiltonian. The classical specific intensity is calculated as the quantum average of the Wigner function. Of course, the quantum radiative transfer equation is more general as it allows one to compute fluctuations of the specific intensity operator. Classically, these fluctuations are negligible. Examples of where such an approach has proved useful are in deriving radiative transfer equations with polarization from first principles, quantum optics and NLTE plasmas.

3. The Diffusion Approximation

The radiative transfer equation [4], being in general a six dimensional non-linear integro-differential equation, is not conducive to closed form solutions. In fact it is a current challenge in the astrophysics, atmospheric physics, high energy density physics and nuclear engineering communities to solve it numerically. Therefore, borrowing a page from the kinetic theory community, a method commonly used to simplify the radiative transfer equation is to construct moments of the equation. The first type of moments that can be constructed of the specific intensity are arrived at by multiplying equation [4] by various powers of Ω, and integrating both sides of the equation over all solid angle. Defining,

$$\varepsilon_v = \frac{1}{c} \int d\Omega I_v(r, \Omega, t)$$

$$\phi_v = \int d\Omega \Omega I_v(r, \Omega, t) \tag{18}$$

$$\Pi_v = \frac{1}{c} \int d\Omega \Omega \Omega I_v(r, \Omega, t)$$

These quantities represent the zero, first and second order moments of the specific intensity. Physically, they are the energy density, flux and pressure tensor per frequency. We will refer to ε_v as the spectrum. Using these definitions, the radiative transfer equation can be written as a set of coupled partial differential equations

40

$$\frac{\partial \varepsilon_v(r,t)}{\partial t} + \nabla \bullet \phi_v(r,t) = \left[4\pi\sigma_v B_v(T) - c\sigma_v \varepsilon_v(r,t)\right]$$

$$\frac{1}{c}\frac{\partial \phi_v(r,t)}{\partial t} + c\nabla \bullet \Pi_v(r,t) = -\sigma_v \phi_v(r,t) \tag{19}$$

$$\frac{\partial \Pi_v(r,t)}{\partial t} + \text{"Fourth Order Moment"}.......$$

This set of moment equations exhibit the classic closure problem. That is, zero order moments are coupled to first order moments; first order moments are coupled to second order moments, ad infinitum. The central question in all approximations to the radiative transfer equation is coming up with a suitable closure scheme. Three related topics are examined here; variable Eddington factors, diffusion, and the telegraphers equation.

3.1. *Variable Eddington Factors*

Variable Eddington factors is less a closure scheme than a way of recasting the coupled moment equations. Define a tensor quantity, called the Eddington factor, by

$$\chi_v(r,t) = \frac{\int d\Omega \Omega \Omega I_v(r,\Omega,t)}{\int d\Omega I_v(r,\Omega,t)} = \frac{\Pi_v(r,t)}{\varepsilon_v(r,t)} \tag{20}$$

Physically, this quantity represents the mean of the tensor $\Omega\Omega$ over all directions weighted by the specific intensity. Substituting equation [20] into equation [19], yields

$$\frac{\partial \varepsilon_v(r,t)}{\partial t} + \nabla \bullet \phi_v(r,t) = \left[4\pi\sigma_v B_v(T) - c\sigma_v \varepsilon_v(r,t)\right]$$

$$\frac{1}{c}\frac{\partial \phi_v(r,t)}{\partial t} + c\nabla \bullet \chi_v \varepsilon_v(r,t) = -\sigma_v \phi_v(r,t) \tag{21}$$

Although the above set of equations is seemingly closed, it is in fact not due to the presence of the Eddington factor. However, by writing the coupled moment equations is this fashion, the Eddington factor provides a degree of freedom to characterize the radiation field. For example, it is possible to solve a transport problem every 10 or 20 cycles or so and construct the Eddington factor, This Eddington factor would then be used in the solution of the coupled moment equations and would be updated when the transport solve was performed. This is essentially what is done in ZEUS-2D[13].

3.2. *Multi-Group Diffusion*

For an isotropic radiation field, the Eddington factor simplifies considerably to

$$\chi_v = \frac{1}{3}\begin{pmatrix} 1 & 0 & 0 \\ 0 & 1 & 0 \\ 0 & 0 & 1 \end{pmatrix} \tag{22}$$

If it is assumed that interactions will matter dominate the flow, that is,,

$$\frac{1}{c}\frac{\partial \phi_v(r,t)}{\partial t} << -\sigma_v \phi_v(r,t) \tag{23}$$

Then the flux equation becomes simply a Fick's type law $\phi_v = -\dfrac{c}{3\sigma_v}\nabla \varepsilon_v$

and the energy density equation becomes the familiar diffusion equation

$$\frac{\partial \varepsilon_v(r,t)}{\partial t} - \nabla \bullet \left(\frac{c}{3\sigma_v}\nabla \varepsilon_v(r,t)\right) = \left[4\pi\sigma_v B_v(T) - c\sigma_v \varepsilon_v(r,t)\right] \tag{24}$$

This equation along with the material temperature energy balance equation

$$\frac{\partial\left[\rho C_v T(r,t)\right]}{\partial t} = -\int dv \sigma_v \left[B_v(T) - \varepsilon_v(r,t)\right] \tag{25}$$

42

form the basis for the multi-group diffusion equations. It should be remarked that the above analysis is equivalent to assuming that the specific intensity is nearly isotropic

$$I_v(r,\Omega,t) = \frac{c\varepsilon_v(r,t)}{4\pi} + \frac{3}{4\pi}\Omega\bullet\phi_v(r,t) \qquad (26)$$

To see how this arises physically, assume LTE and ignore scattering and rearrange the terms in the radiative transfer equation such that

$$I_v(r,\Omega,t) = B_v(T) - \frac{1}{\sigma_v}\left[\frac{1}{c}\frac{\partial I_v(r,\Omega,t)}{\partial t} + \Omega\bullet\nabla I_v(r,\Omega,t)\right] \qquad (27)$$

Perform an expansion to first order in $1/\sigma_v$ to obtain

$$I_v(r,\Omega,t) = B_v(T) - \frac{1}{\sigma_v}\left[\frac{1}{c}\frac{\partial B_v(T)}{\partial t} + \Omega\bullet\nabla B_v(T)\right] \qquad (28)$$

Note that an expansion in $1/\sigma_v$ is equivalent to an expansion in Ω. Constructing the flux directly from equation [28], yields Fick's Law! That is

$$\phi_v(r,t) = \int d\Omega\,\Omega I_v(r,\Omega,t) = -\frac{4\pi}{3\sigma_v}\nabla B_v(T) \qquad (29)$$

Therefore, Fick's Law arises directly from an inverse mean free path expansion or equivalently, an expansion in isotropy.

The multi-group diffusion approximation works well where the radiation field is nearly isotropic yet the medium is transparent enough that photon mean free paths are long and therefore, the radiation temperature is not determined locally but rather is determined by sources that are far away. The multi-group diffusion approximation is frequently applied in astrophysical and inertial confinement fusion (ICF) applications. For example, for an ICF capsule bathed in radiation, the multi-group approximation works very well in predicting capsule performance[14].

It is clear from the functional form of the Planck function that the above equations forms a set of coupled non-linear integro-differential equations. It is this fact that makes their solution a challenge. As in the transport problem, the presence of a material temperature couples all groups together and it is this fact that makes their solution a challenge. This paper will discuss several methods for their solution including partial temperatures, grey methods such as Lund-Wilson and source iteration, and full matrix methods.

3.3. *The Planckian Diffusion Equation*

An additional simplification arises if it is assumed that the photon spectrum is Planckian. This will occur in optically thick media where the radiation field is determined by many absorptions and re-emissions. In this case the radiation field rapidly becomes Planckian at a temperature not necessarily that of the material. Assuming the radiation spectrum is Planckian with a characteristic temperature T_R (i.e. $\varepsilon_v = B_v(T_R)$), equation [24] can be integrated over frequency to yield

$$\frac{\partial T_R^4}{\partial t} - \nabla \bullet \left(\frac{c}{3\sigma_R(T,T_R)} \nabla T_R^4 \right) = c\left[\sigma_P(T,T)T^4 - \sigma_P(T,T_R)T_R^4 \right] \quad (30)$$

Where the Planck and Rosseland averaged opacities are defined by

$$\sigma_P(T,T_A) = \frac{\int dv\,\sigma_v B_v(T_A)}{\left(acT_A^4 \middle/ 4\pi \right)} \qquad \sigma_R(T,T_A) = \frac{\int dv \frac{1}{\sigma_v} \frac{\partial B_v(T_A)}{\partial T_A}}{\left(acT_A^4 \middle/ \pi \right)} \quad (31)$$

The material energy balance equation is just

$$\frac{\partial \left[\rho C_V T(r,t) \right]}{\partial t} = -ac\left[\sigma_P(T,T)T^4 - \sigma_P(T,T_R)T_R^4 \right] \quad (32)$$

The Planck opacity σ_P sets the time scale for local energy exchange between matter and radiation based on emission and absorption. The Rosseland opacity σ_R determines average transport properties of the radiation flow.

3.4. *Some Observations Regarding Diffusion*

It is clear that by going to the diffusion approximation, the equations of radiation transport have been fundamentally changed. From a mathematical standpoint the radiative transfer equation is a hyperbolic first order equation requiring for boundary conditions that the initial specific intensity be specified along with for example an incoming value of the specific intensity at a boundary. The diffusion equation however, is a parabolic equation needing two boundary conditions to be specified along with an initial condition. From a physical standpoint, the diffusion approximation is acausal. To see this, consider the multi-group diffusion equation with no sink or source terms

$$\frac{\partial \varepsilon_v(r,t)}{\partial t} - \nabla \bullet \left(\frac{c}{3\sigma_v} \nabla \varepsilon_v(r,t) \right) = 0 \tag{33}$$

This equation describes a signal propagating with a velocity $V \approx \dfrac{c}{\sqrt{ct\sigma_v}}$

which is obviously in violation of the speed of light restriction. It is clear that applying the diffusion approximation to transparent media where a signal can propagate at or near the speed of light is in violation of the approximations used in deriving equation [33] so it is not too surprising that the acausal nature has reared its ugly head. However, all is not lost. Flux limiters, which will be discussed next save the day and allow one to apply the diffusion equation to applications where normally only a transport description based on the radiative transfer equation would do.

Before moving on, a question naturally arises which is; what has been lost by throwing out the time derivative of the flux term? And what happens if it is restored? Consider for simplicity the coupled set of moment equations in a vacuum

$$\frac{\partial \varepsilon_v(r,t)}{\partial t} + \nabla \bullet \phi_v(r,t) = 0$$

$$\frac{1}{c}\frac{\partial \phi_v(r,t)}{\partial t} + \frac{c}{3}\nabla \bullet \varepsilon_v(r,t) = 0 \tag{34}$$

Taking the time derivative of the first equation and using the second equation to close the set yields

$$\frac{\partial^2 \varepsilon_v(r,t)}{\partial t^2} - \frac{c^2}{3} \nabla^2 \varepsilon_v(r,t) = 0 \tag{35}$$

This is the wave equation that describes ingoing and outgoing waves moving with a velocity of $c/\sqrt{3}$. Hence, causality has been restored; however, the finite light speed is too small by a factor of .577. The coupled moment equations in the presence of matter yield the so-called Telegrapher's equation[15].

3.5. *Transport Corrected Diffusion: Flux Limiters*

In a vacuum, the radiative transfer equation predicts that the energy density propagates with a velocity c and the flux is given by $\phi_v = c\varepsilon_v \Omega_0$. Where Ω_0 is the solid angle subtended by the ray. This is the maximum flux allowed physically since it represents photons moving unimpeded. Diffusion predicts a very different behavior. As shown above the flux for the diffusion equation (in the absence of sink and source terms) is $\phi_v = \left(\dfrac{c}{3\sigma_v} \nabla \varepsilon_v(r,t) \right)$. However, it is clear that for very strong gradients and/or when $\sigma_v \to 0$ the diffusive flux can be greater than the maximum flux allowed.

Jim Wilson first proposed limiting the diffusive flux to correct for the fact that diffusion predicts faster than light flow speeds in transparent media. Define a new flux with a generalized diffusion coefficient which interpolates between the diffusive flux in optically thick media and the transport flux in optically thin media.

$$\phi_v = c\Lambda\left(\sigma_v, \frac{|\nabla\varepsilon_v|}{\varepsilon_v}\right)\nabla\varepsilon_v$$

$$\Lambda\left(\sigma_v, \frac{|\nabla\varepsilon_v|}{\varepsilon_v}\right) = \frac{1}{3\sigma_v} \quad \text{for short photon mean free paths} \qquad (36)$$

$$\Lambda\left(\sigma_v, \frac{|\nabla\varepsilon_v|}{\varepsilon_v}\right) = \frac{\varepsilon_v}{|\nabla\varepsilon_v|} \quad \text{for long photon mean free paths}$$

The literature[16], contains a variety of choices for the flux limiter. As an aside, Levermore and Pomraning[17] have derived the flux limiter in a rigorous fashion by performing a Chapman-Enskog like expansion of the transport equation. Their result is not quoted here but interested readers are urged to read their papers.

$$\Lambda\left(\sigma_v, \frac{|\nabla\varepsilon_v|}{\varepsilon_v}\right) = \frac{1}{3\sigma_v + \dfrac{|\nabla\varepsilon_v|}{\varepsilon_v}} \quad \text{Wilson sum flux limiter}$$

$$\Lambda\left(\sigma_v, \frac{|\nabla\varepsilon_v|}{\varepsilon_v}\right) = \frac{1}{MAX\left[3\sigma_v, \dfrac{|\nabla\varepsilon_v|}{\varepsilon_v}\right]} \quad \text{Maximum flux limiter} \qquad (37)$$

$$\Lambda\left(\sigma_v, \frac{|\nabla\varepsilon_v|}{\varepsilon_v}\right) = \frac{1}{\left[(3\sigma_v)^n + \left(\dfrac{|\nabla\varepsilon_v|}{\varepsilon_v}\right)^n\right]^{1/n}} \quad \text{Larsen flux limiter}$$

Figure 1 shows a comparison of a variety of choices including the Levermore-Pomraning result $R = \dfrac{|\nabla\varepsilon_v|}{\sigma_v\varepsilon_v}$ is plotted on the horizontal axis and $\Lambda\sigma_v$ is plotted on the vertical axis.

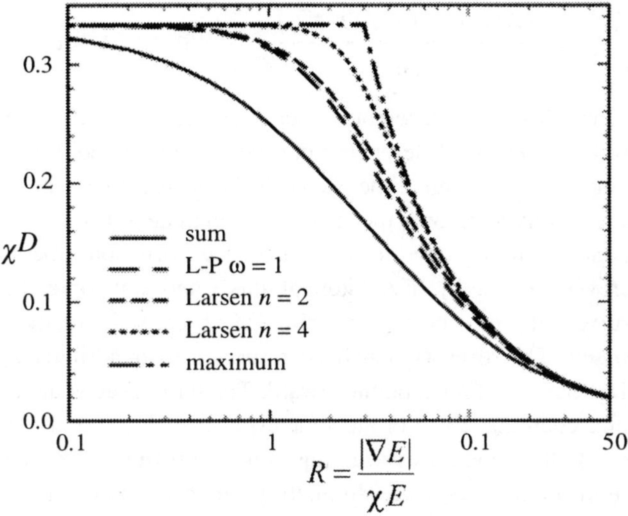

Figure 1. Comparisons of several flux limiter choices (Taken from Olson, Auer, and Hall JQSRT V.56 1996)

Figure 1 shows the Levermore-Pomraning result (long dash-short dash) as derived from the transport equation. Clearly, the Wilson sum flux limiter tends to restrict the flux too much while the MAX flux limiter does not restrict the flux enough. Interestingly, the Larsen flux limiter with n=2 does a very good job of matching the transport solution.

4. Numerical Methods for Diffusion

Complex applications and the nature of the coupled radiation matter equations means that analytic solutions are almost impossible to find. Instead, a numerical approach is sought. The discretization of the multi-group or Planckian diffusion equations means dealing with several issues. The first is time discretization. It is assumed that the coupled radiation and material equations are to be solved implicitly. Explicit time differencing yields a Courant stability condition which severely restricts the time step and therefore is not widely used. There are unconditionally stable explicit schemes for diffusion that have been explored[18]. However, the accuracy of these schemes beyond several Courant time steps is poor. In this paper and in most applications codes, the simple backward Euler difference is used. That is,

$$\frac{\partial A(r,t)}{\partial t} \Rightarrow \frac{\left(A(r,t_{n+1}) - A(r,t_n)\right)}{\Delta t} \qquad \text{where } \Delta t = t_{n+1} - t_n \qquad (38)$$

In spite of higher order time differencing schemes such as Crank-Nicholson, the simple first order backward differencing in time is widely used due to its robust behavior in the limit of large time steps. This is due to its non-oscillatory behavior and its ability to recover the static solution as $\Delta t \rightarrow \infty$.

The second issue is how to discretize the diffusion operator. Space limitations prevent a lengthy discussion of this important topic; however, an incomplete survey of methods such as finite difference, finite element, and Pert will be discussed. The discussion will be restricted to quadrilateral meshes in 2D. Generalizations to 3D are straightforward. The third issue is the treatment of radiation-matter coupling. Even in the simplest case of Planckian diffusion, this problem arises as the equations governing radiation diffusion are coupled non-linearly to the material temperature. In multi-group diffusion this manifests itself as the material temperature being coupled to all photon frequencies

4.1. *Spatial Discretization*

4.1.1. *Finite Difference*

The type of spatial discretization performed on the diffusion operator is intimately connected with the mesh type. The simplest mesh often encountered is the static (Eulerian) orthogonal type. Finite difference methods are frequently applied in this case yielding a discrete representation of the diffusion or div-grad operator. Consider for example the 2D zone

(I, J+1/2)

(I-1/2, J)   (I, J)   (I+1/2, J)

(I, J-1/2)

Assume the independent variable (T_R or ε_v) is defined at the zone center.

Consider the diffusion operator as the gradient of the flux and integrate it over the cell volume of the (i,j) zone. The following result is obtained.

$$\frac{1}{V_{i,j}}\int_{(i,j)}dV\nabla\bullet\Phi = \frac{1}{V_{i,j}}\int_A dA\bullet\Phi$$

$$= \frac{1}{V_{i,j}}\left(\begin{array}{c} A_{i+1/2,j}\Phi_{i+1/2,j} + A_{i,j-1/2}\Phi_{i,j-1/2} \\ + A_{i-1/2,j}\Phi_{i-1/2,j} + A_{i,j+1/2}\Phi_{i,j+1/2} \end{array}\right) \quad (39)$$

Where A_{ab} and Φ_{ab} are the surface area and flux, respectively, at the (a, b) face. The flux at the face is just a function of the difference of the zone centered quantities. For example,

$$\Phi_{i+1/2,j} = \frac{D_{i+1/2,j}\left(\varepsilon_{i+1,j} - \varepsilon_{i,j}\right)}{\Delta x_{i+1/2,j}}$$

$$\Phi_{i,j+1/2} = \frac{D_{i,j+1/2}\left(\varepsilon_{i,j+1} - \varepsilon_{i,j}\right)}{\Delta y_{i,j+1/2}} \quad (40)$$

Rearranging terms yields

$$\frac{1}{V_{i,j}}\int_{(i,j)}dV\nabla\bullet\Phi$$

$$= \frac{1}{V_{i,j}}\left(\alpha_{i+1/2,j}\left[\varepsilon_{i+1,j} - \varepsilon_{i,j}\right] - \alpha_{i-1/2,j}\left[\varepsilon_{i,j} - \varepsilon_{i-1,j}\right]\right) \quad (41)$$

$$+ \frac{1}{V_{i,j}}\left(\alpha_{i,j+1/2}\left[\varepsilon_{i,j+1} - \varepsilon_{i,j}\right] - \alpha_{i,j-1/2}\left[\varepsilon_{i,j} - \varepsilon_{i,j-1}\right]\right)$$

Where $\alpha_{i+1/2,j} = \dfrac{A_{i+1/2,j}D_{i+1/2,j}}{\Delta x_{i+1/2,j}}$ and $\alpha_{i,j+1/2} = \dfrac{A_{i,j+1/2}D_{i,j+1/2}}{\Delta x_{i,j+1/2}}$

The time dependent diffusion equation in its simplest form (without sink or source terms) is almost ready to be solved. The last issue is the time stamp associated with the diffusion coefficients. In general these quantities depend on the matter temperature which depends on time. The simplest approach is to evaluate all diffusion coefficients (i.e. opacities) at the old time step value. This means that the discrete representation of the diffusion equation is semi-implicit. . That is, the solution at the advance time step can be written

$$\hat{\Phi}(t_{n+1}) = \Delta t \hat{M} \Phi(t_{n+1}) + \Delta t \hat{\Phi}(t_n) \tag{42}$$

The diffusion matrix M is symmetric and positive definite. In addition, it is an M-matrix meaning that any given diagonal element is the negative sum of the corresponding off-diagonal elements. This has important consequences in that it implies that the solution vector $\hat{\Phi}$ is positive. The structure of M is tri-diagonal with sub and super diagonals representative of the 5-point stencil in equation [41]. The time lag of the diffusion coefficient means that some type of time step control must be enforced. This is to ensure accuracy. Finally, the semi-implicit equation for $\hat{\Phi}$ can be made implicit simply by wrapping equation [41] in an iteration loop. It is of course required that the diffusion coefficient be continually updated with each iteration. Methods that do this will be discussed at the end.

4.1.2. *Finite Element and Pert Operators*

When the mesh is not aligned with the coordinate system, which can occur for example in radiation-hydrodynamic codes using Lagrangian or ALE (Arbitrary Lagrangian Eulerian) methods, complications can arise when constructing a discrete representation of the diffusion operator. In particular, preserving second order accuracy and positive-definiteness on non-uniform grids continues to be a challenge. In this section, two related methods (finite element and Pert), which have found wide use in codes using complex zoning will be discussed. The discussion will be based on variational methods and is based largely on unpublished notes by R. Tipton[19].

To begin, the time discretized diffusion equation with no sink or source terms can be derived from a variation of the "action"

$$\Theta = \int dV \left[\frac{1}{2} \frac{\Phi(t_{n+1})^2}{\Delta t} + \frac{1}{2} D \left| \nabla \Phi(t_{n+1}) \right|^2 - \frac{\Phi(t_{n+1})\Phi(t_n)}{\Delta t} \right] \tag{43}$$

Changing notation slightly in order to minimize indices, the action in discrete form yields

$$\Theta = \frac{1}{2} \sum_a C_a \Phi(a, t_{n+1})^2 - \frac{1}{2} \sum_{a,b} K_{ab} \Phi(a, t_{n+1}) \Phi(b, t_{n+1})$$
$$- \sum_a E_a \Phi(a, t_{n+1}) \tag{44}$$

Where a denotes node number (i.e. a= 1,2,3,4) and where

$$C_a = \int_{Aa} dV \frac{1}{\Delta t} \quad \text{and} \quad E_a = \int_a dV \frac{\Phi(a, t_n)}{\Delta t} \tag{45}$$

Note that

$$K_{ab} = \frac{\delta^2 \Theta}{\delta \Phi_a \delta \Phi_b} \tag{46}$$

The K matrix represents the spatial discretization of the diffusion operator. The choice of this operator determines the discrete form of the diffusion operator.

Consider the continuous form of the 'action". The finite element method assumes there exists a set of basis functions such that

$$\Phi(x, y) = \sum_a \varphi_a N_a(x, y) \tag{47}$$

Substituting this representation into the div-grad contribution to the action yields

$$\Theta_{Div-grad} = -\frac{1}{2} \sum_{a,b} K_{ab} \varphi_a \varphi_b \tag{48}$$
$$K_{ab} = -\int dV D \nabla N_a \bullet \nabla N_b$$

Tipton's application of finite elements requires two additional ingredients. One is the fact that finite elements like to have the unknowns live at the nodes whereas it has been assumed here that the unknowns live at the zone centers. Tipton defines a dual mesh whose nodes live at the zone centers of the regular

52

mesh. In addition, instead of using quadrilateral basis functions, he splits the quadrilateral into triangles and uses basis functions associated with them. Since there are two ways in 2D of splitting a quadrilateral into triangles, the K matrix is constructed as the result of averaging the two splittings. For a triangular basis function in an r-z cylindrical geometry,

$$K_{ab} = \frac{-2\pi r D s_a \bullet s_b}{2|s_a \times s_b|} \tag{49}$$

Where s_a is a vector defined on the node opposite node a. If the other two nodes are denoted by b and c, where b and c do not equal a, then s_a lies normal to the bc leg and its magnitude is equal to the length of bc. Defining θ_{ab} as the angle opposite nodes a and b, the K matrix becomes for a single triangular basis function,

$$K_{ab} = -\frac{2\pi r}{2} D \cot(\theta_{ab}) \quad \text{in r - z geometry}$$

$$K_{ab} = -\frac{1}{2} D \cot(\theta_{ab}) \quad \text{in planar geometry} \tag{50}$$

Whereas the finite element described above relies on triangular basis functions, the Pert[20] operator relies on bi-linear quadrilateral elements with two point quadrature. A single point quadrature with bi-linear quadrilateral elements will not work due to the fact that for square or rhomboid meshes, couplings between neighboring zones occurs only through corner couplings and not through faces. Hence, the mesh can develop the so-called checkerboard instability.

Consider the mesh in logical coordinates. Consider a specific zone whose nodes are identified with the iso-parametric coordinates $(\xi, \eta) = (\pm 1, \pm 1)$. The bi-linear basis functions are simply

$$N_1 = \frac{1}{4}(1 + \xi)(1 - \eta) \quad N_2 = \frac{1}{4}(1 + \xi)(1 + \eta)$$

$$N_3 = \frac{1}{4}(1 - \xi)(1 + \eta) \quad N_4 = \frac{1}{4}(1 - \xi)(1 - \eta) \tag{51}$$

The div-grad contribution to the action comes about by substituting equation [51] into equation [48]. The result is

$$\Theta_{Div-grad} = 2\pi \int \frac{d\xi \, d\eta \, Dr}{2J}\left[g_{\xi\xi}\left(\frac{\partial\varphi}{\partial\xi}\right)^2 + g_{\eta\eta}\left(\frac{\partial\varphi}{\partial\eta}\right)^2 + g_{\xi\eta}\left(\frac{\partial\varphi}{\partial\xi}\frac{\partial\varphi}{\partial\eta}\right) \right]$$

The g's are the metric tensor and it represents the coordinate transformation between the logical coordinates and physical coordinates. For example,

$$g_{\xi\eta} = \frac{\partial x}{\partial\xi}\frac{\partial x}{\partial\eta} + \frac{\partial y}{\partial\xi}\frac{\partial y}{\partial\eta} \tag{52}$$

The Pert representation for the diffusion operator comes from a two-point quadrature approximation to the integral in the above equation for $\Theta_{Div-grad}$. This is done by taking the dual mesh quadrilateral and decomposing it into four equal area sections. Therefore, the K matrix becomes

$$\begin{pmatrix} -\frac{4\pi Dr}{A}\left(g_{\xi\xi}+g_{\eta\eta}+g_{\xi\eta}\right) & \frac{4\pi Dr}{A}g_{\eta\eta} & \frac{4\pi Dr}{A}g_{\xi\eta} & \frac{4\pi Dr}{A}g_{\xi\xi} \\ \frac{4\pi Dr}{A}g_{\eta\eta} & -\frac{4\pi Dr}{A}\left(g_{\xi\xi}+g_{\eta\eta}+g_{\xi\eta}\right) & \frac{4\pi Dr}{A}g_{\xi\xi} & -\frac{4\pi Dr}{A}g_{\xi\eta} \\ \frac{4\pi Dr}{A}g_{\xi\eta} & \frac{4\pi Dr}{A}g_{\xi\xi} & -\frac{4\pi Dr}{A}\left(g_{\xi\xi}+g_{\eta\eta}+g_{\xi\eta}\right) & \frac{4\pi Dr}{A}g_{\eta\eta} \\ \frac{4\pi Dr}{A}g_{\xi\xi} & -\frac{4\pi Dr}{A}g_{\xi\eta} & \frac{4\pi Dr}{A}g_{\eta\eta} & -\frac{4\pi Dr}{A}\left(g_{\xi\xi}+g_{\eta\eta}+g_{\xi\eta}\right) \end{pmatrix}$$

4.1.3. *Local Support Operators*

Recently, Jim Morel[25] and co-workers have developed a promising 2D and 3D diffusion discretization scheme based on local support operators. Like the methods discussed above, it yields a sparse matrix representation for the div-grad operator. The method has definite advantages. It is second order accurate on distorted meshes, rigorously treats material discontinuities, and has a symmetric positive definite matrix. The disadvantage of the method is that it requires face center as well as cell center unknowns. There is some subtlety regarding implementing flux limiters into the local support scheme. However, David Miller[26] has done this successfully

4.2. *Multi-group and Planckian Diffusion*

To summarize, the solution of the material temperature and spectrum needs to be found from the following coupled equations

$$\frac{\partial \varepsilon_v(r,t)}{\partial t} - \nabla \bullet \left(\frac{c}{3\sigma_v} \nabla \varepsilon_v(r,t) \right) = \left[4\pi\sigma_v B_v(T) - c\sigma_v \varepsilon_v(r,t) \right] \quad (53)$$

$$\frac{\partial \left[\rho C_V T(r,t) \right]}{\partial t} = -\int dv \sigma_v \left[B_v(T) - \varepsilon_v(r,t) \right] \quad (54)$$

The inherent difficulty is solving this set of equations is that even though the equation obeyed by the spectrum looks like it is independent of frequency group coupling, the material equation requires knowledge of the spectrum and emission at all frequencies. Hence, the emission function also requires knowledge of the spectrum at all frequencies. Therefore, the equation for the spectrum is inherently non-linear due to an effective group to group coupling. The partial temperature and grey methods are examples of techniques which have found success in dealing with this problem. These are discussed next. Planckian diffusion can be thought of as a subset of multi-group diffusion. Physically of course, Planckian diffusion arises from assuming the spectrum is Planckian and then integrating over all frequency. Numerically, we will see that in the partial temperature scheme it can be merely thought of as a special case.

In all of these methods, heat capacities and opacities are time lagged so even though the solution methods are not fully implicit. In addition, due to this fact, all of the methods considered will require some sort of time step control on the material and/or radiation temperature. As we will see, the partial temperature method requires a very specific type of time step control. In addition, the methods here should be though of in the wider context of a multi-physics code. Traditionally, physics packages are operator split. Whether this is done first order or second order in time, this fact alone limits the accuracy even in cases where the radiation package is solved fully implicitly. Examples of fully implicit methods will be discussed in section 4.2.3.

The implicit time differencing and the spatial discretization mean that the coupled radiation-material equations will form a system of equations. In general these systems will be non-linear. Due to a variety of techniques, to be discussed next, this non-linear set can be approximated by a system of linear equations. Therefore, linear solvers play an important role in the methods presented here. It is due to advances in linear solvers and their preconditioners that have made radiation diffusion problems in more complex geometries and in 3D possible. This is due in no small part to scaleable (both in the problem size and parallel sense algorithms[21]).

4.2.1. *Multi-group Diffusion: Partial Temperature*

The coupled set of equations is differenced implicitly assuming that the heat capacities and opacities are evaluated at the old time step. Ignoring for the moment issues related to spatial discretization and setting all internal and external sources to zero, we obtain

$$\frac{\left(\varepsilon_v^{n+1} - \varepsilon_v^n\right)}{c\Delta t} = \nabla \bullet \left(\frac{1}{3\sigma_v}\nabla \varepsilon_v^{n+1}\right) + \sigma_v\left(4\pi B_v(T^{n+1}) - \varepsilon_v^{n+1}\right) \qquad (55)$$

$$\rho C_V \frac{\left(T^{n+1} - T^n\right)}{\Delta t} = -\sum_v \Delta v \sigma_v \left(4\pi B_v(T^{n+1}) - \varepsilon_v^{n+1}\right) \qquad (56)$$

The summation is taken over the group index and runs from zero to N_g . Notice that the Planckian diffusion scheme is just a special case of $v = 1$ where $B_v(T) \propto T^4$. Now assume there exist a partial material temperature contribution for each group. That is,

$$T^{n+1} - T^n = \sum_v \left[\tau_{v+1}^{n+1} - \tau_v^{n+1}\right] \text{ where } \tau_{N_g+1}^{n+1} = T^{n+1} \text{ and } \tau_0^{n+1} = T^n$$

$$\frac{\rho C_V}{\Delta t}\left[\tau_{v+1}^{n+1} - \tau_v^{n+1}\right] = -\left(\sigma_v\left(4\pi B_v(T^{n+1}) - \varepsilon_v^{n+1}\right)\right) \qquad (57)$$

We now approximate the emission or Planck function as if it were coming from each group. That is,

$$B_v(T^{n+1}) \Rightarrow B_v(\tau_{v+1}^{n+1}) \approx B_v(\tau_v^{n+1}) + B_v'(\tau_{v+1}^{n+1})\left[\tau_{v+1}^{n+1} - \tau_v^{n+1}\right] \qquad (58)$$

The following linear system results,

$$\left[\tau_{v+1}^{n+1} - \tau_v^{n+1}\right] = \frac{\left(-4\pi\sigma_v B_v(\tau_v^{n+1}) + \sigma_v \varepsilon_v^{n+1}\right)}{\left(\frac{\rho C_V}{\Delta t} + 4\pi\sigma_v B_v'(\tau_{v+1}^{n+1})\right)} \qquad (59)$$

56

Upon approximating the emission or Planck function as before, the multi-group photon equation becomes

$$\frac{\left(\varepsilon_v^{n+1} - \varepsilon_v^n\right)}{c\Delta t} = \nabla \bullet \left(\frac{1}{3\sigma_v} \nabla \varepsilon_v^{n+1}\right)$$

$$+ \sigma_v \left(4\pi B_v (\tau_v^{n+1}) + 4\pi B_v' (\tau_{v+1}^{n+1})\left[\tau_{v+1}^{n+1} - \tau_v^{n+1}\right] - \varepsilon_v^{n+1}\right) \tag{60}$$

Substituting equation [59] into equation [60] yields a linear implicit solution for ε_v^{n+1}. This may be solved via a variety of preconditioned linear solvers such as multi-grid preconditioned conjugate gradient[21].

A word of caution concerning partial temperatures is in order. The method is stable and robust. However, it can suffer in accuracy unless the partial temperature swings for each group from cycle to cycle are limited via a time step control. In Planckian diffusion, since there is just one group, the temperature swing is just the change in matter temperature from cycle to cycle.

4.2.2. Multi-group Diffusion: Iterative Grey Methods

There exists another class of methods that attempt to solve the matter-radiation coupling problem via an iterative procedure. These methods require a grey diffusion accelerator in order to speed-up convergence. We will discuss the fully implicit method of Lund-Wilson[22] and its variant due to Eppley[22]. We will then discuss a semi-implicit method due to Morel, Larsen, and Matzen[22] and its fully implicit extension due to Graziani[22].

The Lund-Wilson method replaces the set of multi-group diffusion equations with the set

$$\frac{\left(\varepsilon_v^{l+1} - \varepsilon_v^n\right)}{c\Delta t} = \nabla \bullet \left(\frac{1}{3\sigma_v} \nabla \varepsilon_v^{l+1}\right)$$

$$+ \sigma_v \left(4\pi\left[B_v (T^l) + B_v' (T^l)\left(T^{l+1} - T^l\right)\right] - \varepsilon_v^{l+1}\right) \tag{61}$$

$$\rho C_v \frac{\left(T^{l+1} - T^n\right)}{\Delta t} =$$

$$- \sum_v \Delta v \sigma_v \left(4\pi\left[B_v (T^l) + B_v' (T^l)\left(T^{l+1} - T^l\right)\right] - \varepsilon_v^{l+1}\right) \tag{62}$$

The index I is an iteration index where I=0 corresponds to the old time step values. Note that upon convergence, $T^{I+1} \rightarrow T^I$ and the coupled multi-group equations are recovered. Define the grey coefficients,

$$\Sigma_1^I = 4\pi\Delta t \sum_v \Delta v \sigma_v B_v(T^I)$$

$$\Sigma_2^I = 4\pi\Delta t \sum_v \Delta v \sigma_v B_v'(T^I)$$

$$\mathrm{E}^I = \frac{1}{c} \sum_v \Delta v \varepsilon_v^I \tag{63}$$

$$\alpha^I = \frac{\Delta t \sum_v \Delta v \sigma_v \varepsilon_v^I}{\mathrm{E}^I}$$

E^I is proportional to an effective radiation temperature raised to the fourth power. α^I is spectrum averaged opacity. In the matter equation, we make the approximation

$$\Delta t \sum_v \Delta v \sigma_v \varepsilon_v^{I+1} = \left(\frac{\Delta t \sum_v \Delta v \sigma_v \varepsilon_v^{I+1}}{\mathrm{E}^{I+1}} \right) \mathrm{E}^{I+1} \approx \alpha^I \mathrm{E}^{I+1} \tag{64}$$

The latter approximation is exact upon convergence. Using the grey coefficients, the, matter equation becomes

$$T^{I+1} - T^I = \frac{\left[(T^n - T^I) - \dfrac{\left[\Sigma_1^I - \alpha^I \mathrm{E}^{I+1}\right]}{\rho C_v} \right]}{\left[1 + \dfrac{\Sigma_2^I}{\rho C_v} \right]} \tag{65}$$

This equation can be solved provided we know E^{I+1}. This quantity is obtained by summing the multi-group photon equation over frequency. There are two ways of doing this depending on how the diffusion coefficients are averaged.

Consider the diffusion contribution to the multi-group photon equation. Also, we assume a simple 1D finite difference representation of the div-grad operator. We can form an equation for E^{I+1} by summing the photon multi-group equation over frequency group.

$$\int dv \nabla \bullet \left(\frac{1}{3\sigma_v} \nabla \varepsilon_v^{l+1} \right) \Rightarrow \sum_v \Delta v D_v (i+1/2) \left[\varepsilon_v^{l+1}(i+1) - \varepsilon_v^{l+1}(i) \right]$$

$$- \sum_v \Delta v D_v (i-1/2) \left[\varepsilon_v^{l+1}(i) - \varepsilon_v^{l+1}(i-1) \right] \tag{66}$$

Lund and Wilson approximate this expression as follows,

$$\sum_v \Delta v D_v (i+1/2) \left[\varepsilon_v^{l+1}(i+1) - \varepsilon_v^{l+1}(i) \right]$$

$$= \left(\frac{\sum_v \Delta v D_v (i+1/2) \left[\varepsilon_v^{l+1}(i+1) - \varepsilon_v^{l+1}(i) \right]}{\sum_v \Delta v \left[\varepsilon_v^{l+1}(i+1) - \varepsilon_v^{l+1}(i) \right]} \right) \left[E^{l+1}(i+1) - E^{l+1}(i) \right] \tag{67}$$

$$\equiv \gamma^{l+1}(i+1/2) \left[E^{l+1}(i+1) - E^{l+1}(i) \right]$$

$$\approx \gamma^l (i+1/2) \left[E^{l+1}(i+1) - E^{l+1}(i) \right]$$

The equation for E^{l+1} becomes,

$$E^{l+1} - E^n = c\Sigma_1^l + c\Sigma_2^l \left[T^{l+1} - T^l \right] - c\alpha^l E^{l+1}(i)$$

$$+ \gamma^l (i+1/2) \left[E^{l+1}(i+1) - E^{l+1}(i) \right] - \gamma^l (i-1/2) \left[E^{l+1}(i) - E^{l+1}(i-1) \right] \tag{68}$$

This is the so-called grey equation. The above equation is a tri-diagonal equation for E^{l+1} which can be solved via back-substitution methods. In two or three dimensions, the above equation is a matrix which can be solved via preconditioned conjugate gradient methods[21]. In solving the coupled multi-group equations, the following steps must be performed

1. Evaluate the grey coefficients [63] using the last available spectrum
2. Solve the grey equation for E^{l+1}
3. Compute the updated temperature T^{l+1}
4. Knowing T^{l+1}, compute the updated spectrum ε_v^{l+1}
5. Is the matter temperature converged $\left| T^{l+1} - T^l \right| < \delta$?
6. If no, repeat steps 1-5 using the latest spectrum to compute the grey coefficients.

7.   If yes, values for matter temperature and spectrum are accepted.

8.

The method due to Lund and Wilson seems to work well in most cases. It does have one drawback however. The grey averaged diffusion coefficients are not guaranteed to be positive. This can cause havoc for matrix solvers. Eppley's method tries to circumvent this problem by defining new grey averaged diffusion coefficients. In Eppley's scheme, instead of one type of grey diffusion coefficient, he defines two. Namely,

$$
\sum_\nu \Delta \nu D_\nu (i+1/2)\varepsilon_\nu^{l+1}(i+1)
$$

$$
= \left(\frac{\sum_\nu \Delta \nu D_\nu (i+1/2)\varepsilon_\nu^{l+1}(i+1)}{\sum_\nu \Delta \nu \varepsilon_\nu^{l+1}(i+1)} \right) \mathrm{E}^{l+1}(i+1)
$$

$$
\equiv \delta^{l+1}(i+1/2)\mathrm{E}^{l+1}(i+1) \approx \delta^{l}(i+1/2)\mathrm{E}^{l+1}(i+1) \tag{69}
$$

$$
\sum_\nu \Delta \nu D_\nu (i+1/2)\varepsilon_\nu^{l+1}(i)
$$

$$
= \left(\frac{\sum_\nu \Delta \nu D_\nu (i+1/2)\varepsilon_\nu^{l+1}(i)}{\sum_\nu \Delta \nu \varepsilon_\nu^{l+1}(i)} \right) \mathrm{E}^{l+1}(i) \equiv \Delta^{l+1}(i+1/2)\mathrm{E}^{l+1}(i) \tag{70}
$$

$$
\approx \Delta^{l}(i+1/2)\mathrm{E}^{l+1}(i)
$$

The grey equation is now slightly modified from before,

$$
\mathrm{E}^{l+1} - \mathrm{E}^{n} = c\Sigma_1^{l} + c\Sigma_2^{l}\left[T^{l+1} - T^{l}\right] - c\alpha^{l}\mathrm{E}^{l+1}(i)
$$

$$
+ \delta^{l}(i+1/2)\mathrm{E}^{l+1}(i+1) - \Delta^{l}(i+1/2)\mathrm{E}^{l+1}(i)
$$

$$
- \left[\delta^{l}(i-1/2)\mathrm{E}^{l+1}(i) - \Delta^{l}(i-1/2)\mathrm{E}^{l+1}(i-1)\right]
$$

Eppley's method produces positive definite grey diffusion coefficients. However, the cost is a non-symmetric matrix for the grey equation. In practice, the robustness of Eppley's variant of Lund and Wilson seems to work well. Methods such as GMRES are satisfactory for solving the non-symmetric grey equation. Methods that attempt to solve the grey equation by splitting the matrix into symmetric and asymmetric contributions and including the asymmetric parts in the overall iteration loop do not seem to be robust. Although they have the advantage of allowing the one to use symmetric matrix solver methods, the method occasionally fails to converge.

Even though it has not been explicitly mentioned as such, the solution of the grey equation in the Lund-Wilson and Eppley variant accelerates the iterative process. Without the grey solution step, the number of iterations can grow into the thousands in regimes of the problem where radiation and matter are tightly coupled. This is certainly true in the next method where a grey accelerator is explicitly introduced.

The starting point of the source iteration method of Morel, Larsen, and Matzen[22] is to expand the Planck or emission function, evaluated at the updated temperature, about its value of the previous time steps' temperature. That is

$$B_v(T^{n+1}) \approx B_v(T^n) + B_v(T^n)\left[T^{n+1} - T^n\right] \tag{72}$$

This is performed in both the photon and the material energy balance equations. By doing this expansion, the coupled set of multi-group equations is effectively linearized. Substituting the above equation into the material energy balance equation and solving for $T^{n+1} - T^n$ yields

$$T^{n+1} - T^n = \frac{\sum_v \Delta v \sigma_v \left[\varepsilon_v^{n+1} - 4\pi B_v(T^n)\right]}{\left[\dfrac{\rho C_v}{\Delta t} + \sum_v \Delta v \sigma_v B_v'(T^n)\right]} \tag{73}$$

Substituting this expression into the photon equation yields

$$-\nabla \bullet \left(\frac{1}{3\sigma_v} \nabla \varepsilon_v^{n+1}\right) + \left(\frac{1}{c\Delta t} + \sigma_v\right)\varepsilon_v^{n+1}$$

$$= Q_v^n + \frac{\varepsilon_v^n}{c\Delta t} + \eta \chi_v \sum_v \Delta v \sigma_v \varepsilon_v^{n+1} \tag{74}$$

Where a new set of grey coefficients arises,

$$\eta = \frac{4\pi \sum_{v} \Delta v \sigma_v B'_v (T^n)}{\dfrac{\rho C_v}{\Delta t} + 4\pi \sum_{v} \Delta v \sigma_v B'_v (T^n)}$$

$$\chi_v = \frac{\Delta v \sigma_v B'_v (T^n)}{\sum_{v} \Delta v \sigma_v B'_v (T^n)} \tag{75}$$

$$Q_v = \sigma_v B_v (T^n) - 4\pi \eta \chi_v \sum_{v} \Delta v \sigma_v B_v (T^n)$$

This is a linearized form for the photon multi-group equation. At this point the group to group coupling still exists. The source iteration process involves replacing the above by

$$-\nabla \bullet \left(\frac{1}{3\sigma_v} \nabla \varepsilon_v^{I+1} \right) + \left(\frac{1}{c\Delta t} + \sigma_v \right) \varepsilon_v^{I+1}$$

$$= Q_v^I + \frac{\varepsilon_v^n}{c\Delta t} + \eta \chi_v \sum_{v} \Delta v \sigma_v \varepsilon_v^I \tag{76}$$

Therefore, when $\varepsilon_v^{I+1} \approx \varepsilon_v^I$, the above equation converges to the linearized form given by equation [74]. Note that whereas with the Lund-Wilson method we converged on matter temperature, here we converge on the spectrum. In practice, a convergence criteria based on converging the spectrum over all groups is not necessary. It seems to be sufficient to converge on the radiation temperature or the radiation energy density. The benefit is a smaller number of iterations with very little loss of accuracy. Therefore, the procedure is

1. Evaluate the grey coefficients using the old time step values for temperature and spectrum
2. Solve the photon equation [76] via iteration
3. Once ε_v^{I+1} is converged, compute the material temperature

Notice that the grey coefficients are evaluated once, at the start of the cycle, as opposed to Lund-Wilson where they are evaluated at each iteration. . In addition, here the material temperature is evaluated once the spectrum is converged whereas in Lund-Wilson, it is updated every iteration. In practice the source iteration method works well except the number of iterations to converge the photon equation rises steeply in optically thick regimes. In order to correct this deficiency, the iteration process is accelerated via a grey equation.

In the paper by Morel, Larsen, and Matzen, besides the spectrum, they also define a quantity which is the difference of the spectrum between the exact solution and the latest guess from iterate I. We denote this quantity as,

$$\delta \varepsilon_v = \varepsilon_v - \varepsilon_v^I \tag{77}$$

In terms of this variable, the multi-group photon equation becomes

$$-\nabla \bullet \left(\frac{1}{3\sigma_v} \nabla \delta \varepsilon_v^I \right) + \left(\frac{1}{c\Delta t} + \sigma_v \right) \delta \varepsilon_v^I \\ = \eta \chi_v \sum_v \Delta v \sigma_v \left[\delta \varepsilon_v^I + \varepsilon_v^I - \varepsilon_v^I \right] \tag{78}$$

In this method, the grey equation is derived by assuming that the multi-group spectrum is given by the equilibrium spectrum. The equilibrium spectrum in turn is given by the solution to equation [74] where the gradient term vanishes. That is,

$$\delta \varepsilon_v \big|_{eq} = \frac{E^I \chi_v \Gamma}{\dfrac{1}{c\Delta t} + \sigma_v} \tag{79}$$

Where the grey coefficient is defined by

$$\Gamma = \left[\sum_v \Delta v \left(\frac{\chi_v}{\dfrac{1}{c\Delta t} + \sigma_v} \right) \right]^{-1} \tag{80}$$

Substituting this expression into equation [78] and integrating over groups yields,

$$-\nabla \bullet \left(\langle D \rangle \nabla E' + \langle \vec{D} \rangle E' \right) + \left[\frac{1}{c\Delta t} + (1-\eta)H \right] E' = \eta \sum_v \sigma_v \left(\varepsilon_v^{l+1} - \varepsilon_v^{l-1} \right) \tag{81}$$

The grey coefficients are given by

$$\langle D \rangle = H \sum_v \Delta v \left(\frac{\chi_v}{\frac{1}{c\Delta t} + \sigma_v} \right) \left(\frac{1}{3\sigma_v} \right) \tag{82}$$

$$\langle \vec{D} \rangle = H \sum_v \Delta v \left(\frac{1}{3\sigma_v} \right) \nabla \left(\frac{\Gamma \chi_v}{\frac{1}{c\Delta t} + \sigma_v} \right) \tag{82}$$

$$H = \Gamma \sum_v \Delta v \frac{\chi_v \sigma_v}{\frac{1}{c\Delta t} + \sigma_v}$$

The multi-group solution procedure is similar to the unaccelerated procedure

1. Evaluate the grey coefficients using the old time step information
2. Update the spectrum using the photon multi-group equation
3. Solve the grey equation for the integrated spectrum
4. Correct the spectrum by adding the correction term $\delta \varepsilon_v \big|_{eq} = \dfrac{E' \chi_v \Gamma}{\frac{1}{c\Delta t} + \sigma}$ to the result from step 2
5. Repeat steps 2-4 until convergence is reached

It would seem that here again the convergence criterion must be based on converging the spectrum in all groups. In practice, however, it turns out to be sufficient to converge on the radiation temperature or the radiation energy density. The number of iterations can be substantially reduced with little loss of accuracy. A more important issue is the grey equation. Unfortunately, the grey equation, like Eppley, does not have a symmetric matrix. This is due to the $\langle \tilde{D} \rangle$ term. Without any justification, dropping this term does not seem to harm the acceleration process significantly. This is frequently done in practice. However, this issue needs to be looked at closer.

There exists a variant of the Morel, Larsen, and Matzen method due to Graziani that solves the non-linear coupled multi-group equations. The starting point is to expand the emission or Planck function not around the old time step but rather the old iterate. That is,

$$B_v(T^{l+1}) \approx B_v(T^l) + B_v(T^l)\left[T^{l+1} - T^l\right] \tag{83}$$

We go through the same algebra as previously done however, at the end we end up with a slightly different equation for the spectrum.

$$-\nabla \bullet \left(\frac{1}{3\sigma_v}\nabla \varepsilon_v^{l+1}\right) + \left(\frac{1}{c\Delta t} + \sigma_v\right)\varepsilon_v^{l+1}$$
$$= \tilde{Q}_v^l + \frac{\varepsilon_v^n}{c\Delta t} + \eta \chi_v \sum_v \Delta v \sigma_v \varepsilon_v^l \tag{84}$$

The grey coefficients are defined by

$$\eta = \frac{4\pi \sum_v \Delta v \sigma_v B_v'(T^n)}{\frac{\rho C_v}{\Delta t} + 4\pi \sum_v \Delta v \sigma_v B_v'(T^n)}$$

$$\chi_v = \frac{\Delta v \sigma_v B_v'(T^n)}{\sum_v \Delta v \sigma_v B_v'(T^n)} \tag{84}$$

$$\tilde{Q}_v = \sigma_v B_v(T^n) - 4\pi \eta \chi_v \left\{\frac{\rho C_v \left(T^l - T^n\right)}{\Delta t} - \sum_v \Delta v \sigma_v B_v(T^n)\right\}$$

Therefore, the "Q" term is slightly modified. The procedure for solving the non-linear multi-group variant follows the same steps as the linear method. However, the convergence criterion is now based on

$$\left|T^{l+1} - T^l\right| < \delta$$

The relative merits of solving the non-linear variant of Morel, Larsen, and Matzen has not been investigated. Further work needs to be done.

4.2.3. *Comments on Full Matrix Methods*

With the advent of increased memory and the progress in preconditioned linear solvers, the idea for fully implicit methods has started looking like an attractive alternative the methods discussed above. We give here some reference to a sampling of the research in this area. Rider, Knoll, and Olson[23] used Newton-Krylov methods along with multi-grid preconditioning to 1D and 2D one temperature Planckian diffusion. They showed that the fully implicit method gave increased accuracy over the usual semi-implicit method where opacities and heat capacities were lagged. Simultaneously, building on earlier work using the ODE integrator methods of Axelrod, Dubois, and Rhodes[24], Brown, Chang, Graziani, and Woodward[24] applied these techniques to 3D multi-group diffusion. Mousseau, Knoll, and Rider[24] extended their earlier work to two temperature Planckian diffusion. Recently, Brown, Shumaker, and Woodward[24] considered fully implicit versus semi-implicit methods where tabulated opacities are used and external sources coming from thermonuclear fusion are included. This latter issue is important as the source term in the radiation equations coming from fusion source terms is a very strong function of temperature and hence places a constraint of time steps and accuracy. There conclusions were that a fully implicit method can achieve more accurate solutions than semi-implicit methods at a cost comparable to semi-implicit methods. In addition, their method is scales very well on parallel machines.

Acknowledgments

The author wishes to thank Tony Mezzacappa for his kind invitation to write a paper for this volume.

References

1. F. Graziani and G. Olson, "Transport Methods: Conquering the Seven Dimensional Mountain", SCaLeS report, UCRL-JC-154432.
2. D. Mihalas and B. Mihalas, "Foundations of Radiation Hydrodynamics", Oxford University Press (1984).
3. G. Pomraning, "The Equations of Radiation Hydrodynamics", Pergamon Press (1973).
4. R. Bowers and J. Wilson, "Numerical Modeling in Applied Physics and Astrophysics", Jones and Bartlett Publishers (1991).

5. J. Castor, "Radiation Hydrodynamics", Cambridge University Press (2004)
6. S. Chandrasekhar, "Radiative Transfer", Dover Publications Inc. (1960).
7. H. Griem, "Principles of Plasma Spectroscopy", Cambridge University Press (1997).
8. M. Boulos, P. Fauchais, and E. Pfender, "Thermal Plasmas", Plenum Press (1994).
9. R. Gelinas and R. Ott, *Ann. Phys* **59**, 323 (1970).
10. C. J. Cannon, "The Transfer of Spectral Line Radiation", Cambridge University Press (1985).
11. Degl'Innocenti
12. F. Graziani, *JQSRT* **83**, 711 (2004).
13. J. Stone, M. Norman and D. Mihalas, *Astrophys. J. Suppl.* **80**, 819 (1992).
14. S. Atzeni and J. Meyer-Ter-Vehn, "The Physics of Inertial Confinement Fusion", Oxford Science Publications (2004).
15. Telegrapher
16. G. Olson, L. Auer, and M. Hall, JQSRT 64, 619 (2000).
17. D. Levermore and G. Pomraning, *Astrophys. J.* **248**, 321 (1981).
18. E. Livne and A. Glasser, *J. Comp. Phys.* **58**, 59 (1985); Richardson, Ferrell, and Long, *J. Comp. Phys.* **104**, 69 (1993); F. Graziani, *J. Comp. Phys.* **118**, 9, (1995).
19. R. Tipton, private notes LLNL (2004).
20. G. Pert, *J. Comp. Phys.* **42**, 20 (1981).
21. C. Baldwin, P. Brown, R. Falgout, F. Graziani and J. Jones, *J. Comp. Phys.* **154**, 1 (1999).
22. C. Lund and J. Wilson in "Numerical Astrophysics", Jones and Bartlett Publishers (1985); K. Eppley, private notes (1981); J. Morel, E. Larsen, and M Matzen, *JQSRT* **34**, 243 (1985); A. Winslow, *J. Comp. Phys.* **117**, 262 (1995); F. Graziani, private notes (2003).
23. D. Knoll, W. Rider, and G. Olson, *JQSRT* **63**, 15 (1999); W. Rider, D. Knoll, and G. Olson, *J. Comp. Phys.* **152**, 164 (1999); V. Mousseau, D. Knoll, W. Rider, *J. Comp. Phys.* **160**, 743 (2000).
24. T. Axelrod, P. Dubois, and C. Rhodes, *J. Comp. Phys.* **54**, 205 (1984); P. brown, B. Chang, F. Graziani, C. Woodward, in, "Iterative Methods in Scientific Computation IV", 343 (1999); P. Brown and C. Woodward, *SIAM J. Sci. Comput.* **23**, 499 (2001); P. Brown, D. Shumaker and C. Woodward, "Fully Implicit Solution of Large-Scale Non-Equilibrium Radiation Diffusion with High order Time Integration", preprint (2004).
25. J. Morel, R. Roberts and M. Shashkov, *J. Comp. Phys.* **144**, 17 (1998); J. Morel, M. Hall and M. Shashkov, J. Comp. Phys. 170, 338 (2001)
26. D. Miller, private notes LLNL (2003).

THE MAGNETOROTATIONAL INSTABILITY

JOHN F. HAWLEY

Department of Astronomy
University of Virginia
PO Box 3818
Charlottesville, VA 22903, USA
E-mail: jh8h@virginia.edu

The magnetorotational instability (MRI) has proven to be of fundamental importance to accretion disk systems. If core collapse supernovae contain weak magnetic fields and differential rotation then this instability should again be present. This paper reviews the basic properties of the MRI and what has been learned in disk studies. While attention is often focused on the exponential growth of the magnetic field from the MRI, the more general point is that a magnetic field fundamentally alters the basic stability criteria in a rotating plasma, regardless of the field's strength. Thus, magnetic fields are likely to be important in a core collapse supernova regardless of whether or not they are amplified to equipartition with the thermal, rotational, or gravitational energy densities.

1. Introduction

Astronomers seem to have a long tradition of ambivalence toward magnetic fields. While it has long been acknowledged that magnetic fields are present in the universe, there has been a prevailing assumption (or at least fond hope) among some astrophysicists that hydrodynamics alone would suffice to describe the basic dynamical properties of most astronomical systems. Theorists would acknowledge the potential importance of magnetic fields, usually in brief asides or parenthetical comments, but more or less assert that unless the magnetic energy was clearly the dominant component of the system the fields generally played a minor secondary dynamical role, such as changing the effective equation of state.

This impression, of course, slightly exaggerates the situation. In my own area of study, namely accretion disks, there has always been a significant minority of workers who emphasized a variety of magnetohydrodynamic (MHD) processes within accreting systems; the review of Blandford (1989) summarizes much of this work. A further confounding factor is that even if the importance of magnetic fields had been recognized 20 years ago, the equations are very difficult to solve except in the most simplified manner. The last decade, however, has seen a significant transition within the disk community, away from a hydrodynamic paradigm to one that is fundamentally MHD. There are at least two reasons for this. First, the elucidation of the accretion disk magnetorotational instability (MRI; Balbus & Hawley 1991) has led to an emerging consensus that magnetic fields are (literally) the driving force behind high-energy accretion processes, whether the fields are weak or strong. The second reason is the exponential increase in computational power available to astrophysicists, and the development of multidimensional MHD simulation codes. While it is not yet true that any arbitrary flow can now be simulated, numerical modeling has truly become an indispensable theoretical tool, allowing us to make headway into some very complex problems that simply cannot be approached by other methods with nearly the same degree of confidence.

One of the questions considered at this workshop is, how important might magnetic fields be for core collapse supernovae? Again, while attention has been directed toward possible roles for magnetic fields for over 30 years (e.g., LeBlanc & Wilson 1970; Bisnovatyi-Kogan 1971) as usual the issue has been whether the magnetic field becomes strong enough to have a significant dynamical effect. Craig Wheeler (these proceedings; also Akiyama et al. 2003) has emphasized that the MRI provides a possible route through which weak magnetic fields can play a significant role. In this paper I shall review results from accretion disk studies that have led to two important conclusions. First, as a linear instability the MRI can amplify magnetic field strength exponentially; what was a weak field initially could then become quite strong. Second, magnetic fields fundamentally alter the basic stability criteria within rotating systems, even when those fields are weak.

I recall hearing a quip about about twenty years ago (and it is probably much older than that) that stated "The magnetic field strength is proportional to our ignorance." To me this expressed quite clearly an attitude that magnetic fields were invoked only as a last measure by desperate people who could not otherwise explain an astrophysical system. It is possible that some in the supernova community are beginning to feel a bit desperate since their models aren't exploding. Is it time to invoke magnetic fields? It is certainly not the case that magnetic fields will play a decisive role in every astrophysical system that we don't currently understand, or even in core-collapse supernovae. But history suggests that if one *assumes* that magnetic fields play no role, one may well be on dangerous ground.

2. MRI Basics

The astrophysical importance of the MRI was first discovered in the context of accretion disks, and in this section I will review the basic properties of the instability within that context. Accretion disks are powered by the release of gravitational energy as gas spirals down onto a compact star. The process of accretion requires a torque to remove angular momentum from the orbiting fluid. The classic accretion disk puzzle has been that the gas in astrophysical disks is not sufficiently viscous to allow for accretion: the Reynolds number (the ratio of the characteristic velocity times length divided by the fluid's viscosity) of an accretion disk is huge. In terrestrial contexts, however, fluid systems with high Reynolds number are often turbulent. This led to the widespread assumption that disks should be as well and that the resulting turbulent stresses would transport angular momentum outward at a sufficient rate to account for the observed accretion. Interestingly, this assumption has been in place from the very beginnings of disk theory. Turbulence is invoked in the seminal accretion paper of Crawford & Kraft (1956), who wrote "The Reynolds number is extremely high, and it is quite certain that the gas is turbulent....The velocity gradient will cause a very rapid turbulent exchange of angular momentum, so that the inner part of the ring will lose angular momentum and move in closer, while the outer part gains angular momentum and spreads out." That, in a nutshell, was the state of disk angular momentum transport theory for the next 35 years.

The problem was that although everyone was certain of the existence of disk turbulence, nobody knew quite how to produce it. The reason is that disks are hydrodynamically stable to the Rayleigh criterion, which simply requires that the angular momentum increase in value radially outward. Indeed, the Keplerian angular momentum distribution, $l \propto R^{1/2}$, is not just stable, it is *stabilizing*. In hydrodynamics the angular momentum is a conserved quantity. A fluid element will conserve its angular momentum unless subjected to nonaxisymmetric pressure forces. Displacing a fluid element radially while conserving its angular momentum results in counter-rotating epicyclic oscillations around the center of the orbit. The flow is stable.

Considerable effort has been devoted to a quest for a hydrodynamic mechanism that would produce turbulence. These efforts can not be de-

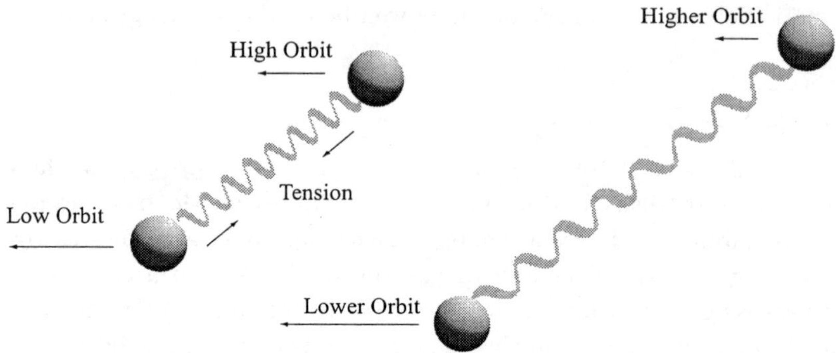

Figure 1. The action of the MRI is analogous to that of a spring connecting two orbiting masses. Spring tension transfers angular momentum from the mass in low orbit to the mass in high orbit, increasing the difference in their angular velocities, their separation, and hence the tension force.

scribed as successful. There are global instabilities in certain (somewhat specialized) types of disk, and local instabilities created by unfavorable gradients in quantities other than angular momentum (e.g., convective instabilities). But no instabilities have been found to arise under general circumstances and to be sustained by tapping into the free energy of differential rotation.

However, the addition of magnetic fields changes everything, for a magnetic field creates new paths by which the disk may access the energy of rotation (Balbus & Hawley 1991). To illustrate the basic mechanism, consider a fluid element in orbit. The counter-intuitive property of orbital dynamics is that if a fluid element loses angular momentum from a deceleration in the azimuthal direction, it drops to a lower orbit which results in an *increase* in its angular velocity. Conversely, a gain in l from an acceleration will decrease Ω. In other words, accelerations cause a slowdown and decelerations cause a speedup. Now consider two fluid elements connected by a magnetic field. The key point is that a magnetic field provides a tension force through the Maxwell stress component $-B_R B_\phi/4\pi$ that allows, for example, two orbiting fluid elements to exchange angular momentum. The transfer of angular momentum from one element to the other causes them to move to lower and higher orbits, respectively. The relative angular velocity between the two fluid elements *increases*, thereby increasing the separation, the magnetic tension, and the angular momentum transfer. The mechanism is exactly analogous to a pair of orbiting masses connected by a spring (Fig. 1; Balbus & Hawley 1992a). What matters is the existence of the tension force, which is what the magnetic field supplies and which is absent in a purely hydrodynamic fluid.

A more complete description of the physics of the MRI in accretion disk systems is provided in the review of Balbus & Hawley (1998). Here I will simply summarize a few of the most salient main points. First, the basic stability requirement is

$$(\mathbf{k} \cdot \mathbf{v_A})^2 > -d\Omega^2/d\ln R, \tag{1}$$

where $\mathbf{k}$ is the wavenumber and $\mathbf{v_A} = \mathbf{B}/4\pi\rho$ is the Alfvén speed. For any Alfvén speed one can find an unstable wavenumber if the angular velocity decreases outward. The maximum unstable growth rate of the instability is

$$|\omega_{max}| = \frac{1}{2}\left|\frac{d\Omega}{d\ln R}\right| \tag{2}$$

and the maximum growth rate occurs for wavenumbers

$$(k \cdot v_A)^2_{max} = -\left(\frac{1}{4} + \frac{\kappa^2}{16\Omega^2}\right)\frac{d\Omega^2}{d\ln R}, \tag{3}$$

where κ is the *epicyclic frequency*, which is given by

$$\kappa^2 = \frac{1}{R^3}\frac{dl^2}{dR}. \tag{4}$$

The MRI is both local and linear, and its presence in differentially rotating systems is independent of both magnetic field strength and orientation. The strength of the magnetic field simply establishes the length of the fastest growing mode.

As a specific example, in a Keplerian accretion disk the angular velocity Ω decreases with radius like $R^{-3/2}$, so these values are

$$|\omega_{max}| = \frac{3}{4}\Omega \tag{5}$$

and

$$(k \cdot v_A)_{max} = \frac{\sqrt{15}}{4}\Omega. \tag{6}$$

Basically, the characteristic wavelength of the ideal MHD MRI is approximately the distance that an Alfvén wave travels in one orbital period, i.e., $\lambda_{MRI} = 2\pi v_A/\Omega$; the growth rate is on order Ω^{-1}. All longer wavelengths are also unstable, with reduced growth rates comparable to the Alfvén crossing time.

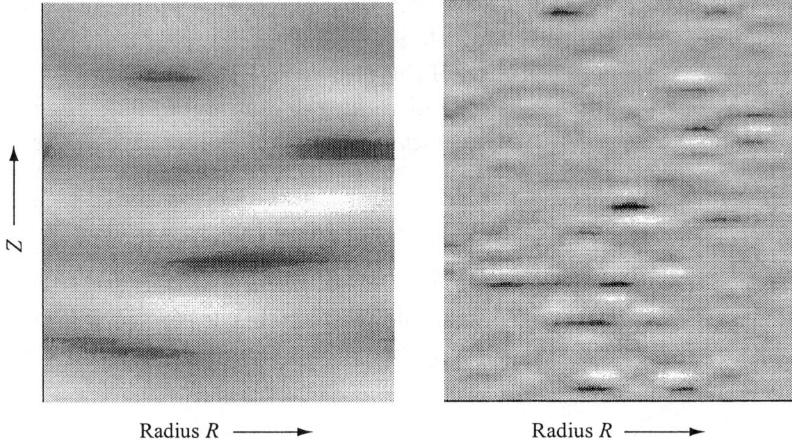

Radius R ⟶   Radius R ⟶

Figure 2. Linear MRI perturbations (in B_R) in a cross-section of a three-dimensional shearing box simulation. On the left the background field is vertical; the modes are nearly axisymmetric, nearly radial, and with a vertical wavelength equal to one-quarter the vertical box size. On the right the background field has the same strength but is azimuthal. In this case the fastest growing modes have an azimuthal wavelength comparable to v_A/Ω, but the transverse vertical and radial wavelengths are very small.

The peak growth rate of the MRI is independent of the strength of the magnetic field. This point is often regarded as one of the more surprising aspects of the MRI, but it illustrates a basic and important fact. The concept of a "weak" magnetic field is generally expressed in the ratio of the total field energy to other energies of the system, e.g., internal or kinetic. But such a concept misses the point: what is important is the existence of a tangential Maxwell stress force, a force that has no hydrodynamic analogue. The magnetic field enables degrees of freedom in the plasma that are simply unavailable with hydrodynamics alone. With the MRI the magnetic field strength sets the wavelength of the most unstable mode, but does not determine whether or not the system is unstable.

Many people intuitively feel that this cannot possibly be correct, that an arbitrarily weak field cannot have such a profound effect on a rotating system. In fact the action of the MRI, as described above, holds so long as the ideal MHD limit is valid. If there is finite resistivity, or other nonideal effects such as ambipolar diffusion or Hall currents, the stability criteria become a bit more complex. The simplest example is the addition of resistivity; if the field diffuses through the plasma faster than it can transfer angular momentum, then the MRI can be stabilized.

While all magnetic field orientations can be unstable in a differentially rotating system such as an accretion disk, background magnetic fields aligned with the rotation axis (defined for convenience as the z-axis) have a different character. For vertical fields the fastest growing mode has $k_z \sim \Omega/v_A$, with the perpendicular wavenumbers equal to zero. The fastest growing modes for other field orientations again have wavenumbers along the field that are on order Ω/v_A, but with transverse wavenumbers $k \rightarrow \infty$ (Balbus & Hawley 1992b). As a practical matter this means that in simulations that use vertical initial fields the instability first appears in a well-resolved wavelength (assuming an appropriate magnetic field strength), whereas with toroidal initial fields the fastest growing modes are always underresolved. The difference in appearance of the linear modes that arise for these two field orientations is illustrated in Figure 2, using results from simple shearing box simulations (e.g., Hawley, Gammie, & Balbus 1995).

3. Disks and Stars

Because the MRI plays such an essential role in accretion disks, and because it requires only differential rotation and a weak magnetic field, the question naturally arises as to whether it might be important in other astrophysical systems. Two such systems come immediately to mind: galactic disks and stellar interiors. Here, since we are concerned with Type II supernovae, we will focus on stars. Stars differ from disks in several fundamental ways. First, disks are supported primarily by rotation, whereas stars are supported primarily by pressure gradients. Second, entropy gradients (stable or unstable) are generally considered to be of secondary importance within disks, but are important to understanding stellar structure, distinguishing, as they do, a strongly stable radiative zone from a convective zone. Third, stars can come into a state of solid body rotation, whereas disks cannot; orbital dynamics simply preclude reaching a state where $d\Omega/dR = 0$ in a disk. In stars, however, the MRI may bring the radiative zone into just such a state (Balbus & Hawley 1994).

The general stability criteria for a rotating magnetofluid were laid out in a series of important papers by Balbus (1995; 2000; 2001), who considered the local axisymmetric stability of a weakly magnetized, stratified system that is differentially rotating with angular frequency $\Omega(R, z)$. For the case of an unmagnetized adiabatic fluid with an equation of state $P \propto \rho^\gamma$, the stability criteria are the well-known classical *Høiland criteria* (Tassoul

1978):

$$\frac{-1}{\gamma\rho}(\nabla P)\cdot\nabla\ln P\rho^{-\gamma} + \frac{1}{R^3}\frac{\partial l^2}{\partial R} > 0, \tag{7}$$

$$\left(-\frac{\partial P}{\partial z}\right)\left(\frac{\partial l^2}{\partial R}\frac{\partial\ln P\rho^{-\gamma}}{\partial z} - \frac{\partial l^2}{\partial z}\frac{\partial\ln P\rho^{-\gamma}}{\partial R}\right) > 0, \tag{8}$$

where $l^2 \equiv R^4\Omega^2$ is the square of the specific angular momentum. Instability can arise from unfavorable entropy gradients (the Schwarzschild criterion) or from unfavorable angular momentum gradients (the Rayleigh criterion). In the presence of a magnetic field, however, Balbus (1995) showed that the Høiland criteria must be replaced by the following:

$$\frac{-1}{\gamma\rho}(\nabla P)\cdot\nabla\ln P\rho^{-\gamma} + \frac{1}{R^3}\frac{\partial\Omega^2}{\partial R} > 0, \tag{9}$$

$$\left(-\frac{\partial P}{\partial z}\right)\left(\frac{\partial\Omega^2}{\partial R}\frac{\partial\ln P\rho^{-\gamma}}{\partial z} - \frac{\partial\Omega^2}{\partial z}\frac{\partial\ln P\rho^{-\gamma}}{\partial R}\right) > 0. \tag{10}$$

Two features of the Balbus criteria are particularly striking. First, the criteria are the same as the Høiland criteria with the substitution of the angular velocity Ω for the angular momentum l. This seemingly minor change has enormous significance. In a Keplerian disk $l \propto R^{1/2}$, but $\Omega \propto R^{-3/2}$, so the disk is linearly stable by criterion (7), but *unstable* by (9). Second, the stability criteria for the magnetized plasma make no reference to the magnetic field! It is the presence of the magnetic field, not its amplitude, that changes the fundamental stability properties of the system.

Figure 3 provides an illustration of how dramatically the stability criteria can change in the presence of a magnetic field. In this plot the x-axis is the negative of the square of the Brunt-Väisälä frequency, $N^2 = (1/\gamma\rho)(\partial P/\partial R)(\partial\ln P\rho^{-\gamma}/\partial R)$, normalized to the orbital frequency Ω^2. The y-axis is the square of the Alfvén frequency normalized to orbital frequency; a Keplerian distribution of angular momentum is assumed. With just hydrodynamics the system is unstable for all wavenumbers according to the Høiland criterion, namely, $N^2 + \kappa^2 < 0$, where κ is the epicyclic frequency (eq. 4; $\kappa = \Omega$ for a Keplerian disk) . A magnetized system with no entropy gradient is MRI unstable wherever $(kv_A)^2/\Omega^2 < 3$. With both a magnetic field and an entropy gradient things are more complex. Everything to the right of the thick line is unstable. In the upper right portion of the diagram, the magnetic field is relatively strong (or, more precisely, the product kv_A is large). The flow would be stable but for the presence of an

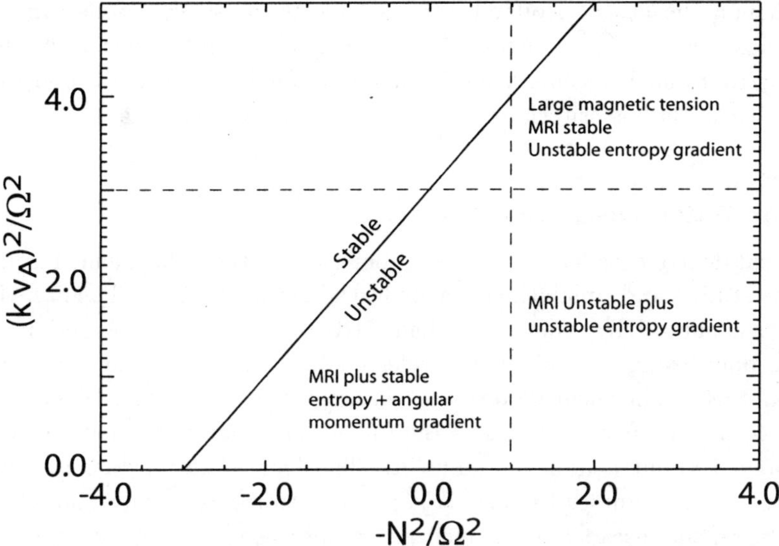

Figure 3. The domain of instability for a magnetized system with a Keplerian angular velocity distribution and radial entropy gradients.

adverse entropy gradient; a sufficiently strong field (or large wavenumber) will stabilize even that. Similarly, in the lower left quadrant the presence of a stable entropy gradient can, if sufficiently strong, stabilize the MRI.

Another demonstration of the fundamental differences between MHD and hydrodynamics is provided by an analysis of a differentially rotating, magnetized system with finite thermal conductivity along the magnetic field lines (Balbus 2000, 2001). The stability criteria then become:

$$\frac{-1}{\rho}(\nabla P) \cdot \nabla \ln T + \frac{1}{R^3}\frac{\partial \Omega^2}{\partial R} > 0, \tag{11}$$

$$\left(-\frac{\partial P}{\partial z}\right)\left(\frac{\partial \Omega^2}{\partial R}\frac{\partial \ln T}{\partial z} - \frac{\partial \Omega^2}{\partial z}\frac{\partial \ln T}{\partial R}\right) > 0. \tag{12}$$

Again, the mere presence of a magnetic field makes an enormous difference. This time, entropy gradients have been replaced by temperature gradients, independent of the size of the coefficient of thermal conductivity. As Balbus emphasizes, the angular velocity gradients and temperature gradients are sources of free energy in the system. By providing a means of tapping into that free energy, a means that is absent in a hydrodynamic system, the magnetic field profoundly alters the stability properties of the plasma.

Although the classic Høiland criteria have been widely applied in astrophysical situations, particularly to rotating stars and supernovae, almost all astrophysical plasmas contain magnetic fields. In general one should expect the Balbus criteria to apply, not the Høiland criteria.

4. Accretion Disk Simulations

A great deal of detailed information about the MRI can be gleaned from the many studies and simulations devoted to accreting systems. The first three-dimensional, MHD, time-dependent MRI simulations of accretion disks were done using a local model, referred to as the *shearing box* (Hawley et al. 1995). The shearing box is a local region of an accretion disk, viewed in a corotating frame, that uses the full set of dynamical equations locally expanded about a fiducial radius R. The local shearing box is a useful laboratory for studies because it is possible to resolve many scales within the turbulent cascade, and to incorporate increasingly complex physics into the system. Shearing box simulations have been done for both unstratified (no vertical gravity; Hawley et al. 1995, 1996; Matsumoto & Tajima 1995) and stratified (Brandenburg et al. 1995; Stone et al. 1996) disk sections, with a variety of initial magnetic field topologies and strengths and different background rotation laws (Hawley, Balbus & Winters 1999), and for a number of different box sizes, resolutions, and initial conditions (Sano et al. 2004), and with a variety of numerical algorithms. The MRI has been simulated in the local shearing box for a wide variety of physics in addition to ideal MHD, including ion-neutral plasmas (Hawley & Stone 1998), resistive plasmas (Fleming et al. 2000), plasmas with Hall currents (Sano & Stone 2002), and radiative plasmas (Turner et al. 2002). In all cases the results are in complete agreement with the linear stability analyses (where applicable), and they have demonstrated the importance of the MRI for many different plasma states.

In the linear regime the amplitude of the instability grows exponentially, but what is amplified is not the total magnetic energy, but the energy in the magnetic field *perturbations*. The initial background field is not noticeably perturbed until several e-folding times have passed. Exponential growth typically ceases shortly after the energy in the perturbed field exceeds that in the initial field. This usually happens within 2 to 3 rotation periods (depending upon the amplitude of the initial perturbations). The nonlinear state that results from the saturation of the linear modes is quite generally MHD turbulence. In these simulations saturation of the MRI comes about from reconnection and turbulent dissipation rather than by adjusting the overall background angular velocity distribution (see, e.g., Sano et al. 2004). The MRI continues to operate, however, sustaining the turbulence with free energy from the differential rotation. This is possible because the turbulence has strongly correlated velocity and magnetic field fluctuations with a significant nonzero R–ϕ stress, from both a Maxwell, i.e., $-B_R B_\phi/4\pi$, and a Reynolds, $\rho \delta v_R \delta v_\phi$, stress component. The Maxwell component of the stress always exceeds the kinematic Reynolds stress by a factor of several. In disk simulations the overall increase in magnetic energy tends to be modest. The magnetic pressure typically remains subthermal, with a ratio of $\beta = P_{gas}/P_{mag} \sim 10$.

Because of the greatly reduced computational expense, axisymmetric simulations are preferable to fully three-dimensional simulations whenever they are sufficient. Axisymmetric simulations have been useful for MRI accretion studies, particularly in global simulations. There are some significant limitations, however. First, Cowling's antidynamo theorem tells us that a sustained magnetic dynamo is not possible in axisymmetry. Second, the character of turbulence is different in two dimensions compared to three. Third, in axisymmetry the toroidal field MRI cannot develop. This limits the initial background magnetic field to vertical fields. Radial fields can also be unstable to the MRI, but the simple linear growth of toroidal field due to the differential rotation tends to dominate, at least initially, because shear can work directly upon the background field whereas the MRI amplifies the field perturbations. And finally, the axisymmetric vertical field MRI displays a unique characteristic: background z-fields produce coherent radially streaming flows of considerable amplitude (the "channel solution," Hawley & Balbus 1992). These channel solutions can be seen emerging in the linear mode pattern in the left hand frame of Figure 2. In three dimensions the channel solution is subject to nonaxisymmetric instabilities (Goodman & Xu 1994) that cause it to lose coherence and rapidly break down into turbulence.

Even in three dimensional simulations there are observable differences between initial vertical and toroidal background fields. Local simulations with uniform vertical field tend to have the strongest turbulence, and time-dependent quantities such as stress or magnetic energy often show significant spikes due to intermittent appearance of the channel mode. In global simulations these streaming modes can rapidly change the structure of a disk initially threaded by a vertical field. When the background field is toroidal the modes that first appear are those with small spatial scales; the dominant perturbations in the flow gradually increase in size as the system evolves. The more general case of a tangled initial magnetic field shows rapid growth of perturbations on a variety of lengthscales; this rapidly leads to turbulence.

5. Implications for Supernovae

The basic question is: are magnetic fields important for supernovae? It seems to be somewhat uncontroversial that a *strong* magnetic field could have a profound impact on the evolution of a core-collapse supernova. For example, in their review Wheeler et al. (2002) discuss the arguments supporting the idea that strong magnetic fields could produce the asymmetries seen in supernova ejecta through the formation of bi-polar jets. The general LeBlanc-Wilson mechanism has been invoked as a potential model for the strong jets produced in gamma ray burst systems. Within the accretion disk and AGN community there seems to be a growing consensus that jets *require* the combination of rotation and magnetic field. All of this contributes to the feeling that magnetic fields must be important in at least some classes of supernovae.

The issue as always is just how large must the field be, and do processes occur naturally that could amplify a pre-existing field to the required level? Arguments based on traditional dynamo mechanisms seemed unpromising, but Akiyama et al. (2003) specifically drew attention to the possibility that the MRI could provide exponential amplification of fields. A stellar interior would be expected to be unstable to the MRI if there is differential rotation and weak magnetic field. Assuming that the angular velocity is proportional to R^{-q}, then the system will be unstable for $q > 0$ with a maximum growth rate of $q/2$ (eq. 2). Disk simulations of the MRI have generally begun from a state of equilibrium, so the behavior of the MRI in a dynamic system such as a collapsing stellar core may be somewhat different from that seen so far in numerical models. The general insights found in accretion disk studies should largely carry over to this new context.

There is little need for speculation since the issue will ultimately be addressed by simulations. How much work would it be to include MHD in core collapse supernova simulations? MHD is considerably more complicated than hydrodynamics, but the degree of complexity pales compared to the problems associated with neutrino and radiation transport with which modelers must already contend. Although the MRI and its associated physics are fundamentally three-dimensional phenomena, axisymmetric simulations should be sufficient to provide some initial indication as to the importance of magnetic fields and the MRI. Even in axisymmetry, however, it may prove difficult to use sufficient grid resolution. Sano et al. (2004) argue that at least 6 grid zones are required to resolve a given MRI wavelength sufficiently to allow for mode growth, and if the field is weak the most unstable wavelengths will be small.

Once sufficiently well-resolved MHD simulations are carried out, how would one know that the MRI is operating? One indicator would be exponential growth of the magnetic field energy, although this is not a definitive test. Field amplification is generally described as a dynamo process, and amplification could be achieved either through a kinematic mechanism, meaning that the magnetic Lorentz forces play no role, or through fully dynamical process. The wrapping up of radial field lines in a differentially rotating system is an example of a simple kinematic mechanism. The MRI, on the other hand, is a fully MHD process and operates because of the presence of Lorentz forces. Thus if a simulation produces field amplification even with the Lorentz forces not included, then the MRI is not the cause.

Ultimately, what the MRI does best is transport angular momentum, extracting energy from the differential rotation and putting it into poloidal fluid motions and magnetic field energy. Hence field growth accompanied by significant angular momentum transfer would be another indicator of the presence of the MRI in a simulation. This point is worth emphasizing: because the MRI obtains its power from differential rotation, the question is not the strength of the magnetic field but how much rotation energy there is in the collapsing star. This is a difficult question, depending as it does on uncertain details of the progenitor. However, *if* the rotational energy is significant, then the MRI can tap that energy and put it to other uses,

Magnetic fields may prove to be significant even if they are not amplified to large levels in some types of systems. If one were to draw a general lesson from the experience with the MRI and accretion, it would be that a magnetofluid has fundamentally different properties from a hydrodynamic fluid even if the total energy density in the magnetic field is less than other forms such as the thermal, gravitational, or rotational energy. The basic stability criteria (i.e., the Høiland criteria) are changed and new avenues for energy and angular momentum transport are enabled. Supernova modelers, like the disk community before them, ignore these fields at their peril.

80

Acknowledgments

I would like to thank the organizers for their invitation to participate in the workshop on core collapse supernovae, and Steve Balbus for useful discussions. This work was partially supported by NSF grant PHY-0205155, and NASA grant NNG04-GK77G.

References

1. Akiyama, S., Wheeler, J. C., Meier, D. L., & Lichtenstadt, I., Astrophys. J., 584, 954 (2003).
2. Balbus, S. A., Astrophys. J., 453, 380 (1995).
3. Balbus, S. A., Astrophys. J., 534, 420 (2000).
4. Balbus, S. A., Astrophys. J., 562, 909 (2001).
5. Balbus, S. A., & Hawley, J. F., Astrophys. J., 376, 214 (1991).
6. Balbus, S. A., & Hawley, J. F., Astrophys. J., 392, 662 (1992a).
7. Balbus, S. A., & Hawley, J. F., Astrophys. J., 400, 610 (1992b).
8. Balbus, S. A., & Hawley, J. F., MNRAS, 266, 769 (1994)
9. Balbus, S. A., & Hawley, J. F., Rev. Mod. Phys., 70, 1 (1998).
10. Bisnovatyi-Kogan, G. S., Soviet Astron., 14, 652, (1971).
11. Blandford, R. D., in *Theory of Accretion Disks*, eds. F. Meyer et al., (Dordrect: Kluwer) (1989).
12. Brandenburg, A., Nordlund, Å., Stein, R. F., & Torkelsson, U., Astrophys. J., 446, 741 (1995).
13. Crawford, J. A., & Kraft, R. P., Astrophys. J., 123, 44 (1956).
14. Fleming, T., Stone, J. M., & Hawley, J. F., 530, 464 (2000).
15. Goodman, J., & Xu, G., Astrophys. J., 432, 213 (1994).
16. Hawley, J. F., & Balbus, S. A., Astrophys. J., 400., 595 (1992).
17. Hawley, J. F., Gammie, C. F., & Balbus, S. A., Astrophys. J., 440, 742 (1995);
18. Hawley, J. F., Gammie, C. F., & Balbus, S. A., Astrophys. J., 464, 690 (1996);
19. Hawley, J. F., Balbus, S. A., & Winters, W. F., Astrophys. J., 518, 394 (1999).
20. Hawley, J. F., & Stone, J. M., Astrophys. J., 501, 758 (1998).
21. LeBlanc, L. M., & Wilson, J. R., Astrophys. J., 161, 541 (1970).
22. Matsumoto, R., & Tajima, T., Astrophys. J., 445, 767 (1995).
23. Sano, T., & Stone, J. M., Astrophys. J., 577, 534 (2002).
24. Sano, T., Inutsuka, S., Turner, N. J., & Stone, J. M., Astrophys. J., 605, 321 (2004)
25. Stone, J. M., Hawley, J. F., Gammie, C. F., & Balbus, S. A., Astrophys. J., 463, 656 (1996).
26. Tassoul, J.-L., *Theory of Rotating Stars* (Princeton University, Princeton) (1978)
27. Turner, N. J., Stone, J. M., & Sano, T., Astrophys. J., 566, 148 (2002).
28. Wheeler, J. C., Meier, D. L., & Wilson, J. R., Astrophys. J., 568, 807 (2002).

Section 3
The Core Collapse Supernova Mechanism

SUPERNOVAE MODELING: A PERSONAL HISTORY

JAMES R. WILSON*

This paper will be an account of the history of the author's progress in modeling collapse supernovae explosions. Many people have contributed to our understanding of the collapse problem, but I will only present my own steps along the way. This paper is not a full history of the evolution of our understanding of the stellar collapse process.

1. Ancient History

1.1. *1968 Calculation for Colgate*

Sterling Colgate had a hypothesis for an energization of the envelop above a proto neutron star by interactions of neutrinos with hot matter behind the shock. I ran several calculations with a hydrodynamic computer program in which the photon transport section was modified to treat neutrinos. With the parameters used the shock was somewhat broadened but not enough to be interesting as a source of explosion energy. Results were not published.

1.2. *1970 Collapse of a Rotating Magnetized Star*

James LeBlanc and I took a magneto-hydrodynamic computer code that we had and modified it to study the collapse of star. A gravitational potential was added as well as a simple model for neutrino diffusion. The 1965 Harrison Wheeler equation of state was used for the nuclear matter. An EOS for photons and electron pairs was included . The thermal part of the nuclear EOS was treated as a perfect gas with a 5/3-1 pressure coefficient. Only electron neutrinos were included and an opacity proportional to temperature squared was used.

The calculation started with a 7 solar mass star with a central density of 10^8 g/cc in static equilibrium. The star was given an initial angular

---

*This work was written under the auspices of U. S. Department of Energy under contract W-7405-ENG-48.

velocity of 0.7 radians/sec and a polar magnetic field whose energy was equal to 0.00025 times the gravitational potential energy.

The star collapsed and twisted up the magnetic field to put a large amount of energy into the field. A strong axial jet resulted. The magnetic field energy rose to 10^{52} ergs. The jet had a velocity of 6.0×10^9 cm/sec[1].

Later calculations with more realistic stellar models only gave weak jets.

1.3. *1971 First Spherical Collapse Calculations*

A full general relativistic computer program was written. It solved the Boltzmann transport equation for the neutrinos in both the angle and the energy variables. When this project was started full stellar evolution calculations that have since been made were not available so a range of initial stellar masses were studied. Masses ranged from 1.25 to 4.30 solar mass. General relativistic effects were quit important for the high mass stars. Only the 1.25 solar mass star yielded an explosion. For this case the neutrino losses were small. It was only a bounce explosion. The ejected matter had a kinetic energy of about 3×10^{50} ergs. The model had no envelope. Since the energy would be dissipated in the envelope, in a more realistic calculation probably no explosion would obtain[2].

1.4. *1974 Coherent Neutrino Scattering*

D. Z. Freedman introduced the idea of using coherent neutrino scattering of heavy nuclei by the recently established neutral-current theory to produce an explosion.

The proposal was that the coherent scattering of post-bounce neutrinos on heavy nuclei would be strong enough to expel the matter outward by momentum exchange.

For nuclei the cross-section for scattering per nucleon was multiplied by the atomic weight of the matter. Since only scattering was involved, the matter was only slightly heated. The nuclei remained intact. A properly evolved initial model was used. An explosion of 10^{50} ergs resulted. Later refinements of the coherent scattering process reduced the scattering cross-section sufficiently that no explosion resulted[3].

1.5. *Burn of in-falling matter*

Since the coherent neutrino scattering mechanism disappeared it was thought that, perhaps, the burning of the matter falling in on to the proto-neutron star might be enough to start an explosion. The energy supplied by the nuclear burning was not enough to make an explosion. Again we were left with no explosion[4].

1.6. *Improved Neutrino Model*

Previous calculations only included electron and anti-electron neutrinos. The equations for electron-positron annihilation to form neutrino pairs were added to the calculations. This allowed for the inclusion of mu neutrinos, which can only be produced in large number by annihilation of electrons by positrons.

The neutrino equations now used treated neutrino degeneracy correctly. A "flux limited" diffusion model was introduced for the neutrino transport. All subsequent calculations used this method of transport. The model now used Newtonian equations. No explosion resulted[5].

1.7. *Inclusion of Neutrino Electron Scattering Energy Exchange*

When the envelop of the proto-neutron star that we wish to expel becomes hot enough the scattering of neutrinos on electrons and positrons becomes the principle energy depositing mechanism. To model this effect (the energy exchange between neutrinos and electrons or positrons) a Fokker-Planck type equation for diffusion in momentum was implemented. Because the fractional energy exchange per collision in scattering is large, a calibration was made by comparing with Monte Carlo calculations made by D. Tubbs. The Fokker-Planck algorithm was found to be adequate. However, still no reliable explosions occurred, but the system now appeared very close to being successful[6].

1.8. *Neutron Fingers*

A two dimensional combined neutrino radiation diffusion and hydrodynamics calculation of non-spherical instabilities in the collapsed core of a massive star was made. The core initial properties were taken from a one-dimensional collapse calculation with detailed microphysics and neutrino transport. The core was found to be unstable even though the entropy increased going outward in the core. We found out about the process called "salt fingers" so we called our process "neutron fingers". The instability occurs because a decreasing gradient in electron fraction overcomes the rising entropy. The geometry of this instability is of the form of fingers due to the high core viscosity. From these calculations we also observed an entropy- driven instability in the region above the proto-neutron star. This later instability proved subsequently to be less important than the neutron fingers instability[7].

1.9. *Detailed Description of the Core-Collapse Model*

The computer program for the study of core collapse had evolved slowly over the previous decade. The additions were partially described in the various papers discussed above. We wrote a full description of the model as of 1982. The reference below covered the following:

Thermonuclear burn;

Electron capture;

Equation of state for electrons, positrons, photons, nuclei, and free nucleons;

Hydrodynamics shock treatment and neutrino hydrodynamic effects;

Neutrino-electron elastic scattering;

Neutrino transport electron neutrinos and mu and tau neutrinos;

Neutrino matter interactions;

Automatic rezoning[8,9].

1.10. *Equation of State Near Nuclear Density*

The question is: Is the transition from subnuclear density to supranuclear density smooth or abrupt? First, I modeled heavy nuclei in a two-dimensional liquid-drop model and found that at high density non-spherical drops were the most stable. With Ravenhall and Pethick we did the study in more detail and found about one MeV energy difference in energy for non-sphericity. Later, I calculated the electrostatic energy using a Thomas-Fermi approach for the electrons. This gave an additional subnuclear binding of about one MeV. The conclusion is that the transition from subnuclear to supranuclear density should be smooth. No simple phase transition. Further work on this region of the equation of state is discussed in the following reference[10].

1.11. *Equation of State at Very High Density*

Many cold equations of state at very high density have P/density greater than c^2. Some even have sound speed greater than c. We use a bag model above a few times nuclear density where the bag model P meets the lower density P equation tangentially. The part of the EOS affected only occurs very late in the calculations.

1.12. *Revival of Bounce Shock by Neutrino Heating*

In 1984 I noticed that in the matter well above the neutrino photospheric radius, the entropy had risen slightly. I first thought it might be due to a fault in the computer program. After looking at the output closely I decided that the effect was real. The effect was so slight that I needed to make the program more efficient so that it could be run far enough in time to see whether the entropy rise was in fact important. As is now known, it turned out to be quit important. Bethe and I subsequently studied the effect carefully and derived a simple equation to explain it[11].

1.13. *Numerical Solution of the General Relativistic Boltzmann Equation*

Back in 1971 I gave results for a collapse calculation in which the full Boltzmann equation in angle and energy was solved in the general relativistic framework. I had since learned how to use flux limeters properly so that the detail of an angular resolved solution was not needed. I was asked how it was done numerically hence[12]:

1.14. *Neutrinos From Gravitational Collapse*

In 1987 Mayle and I published a paper giving details on the spectra and luminosities of the various neutrinos produced in a collapse calculation. Then conveniently supernova 1987a occurred. We did an analysis of the neutrinos from SN 1987a using a pre- collapse model for the event made by S. Woosley. The paper giving the results of our calculations was rejected for publication.

I will give some of the figures from the rejected paper. In Figure 1 the luminosity for a calculation made with an iron core of 1.64 solar mass and a soft equation of state, a calculation with the same mass and hard EOS, and a calculation with a small iron core, 1.27 mass and a soft EOS are given. Since the observed neutrino signal lasted 12 sec. only the first calculation is relevant. Figure 2 shows the detector efficiencies used for the analysis given in Figure 3. Figure 3 shows the comparison of calculated spectra and observed spectra from the IMB and Kamiokanda detectors. Figure 4 shows the accumulated emitted energy versus time for observations

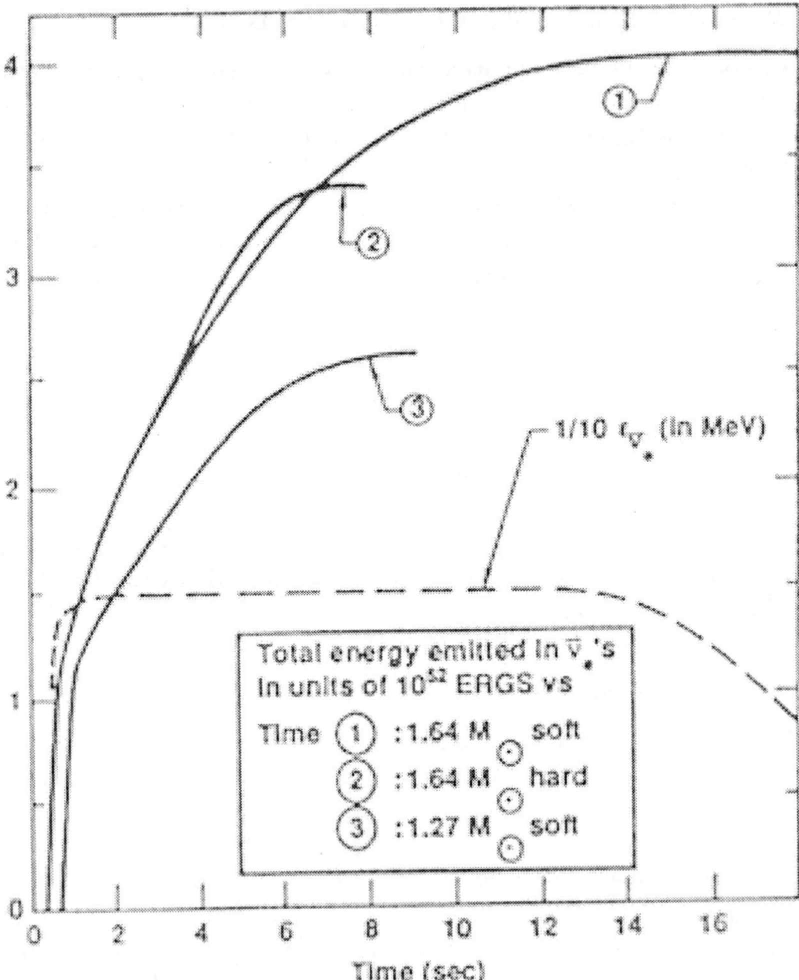

Figure 1. The luminosity for a calculation made with an iron core of 1.64 solar mass and a soft equation of state, a calculation with the same mass and hard EOS, and a calculation with a small iron core, 1.27 mass and a soft EOS are given. Since the observed neutrino signal lasted 12 sec. only the first calculation is relevant.

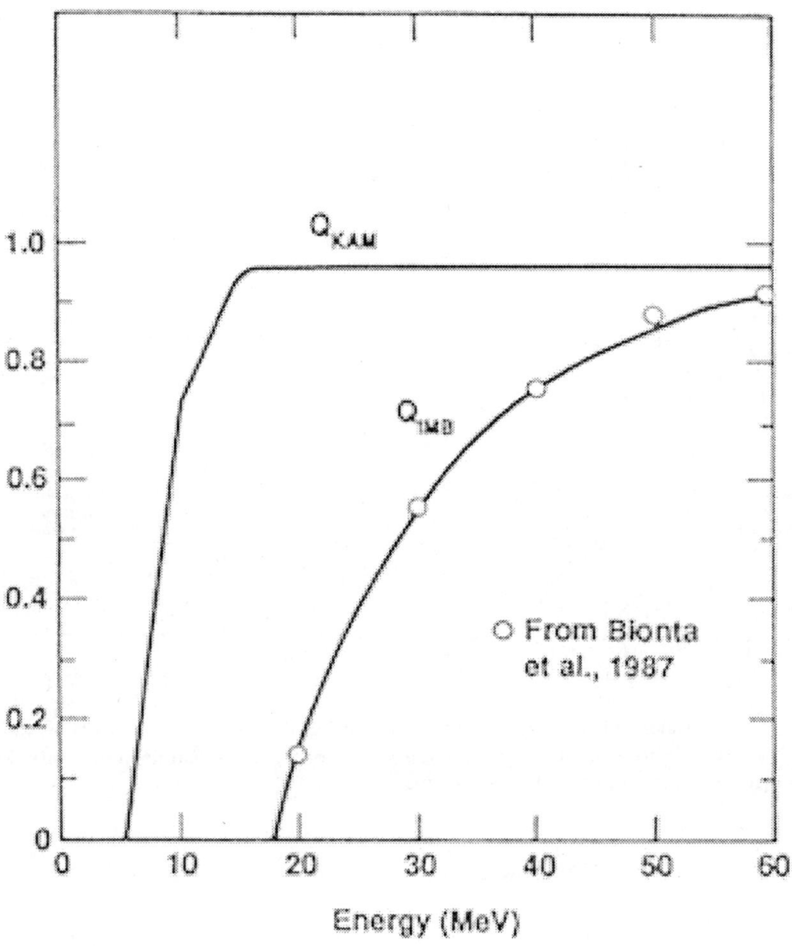

Figure 2.   The detector efficiencies used for the analysis given in Figure 3.

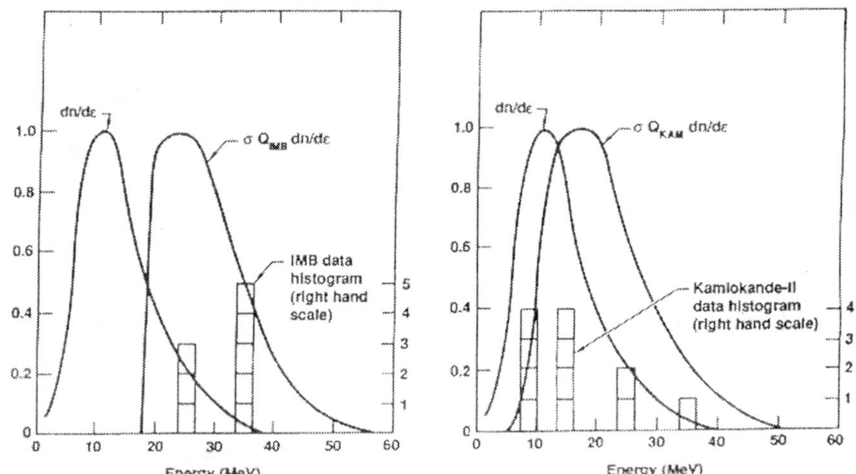

Figure 3. Anti electron neutrino number distriution dn/dE and cross-section, detector efficiency weighted number distribution versus energy. 4a Kamiokande. 4b IMB. Observations have been put into histograms.

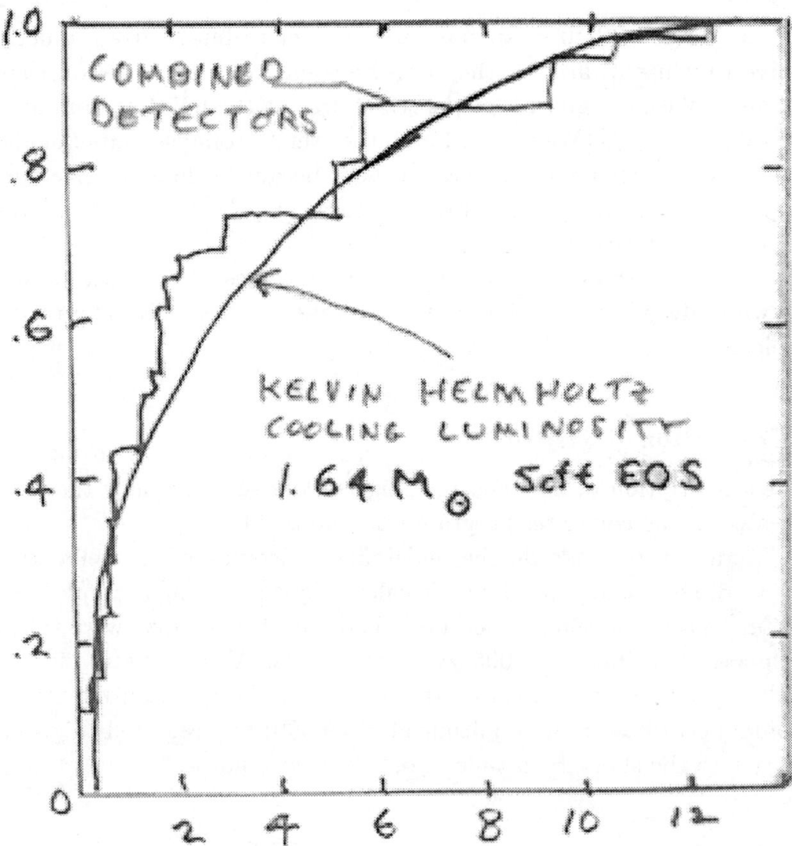

Figure 4.   The accumulated emitted energy versus time for observations and calculation. The observational curve is made by weighting each event by its energy and the detector's efficiency. In our analysis three of the Kamiokanda events would be electron scattering events.

and calculation. The observational curve is made by weighting each event by its energy and the detector's efficiency. In our analysis three of the Kamiokanda events would be electron scattering events[13].

1.15. *Oxygen-Magnesium-Neon Collapse Supernova*

Stars in the range of 8-10 solar mass are thought to burn carbon quietly and evolve to white dwarfs. If they acquire some mass by accretion they may collapse. We were given a model which was at the point of imminent collapse by K. Nomoto. We indeed found the star to collapse and explode. However, it would not be an observable event because it had no envelope. The luminosity by Ni decay was estimated to be only 10^{39} ergs/sec. It was dubbed a "silent supernova".

This is an important type of supernova since, as later work shows, OMgNe white dwarfs may be induced to collapse by close encounters with black holes[14].

1.16. *Convection Revisited*

A detailed description of how neutron fingers work and also how they are implemented in the computer program was published.

New studies were made on the instabilities encountered in stellar collapse. Two-dimensional hydrodynamic calculations including a simple neutrino diffusion scheme were carried out. The initial conditions were taken from a spherical calculation 0.005 sec after bounce. A perturbation in the infall velocity was imposed to start the instability. To get a strong effect, a sinusoidal perturbation of amplitude about 0.0001 was required to give a strong drive to the shock by a time 0f 0.3 sec post bounce[15].

1.17. *Pions in Supranuclear Matter*

We modeled the heavy ion collision experiments that were performed at the Bevalac in Berkeley using hydrodynamics and an in-medium formulation of pions. Pions were treated by a Monte-Carlo algorithm. Nuclear viscosity was included. We achieved very good agreement with the experimental results. The pion formulation we used was then incorporated in our equation of state. The inclusion of pions increased the temperature in the core with the final effect of increasing the energy of the explosion. The calculations now yielded more than 10^{51} ergs of kinetic energy[16].

1.18. *Nucleosynthesis by the r-process*

In our study of OMgNe supernovae we did a simple nucleosynthesis calculation in order to see what might be observable. Now we had a high mass star model that gave the proper explosion energy. It seemed timely to calculate the nucleosynthesis that would arise therefrom. The results of the r-process calculation was very encouraging. The calculation over produced elements in the atomic mass range up to about 90, and under produced elements in the mass range above 200. In the middle range of nuclear mass the code well matched the solar abundances[17].

1.19. *Gravitational Bending of Neutrino Orbits*

Neutrino orbits are strongly bent near to the surface of a proto-neutron star late in time, several seconds after bounce. This bending greatly increases the neutrino-antineutrino annihilation above the star since the cross-section for annihilation is equal to the square of the center- of-mass energy. The bending increases the annihilation rate over the straight-line orbits by up to a factor of 30 near the stellar surface. This effect is also very important for our neutron-star model for Short-duration gamma-ray bursts[18].

1.20. *Recent Summary of Our Work*

To finish off ancient history I want to mention the latest write-ups of our ancient supernova work[19].

2. Modern History

2.1. *Supernovae Explosions Induced By White Dwarf Passing a Black Hole*

In 1996 G. Mathews and I published a paper on neutron star binaries in which we described the increase in density of the neutron stars as the stars came closer together[20]. Later we realized that the effect was related to the stars being in a strong gravitational field and the dominant term in compression effect was proportional to the square of the spatial part of the four velocity. A code was written to study white dwarf stars in orbit about a black hole. The stars were found to increase in density in a manner very similar to neutron stars in binary orbits. We studied three sizes of black holes; a few tens of solar mass, a few thousand solar mass, and millions or more solar mass. For the low-mass black holes the effect was too weak to bring even a 1.2 solar-mass white dwarf up to ignition density. For a thousand solar-mass black hole, even a 0.6 solar mass white dwarf was brought up to a density 0f 3.5×10^9 g/cc. The white dwarf has under gone a compression of a factor of one thousand. Globular clusters are thought possibly to contain black holes in the thousand solar mass range. If globular clusters contain massive black holes we estimate a possible rate for these events out to a red shift of one of a few/10^5 per year. For supermassive black holes we estimate a similar rate. Subsequent work by D. Dearborn has shown that these black-hole-ignited white dwarfs would burn 98% of the star into Ni. The energy of a 0.6 solar mass dwarf would be 10^{51} ergs like a normal type 1 supernova, but the spectra would be much different[21].

2.2. *A Model For The Supernova Remnant Near Sagittarius Black Hole*

A few years ago[22], a supernova remnant was found by radio and X-rays to be about 2 pc from the supermassive black hole in Sagittarius. We (Mathews, Dearborn and I) have modeled this event. Djehuty, a 3D stellar burn computer program was modified to model the explosion of a white dwarf of 0.60 solar masses passing by the Galactic-center black hole. The gravitational potential in the burn code which is Newtonian otherwise was modified by multiplying the potential by $(1.0 + 2.0 \times U^2/c^2)$. U^2 the square of the spatial four velocity was calculated for appropriate stellar trajectories as a function of time. The time scale for the final factor of two in compression is the order of a few seconds. We find that a detonation of the whole star occurs even though the burn starts slightly off center from tidal distortion. The white dwarf was burned to 96.5% Ni, 1.3% Si, 0.2% Mg, 0.1% O and 1.8% He. If the white dwarf is only a few million years old it will ignite early and produce 90.0% Ni and 7.1% He. and the rest as before. The estimated rate for this type of event to occur at the Sagittarius black hole is a few/10^5 per year. A weakness of the model is that the observed remnant is a metal-rich so called "mixed morphology" type remnant, while our model yields mostly Ni. Two possibilities are currently under study. The first is that the explosion is induced in the dense core of a red giant whose outer layers had been stripped off by tidal forces earlier on in the stars approach. A bare core of a red giant has been placed on a trajectory close to the hole and it ignited. A more realistic calculation is planned for the future. In Sec. 1.14 we discussed the collapse of a O,Ne,Mg star to produce a supernova explosion. Heavy white dwarfs 1.0-1.2 solar mass are thought to be O,Ne,Mg stars[23]. We plan to use our standard supernova collapse program with the gravitational constant modified by the factor $(1.0+2.0\times U^2/c^2)$ to study this possibility of a collapse type explosion.

2.3. *Core Instabilities In Collapse Supernovae*

In most supernova models with pure spherical symmetry, after a massive star collapses due to the exhaustion of its nuclear fuel, the neutrino luminosity from the proto-neutron star is too low to heat the in-falling material sufficiently to expel matter from the star.

One important mechanism for such heating, for example, is that the proto-neutron star can become hydrodynamically unstable a few hundred milliseconds after the core bounce due to the so-called "neutron-finger" instability noted above. This instability results from the build up of material with a large neutron-to-proton ratio near the surface of the proto-neutron star. Sufficiently neutron-rich material can overcome the non-buoyancy caused by the high entropy near the surface. As surface material sinks downward neutrino-rich material rises to the surface. This enhances the neutrino luminosity and produces enough heating of material behind the shock to produce an explosion. See Sec. 1.8.

In this section we discuss the possible roles of magnetohydrodynamics in rotating core-collapse supernovae. While the models we utilize are adequate to illustrate the order of magnitude of these effects, it should be emphasized that it will be necessary to do these calculations in two or three dimensional magneto-hydrodynamics (MHD) to prove that the ideas presented here are correct.

Such calculations, however, present an exceedingly difficult computational challenge. A much simpler model is employed which involves the average spherical effects of the inherently multidimensional rotation and magnetic effects as a perturbation on the one-dimensional hydrodynamics. This is a reasonable approximation as a means of exploring the parameter space and extracting the essential physics as long as we are considering small rotation rates and relatively weak but realistic magnetic fields.

As we shall see, the main effects in our perturbation analysis are the build up of magnetic turbulence and field energy above the proto-neutron star shortly after collapse and the formation of magnetic bubbles that lead to magnetic-driven convection below the surface of the proto-neutron star. Both of these effects together can significantly impact the explosion process.

Neutron stars have magnetic fields from 10^{12} to 10^{13} gauss so we consider initial uniform magnetic fields that will result in final fields in the cool neutron star in the above range. Massive stars tend to be rapidly rotating so we assume initial uniform angular velocity such that the final period will be longer than a few milliseconds. We assume that the magnetic fields will slow the stars down to the observed neutron star periods.

We first describe the magnetic turbulence model for explosion initiation. This process was suggested by John Hawley at the Seattle meeting. Above the proto-neutron star for a few tenths of a second after bounce is a region below the bounce shock front and above the almost static proto-neutron star radius where matter is slowly accreting. At a post bounce time tpb=200ms the neutron star radius is 40 km and the shock radius is at 170km.

Later at tpb=300ms the proto-neutron star radius is 32 km and the shock radius has contracted to 140km. The Mach number in the subshock region is about 0.1.

In the sub-shock region large magnetic field amplification is possible for a rapidly rotation magnetized collapsing star. To model the evolution of the magnetic field, we follow the general principles given in Balbus and Hawley[24]. They studied the magnetic field instability in accretion flows. They found an exponential growth in turbulent magnetic field that has a growth rate proportional to the logarithmic radial gradient of the angular velocity. In our case the exponentiation rate is about $0.9 \times$ angular velocity. We integrate the buildup of magnetic field in matter after the matter falls below the outer shock position starting with the field present in the unperturbed calculation. The pressure in the magnetic field is added to the matter pressure by a solid angle weighting. This additional pressure is enough to increase the luminosity sufficiently to induce an explosion. For an initial magnetic field that gives a final 10^{12} gauss the minimum initial angular velocity required for explosion was 0.07 rad/sec. This rotation speed leads to a final neutron star period of a few ms.

The formation of a magnetic bubble arises due to the non-homologous nature of the collapse. We start with a precollapse star in uniform rotation threaded by a uniform field in the axial direction. The differential rotation that ensues leads to two toroidal magnetic field bubbles in in the shape of a doughnuts. The toroidal magnetic bubble is buoyant and its motion brings neutrino rich matter up to the proto-neutron surface.

With an initial magnetic field that yields a final 10^{12} gauss a rotation rate of 0.30 radians/sec. is required. This rotation rate yields a final rotation period of only a little more than one millisecond. However with an initial field strength leading to 10^{13} gauss an initial rotation rate of 0.03

radians/sec is sufficient to start the explosion.

Magnetic fields of the strength we consider certainly exist. The observation that most collapse supernova remnants emit polarized optical radiation hints to the existence of large early rotation rates. Our calculations being perturbation calculations on top of a spherical computer model means the calculations may not be good quantitatively. Since we now have three sources of luminosity enhancement that can work in parallel even if the effects were somewhat weaker they should yield an explosion.

2.4. *Recent Collapse Supernova Calculations and the R-process*

In our calculations of the supernova explosions at late times we have obtained high entropies at late times in the bubble that forms above the proto-neutron star. People who make so-called— "wind" calculations of the heating of matter above the proto-neutron star obtain low entropies. A few years ago I developed a wind computer program that used the time history of the neutrino emission from the proto-neutron star and applied it to heat the matter above the star. The outer material was taken from a collapse calculation shortly after bounce. The material was then heated by the neutrinos and the resulting hydrodynamics was followed explicitly. The first calculation was made three years ago and used the neutrino output from the full code as calculated then. The maximum entropy found in the wind calculation at a time 15 seconds after bounce was 1000. The full supernova program calculation only gave an entropy of 500. at a time of 15 seconds. The most recent full code calculation now gives an entropy close to 1000 at a time of 15 seconds post bounce. The wind calculation using the neutrino output from the newest full calculation still gives an entropy of about 1000 at 15 seconds. The time histories of the maximum entropies are considerably different. The evolution of the electron fraction, Ye, is not followed in the wind model at present. The electron fraction evaluated at the point of maximum entropy in the full calculation falls from 0.46 to about 0.35 20 seconds later. The combination of a higher entropy and a lower Ye should give a much higher yield of transuranic elements than the calculation reported in 1994. See Sec. 1.18.

Currently G. Mathews and K. Otsuki are using the output of the supernova explosion to calculate the yield of r-process elements.

98

2.5. *Conclusions*

It takes more than 37 years to understand supernovae!

References

1. J. M. LeBlanc and J. R. Wilson, *Ap. J.* **161**, 541 (1970).
2. J. R. Wilson, *Ap. J.* **163**, 209 (1971).
3. J. R. Wilson, *Phys. Rev. Lett.* **32**, 15 (1974).
4. Barkat, Rakavy, Reiss, and Wilson, *Ap. J.* **196**, 633 (1975).
5. J.R. Wilson, Couch, Cochran, and LeBlanc, *Annals Of The New York Academy Of Science* **262**, 54 (1975).
6. *Proceedings of the International School of Physics "Enrico Fermi", 1975*, Published 1978.
7. L. Smarr, J. Wilson, R. Barton, and R. Bowers, *Ap. J.* **246**, 515 (1981).
8. R. L. Bowers and J. R. Wilson, *Ap. J. Supp.* **50**, 115 (1982).
9. R. L. Bowers and J. R. Wilson, *Ap. J.* **263**, 366 (1982).
10. H. A. Bethe, G. E. Brown, J. Cooperstein, J. R. Wilson, *Nuclear Physics* **A403**, 625 (1983).
11. H. A. Bethe and James R. Wilson, *Ap. J.* **295**, 14 (1985).
12. J. R. Wilson, *Astrophysical Radiation Hydrodynamics*, Winkler and Norman (eds.) Reidel Publishing Company, 477 (1986).
13. R. Mayle, J. R. Wilson, and D. Schramm, *Ap. J.* **318**, 288 (1987).
14. R. Mayle and J. R. Wilson, *Ap. J.* **334**, 909 (1988).
15. J. R. Wilson and R. W. Mayle, *Physics Reports* **163**, 63 (1988); D. S. Miller, J. R. Wilson and R. W. Mayle, *Ap. J.* **415**, 278 (1993).
16. T. L. McAbee, J. R. Wilson, *Nuclear Physics A* **576**, 626 (1994); R. W. Mayle, M. Tavani, and J. R. Wilson, *Ap. J.* **418**, 398 (1993).
17. S. E. Woosley, J. R. Wilson, G. J. Mathews, R. D. Hoffman, and B. S. Meyer, *Ap. J.* **433**, 229 (1994).
18. *Ap. J.* **517**, 859 (1999).
19. J. R. Wilson and H. E. Dalhed, *World Scientific* (2002), J. R. Wilson G. J. Mathews, *Relativistic Numerical Hydrodynamics* Cambridge University Press.
20. J. R. Wilson, G. J. Mathews, and P. Marronetti, *Phys. Rev. D* **54**, 2 (1996).
21. J. R. Wilson and G. J. Mathews, *Ap. J.* **610** (2004).
22. Y. Maeda, *et al.*, *Ap. J.* **570** 671 (2002).
23. D. S. P. Dearborn, G. J. Mathews, and J. R. Wilson, Submitted *Ap. J.* (2004).
24. S. A. Balbus and J. F. Hawley, *Rev. Mod. Phys.* **70** (1988).
25. G. J. Mathews and J. R. Wilson, Submitted *Ap. J.* (2004).

ISSUES WITH CORE-COLLAPSE SUPERNOVA PROGENITOR MODELS

S. W. BRUENN *

Florida Atlantic University,
777 West Glades Road,
Boca Raton, FL 33431, USA
E-mail: bruenn@fau.edu

The status of core collapse supernoova progenitor models is reviewed with a focus on some of the current uncertainties arising from the difficulties of modeling important macrophysics and microphysics. In particular, I look at issues concerned with modeling convection, the implications of the still uncertain ^{12}C$(\alpha, \gamma)^{16}$O reaction rate, the uncertainties involved with the incorporation of mass loss, rotation, and magnetic fields in the stellar models, and the possible generation of global instabilities in stellar models at the late evolutionary stages.

1. Inroduction

The core-collapse supernova mechanism is still an unsolved problem. The failure of "state-of-the-art" one-dimensional core collapse simulations, utilizing multienergy-multiangle neutrino transport schemes and realistic opacities, and the plethora of evidence that the phenomenon is inherently multidimensional has inaugurated a new era of supernova modeling. Supernova codes are now beginning to couple multidimensional hydrodynamics to multidimensional neutrino transport, or at least neutrino transport along radial rays. Furthermore, spectacular advances are being made in the microphysics, particularly in the equation of state and the neutrino opacities. Many of the issues and prospect of both the microphysics and macrophysics involved in realistic core collapse supernova modeling are discussed in this volume.

Here I discuss the input data to these core collapse simulations, namely the progenitor models. Supernova modelers (myself included) tend to take

---

*Work partially supported by grant from the DOE Office of Science Scientific Discovery through Advanced Computing Program.

delivery of these progenitor models without being fully cognizant of the approximations that are made in order to evolve these models from the main sequence to the point of core-collapse. Phenomena such as convection, rotational instabilities, and mass loss involve huge ranges of spatial and temporal scales and/or uncertain physics and require some sort of parameterization. Some of the macrophysics is inherently multidimensional. The purpose here is gather together most of the salient approximations and parameterizations that are made in computing progenitor models, and to increase thereby the awareness of the supernova community to the many aspects of current progenitor models open to future revision. Unfortunately, without a viable model of the core-collapse supernova mechanism, it is difficult to assess the effect on the core-collapse supernova scenario of variations within the uncertainties of the progenitor models. Conventional wisdom would point to the mass of the precollapse iron core, the density profile in the adjacent silicon and oxygen layers, and its rate of rotation as being particularly important. The reader will find many details of recent progenitor models reviewed by Maeder & Meynet (2000) and Woosley et al. (2002).

2. Convection

Convection at the high Reynolds number characterizing flows inside stars is highly turbulent and chaotic characterized by eddies on a vast spectrum of scales. A great source of uncertainty in current stellar evolutionary calculations is how to model the thermal and compositional mixing at convective boundaries, and how to model the reactive flows that occur during late evolutionary phases when convective and nuclear time scales become comparable. First-principled numerical simulations of turbulence with the necessary resolution are not yet practicable, and needless to say it has been impossible to couple a first-principled calculation of turbulence with a stellar evolution code. Almost all stellar evolution codes model convection with some variant of "mixing length theory" (MLT) (Böhm-Vitense, 1958) which attempts to capture the effects of convection by an essentially one parameter diffusion process. The convective diffusivity is taken to be $K_{\text{conv}} = \frac{1}{3} v_{\text{conv}} \ell$ where v_{conv} is the mean velocity of a typical convective eddy as it traverses a mean free path, ℓ. The mean velocity is computed from the buoyancy of the eddy and Newton's laws, and the mean free path, ℓ, referred to as the mixing length, is the free parameter of the theory. It is typically taken to be some fraction of the pressure scale height. A number of uncertainties attend this attempt to model convection and an attempt will be made to describe these below.

Some sort of convective motions will occur if a fluid is unstably stratified, that is, if a displaced fluid element finds itself subjected to a buoyancy force tending to amplify the displacement. Whether a fluid element will be unstable, and if so the mean velocity that it will acquire, will depend on the assumptions made as to how the fluid element is displaced. If it is displaced adiabatically (the typical assumption) and at constant composition, then the resulting convection if it occurs is referred to as Ledoux convection. If it moves adiabatically but maintains the same composition as the background, then the resulting convection is referred to as Schwarzschild convection. If the composition gradient is zero the criterion for the two is the same. As the background composition gradient in a star, when nonzero, almost always goes from heavier to lighter elements as a function of radius (*e.g.*, in the wake of a retreating convective region), the composition gradient tends to be stabilizing. Thus Ledoux convection is more restrictive, in the sense that a fluid can be unstable to Schwarzschild convection but stable to Ledoux convection.

Regions unstable to Schwarzschild convection but stable to Ledoux convection can be doubly diffusive unstable (Kato, 1966), a phenomenon usually referred to as semiconvection, although this term has been used to refer to a multitude of sins. A fluid element perturbed outward under these conditions will find itself hotter than the background and therefore tend to continue the displacement, but will be stabilized by its heavier composition. Thermal diffusion, if faster than compositional diffusion, will tend to thermally equilibrate the fluid element with the background while leaving it with a compositional difference tending to drive it back. What can result is an oscillation of the fluid element with growing amplitude. It is unclear how to mix the material under these conditions. Two dimensional numerical simulations (Merryfield, 1995) suggest a complicated situation. Large-amplitude standing waves which break and mix over a distance of the order of a wavelength will arise if the instability is strongly driven. If the instability is weakly driven short waves arise initially and then organize themselves into longer waves which occasionally overturn and mix, and ultimately come to resemble horizontally propagating solitary waves. It is not clear how to connect the results of these simulations, which were unable to reach steady state, with the statistical steady state that presumably develops over the evolutionary time scales of stars. Extreme assumptions among stellar evolution modelers are that semiconvective mixing is fast and the use the Schwarzschild criterion for convection is therefore appropriate, or that it is slow and the use of the Ledoux criterion is therefore appropriate.

Semiconvection originally referred to another ambiguity that arises when a hydrogen burning core moves outward in mass as its helium content grows (Schwarzschild & Härm, 1958). This happens in some massive star models as the pressure in the convective core becomes more dominated in time by radiation and convective instability is more easily achieved. A chemical discontinuity arises at the convective core boundary. If electron scattering dominates the opacity, as is the case for massive stars, then the opacity increases across the convective core boundary and a problem arises as to the placement of this boundary. As the boundary is approached from the inside the radiative gradient becomes equal to the adiabatic gradient. But just outside the boundary the opacity increase implies that the radiative gradient must exceed the adiabatic gradient. Hence the boundary should be moved outward. Doing so removes the composition gradient, however, and hence removes the need to move the boundary outward in the first place. This poses a dilemma, and a number of schemes have been proposed for dealing with it. These have been summarized by Stothers (1970).

A related mixing ambiguity, also referred to as semiconvection, can happen in stars with expanding helium burning cores (Schwarzschild & Härm, 1969; Paczyński, 1970; Castellani et al., 1971b,a; Robertson, 1971; Robertson & Faulkner, 1972). In this case the electron scattering is the same just inside and just outside the core, but the carbon rich mixture inside the core has a higher free-free opacity. This by itself does not prevent the boundary of the convective core from being located unambiguously, as curves a to c in Fig. 1 illustrate. The convective boundary occurs where the radiative gradient becomes equal to the adiabatic gradient, and curve b has correctly located this boundary. As the helium burning core grows, however, a point is reached where the opacity develops a minimum inside the convective core and then increases outward to the core boundary. In this case the attempt to find the convective core boundary leads to the possibilies illustrated by curves d to f. If curve d is chosen to locate the core boundary, the region between i and j is not convective, contradicting the choice. If curves e or f are chosen, the material at the edge of the core will be unstable to further convection since $\nabla_{\mathrm{rad}} > \nabla_{\mathrm{rad}}$ there. What is frequently done is to assume that curve e, modified by the horizontal segment connecting points m and n represents the correct choice. This is achieved by assuming that the requisite "semiconvective" compositional mixing takes place between points m and n causing $\nabla_{\rm rad} = \nabla_{\rm rad}$ there.

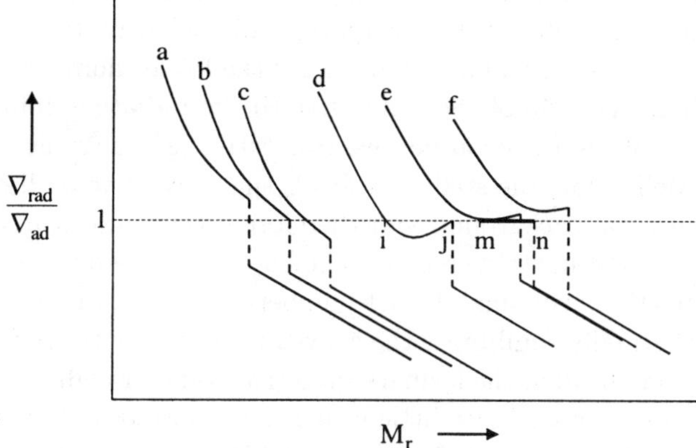

Figure 1. The curves illustrate profiles of $\nabla_{\rm rad}/\nabla_{\rm ad}$ in the vicinity of the boundary between a convective helium burning core and a radiative hydrogen envelope. Curves a to c illustrate an unambiguous determination of the convective core boundary (*e.g.*, curve b). Curves d to f illustrate the ambiguity that arises when a minimum in the opacity occurs within the convective core. (Figure adapted from Fig. 7 of (Paczyński, 1970).)

Another problem with the MLT parameterization of convection is overshooting, which refers to the tendency of convective eddies to penetrate the radiative layers surrounding a convective zone and hence induce a mixing of a region larger (in mass) than classically allowed by the strict adoption of the Schwarzschild or Ledoux criterion. This is a consequence of the fact that while the acceleration of the convective motions cease at the boundary of a convective region, there velocities do not. Thus convective overshooting may be present at the border of any convective region, and is not confined to any particular evolutionary phase. The effect of overshooting is to increase the mass in the convective region that is mixed which, in turn, can have a number of consequences for stellar ages, nucleosynthesis, and pre-supernova structure. Unfortunately, is not a natural outcome of MLT, due to the local nature of the theory (Renzini, 1987).

The radial extent, $\ell_{\rm OV}$, of the thermal and chemical mixing from the formal convective core boundary is typically parameterized by an ad hoc formula such as $\ell_{\rm OV} = \alpha_{\rm OV} \min(r, H_{\rm p})$, where $H_{\rm p}$ is the pressure scale height, r is the core radius, the distance from the core edge to the surface, or some other such scale that naturally limits the extent of overshooting, and $\alpha_{\rm OV}$ is the parameter of the theory, typically below unity. The extent of convective overshooting will be a function of the Péclet number (*e.g.*, Zahn, 1999), which is the ratio of the convective to the radiative diffusivity. For large Péclet numbers at the border of a convective region (typical of convective regions well below the stellar surface), the convective eddies exchange little heat with the background and therefore establish a nearly adiabatic gradient beyond the unstable region. They are therefore decelerated by the stable stratification. For small Péclet numbers, however, radiation diffusion will tend to thermally equilibrate the convective eddies with the background as they penetrate beyond the formal convective boundary which will weaken their deceleration. In this case little heat is transported, but chemicals and momentum can be transported an appreciable distance. (Technically, the former (large Péclet number) case is referred to as convective penetration, the latter is referred to as convective overshooting (Zahn, 1991).)

Some observational constraints suggesting a value of ~ 0.2 for $\alpha_{\rm OV}$ are provided by the size of gaps (blue loops) in open star cluster color-magnitude diagrams (Maeder & Mermilliod, 1981; Stothers & Chin, 1991a,b; Stothers, 1991; Nordström et al., 1997), the asteroseismology of η Bootis (Di Mauro et al., 2003), accurate stellar dimensions derived from well-detached double-lined eclipsing binaries (Giménez et al., 2004) (which suggests a somewhat larger value of $\alpha_{\rm OV}$ for massive stars). Beyond this the value of $\alpha_{\rm OV}$ must be inferred from numerical simulations or guessed at.

A number of numerical simulations investigating the nature of turbulent compressible convection and convective overshooting have been performed. These include two dimensional simulations (Hurlburt et al., 1986, 1994; Dintrans et al., 2003), three dimensional simulations (Cattaneo et al., 1991; Muthsam et al., 1995; Singh et al., 1998; Stein & Nordlund, 1998; Brummell et al., 2002), three dimensional simulations with rotation (Brummell et al., 1996; Browning et al., 2004), three dimensional simulations with ionization (Rast et al., 1993; Rast & Toomre, 1993a,b), and two dimensional simulations (Bazan & Arnett, 1994, 1998). and they reveal a rather complicated picture. Depending on the density contrast, upward-moving flows are typically broader and slower moving than downward-moving flows (a trend seen in the neutrino driven convecting regions in post collapse stellar cores). Ionization regions can exagerate this trend. The downward flows traverse multiple scale heights and penetrate the stable layers below by a significant fraction of the local pressure scale height. Because of the low filling factor of the plumes, however, they do not establish an adiabatic gradient there. Convective overshooting can excite gravity waves which leads to further mixing. The use of MLT during shell oxygen burning and later nuclear burning stages is particularly problematic, as nuclear burning timescales at the base of the convecting region and convective timescales across the convective region become comparable. The simulations show inhomogeneities in the composition and strong fluctuations in space and time unlike the smooth, steady flow presupposed by one-dimensional stellar evolutionary calculations with MLT.

In conclusion we observe that MLT is a phenomenological parameterization of convection which is applied to a variety of convective phenomena in a physically motivated but crude way. Different prescriptions for MLT convection can lead to substantial differences in the interior structure of massive stars in their late evolutionary phases. We note just one example. The nonrotating models computed by Hirschi et al. (2004), who used the Schwarzschild criterion for convection with overshooting, and with convective diffusion beyond He burning, develop considerably larger Si core masses than the models computed by Rauscher et al. (2002), who used the Ledoux criterion for convection with semiconvection and without overshooting, and convective diffusion beyond He burning.

2.1. $^{12}C(\alpha, \gamma)^{16}O$ Reaction Rate and Convection

The structure and explosive yields of massive stars depends on the mass fraction, X_C, of ^{12}C left after He burning, and this depends both on the combined effects of the $^{12}C(\alpha, \gamma)^{16}O$ reaction rate and the treatment of overshooting and semiconvection which governs the growth of the helium burning core (Weaver & Woosley, 1993; Thielemann et al., 1996; Imbriani et al., 2001). The triple-α reaction and the $^{12}C(\alpha, \gamma)^{16}O$ reaction compete with each other, and the ratio of the two rates determines directly the ratio of ^{12}C and ^{16}O produced during core helium burning. The mixing of fresh He fuel into the He-burning core at late stages and high temperatures, when a core without growth by semiconvection and overshooting of convection eddies would have already ceased to process any He fuel, probes the ^{12}C rate at higher temperature with the effect of turning much of the remaining ^{12}C into ^{16}O. However, despite years of effort, the ^{12}C and ^{16}O cross section is still unknown to within a factor of a few (Buchmann et al., 1996; Angulo et al., 1999). Furthermore, as discussed above the treatment of convection in stellar evolutionary codes is by means of MLT, which is rudimentary and phenomenological, and cannot address the question of convective overshooting.

The implication of the uncertainty in the value of X_C left after helium burning is that its value affects the later structure of the star principally through its effect during the time interval that elapses between the end of the central C burning and the beginning of the central Ne burning. During this time the CO core experiences a phase of gravitational contraction which is partially alleviated by the formation of one (or more) convective C shell episodes. These convective episodes stop for a while the outwardly advancing C-burning front while the reservior of fuel contained in the convective shell is consumed. During this time the C-burning front remains essentially fixed in mass and slows down the contraction of the region above the front. A larger value of X_C after core carbon burning allows a more effective support of the layers above the C-burning front during C-shell burning and hence the formation of a less steep mass-radius relation. These differences in the mass-radius relations that form before the Ne ignition remain through later core contractions until the final explosion.

The situation is shown schematically in Fig. 2, which is adapted from Fig. (12) of Imbriani et al. (2001). The lines denoted by X_C^{high} and X_C^{low} represent the mass-radius relations for a star at the onset of core collapse having a high and a low value of X_C, respectively, after core He-burning. Once the explosion commences and the shock wave moves outward, it is radiation dominated, gains or loses only a small fraction of its energy to the matter, and therefore expands essentially adiabatically (Weaver & Woosley, 1980). The temperature behind the shock is therefore a function only of its radius and the explosion energy. At the same time, the matter which is subject to complete silicon burning, incomplete silicon burning, or oxygen burning, and whose final composition therefore depends only on its initial proton fraction, Y_e, is determined only by the peak post shock temperatures, and therefore by the geometrical distance of the matter from the core center. Because of the mass-radius relation (Fig. 2), the mass of material undergoing incomplete explosive Si burning and explosive O burning, which produce the bulk of the elements from ^{28}Si to ^{55}Mn, is greater for small X_C. On the other hand, the lighter elements from ^{20}Ne to ^{27}Al are

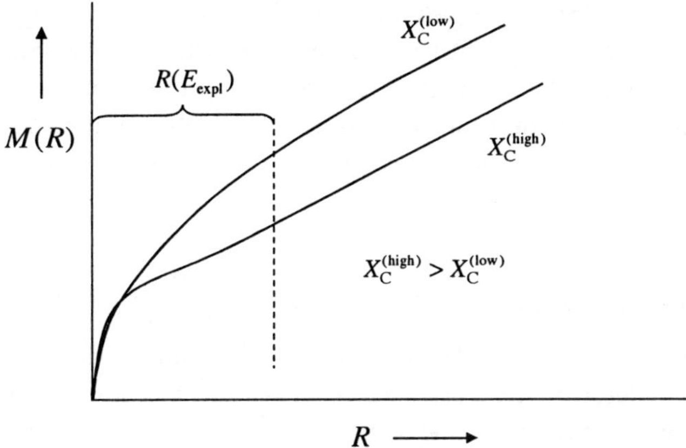

Figure 2. The enclosed mass as a function of radius, R, of two initially identical main sequence stars at the onset of core collapse. The X_C^{low} and X_C^{high} curves represent the case of a low and a high mass fraction of ^{12}C after core He-burning.

produced in the C convective shell and partial destroyed by the shock, and their production therefore scales with X_C. Ignoring the subtleties of many of the production chains, it is seen that a large X_C favors the production of elements at the lighter end of the ^{20}Ne to ^{55}Mn range, while the opposite is true of a small X_C. A dramatic illustration of this trend is shown in Fig. 4 of Weaver & Woosley (1993), who tried to put limits on the ^{12}C$(\alpha,\gamma)^{16}$O reaction rate by requiring that the final explosive yields to have a scaled solar relative abundance. (Arnett (1971) made an analogous attempt to fix the ^{12}C$(\alpha,\gamma)^{16}$O reaction rate by using the observed ^{12}C to ^{16}O ratio.) It must be remembered that these attempts to fix the ^{12}C$(\alpha,\gamma)^{16}$O reaction rate by using the results of stellar evolutionary calculations actually fix a combination of this rate and the particular MLT scheme used.

Further problems with MLT convection for the evolution of massive stars, in brief, are the fact that it fails to deal with the interaction between convection and rotation, and the generation and transport of magnetic fields (Zahn, 1999), and convective nuclear burning (Bazan & Arnett, 1994, 1998; Asida & Arnett, 2000). Attempts have been made to overcome the limitations of MLT by a first principles approach (*i.e.*, direct numerical solutions of the fundamental equations) (*e.g.*, Stein & Nordlund, 1998; Singh et al., 1998; Deupree, 1998; Asida & Arnett, 2000), or by more sophisticated convection models (*e.g.*, Canuto & Mazzitelli, 1991, 1992), but these have not made there way into evolutionary calculations of massive stars to core collapse.

2.2. *Weak Interactions*

Weak interactions affect Y_e, the proton fraction, and therfore play an important role in determining both the presupernova stellar structure and the nucleosynthesis. They affect the structure because, at all times, the pressure is mostly due to electrons - at first, nonrelativistic and nondegenerate, but later neither. They affect the nucleosynthesis because the synthesis of all nuclei except those with equal numbers of neutrons and protons is sensitive to Y_e.

The weak interaction rates after oxygen burning are particularly difficult to calculate as a large number of excited states with uncertain properties become populated so that their decay must be dealt with statistically. Early attempts in this direction were made by Hansen (1968); Mazurek et al. (1974); Takahashi et al. (1973). However, it was (Fuller et al., 1980, 1982b,a; Fuller, 1982; Fuller et al., 1985) who recognized the key role played by the Gamow-Tellar resonance and noted that measured decay rates exploited only a small fraction of the available strength. More recently new shell-model calculations of the distribution of Gamow-Tellar strength have resulted in an improved—and often reduced—estimate of its strength (Martnez-Pinedo & Langanke, 1999; Langanke & Martnez-Pinedo, 2000; Martnez-Pinedo et al., 2000; Langanke & Martnez-Pinedo, 2003). Inclusion in presupernova evolutionary models of these new rates for electron capture and beta dacay (Heger et al., 2001) lead to slightly higher central proton fractions and smaller outer core entropies at the time of core collapse, leading to slightly smaller iron core masses. Incorporation of the new rates in core collapse simulations (Langanke et al., 2003; Hix et al., 2003) lead to an increased importance of nuclear vs free proton electron capture and reduced initial mass behind the shock with lower densities, proton fractions, and entropies. However, the reduced electron capture in the outer layers slows their collapse and allows the shock to reach a slightly larger maximum radius. The collapsing core encounters a range of large and neutron rich nuclei whose beta strengths have not yet been calculated in detail, underscoring the need for more work in this area.

2.3. *Rotation*

Massive stars are observed to be rapid rotators, with equatorial velocities spanning the range $100-400$ km s^{-1} (Fukuda, 1982; Halbedel, 1996; Penny, 1996; Howarth et al., 1997). As a consequence, a number of instabilities leading to composition mixing and angular momentum transport are predicted to occur within these stars as they evolve, leading to differences in the structure of supernova progenitors. Furthermore, the rotation rate of progenitor cores may play a role in the supernova mechanism and is dependent on the degree to which angular momentum transport has occurred during the course of prior evolution.

A number of observations point to the operation of rotationally induced mixing processes in massive stars. The ratio B/R, the number of blue to red supergiants, increases with the metalicity, Z, (*e.g.*, Langer & Maeder, 1995; Maeder & Meynet, 2000), and this cannot be accounted for by mass loss or convection. For a number of reasons (Maeder & Meynet, 2001) rotation favors the development of the red supergiant structure. The increase of the B/R ratio with Z results from the increase of the mass loss rate with Z, and with it the loss of angular momentum of the star, rapidly reducing its rotation rate during the MS phase and thus reducing its propensity to become a red supergiant during later phases. Rotational mixing in the radiative envelopes of massive stars will modify their surface abundances. One would naively expect a depletion of an initial surface abundance of fragile nuclei, such as ^{3}He, ^{6}Li, ^{7}Li, ^{9}Be, ^{10}B, and ^{11}B, mixed down and destroyed by proton capture at higher interior temperatures. At the same time, hydrogen burning in massive stars is governed by the CNO cycle, and this has the effect of converting most of the initial ^{12}C and ^{16}O into ^{14}N. Rotational mixing to the surface of material in which the CNO cycle was operative should be manifested as a depletion of ^{12}C and ^{16}O and an enhancement of ^{14}N. These effects have been observed. For example, Proffitt & Quigley (2001); Venn et al. (2002) have observed boron depletions in B type stars in OB associations, consistent with the predictions of Fliegner et al. (1996) and the rotating models of Heger & Langer (2000). Some non-supergiant B stars show a moderate increase in N abundance (Gies & Lambert, 1992; Lennon et al., 1996).

Ideally, the evolution of rotating stars should be calculated multi-dimensionally, with the composition and angular momentum transport arising directly from the calculation itself. This program cannot be carried out with current computer resources. Rather, the equations of stellar structure are kept one-dimensional. Initially this was accomplished by replacing the usual spherical coordinates by new coordinates characterizing the equipotentials (which have cylindrical symmetry) (Kippenhahn et al., 1970). More recently, this is accomplished by making the assumption (Zahn, 1992) of highly anisotropic turbulence in radiative layers. In particular, turbulence generated by, say, shear in the presence of differential rotation is much stronger in the direction perpendicular to gravity ("horizontal direction") than in the vertical direction, the latter being suppressed by the stable vertical stratification. If true, the strong horizontal turbulence makes the angular velocity Ω and the composition nearly constant on isobaric surfaces, rather than cylinders, giving rise to a "shellular" rotation law. In this case, the motion is not cylindrical. Nevertheless, a consistent 1-D scheme has been formulated (Meynet & Maeder, 1997, 2000).

The critical assumption of highly anisotropic turbulence in radiative stellar zones has indirect observational support, both in the fact that turbulent motions caused by shear stresses in the Earth's atmosphere are highly anisotropic in those regions where the stratification is stable, and in the study of the solar tachocline (Spiegel & Zahn, 1992). (The tachocline is the transition zone between the rigid rotation in the radiative interior and the external convective zone, where rotation varies with latitude.) If the horizontal turbulence is intense, then the tachocline is very thin, and the latter is supported by helioseismological observations.

Keeping the equations of stellar structure one-dimensional allows stellar evolutionary calculations to be performed, but requires that various instabilities leading to angular momentum transport and the mixing of chemical elements, which play a major role in massive star evolution, be parameterized. Since the diffusion and advection of composition and angular momentum operate on Ω and $(r\sin\theta)^2\Omega$, respectively, their vertical transport rates are different, being much smaller for the composition. Gradients in composition (μ-gradients) that develop during the evolution of the star tend to reduce the vertical transport rates, so the effect of these μ-gradient effects must either also be parameterized (Heger et al., 2000) or attempt to incorporate them more consistently in the instability and mixing algorithms (Maeder & Zahn, 1998).

Rotation in convective zones is relatively easy to handle until oxygen shell and particularly silicon burning. Chemical homogeneity can be assumed and, if the high viscosity associated with turbulence tends to solid-body rotation, then rigid body rotation can also be assumed. An alternative (Endal & Sofia, 1976) is that convection preserves the angular momentum of the convective elements leading to equalization of the specific angular momentum. Supporting the tendency towards rigid body rotation over alternatives, however, is the observation that the solar convection zone deviates from solid body rotation by less than 5% (Antia et al., 1998). Complications in handling convective zones begin with oxygen shell burning. The times scales for convective mixing, nuclear burning and angular momentum transport become similar, and the feasibility of constructing self-consistent models with one-dimensional evolution equations becomes highly suspect.

At convective boundaries and in radiative zones a number of instabilities can lead to significant transport of composition and angular momentum (*e.g.*, Endal & Sofia, 1978; Heger et al., 2000; Meynet & Maeder, 2000; Maeder & Meynet, 2000, and many others). These include the Eddington-Sweet circulation (von Zeipel, 1924; Eddington, 1925; Vogt, 1925) (a circulation that arises because a component of the radiation stress is directed along equipotential surfaces), shear instabilities (Spiegel & Zahn, 1970; Zahn, 1974) (dynamical: arising when the free energy in differentially rotating layers exceeds the work against restoring forces required to adiabatically overturn the fluid; secular: as above but allowing thermal diffusion in the overturning fluid to remove a stabilizing temperature gradient), the Solberg-Høiland instability (Tassoul, 2000) (analogous to the Ledoux criterion but including the angular momentum gradients), and the Goldreich-Schubert-Fricke instability (Goldreich & Schubert, 1967; Fricke, 1968) ((1) a secular analogue to the Solberg-Høiland stability criterion, and (2) a criterion for the generation of meridional flows for nonconservative rotation profiles).

During the main-sequence evolution of rapidly rotating massive stars, angular momentum transport in the outer radiative regions, principally by the Eddington-Sweet circulation, quickly establishes a steady state in which the diffusion of angular momentum is balanced by advection of angular momentum due to circulation (Zahn, 1992; Urpin et al., 1996; Talon et al., 1997). Neglecting angular momentum loss at the surface, this leads to a differential rotation in which the angular velocity at the convective core boundary is about a factor of 1.15 that at the surface.

Concerning the late evolutionary stages, which are of most interest to supernova modelers, there have been two recent stellar evolutionary calculations of rotating stars that have been carried out to these stages (without magnetic fields). These are by Heger et al. (2000) (HLW), who evolve to core collapse, and Hirschi et al. (2004) (HMM), who evolve through central oxygen or silicon burning. The two groups employed different numerical methods of incorporating the effects of rotation, and different parameterizations of convection.

HMM find that for stars with zero age main-sequence masses (M_{MS}) < 30 $M_\odot$ rotation tends to increase the mass of the iron core prior to collapse. This trend is confirmed by HLW who confine their study to $M_{MS} < 25~M_\odot$. The larger iron cores in the rotating models result from rotational mixing in prior evolutionary phases, mainly the H-burning phase where the mixing and nuclear timescales are comparable. The larger He cores resulting from the rotational mixing during H-burning cause them to have lower densities and higher temperature during He-core-burning, and this leads to lower C to O ratios at core He exhaustion. HLW, who parameterize the inhibiting effect of μ-gradients on mixing, find on varying this parameter that the final iron core masses are a sensitive functions of this parameter. The more efficient the rotational mixing (or less strong are the inhibiting effects of the μ-gradients), the greater the core masses. (This underscores the fact that the precollapse structure of rotating stars is highly dependent on approximate numerical treatments of complicated physics.) The larger precollapse iron core models resulting from rotation imply, of course, a smaller initial M_{MS} that will lead to core collapse. HMM obtain significantly greater core masses then HLW, due to their different treatments of rotational mixing and convection, again underscoring the effect of different approximations. Both groups find that only convective process are rapid enough to notably redistribute angular momentum after core helium burning The effect is to leave each convective region with a constant angular velocity. The distribution of angular momenta in the final models is therefore characterized by rounded saw-tooth patterns, each saw tooth being the constant angular velocity imprint of a convective zone. For $M_{MS} > 30~M_\odot$, HMM find that rotationally enhanced mass loss causes the star to enter the Wolf-Rayet phase earlier and to therefore spend more time undergoing heavy mass loss. This erodes the star and results in smaller cores at the pre-supernova stage.

The two groups find that the angular momentum of precollapse cores tend to converge to a value that would imply a rotation rate upon collapse to a neutron star of about 1 ms or less, which is near breakup for even the slowest rotators. The corresponding angular momentum is about 100 times the angular momentum of the fastest rotating pulsars observed. The implication of this in unclear, as it is not yet established how rapidly newly formed pulsars rotate. For example, the fastest rotating young pulsar, PSR J053726910 (Marshall et al., 1998), has a period of 16 ms. But with its estimated age of $\sim 5 \times 10^3$ yr an extrapolation to an initial rotation rate of ~ 1 ms, while not the only possibility, is not unreasonable. The Crab (and by implication others), on the other hand, may never have rotated near breakup (Trimble & Rees, 1970) unless this rotational energy ($\sim 3 \times 10^{52}$ ergs) were radiated as gravitational waves. Otherwise this energy would surely have been seen in the optical and manifested now in the expansion velocity of the nebula.

2.4. *Rotation and Mass Loss*

Mass loss from the stellar surface (stellar winds) significantly affects the evolution of massive stars, particularly the Upper MS stars (Chiosi & Maeder, 1986; Maeder & Meynet, 1987) where an appreciable percentage of the mass of these star can be lost during their evolution. Type 1b and 1c supernovae probably arise from stars having suffered extensive mass loss. Mass loss also affects rotating stars, as these winds can transport large amounts of angular momentum out of the star. This is particularly true for equatorial mass loss by anisotropic stellar winds (Maeder, 1999). However, the uncertainties in these mass-loss rates, particularly for red supergiants (Lamers & Cassinelli, 1999) are considerable due both to the uncertain physics involved and the uncertainties in the observational data and their interpretation. (For example, mass loss rates from hot O and B stars have rencently been revised, generally downward (Nugis & Lamers, 2000; Vink et al., 2000, 2001; Bouret et al., 2004), owing to a improved treatments of clumping and multiple scatering.

Stellar evolutionary calculations of nonrotating stars with mass find that stars with solar metalicity initially more massive than $\sim 30 - 35$ M$_\odot$ converge to a hydrogen free star of roughly 5 M$_\odot$ (Maeder, 1990; Schaller et al., 1992; Woosley et al., 1993; Meynet et al., 1994). This assumes the loss of the hydrogen envelope either during the main sequence phase (as luminous blue variables) or as red supergiants), and a mass loss rate for hydrogenless Wolf-Rayet stars being given by a positive power of the remaining mass (Langer, 1989b,a). The latter causes the convergence in mass. However, while the masses of these stars may converge, the thermal and chemical structures of these stars retain some memory of their former masses (Woosley et al., 1993). For example, a helium core undergoing carbon burning and trimming down from a larger mass will have a higher temperature and lower density than a constant mass helium core of the same final mass. It will therefore burn more of its helium to oxygen than its constant mass counterpart. Thus, despite similar final iron core masses, stars that have trimmed down from larger initial masses will have larger carbon-oxygen mantles with larger oxygen to carbon ratios and shallower density gradients in the outer regions than stars of the same final mass that have trimmed down from smaller initial masses. Nonrotating stars that do not succeed in loosing their hydrogen envelopes during hydrogen and core helium burning have internal structures that are little affected by the mass loss, as the interior evolution is largely decoupled from the surface. However, these stars may give rise to different supernova types (Type IIL versus Type IIP, for example) depending on the remaining mass of the hydrogen envelope.

Rotating stars undergoing mass loss can lose considerable angular momentum during evolution (Packet et al., 1980; Heger & Langer, 1998). The mass loss itself is affected by rotation through centrifugal forces, nonradial forces, and gravitational darkening (von Zeipel's theorem). The mass loss rate is increased by rotation, but it is now appreciated that it could be either oblately (*i.e.*, predominantly equatorial) or prolately (*i.e.*, predominantly polar) asymmetrical (Owocki et al., 1998; Petrenz & Puls, 2000), with much less angular momentum being lost by the star for a given amount of mass lost in the latter case.

2.5. *Rotation and Magnetic Fields*

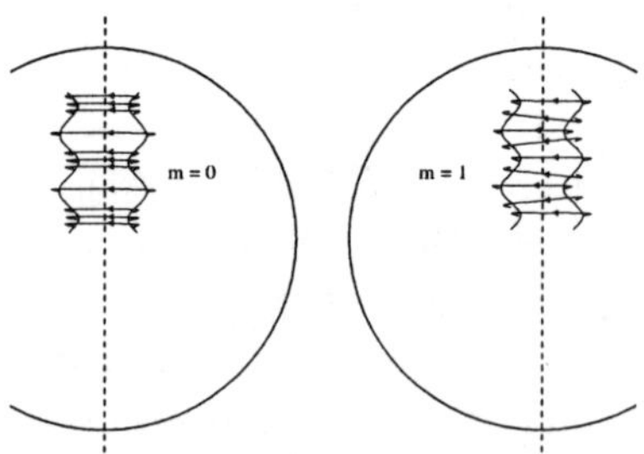

Figure 3. Illustration of $m = 0$ and $m = 1$ instabilities near the axis of a star, where m is the azimuthal mode of number of the perturbation. In stably stratified (radiative) zones, the $m = 1$ instability is more likely to occur as it involves displacements close to equipotential surfaces. (Figure adapted from Fig. 1 of (Taylor, 1973).)

The growth of magnetic fields and their influence on the evolution of rotating stars has recently begun to be considered in stellar evolutionary calculations of stars to core collapse. The prime importance of magnetic fields here is in their potential for transferring angular momentum. As discussed in the previous Section, without magnetic fields the various hydrodynamic instabilities leading to angular momentum transport (as presently understood) are two weak to prevent the cores of supernova progenitors from ending up with about 100 times the angular momentum of the shortest period young pulsar. While this large angular momentum might be appropriate for the collapser model of gamma-ray bursts (MacFadyen et al., 2001), it is likely too high to account for pulsar rotation rates at birth. Magnetic effects might provide additional angular momentum transport during stellar evolution. A possible case in point is the Sun, whose near-uniform rotation of the radiative core and the small difference in rotation rate between the core and the convective envelope, the latter being continuously spun down by the solar wind torque, has been established through helioseismology (Corbard et al., 1997; Schou et al., 1998). On the other hand, it has been known for a long timed that hydrodynamic instabilities alone are incapable of accounting for this (*e.g.*, Spruit et al., 1983).

In convective zones, angular momentum transport by turbulent viscosity is very efficient. Where the transport of angular momentum by magnetic fields may be critical is in radiative zones. Spruit (1999, 2002) has summarized much of the literature pertaining to this issue and described how a magnetic dynamo based on the an instability studied by Taylor (1973) and others and tapping the free energy available from differential rotation could operate in stably stratified zones. The basic picture is that differential rotation will wind up an initially weak field producing a predominantly toroidal (azimuthal) field. This field is subject to a number of instabilities, the m = 1 Taylor instability, exhibited in Fig. 3, being likely the most relevant. The poloidal component generated by this instability will be stretched into a strong toroidal component which will in turn be subjected to instabilities, forming a dynamo. An estimate of the equilibrium strength of the field generated by this "Taylor-Spruit" dynamo is made by equating the growth timescale to the attenuation timescale by mangetic diffusivity. The radial component over the largest unstable lengths is chosen to evaluate these timescales as this determines the maximum saturation field. The result is obtained for the two limiting cases in which the stability is provided either by a thermal or μ gradient.

Comparisons of estimates of the angular momentum transport by the magnetic fields generated by the Taylor-Spruit dynamo with the angular momentum transport by hydrodynamic instabilities indicate that the former could be of major importance (Maeder & Meynet, 2004). A recent stellar evolutionary calculation (Heger et al., 2004) incorporating the estimates of magnetic torques produced by the Taylor-Spruit magnetic fields show a reduction by a factor of 10 in the final iron core angular momentum. It must be observed, however, that angular momentum transport by magnetic fields has the potential of being completely ineffective in a star or so effective as to lead to near uniform rotation throughout the entire star with the result of a much too slow rotation of the remnant (Spruit & Phinney, 1998). Thus, while the above results are encouraging, the fact that magnetic fields have an almost "just so" effectiveness must regarded as extremely preliminary.

2.6. *Global Asymmetries*

A potentially important effect in the late evolutionary stages of massive stars, which would require a multidimensional evolutionary code to follow, is the generation of overstable g-modes driven by shell nuclear burning (Goldreich et al., 1996). The idea is that nuclear burning rates in silicon and oxygen burning shells are extremely temperature sensitive. Consider an $\ell = 1$ perturbation of the shell to the right, for example. The left-hand side of the shell will then be compressed and heated. Nuclear burning in the compressed region will be greatly enhanced and will generate a large local overpressure which will push the shell back to the left. If overstable, this g-mode will oscillate with growing amplitude with the very interesting possibility of generating a significant global asymmetry of the stellar core just prior to collapse.

2.7. *Conclusions*

The evolutionary calculations of massive stars to core collapse are extremely difficult, involving a variety of physics on multiple length and time scales and in multiple dimensions. Teams are incorporating more realistic physics into the stellar codes, but are still constrained to model inherently multi-dimensional phenomena in one dimension. This review has attempted to make supernova modelers aware of the many approximations and parameterizations that are perforce made in the course of evolving a star from the main-sequence to core collapse.

References

Angulo, C., Arnould, M., Rayet, M., Descouvemont, P., Baye, D., Leclercq-Willain, C., Coc, A., Barhoumi, S., Aguer, P., Rolfs, C., Kunz, R., Hammer, J. W., Mayer, A., Paradellis, T., Kossionides, S., Chronidou, C., Spyrou, K., degl'Innocenti, S., Fiorentini, G., Ricci, B., Zavatarelli, S., Providencia, C., Wolters, H., Soares, J., Grama, C., Rahighi, J., Shotter, A., & Lamehi Rachti, M. 1999, Nuclear Physics A, 656, 3

Antia, H. M., Basu, S., & Chitre, S. M. 1998, Monthly Notices of the Royal Astronomical Society, 298, 543

Arnett, W. D. 1971, Astrophysical Journal Letters, 170, 1621

Asida, S. M. & Arnett, D. 2000, Astrophysical Journal, 545, 435

Bazan, G. & Arnett, D. 1994, Astrophysical Journal Letters, 433, L41

—. 1998, Astrophysical Journal, 496, 316

Böhm-Vitense, E. 1958, Zeitschrift für Astrophysik, 46, 108

Bouret, J.-C., Lanz, T., & Hillier, D. J. 2004, astro-ph/0412346v1

Browning, M. K., Brun, A. S., & Toomre, J. 2004, Astrophysical Journal, 601, 512

Brummell, N. H., Clune, T. L., & Toomre, J. 2002, The Astrophysical Journal, 570, 825

Brummell, N. H., Hurlburt, N. E., & Toomre, J. 1996, The Astrophysical Journal, 473, 494

Buchmann, L., Azuma, R. E., Barnes, C. A., Humblet, J., & Langanke, K. 1996, Physical Review C, 54, 393

Canuto, V. M. & Mazzitelli, I. 1991, Astrophysical Journal, 370, 295

—. 1992, Astrophysical Journal, 389, 724

Castellani, V., Giannone, P., & Renzini, A. 1971a, Astrophysics and Space Science, 10, 355

—. 1971b, Astrophysics and Space Science, 10, 340

Cattaneo, F., Brummell, N. H.; andToomre, J., Malagoli, A., & Hurlburt, N. E. 1991, Astrophysical Journal, 370, 282

Chiosi, C. & Maeder, A. 1986, Annual Review of Astronomy and Astrophysics, 24, 329

Corbard, T., Berthomieu, G., Morel, P., Provost, J., Schou, J., & Tomczyk, S. 1997, Astronomy and Astrophysics, 324, 298

Deupree, R. G. 1998, Astrophysical Journal, 499, 340

Di Mauro, M. P., Christensen-Dalsgaard, J., Kjeldsen, H., Bedding, T. R., & Paternò, L. 2003, Astronomy and Astrophysics, 404, 341

Dintrans, B., Brandenburg, A., Nordlund, A., & Stein, R. F. 2003, Astrophysics and Space Science, 284, 237

Eddington, A. S. 1925, The Observatory, 48, 73

Endal, A. S. & Sofia, S. 1976, Astrophysical Journal, 210, 184

—. 1978, Astrophysical Journal, 220, 279

Fliegner, J., Langer, N., & Venn, K. A. 1996, Astronomy and Astrophysics, 308, L13

Fricke, K. 1968, Zeitschrift für Astrophysik, 68, 317

Fukuda, I. 1982, Astronomical Society of the Pacific, 94, 271

Fuller, G. M. 1982, Astrophysical Journal, 252, 741

Fuller, G. M., Fowler, W. A., & Newman, M. J. 1980, Astrophysical Journal Supplement, 42, 447

—. 1982a, Astrophysical Journal Supplement, 48, 279

—. 1982b, Astrophysical Journal, 252, 715

—. 1985, Astrophysical Journal, 293, 1

Gies, D. R. & Lambert, D. L. 1992, Astrophysical Journal, 387, 673

Giménez, A., Claret, A., Ribas, I., & Jordi, C. 2004, in ASP Conference Series, Vol. 173, Stellar Structure: Theory and Test of Connective Energy Transport, ed. A. Giménez, E. A. Guinan, & B. Montesinos (ASP: San Francisco)

Goldreich, P., Lai, D., & Sahrling, M. 1996, in Unsolved Problems in Astrophysics, ed. J. N. Bahcall & J. P. Ostriker (Princeton: Princeton University Press), 269–280

Goldreich, P. & Schubert, G. 1967, Astrophysical Journal, 150, 571

Halbedel, E. M. 1996, Publications of the Astronomical Society of the Pacific, 108, 833

Hansen, C. J. 1968, Astrophysical Journal Supplement, 1, 499

Heger, A. & Langer, N. 1998, Astronomy and Astrophysics, 334, 210

—. 2000, Astrophysical Journal, 544, 1016

Heger, A., Langer, N., & Woosley, S. E. 2000, Astrophysical Journal, 528, 368

Heger, A., Woosley, S. E., Langer, N., & Spruit, H. C. 2004, in Proceedings of the IAU Symposium 215, Stellar Rotation, astro-ph/0301374, ed. A. Maeder & P. Eenens

Heger, A., Woosley, S. E., Martínez-Pinedo, G., & Langanke, K. 2001, Astrophysical Journal, 560, 307

Hirschi, R., Meynet, G., & Maeder, A. 2004, Astronomy and Astrophysics, 425, 649

Hix, W. R., Messer, O. E., Mezzacappa, A., Liebendrfer, M., Sampaio, J., Langanke, K., Dean, D. J., & Martnez-Pinedo, G. 2003, Physical Review

Letters, 91, 201102

Howarth, I. D., Siebert, K. W., Hussain, G. A. J., & Prinja, R. K. 1997, Monthly Notices of the Royal Astronomical Society, 284, 265

Hurlburt, N. E., Toomre, J., & Massaguer, J. M. 1986, Astrophysical Journal, 311, 563

Hurlburt, N. E., Toomre, J., Massaguer, J. M., & Zahn, J.-P. 1994, Astrophysical Journal, 421, 245

Imbriani, G., Limogi, M., Gialanella, L., Terrasi, F., Straniero, O., & Chieffi, A. 2001, Astrophysical Journal, 558, 903

Kato, S. 1966, Publications of the Astronomical Society of Japan, 18, 374

Kippenhahn, R., Meyer-Hofmeister, E., & Thomas, H.-C. 1970, Astronomy and Astrophysics, 155, 155

Lai, D. & Goldreich, P. 2000, Astrophysical Journal, 535, 402

Lamers, J. H. G. L. M. & Cassinelli, P. 1999, Introduction to Stellar Winds (Cambridge/New York : Cambridge University Press)

Langanke, K. & Martnez-Pinedo, G. 2000, Nuclear Physics A, 673, 481

—. 2003, Reviews of Modern Physics, 75, 819

Langanke, K., Martnez-Pinedo, G., Sampaio, J. M., Dean, D. J., Hix, W. R., Messer, O. E., Mezzacappa, A., Liebendrfer, M., Janka, H.-T., & Rampp, M. 2003, Physical Review Letters, 76, 241102

Langer, N. 1989a, Astronomy and Astrophysics, 220, 135

—. 1989b, Astronomy and Astrophysics, 210, 93

Langer, N. & Maeder, M. 1995, Astronomy and Astrophysics, 295, 685

Lennon, D. J., Dufton, P. L., Mazzali, P. A., Pasian, F., & Marconi, G. 1996, Astronomy and Astrophysics, 314, 243

MacFadyen, A. I., Woosley, S. E., & Heger, A. 2001, Astrophysical Journal, 550, 410

Maeder, A. 1990, Astronomy and Astrophysics Supplement Series, 84, 139

—. 1999, Astronomy and Astrophysics, 347, 185

Maeder, A. & Mermilliod, J. C. 1981, Astronomy and Astrophysics, 321, 136

Maeder, A. & Meynet, G. 1987, Astronomy and Astrophysics, 182, 243

—. 2000, Annual Review of Astronomy and Astrophysics, 38, 143

—. 2001, Astronomy and Astrophysics, 373, 555

—. 2004, Astronomy and Astrophysics, 411, 543

Maeder, A. & Zahn, J.-P. 1998, Astronomy and Astrophysics, 334, 1000

Marshall, F. E., Gotthelf, E. V., Zhang, W., Middleditch, J., & Wang, Q. D. 1998, Astrophysical Journal Letters, 499, L179

Martnez-Pinedo, G. & Langanke, K. 1999, Physical Review Letters, 83,

120

4502

Martnez-Pinedo, G., Langanke, K., & Dean, D. J. 2000, Astrophysical Journal Supplement, 126, 493

Mazurek, T. J., Truran, J. W., & Cameron, A. G. W. 1974, Astrophysics and Space Science, 27, 261

Merryfield, W. J. 1995, Astrophysical Journal, 444, 318

Meynet, G. & Maeder, A. 1997, Astronomy and Astrophysics, 321, 465

—. 2000, Astronomy and Astrophysics, 361, 101

Meynet, G., Maeder, A., Schaller, G., Schaerer, D., & Charbonnel, C. 1994, Astronomy and Astrophysics Supplement Series, 103, 97

Muthsam, H. J., Goeb, W., Kupka, F., Liebich, W., & Zoechling, J. 1995, Astronomy and Aastrophysics, 293, 127

Nordström, B., Andersen, J., & Andersen, M. I. 1997, Astronomy and Astrophysics, 322, 460

Nugis, T. & Lamers, H. J. G. L. M. 2000, Astronomy and Astrophysics, 360, 227

Owocki, S. P., Cranmer, S. R., & Gayley, K. G. 1998, Astrophysics and Space Science, 260, 149

Packet, W., Vanbeveren, D., de Loore, C., Sreenivasan, S. R., & de Greve, J. P. 1980, Astronomy and Astrophysics, 82, 73

Paczyński, B. 1970, Acta Astronomica, 20, 195

Penny, L. R. 1996, Astrophysical Journal, 463, 737

Petrenz, P. & Puls, J. 2000, Astronomy and Astrophysics, 358, 956

Proffitt, C. R. & Quigley, M. F. 2001, The Astrophysical Journal, 548, 429

Rast, M. P., Nordlund, A., Stein, R. F., & Toomre, J. 1993, Astrophysical Journal Letters, 408, L53

Rast, M. P. & Toomre, J. 1993a, Astrophysical Journal, 419, 224

—. 1993b, Astrophysical Journal, 419, 240

Rauscher, T., Heger, H., Hoffman, R. D., & Woosley, S. E. 2002, Astrophysical Journal, 576, 323

Renzini, A. 1987, Astronomy and Astrophysics, 51888, 49

Robertson, J. W. 1971, Astrophysical Journal, 170, 353

Robertson, J. W. & Faulkner, D. J. 1972, Astrophysical Journal, 171, 309

Schaller, G., Schaerer, D., Meynet, G., & Maeder, A. 1992, Astronomy and Astrophysics Supplement Series, 96, 269

Schou, J., Antia, H. M., Basu, S., Bogart, R. S., Bush, R. I., Chitre, S. M., Christensen-Dalsgaard, J., di Mauro, M. P., Dziembowski, W. A., Eff-Darwich, A., Gough, D. O., Haber, D. A., Hoeksema, J. T., Howe, R., Korzennik, S. G., Kosovichev, A. G., Larsen, R. M., Pijpers, F. P., Scherrer, P. H., Sekii, T., Tarbell, T. D., Title, A. M., Thompson, M. J., & Toomre, J. 1998, Astrophysical Journal, 505, 390

Schwarzschild, M. & Härm, R. 1958, Astrophysical Journal, 128, 348

—. 1969, Bulletin of the American Astronomical Society, 1, 99

Singh, H. P., Roxburgh, I. W., & Chan, K. L. 1998, Astronomy and Astrophysics, 340, 178

Spiegel, E. A. & Zahn, J.-P. 1970, Comments on Astrophysics and Space Physics, 2, 178

—. 1992, Astronomy and Astrophysics, 265, 106

Spruit, H. C. 1999, Astronomy and Astrophysics, 349, 189

—. 2002, Astronomy and Astrophysics, 381, 923

Spruit, H. C., Knobloch, E., & Roxburgh, I. W. 1983, Nature, 304, 520

Spruit, H. C. & Phinney, E. S. 1998, Nature, 393, 139

Stein, R. F. & Nordlund, A. 1998, Astrophysical Journal, 499, 914

Stothers, R. 1970, Monthly Notices of the Royal Astronomical Society, 151, 65

—. 1991, Astrophysical Journal, 383, 820

Stothers, R. & Chin, C.-W. 1991a, Astrophysical Journal, 374, 288

—. 1991b, Astrophysical Journal Letters, 381, L67

Takahashi, K., Yamada, M., & Kondoh, T. 1973, Atomic Data and Nuclear Data Tables, 12, 101

Talon, S., Zahn, J.-P., Maeder, A., & Meynet, G. 1997, Astronomy and Astrophysics, 322, 209

Tassoul, J.-L. 2000, Stellar Rotation, Cambridge astrophysics series ; 36 (New York : Cambridge University Press)

Taylor, R. J. 1973, Monthly Notices of the Royal Astronomical Society, 161, 365

Thielemann, F.-K., Nomoto, K., & Hashimoto, M. 1996, Astrophysical Journal, 460, 408

Trimble, V. & Rees, M. 1970, Astrophysical Letters, 5, 93

Urpin, V. A., Shalybkov, D. A., & Spruit, H. C. 1996, Astronomy and Astrophysics, 306, 455

Venn, K. A., Brooks, A. M., Lambert, D. L., Lemke, M., Langer, N., Lennon, D. J., & Keenan, F. P. 2002, Astrophysical Journal, 565, 571

Vink, J. S., de Koter, A., & Lamers, H. J. G. L. M. 2000, Astronomy and Astrophysics, 362, .295

—. 2001, Astronomy and Astrophysics, 369, 574

Vogt, H. 1925, Astronomische Nachrichten, 223, 229

von Zeipel, H. 1924, Monthly Notices of the Royal Astronomical Society,

223, 665

Weaver, T. A. & Woosley, S. E. 1980, Texas Symposium on Relativistic Astrophysics, 9th, Munich, West Germany, December 14-19, 1978, Proceedings. Annals of the New York Acadamy of Science, 336, 335

—. 1993, Physics Reports, 227, 65

Woosley, S. E., Heger, A., & Weaver, T. A. 2002, Reviews of Modern Physics, 74, 1015

Woosley, S. E., Langer, N., & Weaver, T. A. 1993, Astrophysical Journal, 411, 823

Zahn, J.-P. 1974, in Stellar instability and evolution; Proceedings of the Symposium, Canberra, Australia, August 16-18, 1973, ed. e. a. Ledoux P., Annals of the NY Academy of Science (Dordrecht: D. Reidel), 185–194

Zahn, J.-P. 1991, Astronomy and Astrophysics, 252, 179

—. 1992, Astronomy and Astrophysics, 265, 115

Zahn, J.-P. 1999, in ASP Conference Series, Vol. 173, Stellar Structure: Theory and Test of Connective Energy Transport, ed. A. Gimenez, A. Guinan, & B. Montesinos (San Francisco: Astronomical Society of the Pacific), 121–132

CAPTURING STELLAR CORE HYDRODYNAMIC INSTABILITIES IN CORE-COLLAPSE SUPERNOVAE

JOHN M. BLONDIN

Department of Physics,
North Carolina State University,
Raleigh, NC 27695-8202, USA
E-mail: John_Blondin@ncsu.edu

Current efforts to model the multidimensional dynamics of core-collapse supernovae show effects of strong turbulence during the post-bounce phase. This motion
may be seeded by convection driven by neutrino heating below the accretion shock,
but this shock itself is capable of generating such turbulence through the spherical
accretion shock instability (SASI). Here we review recent studies of the SASI and
discuss its role in core-collapse supernovae.

1. Introduction

The breadth of physics described in this volume is a testament to the complexity of core-collapse supernovae. Unfortunately a successful explosion
depends sensitively on all of this physics, from the uncertain equation of
state at nuclear densities to the challenging problem of the transport of
an intense flux of neutrinos from the optically thick proto-neutron star
through the optically thin accretion shock region. Nonetheless, it is useful
to remember that the phenomenon of a core-collapse supernova begins with
a hydrodynamic implosion, and must end with an expanding shock wave.
The intermediate steps are critically dependent on a wide range of complex
physical processes, but hydrodynamics must play a key role.

This fundamental role of hydrodynamics has received more attention
over the past decade because of the growing realization that turbulent flows
behind the accretion shock may contribute in a significant way to successful supernova explosions. While the importance of convection has been
recognized for some time[1,2,3], it has become the focus of a flurry of activity
involving multi-dimensional simulations since the ground-breaking work by
Herant, Benz and Colgate[4]. Most two-dimensional supernova simulations
exhibit strong turbulent motions below the stalled accretion shock[4,5,6,7,8],

which has lead some authors to claim that convection plays a key role in core-collapse supernovae[4,9]. So ingrained is this idea that it is referred to as the "convectively supported neutrino-heating mechanism."

1.1. *Thermal Convection*

In this convection paradigm the turbulent motions are driven by the buoyancy of gas just above the gain radius, a characteristic radius above which neutrino heating is more than cooling. Because the energy deposition rate drops off with increasing radius due to the geometric dilution of the neutrino flux, neutrino heating is most efficient just above the gain radius and less efficient farther out. The result is a negative entropy gradient in the post-shock gas. The convective motion driven by this entropy gradient is therefore characterized by high-entropy plumes rising up from the gain radius. This uprising meets the inflow of lower entropy gas flowing in from the shock front, rolling the flow over to form a convective cell bounded by the gain radius on the inside and the accretion shock on the outside.

These characteristics of convection are apparent in the early stages of most 2D supernova simulations. For example, the image of gas entropy at a time of 137 ms after bounce shown in Mezzacappa *et al.*[8] shows about 7 high-entropy plumes at a point roughly half way between the gain radius and the accretion shock. At later times, however, the dynamics always becomes more ambiguous. Complications arise when these plumes reach the vicinity of the spherical accretion shock and begin to distort the shape of the shock; At this point one must also consider the dynamics of the accretion shock.

1.2. *Shock Dynamics*

In contrast to thermal convection driven by neutrino heating, the turbulent motions seen in the supernova models of Blondin, Mezzacappa and DeMarino[10] are driven by the obliquity of the accretion shock. In this case of shock-driven turbulence, entropy differences arise only from changes in the shock obliquity and variations in the shock speed and radius.

Because a local stability analysis finds that a planar adiabatic shock is stable to perturbations[11], small wavelength (i.e., wavelengths much smaller than the radius of the shock) perturbations to the accretion shock front would create small-scale disturbances that advect harmlessly inward with the accretion flow. On length scales comparable to the shock radius, however, these disturbances feed back on the shock itself and one must consider the global stability of the spherical accretion shock. On the basis of 2D hydrodynamic simulations[10], one finds that these global modes are unstable, with the largest wavelength (corresponding to a spherical harmonic of $l = 1$) being the dominant mode. This global instability is referred to as the Spherical Accretion Shock Instability, or SASI.

While previous authors have credited the turbulent motions observed in supernova simulations to the effects of thermal convection, there has been no serious attempt to answer the question of what generates the large-scale asymmetry, and why is it dominated by the longest wavelengths ($l = 1$). Is the turbulence the result of convective motions, shock dynamics, a combination of the two, or perhaps something entirely different?

2. Modeling Supernova Dynamics

How does one advance our understanding of the hydrodynamical instabilities in core-collapse supernovae? One can borrow a lesson from solar physics, where thermally driven convection is an undisputed key process affecting the structure and energy transport in the Sun. A realistic model of the Sun capable of explaining the observed internal rotation profile would have to include 3D turbulent convection in a spherical geometry in the presence of global rotation and shear layers, radial stratification including boundary layers above and below the convective zone, and magnetic fields. While such complex global models are attempted, much of the progress has been made through more narrowly defined numerical experiments that focus on specific aspects of the overall problem. For example, 3D simulations of turbulent compressible convection in Cartesian coordinates provide quantitative measures of the Reynolds stresses in a controlled environment[12], which can then be used to better understand transport processes through the solar convection zone.

A similar, perhaps more dramatic situation occurs in the supernova problem. Given the wealth of physical phenomena affecting the dynamics of core-collapse supernovae, existing multi-dimensional simulations are difficult to interpret[13]. A complementary approach is to develop an analytic model of the dynamics of core-collapse supernovae that can be used to study a specific physical process in supernovae without the complication of many competing physical processes[14]. To this end we follow the work of Janka[13] to describe a dynamical model of the post-bounce accretion phase that is not susceptible to convection, and hence may be used to study the shock dynamics of core-collapse supernovae in a pristine environment.

The post-bounce accretion phase of core-collapse supernovae is characterized by a standing accretion shock at a radius of one to a few hundred kilometers. The shock maintains a roughly constant radius due to a balance between the ram pressure of the outer core falling into the shock at near free-fall velocity, and the high thermal pressure inside the spherical cavity defined by the shock. The gas behind the shock continues to fall radially inward but with an infall velocity decreasing monotonically with increasing depth below the shock. Eventually this post-shock gas settles onto the surface of the protoneutron star, after losing much of its entropy through neutrino emission in the vicinity of a characteristic neutrinosphere[13]. This dynamical model of a standing accretion shock maintained by strong cooling near the surface of the accreting star is a close match to the semi-analytic "fall-back" model described by Chevalier[15].

The thermodynamics of the gas in this cavity bounded by the accretion shock on the outside and the nascent neutron star on the inside is complicated by the changing role of the relativistic electrons versus the non-relativistic nucleons in determining the equation of state, as well as the challenging problem of neutrino heating and cooling. The pressure in the bulk of the volume of this cavity is determined by relativistic electron-positron pairs and radiation[13], allowing one to model this gas with an effective adiabatic index of $\gamma = 4/3$. While the pressure near the neutrinosphere is instead dominated by nonrelativistic nucleons, this region does not directly affect the dynamics immediately behind the shock. Furthermore, because the volume is dominated by the relativistic electron gas, any global dynamical modes that might develop would be primarily determined by the $\gamma = 4/3$ gas. Thus, in order to keep our dynamical model from becoming too complex, we adopt a constant $\gamma = 4/3$ throughout the postshock region.

The neutrino cooling rate per unit volume is approximately[13] proportional to the density, ρ, and to the sixth power of the temperature, T^6. Given a relativistic equation of state with pressure $P \propto T^4$, this can be expressed as $\rho P^{3/2}$. The coefficient of this cooling function is treated as a free parameter that can be adjusted to set the radius of the standing accretion shock[16].

While one could include a neutrino heating term in such a model[14], this would open up the dynamics to the possibility of convection. Since our goal is to separate the effects of the shock-driven turbulence from thermally-driven convection, we do not include any heating term in our model. By assuming a steady flow, all of the shocked gas must have passed through the same accretion shock and hence was raised to the same value of entropy. Initially this gas is roughly adiabatic (the cooling function is only important near the surface of the accreting core), and it flows inwards from the shock with a constant value of entropy. Only when it approaches the surface does the cooling become important and the local entropy decreases. The result is a positive entropy gradient (stable to convection) deep in the post-shock accretion flow, and a zero entropy gradient (marginally stable) closer to the standing shock. There is no convection in this model!

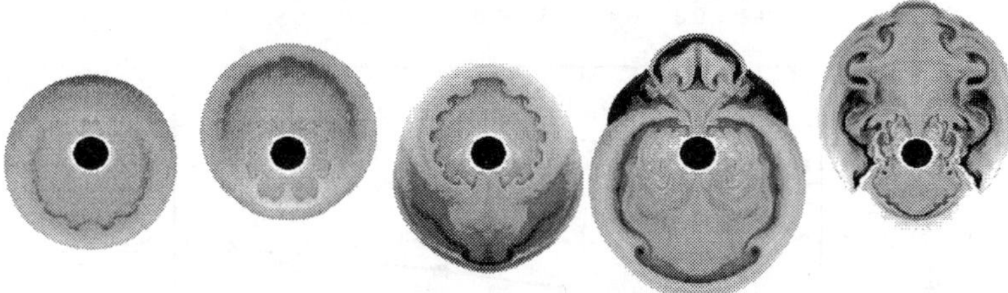

Figure 1.   The time evolution of the SASI, driving a steady, spherical accretion shock into an oscillating, aspherical shock. The shading depicts variations in gas entropy, and each image is separated in time by roughly 22 ms (evolving from left to right).

3. Spherical Accretion Shock Instability

Time-dependent hydrodynamic simulations based on the supernova model described above demonstrate that the accretion shock in core-collapse supernovae is dynamically unstable[10].   An example of the evolution of the SASI is show in Fig. 1, where an initially steady-state, spherically-symmetric accretion shock becomes highly aspherical after only a few flow crossing times. In the linear regime the accretion shock behaves like a resonant cavity, with standing pressure waves filling the interior. The longest wavelength modes are seen to grow exponentially, and eventually the shock front is distorted by a dominant 'sloshing mode' characterized by a spherical harmonic of $l = 1$. Once the spherical symmetry of the accretion shock is broken, the post shock flow is no longer radially inward. This non-radial flow continues to drive the sloshing mode to larger and larger amplitudes, resulting in an increasingly larger shock radius.

The growth of the SASI in the linear regime can be quantified by tracking the volume integral of the gas entropy weighted by Legendre functions[17].

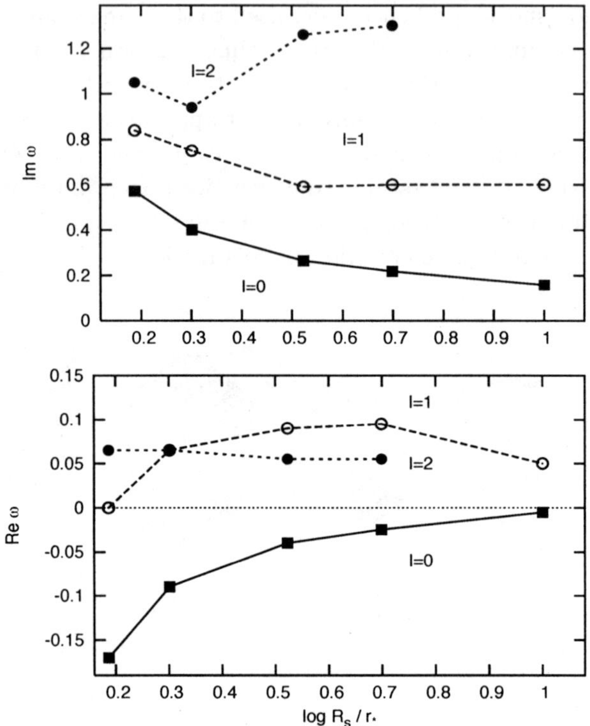

Figure 2. The real and imaginary parts of the growth rate, ω, as a function of the shock height relative to the radius of the accreting star, R_s/r_*, for three different axisymmetric modes: $l = 0$, 1, and 2.

This quantity is found to evolve with a time dependence described by an exponentially-growing sinusoid. This growth curve can thus be fitted to extract a complex growth rate, ω, the real part of which corresponds to the growth of the instability and the imaginary part to the oscillation frequency of the SASI. The extracted growth rates for various modes are shown in Fig. 2 as a function of the shock stand-off distance. In the regime relevant to the post-bounce epoch in core-collapse supernovae, one finds that the accretion shock is always stable to radial modes and the $l = 1$ mode is typically the most unstable.

This result is consistent with the many two-dimensional supernova simulations published to date, which exhibit a strong $l = 1$ morphology of the accretion shock after hundreds of milliseconds after bounce.

An interesting result of this study of linear growth of the SASI is the drop in the growth rate of the dominant $l = 1$ mode when the shock stand-off distance becomes smaller than the radius of the accreting proto-neutron star. This result may have relevance to the models of Mezzacappa *et al.*[8], where some global asymmetry was evident midway through the simulations but ultimately the models ended with a nearly spherical accretion shock only a small distance above the neutron star.

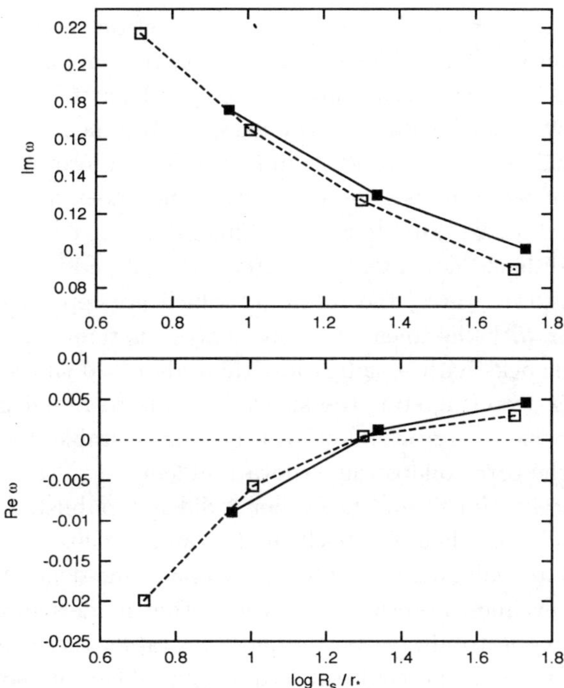

Figure 3. The real and imaginary parts of the growth rate, ω, as a function of the shock height relative to the radius of the accreting star, R_s/r_*, for the fall-back model. Results are shown for the linear stability analysis (solid line) and for 1D numerical simulations using VH-1 (dashed line).

4. Code Validation

An important contribution of this analytic model is the ability to quantitatively validate the time-dependent hydrodynamics codes used in core-collapse simulations. While most such codes used in the astrophysics community are tested against standard problems[18], it is important that generic hydrodynamics codes be validated against test problems that are as close as possible to the problem being studied[19]. The most rigorous testing involves comparison with a known analytic solution. In the absence of such, one can compare simulation results from different numerical codes. The last resort is testing for numerical convergence. However, one must keep in mind that this last method only answers the question of whether spatial resolution is affecting the results of the simulation, not the question of whether the results are correct.

The linear stability analysis of spherical accretion shocks performed by Houck and Chevalier[16] provides a valuable test problem for supernova simulation codes. As we have argued above, their fall-back model is very similar to the supernova model defined herein using the arguments in Janka[13]. Any one-dimensional supernova code should be able to reproduce the complex growth rates of radial modes computed from the linear stability analysis.

Blondin and Mezzacappa[17] ran one-dimensional hydrodynamic simulations of the fall-back model and measured the real and imaginary parts of the complex growth rates, the results of which are reproduced in Fig. 3. The agreement with the linear stability analysis is remarkable. For standing accretion shocks with a radius less than about 20 times the radius of the underlying accreting star, the shock is stable to radial perturbations. While shocks with a large stand-off distance are unstable, this regime does not apply to the core-collapse supernova problem.

Unfortunately, Houck and Chevalier[16] did not publish any results for non-radial modes of their fall-back model, and so this approach cannot (yet) be used to validate the two-dimensional time-dependent hydrodynamic codes. (While the results from their other models show a trend for the spherical mode to always be the most unstable, these results may be affected by their particular choice of zero tangential flow at the outer boundary.) In the absence of an analytic model for validation, we can turn to a comparison of different numerical codes. ud-Doula and Blondin[20] present results from 2D simulations of the SASI using two different hydrodynamics codes: ZEUS-MP and VH-1. The results are illustrated in Fig. 4 in the form of the time evolution of the power in the $l = 1$ mode. Despite significant

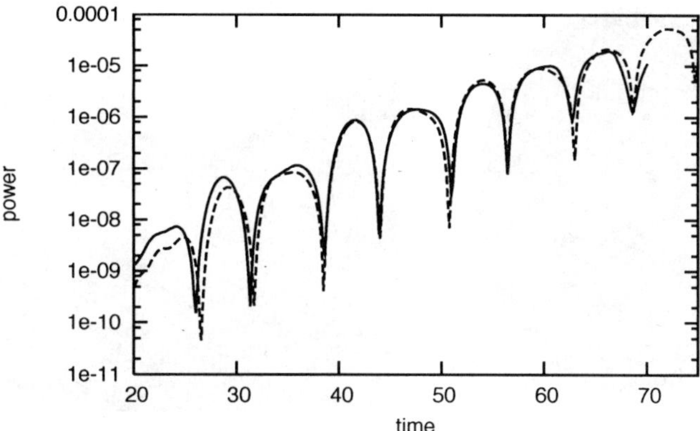

Figure 4. The evolution of the power in the $l = 1$ mode of the gas entropy from a two-dimensional simulation of the SASI using VH-1 (solid line) and Zeus (dashed line).

differences in the algorithms of these codes (VH-1 uses a Riemann solver following the PPMLR scheme[21] while ZEUS-MP[22] uses an artificial viscosity to generate the requisite entropy at the accretion shock), the evolution of the SASI is strikingly similar.

5. Non-Linear Evolution

The transition from a linear growth characterized by nearly radial post-shock flow and a spherically-symmetric shock to non-linear evolution characterized by strong non-radial flows and asymmetric shocks is evidenced by the longer timescale of the evolution. Rather than oscillations on the sound crossing time, variations in the flow now change on the timescale of flow across the diameter of the shock[17]. Thus, the non-linear evolution of the SASI appears to be driven not by growing acoustic modes, but by asymmetric post-shock flow sloshing back and forth. It is this strong non-radial flow, along with the entropy variations induced by changing shock speeds and obliquity, that characterize the turbulent flow seen in virtually all multi-dimensional core-collapse SN simulations[8,6,7].

It is important to understand the origin of this turbulent flow. Entropy variations and post-shock flow are created by the obliquity and evolution of the shock, NOT by buoyancy effects in the gain region. The role of the accretion shock is illustrated in Fig. 5, where the same snapshot of the

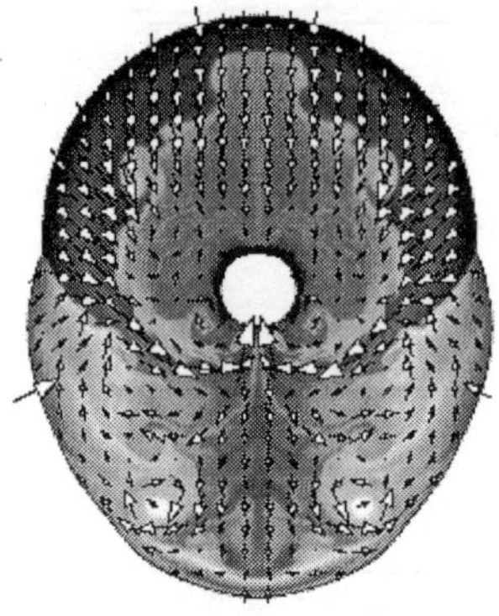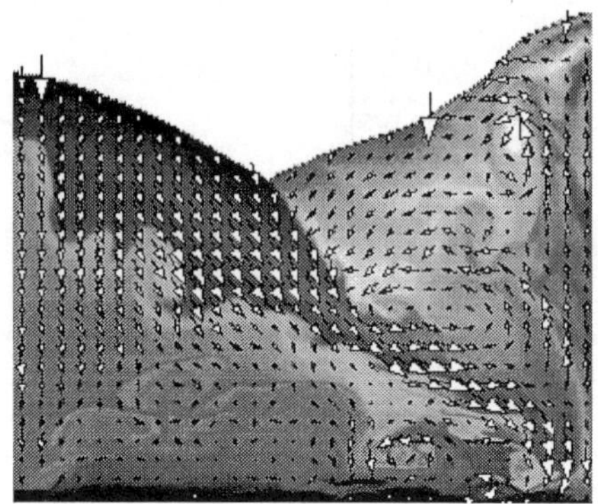

Figure 5. The postshock flow is shown here at a time shortly after the transition to the non-linear stage of the SASI. Both images are from the same data set, but displayed in different geometries. The image on the left shows the true physical structure of the SN shock, while the image on the right shows the same postshock flow but displayed on a grid of radius versus polar angle. The darker shading represents lower entropy gas.

SASI is shown in both normal cartesian coordinates and in polar coordinates (radius vs. polar angle). In the latter image the preshock accretion flow is directed vertically downward, and if the shock were spherical it would describe a horizontal line. From this image it is clear that the downdraft of low entropy gas is a continuous flow stream that starts at the shock in the upper left of the image and flows down and to the right all the way to the surface of the accreting star. The lower entropy of the gas in this downdraft is a result of the lower shock velocity on the left hand side of the image. This half of the shock is retreating in radius such that the effective shock velocity is less than that of a standing spherical accretion shock. Additionally, the obliquity of this retreating half of the shock deflects the lower-entropy gas to the right, driving it underneath the corresponding updraft on the other half of the SN shock.

If one carefully examines the published 2D supernova simulations, one finds that the late time evolution (i.e., more than 200 ms after bounce) is characterized by very fast low-entropy downdrafts originating from an oblique section of the accretion shock. These rapid downflows are driven by the shock dynamics, not by convection. While these models do exhibit rising plumes of high-entropy gas (presumably driven by the buoyancy resulting from a negative entropy gradient), one could argue that their biggest influence on the dynamics of the supernova is through their perturbation of the accretion shock. Once this shock breaks spherical symmetry, the nonlinear SASI dominates the dynamics.

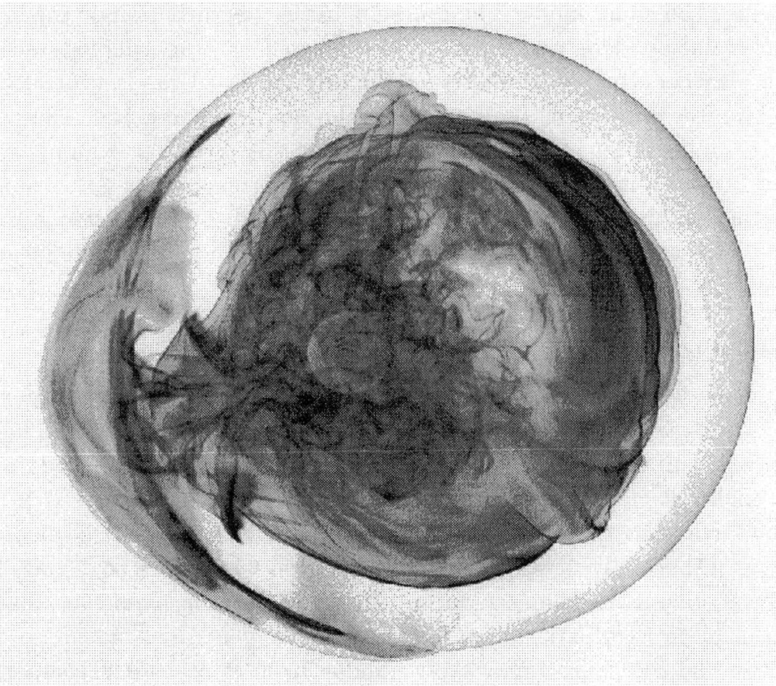

Figure 6. A volume rendering of the gas entropy illustrating the growth of the SASI in a three-dimensional simulation.

6. Three-Dimensional Simulations

One concern regarding the two-dimensional simulations of the SASI and other 2D supernova models is the imposed axisymmetry. Because the SASI is a global instability with perturbations flowing along the symmetry axis, one could imagine that in three dimensions the dominant $l = 1$ mode might not remain coherent. Three-dimensional simulations on a large cartesian grid ($> 10^8$ zones) confirm that the SASI is not just an artifact of axisym-

134

metric simulations[17]. These simulations show that the linear evolution of the SASI is essentially the same in 3D as in 2D, with the $l = 1$ mode dominating the growth. The structure of the 3D accretion shock altered by the SASI is shown in Fig. 6. This image is from a stage where the instability is transitioning from linear to non-linear evolution.

It is interesting to note that in all of the 3D SASI simulations run to date (i.e., the same model but different initial perturbations), the subsequent evolution does not maintain the clean $l = 1$ morphology seen in Fig. 6. Rather, the axisymmetry of the original $l = 1$ mode typically disappears after only a few oscillations in the non-linear regime. This raises a critical issue in supernova modeling in that this is a clear example where 3D evolution is inherently different from the corresponding 2D evolution. Nevertheless, the postshock flow remains turbulent and the average shock radius continues to increase as seen in the corresponding 2D simulations.

7. Conclusions

What have we learned? First and foremost we should, by now, be aware that asymmetry is a natural outcome of a stalled accretion shock. This is a common result of 2D supernova models to date, and the linear study of the SASI makes it clear this is an inescapable result. Second, shock dynamics are important to the evolution of core-collapse supernovae. Even in the absence of thermally-driven convection, we expect the shock to deviate from its initial spherical symmetry due to the SASI. Note in particular that the patterns of entropy variation and flow structure generated by the unstable shock of the SASI are remarkably similar to that seen in 2D supernova simulations. Can the same be said of convection in the absence of an accretion shock?

This work shows that, at a minimum, core-collapse supernova simulations must have two dimensions and cover a polar angle from 0 to π. However, the SASI models also warn us that two-dimensional simulations are ultimately insufficient to capture the full dynamical behavior of the accretion shock. The supernova community must aim for full three-dimensional simulations.

A by-product of these SASI models is the ability to validate the hydrodynamic algorithms used by this community. Using the linear growth of the SASI, we can now show quantitatively that hydrodynamic codes can reproduce the correct time-dependent hydrodynamic evolution of the postbounce accretion phase. We have shown this true for PPM and Zeus-based codes which represent the majority of the supernova modeling community. Can other codes match this test?

Acknowledgments

This work is supported by a SciDAC grant from the U.S. DOE.

References

1. W. D. Arnett, in *IAU Symp. 125, The Origin and Evolution of Neutron Stars*, ed. D. J. Helfand and J. H. Huang (Dordrecht: Reidel), 273 (1986).
2. H. A. Bethe, G. E. Brown and J. Cooperstein, *Astroph. J.*, **322**, 201 (1987).
3. A. Burrows, *Astroph. J. Letters*, **318**, L63 (1987).
4. M. Herant, W. Benz and S. A. Colgate, *Astroph. J.*, **395**, 642 (1992).
5. D. S. Miller, J. R. Wilson and R. W. Mayle, *Astroph. J.*, **415**, 278 (1993).
6. A. Burrows, J. Hayes and B. A. Fryxell, *Astroph. J.*, **450**, 830 (1995).
7. H.-Th. Janka and E. Müller, *Astron. and Astroph.*, **296**, 167 (1996).
8. A. Mezzacappa *et al.*, *Astroph. J.*, **495**, 911 (1998).
9. M. Herant *et al.*, *Astroph. J.*, **435**, 339 (1994).
10. J. M. Blondin, A. Mezzacappa and C. DeMarino, *Astroph. J.*, **584**, 971 (2003).
11. G. B. Whitham, *Linear and Nonlinear Waves* (New York: Wiley), 307 (1974).
12. N. H. Brummell, N. E. Hurlburt and J. Toomre, *Astroph. J.*, **493**, 955 (1998).
13. H.-T. Janka, *Astron. and Astroph.*, **368**, 527 (2001).
14. A. Burrows and J. Goshy, *Astroph. J. Letters*, **416**, L75 (1993).
15. R. A. Chevalier, *Astroph. J.*, **346**, 847 (1989).
16. J. C. Houck and R. A. Chevalier, *Astroph. J.*, **395**, 592 (1992).
17. J. M. Blondin and A. Mezzacappa, *Astroph. J.*, submitted.
18. J. M. Stone *et al.*, *Astroph. J.*, **388**, 415 (1992).
19. A. C. Calder *et. al.*, *Astroph. J. Suppl.*, **143**, 201 (2002).
20. A. ud-Doula and J. M. Blondin, *Astroph. J.*, submitted.
21. P. Colella and P. R. Woodward, *J. Comput. Phys.*, **54**, 174 (1984).
22. J. M. Stone and M. L. Norman *Astroph. J. Suppl.*, **80**, 753 (1992).

OPEN ISSUES IN CORE-COLLAPSE SUPERNOVAE - PROGENITORS AND 3-DIMENSIONAL SIMULATIONS

C. L. FRYER, G. ROCKEFELLER, F. X. TIMMES *

T-6, MS B227
Los Alamos National Laboratory
Los Alamos, NM 87545
E-mail: fryer@lanl.gov, gaber@lanl.gov, timmes@lanl.gov

A. L. HUNGERFORD

CCS-4, D409
Los Alamos National Laboratory
Los Alamos, NM 87545
E-mail: aimee@lanl.gov

K. E. BELLE

X-2, B227
Los Alamos National Laboratory
Los Alamos, NM 87545
E-mail: belle@lanl.gov

Modeling core-collapse supernovae is truly a complex computational and physical problem, requiring a detailed knowledge of a wide variety of physics and the latest techniques in both transport algorithms and hydrodynamics. Just understanding the hydrodynamic flows in core-collapse requires a much more comprehensive set of tests than is usually applied to hydrodynamic codes. In this article, we discuss a number of computational tests (with both analytic and experimental solutions) that should be added to any suite of code tests checking the viability of a code to model core-collapse supernovae. We demonstrate the weaknesses and strengths of various codes on these tests.

---

*C.L.F., G.R., AND F.X.T. are also affiliated with the Physics Department, University of Arizona, Tucson, AZ 87521

1. Computing Core-Collapse Supernovae

Theoretical astrophysics has been divided into two disciplines: studies based on well-chosen (hopefully), simplifying assumptions facilitating analytic (or semi-analytic) derivations and the detailed simulations of complex systems (sometimes also based on simplifying assumptions). Analytic studies have the advantage that, when the calculation is complete, the answer is known. In computational science, even the simplest problems can be plagued with numerical artifacts. Unfortunately, many problems in astrophysics can not be solved using analytic studies. For these problems, the simplifying assumptions required to reach a solution reduce the analytic studies to, at best, order of magnitude estimates, and, at worst, no predictive ability whatsoever.

No problem exemplifies these troublesome issues better than the core-collapse supernova problem. Understanding core-collapse supernovae requires putting together a wide range of physics. Each piece of physics, from the equation of state to the neutrino transport to the possibly important magnetic fields, is sufficiently complex that it can not even be fully tested analytically. To model core-collapse supernovae, these pieces must be put together into a calculation that can solve the full radiation hydrodynamics problem with the cumulative errors from all the pieces held to a minimum. At present, it appears that a 10-20% error in some aspects of this problem (if not all) can alter the fate of a collapsing star from a direct collapse to a black hole to the energetic (10^{51} erg) explosion labeled "supernovae" by astronomers. This problem can NOT be solved using analytic assumptions. Even the assumptions used to make the core-collapse problem computationally tractable[a] introduce errors that make it difficult to believe the final results from current simulations. Core-collapse theorists are pushing the frontier of computational science and have become a sort of jack-of-all trades understanding issues with supercomputing, computational methods, theoretical nuclear and particle physics, and astronomy.

A number of issues exist for those brave, or foolish, enough to pursue the problem of core-collapse: understanding the equation of state at nuclear densities, uncertainties in the progenitors, modeling neutrino transport, modeling hydrodynamics, modeling radiation hydrodynamics (instead of the decoupled solutions most astronomers use), and even the possible effect of magnetic fields. This book should cover these issues several times over.

---

[a] By tractable, we mean that the simulation must run on a supercomputer in less than 1 year.

Here we focus on two issues: those of the progenitors (Section 2) and those of modeling the hydrodynamics (Section 3). We conclude with a discussion of the future in this problem.

2. Issues with Progenitors

In any modeling problem where the goal is to match some observed experiment using simulations, the first set of errors in the computation arises from the initial conditions. Core-Collapse modelers get their initial conditions from theorists studying stellar evolution. These theorists model stars using implicit codes (to avoid Courant condition restrictions) from the onset of hydrogen burning to the collapse of the iron core. These codes make a number of simplifying assumptions from how to model mass-loss in stellar winds, to the effects of rotation, and to the correct modeling of convection (all of which are modeled with recipes in these 1-dimensional implicit codes). Stellar theorists are improving these recipes (e.g. studying convection by using multi-dimensional simulations and modifying their recipes to mimic these studies). But these uncertainties may dominate the uncertainties in supernova modeling.

One of the standard initial conditions we have used is the "s15s7b" progenitor from Woosley & Weaver (1995). This progenitor was produced using the stellar evolution code KEPLER. But KEPLER has gone through its own evolution and newer versions of the code produce different results. Comparing to other codes, the difference in results are even larger. In this section, we compare the results of the most recent version of KEPLER (Rauscher et al. 2003) to recent simulations from the revised version of TYCHO (Arnett 1996) provided by Young & Arnett (private communication). For our comparison, we try to minimize the differences by using non-rotating $20\,M_\odot$ stars (low enough mass to have only a little mass loss from winds). The dominate difference according to the stellar theorists, then, is the recipe of convection[b].

The top panel of Figure 1 shows entropy versus mass profile of this $20\,M_\odot$ star at the time of collapse from both the KEPLER (Rauscher et al. 2003) and TYCHO (Young & Arnett - private communication). The sharp entropy differences in the Rauscher et al. (2003) model mark boundaries in the abundances (see bottom panel of Figure 1). Note that the variation in the entropy is much more gradual with the TYCHO code. This can

---

[b]But it is unclear how well the hydrodynamics and nuclear burning in these codes have been tested so we can not prove that the differences are dominated by convection recipes.

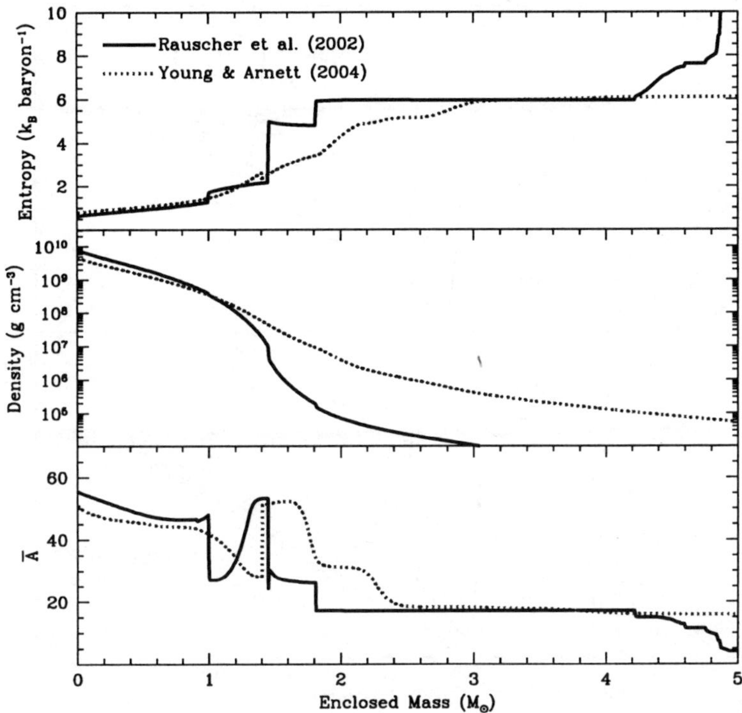

Figure 1. Progenitor Comparison: (Top Panel) Entropy versus enclosed mass from the KEPLER (Rauscher et al. 2003) and TYCHO (Arnett 1996; Young & Arnett) codes at the time of collapse. Note the smooth distribution of entropy in the TYCHO model. This could be due to a difference in the overshooting between these two codes. (Middle Panel) The lower entropy from the TYCHO code leads to higher densities. As we shall see, this may have major repurcussions on the ability to explode such a star. (Bottom Panel) Mean atomic mass of the matter versus mass. Note that the discrete jumps in elemental abundance correspond to the sharp changes in the entropy profile for the Kepler model. The abundance profile has shifts that are almost as sharp in the TYCHO simulation, but the entropy profile has much more gradual variations.

plausibly be explained by a different algorithm for the convective mixing (especially the method for modeling overshoot). But these differences lead to nearly a factor of 2 difference in the entropy at specific parts of the star. On the surface, one might think that the limited spatial region where these entropy differences are large should minimize the structure differences. However, the entropy differences change the structure of the entire star

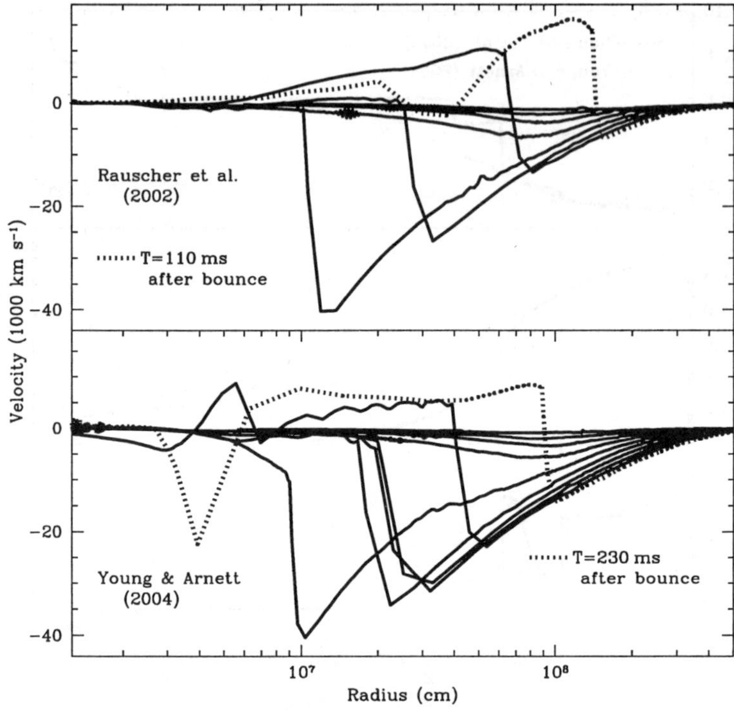

Figure 2.   Progenitor Comparison: Time snapshots (in 50 ms intervals) of the velocity profile (velocity versus radius) for collapse models of the KEPLER (top) and TYCHO (bottom) progenitors.

beyond that radius, leading to over an order of magnitude difference in the density beyond the inner $1.5\,M_\odot$ (Fig. 1).

We have modeled both these stars through collapse and explosion in a 1-dimensional code with an artificially increased neutrino energy (using the technique described in Willems et al. 2005). We tuned the neutrino energy to drive a strong explosion with the KEPLER model. Figure 2 shows a series of time snapshots of the velocity profile from identical collapse simulations of the KEPLER (top) and TYCHO (bottom) progenitors. After 110 ms, the KEPLER progenitor has a strong explosion pushing beyond 1000 km. The TYCHO progenitor takes longer (230 ms) for the explosion to reach this point and the resultant explosion is much weaker (maximum velocities are a factor of 2 less than the KEPLER progenitor).

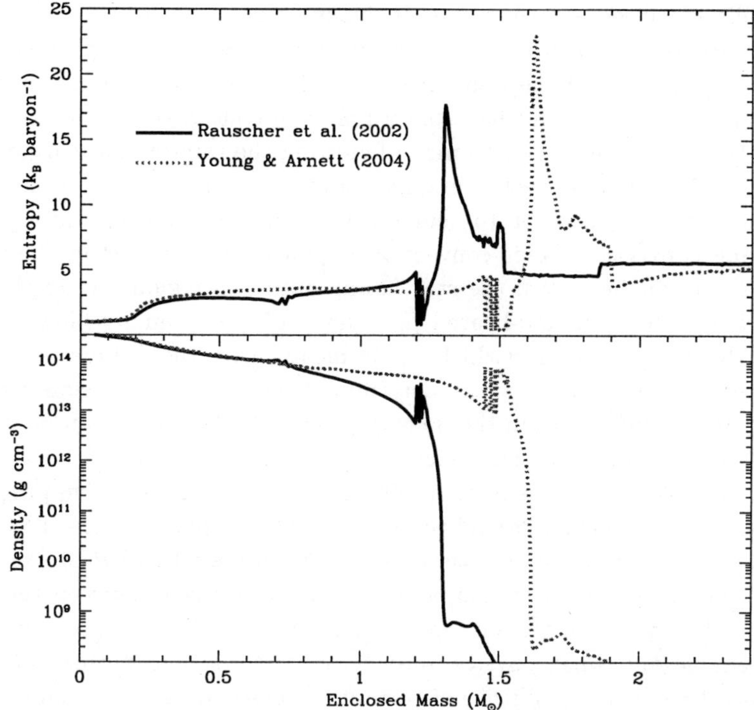

Figure 3. Progenitor Comparison: Entropy (top panel) and Density (bottom panel) versus enclosed mass for the KEPLER and TYCHO progenitors when the explosion has reached 1000 km. Because the explosion takes longer to occur with the TYCHO progenitor, more mass accretes onto the proto-neutron star, driving the explosion further out in mass space. This leaves behind a more massive remnant (If we include fallback, the TYCHO progenitor becomes a black hole). Because more energy was needed to drive the explosion, the entropy in the exploding material for the TYCHO progenitor is slightly higher.

Figure 3 shows the entropy (top) and density (bottom) versus enclosed mass profiles from these two progenitors when the shock reaches 1000 km. Because the TYCHO progenitor took longer to explode, more mass settled onto the proto-neutron star, and the resultant remnant (neglecting fallback) is $1.6\,M_\odot$, more than 20% bigger than the KEPLER counterpart. If we include fallback, the fate of these two stars is even more different. The KEPLER progenitor will produce a neutron star with a gravitational mass of less than $1.4\,M_\odot$. The TYCHO progenitor has considerable fallback and eventually collapses to form a $> 4\,\mathrm{M}_\odot$ black hole.

The differences in progenitors led to differences in the explosion that ranged from a strong explosion with a smallish neutron star remnant to a weak supernova explosion with a black hole remnant. A simplistic explanation of these differences can be made following the ram-pressure argument of Fryer (1999). The convective engine works under the following basic idea: neutrinos leak out of the proto-neutron star and heat the region just above it, driving convection. This convection region is in turn "capped" by the ram-pressure of the rest of the star falling onto this region. An explosion occurs when this neutrino-heated pressure cooker can finally blow off its lid. In this simple picture (which by no means gives the full discussion of the supernova problem), one can determine trends in the explosion just by looking at the differences in the density profiles of these progenitors at the time of collapse(Fig. 1). The denser TYCHO progenitor has a lid that is harder to explode. Just as Fryer (1999) found that more massive progenitors took longer to explode and produced weaker explosions, the TYCHO progenitor also produces a weaker explosion than the KEPLER progenitor.

Up until now, very few comparisons have been made between the different stellar evolution models, so it is difficult to tell which progenitor is more reliable and what physics must be understood and better modeled to reduce these errors. With our current understanding of the supernova engine, these differences can make a huge difference in the fate of the star. The progenitors *must* be understood for us to get a final answer to the supernova problem.

3. Multi-Dimensional Hydrodynamics

A number of hydrodynamics techniques are available for the budding astronomer and a wide variety of techniques are used in core-collapse simulations. Unfortunately, no technique is perfect: as Grant Bazan said at this meeting: "every technique has its poison". For complex problems like supernovae, there may well be no ideal code. Thus, it is important to know where the "poisons", or weaknesses, of each code lie. Given the complex hydrodynamic motions involved in the supernova problem, a single test of the code is not sufficient. Here we start to make a list of all of the physics that the code must model accurately:

- **Convective Motions:** The neutrino-driven core-collapse supernova engine seems to rely upon convection to bolster the efficiency of the neutrino heating. A number of tests exist to check the ability of a code to model Rayleigh-Taylor convection. But these tests do not fully check the specific geometric conditions in core-collapse supernovae. We will not cover this further in this proceedings.

- **Shocks:** Both the bounce shock of the collapsed stellar core and the accretion shock are important in determining the initial entropy of the material in the convective region and can then play a major role in the strength of the convection. Many tests focus on the ability of a code to model the shock jump conditions. But many of this tests are also too simplistic for the supernova problem. We will discuss this in more detail here.
- **Artificial Diffusion:** Following the elemental abundance and heat transfer is also critical in the supernova problem. This effect can be tested in a lot of shock tests and we will discuss this issue along with our shock discussion.
- **Gravity:** The power source for the supernova ultimately arises from the release of potential energy. Any explosion must ultimately overcome the gravitational well of the neutron star. A number of tests also exist to check how well a code models gravitational forces, but we will not discuss these in detail here.
- **Angular Momentum:** If the star is also rotating, then how well the code conserves angular momentum, and how much artificial angular momentum transport exists in the code is also crucial. We will show some of these tests here.

This list is far from complete. Even so, there is too much on this list to cover in this proceedings. We will focus this proceedings to some issues on shock modeling and discuss the angular momentum issue. Much of this work comes from a recently submitted paper by Fryer, Rockefeller, & Warren (2005) and we direct readers to that paper for a broader discussion.

3.1. *Shocks and Diffusion*

Modeling hydrodynamic shocks is the central test of most hydrodynamics codes. Many tests exist, but most of them (e.g. Sedov, shock tube, Noh problem) have severe limitations. The shocks in all these tests have simple structures. This is what allows one to derive analytic solutions to these tests. But it also allows modelers to fine-tune their codes to perform well on these tests. We believe such "tuning" is the wrong way to test a code. Will these tuned codes work well on the more complex structures in supernova collapse, or did the tuning only help the code when modeling the test? Here

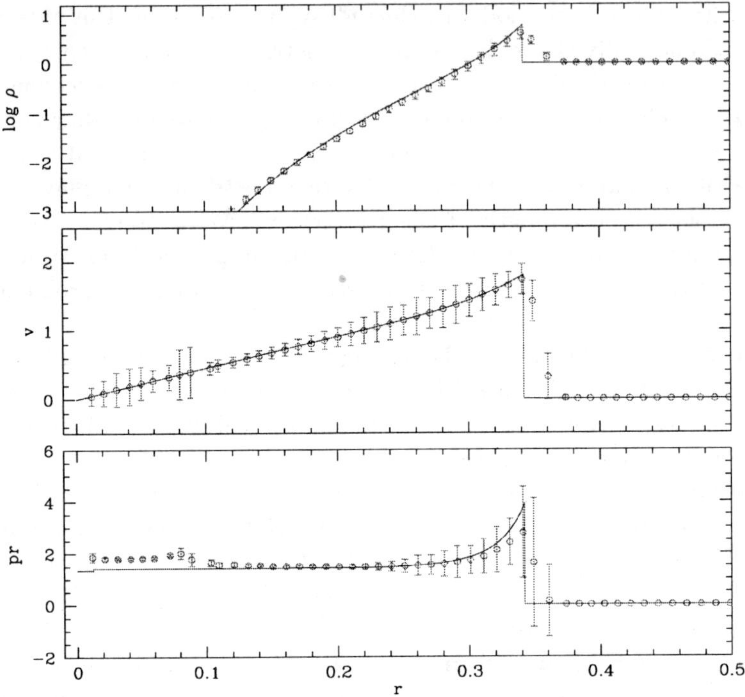

Figure 4. Sedov Shock Problem: The density, velocity and pressure for a Sedov explosion. The line shows the analytic solution. The points with the error bars gives the results from an SPH simulation where the point is the mean value of the particles in a radius bin and the error bars denote the 1-sigma deviation of the particles away from this mean.

we step through a series of increasingly complex shock tests starting with a Sedov shock problem and moving through to the complex experimental problem of the Galactic center. Our hope is that the complexity of the Galactic center experiment will deter even the most nefarious (you know who you are) code tuners.

Figure 4 shows a Sedov shock problem test of the 3-dimensional smooth particle hydrodynamics code "SNSPH" (Fryer, Rockefeller, & Warren 2005). The line shows the analytic result, the points with error bars denote the mean SPH value (binned in radius) with the 1-sigma scatter in the SPH particles. This scatter is dominated by the scatter in the initial

SPH conditions. The set-up for this simulation is a spherically symmetric uniform-density profile with a high energy central point. The particles are set randomly in a series of shells, where we have strived to relax the particles so that they are all roughly equidistant within each shell and this distance is the same as the distance between the shells. However, the error in the separations lead to errors in the densities. This error dominates the scatter in the data plotted in Figure 4. Although there is some scatter, the mean values (especially the velocity) are very close to the correct answer.

Our next step in sophistication is to take a realistic star and model explosions. Figure 5 shows the velocity profile of the explosion of a $15\,M_\odot$ star using a variety of codes: LAHYC - the 1-D lagrangian supernova code used in Herant et al. (1994) and described more fully in Fryer et al. (1999), Chicago's FLASH code, and the LANL-based RAGE code. The FLASH and RAGE codes are both Adaptive Mesh Refinement (AMR) codes using Riemann solvers. In all cases, the Lagrangian technique can outperform both the AMR codes as far as accuracy versus resolution, but bear in mind that the AMR codes are designed to work in multi-dimensions and the Lagrangian code used here does not extrapolate well to more complex structures. We do not yet have a 3-dimensional simulation of this problem comparing SPH and AMR techniques.

Figure 5 shows the energy versus radius for same suite of simulations. The important feature to focus on is the spike in energy where the shock has heated the star. Getting this feature right is important in modeling supernova light curves, and all codes do reasonably well on this part of the problem. But even more important is the nickel distribution (Figure 6). The RAGE calculations have considerable numerical diffusion. This is a problem facing many Eulerian codes, as it occurs in the advection step. At least in the standard set of options, the RAGE code has too much diffusion to accurately model this problem[c]. However, the FLASH code works very well, matching the Lagrangian solution (which, because it is Lagrangian, is the exact solution). Eulerian codes have been plagued by this artificial diffusion for decades, some of it leading to spurious results (e.g. the explosion of sub-Chandrasekhar white dwarfs). But, as the flash simulation demonstrates, the high order differencing scheme in FLASH has overcome this problem. So this problem is not a show-stopper for Eulerian codes. Since SPH is Lagrangian, there is no advection step, so numerical

---

[c]However, we have not explored many integrator options existing in RAGE and there may be a set of parameters that is ideal for the supernova problem.

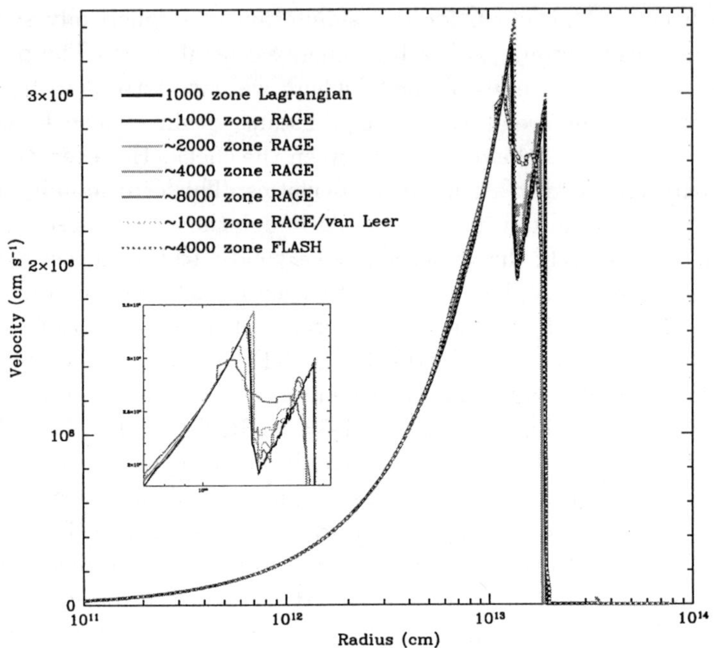

Figure 5.   Supernova Explosion Test: Velocity versus radius of the explosion of a $15\,M_\odot$ star, 39,000 s after the launch of the explosion. The lines show the results from a low resolution Lagrangian simulation, a simulation using moderate resolution with Chicago's FLASH code, and a simulation showing a range of effective resolutions using the RAGE code.

diffusion is not a problem for SPH.

Another strength of SPH is that it is not tuned for the 1-dimensional Sedov test, and the errors in Figure 4 will not increase if the shock does not travel along our shells. Most tests of Riemann codes concentrate on

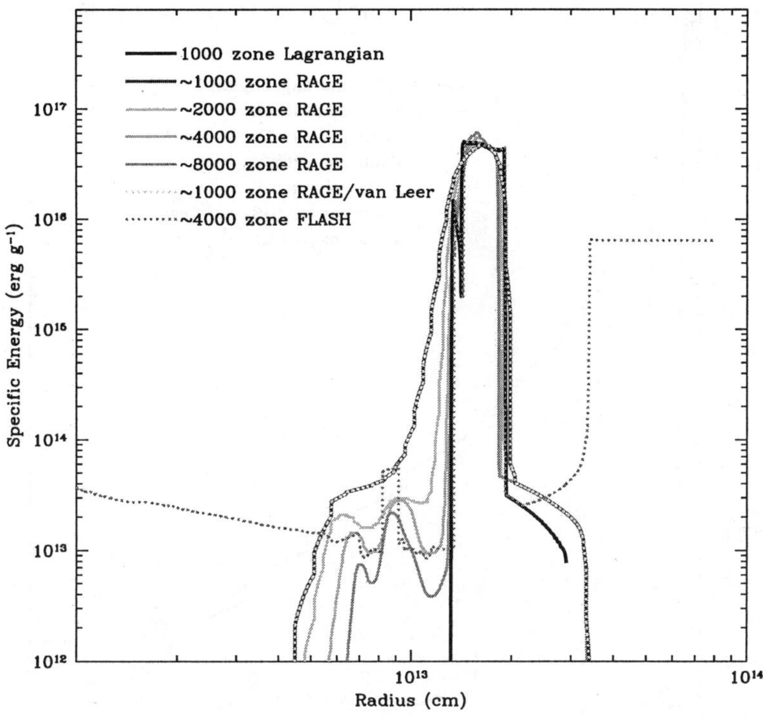

Figure 6.   Supernova Explosion Test: Energy versus radius for the same suite of simulations as Figure 5.

shocks moving along the grids. Those Eulerian codes that were fine-tuned to solve the Sedov (or shock-tube or Sod problem) may not perform well when the shock is not aligned with the grid. One can misalign the shock with respect to the grid, but a final test requires modeling some complex shock structure.

Here we propose an experimental test based in astrophysics: shocks in the Galactic center. The gas in the Galactic center is composed of the wind material from roughly a dozen stars. The winds from all these stars interact and shock against each other, ultimately emitting X-rays. This diffuse X-ray flux is observed, and because it depends upon the square of the gas density, is an ideal diagnostic of the shock jump conditions. Rockefeller et al. (2004) modeled these conditions, finding they could match the diffuse

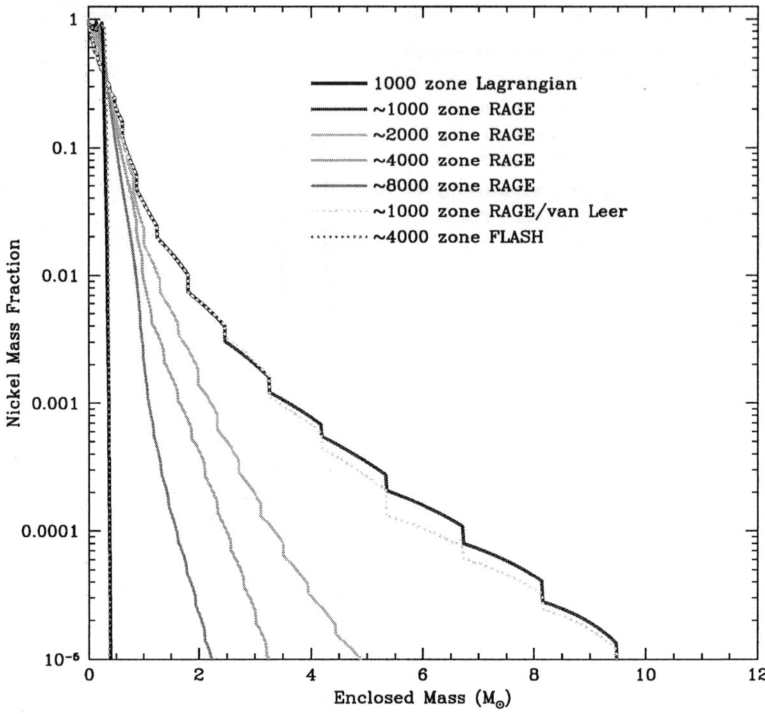

Figure 7. Supernova Explosion Test: Nickel mass fraction versus radius for the same suite of simulations as Figure 5.

X-ray flux in the galactic center with the current mass-loss rates (Figure 8).

The current drawbacks of this test are the standard errors of any experimental test: errors in the initial conditions and errors in the measurements. Unlike most experiments where once the test has been run, you can't check the initial conditions, the long timescales involved in most astronomical phenomena allow us to keep testing our initial conditions with time. In the Galactic center, the dominant error arises from our knowledge of the mass-loss from stars. Within some short-term variation, the mass-loss is the same now (on average) as it was 1000-10,000 years ago, so we can keep on homing in on the correct mass-loss rate. Hence, this uncertainty will gradually become less with time. Likewise, the X-ray flux measurement

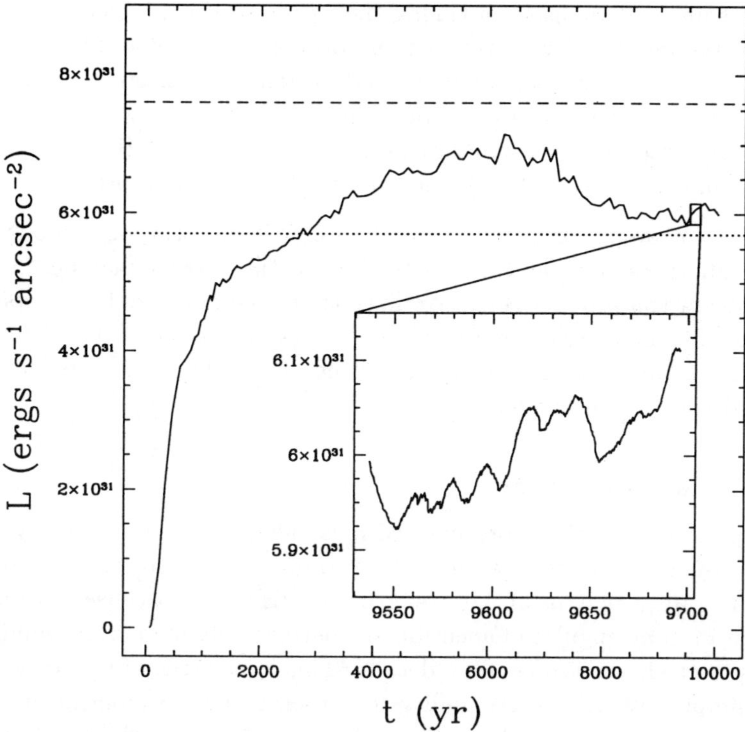

Figure 8. Diffuse X-ray emission (solid line) from the Galactic Center as a function of time (Rockefeller et al. 2004). The initial rise is because we begin the simulation with no gas in the region and allow winds to gradually fill up the region. The dashed, dot-dashed lines show the upper,lower limits respectively from the X-ray observations of the diffuse emission.

will also improve over the next few decades, allowing us to reduce the measurement errors as well. However, even now, this test can confirm that ones code does not have systematic errors in the shock treatment by more than 30-50%.

We have moved quickly through a series of tests of the shock treatment in the code. Let's recap some of the most crucial points these tests bring to light:

- Although the Sedov, shock tube, and Sod problems are important code tests, they do not fully test a codes ability to model shocks if the tests are limited to shocks along the grid. Especially since some coders insist on tuning their code to this problem.

- Testing velocity, density, energy profiles is not sufficient. One must also test to make sure artificial diffusion is not altering other quantities relevant to a given problem.
- A complex shock structure also makes a nice test. Here we propose an experimental test based around the Galactic center.

How well a code performs is not based on how it performs on any one test, but on all these tests and, in particular, to those tests that best mimic the problem you are modeling. We have tried to emphasize here tests that are probably more relevant to testing a supernova code, but tend to be neglected in most code test suites. Other hydrodynamic tests must be included (rarely can one do too few tests).

4. Angular Momentum

If the progenitor star is rotating, we must also test a codes ability to a) conserve angular momentum and b) avoid too much artificial angular momentum transport. One of the strengths of SPH is that it conserves angular momentum (any angular momentum loss derives from numerical round-off). See Fryer et al. (2005) for more details. This is not true of grid codes.

A simple test of a code's ability to conserve angular momentum is the simulation of a binary system. Figure 9 shows time snapshots of a binary system modeled for 18 orbits in SPH. The binary was placed in an orbit slightly smaller than its Roche overflow radius, so the stars slowly expand and begin to lose mass through Roche-Lobe overflow. But not all the expansion is caused by Roche-Lobe overflow. The tidal forces in this problem lead to friction that, with our artificial viscosity, causes the star to heat up and expand. This numerical artifact of codes with artificial viscosity leads to poor energy conservation (Figure 10). Because of the high artificial viscosity in this simulation, the system gains 10% of its total energy after 10 orbits, and another 10% after 18 orbits. The artificial viscosity terms were set to the standard $\alpha = 1.0, \beta = 2.0$ values used to model strong shock situations. At the expense of shock modeling, we could decrease the artificial viscosity to reduce the errors in the energy conservation. But this would be fine-tuning the code to make it work less well on shocks. For the supernova problem, we need to do both, so fine-tuning is not an option (it can only be used as a check).

Figure 11 shows the mass-weighted separation of this binary. The decrease in orbital separation occurs because of the Roche-Lobe overflow of

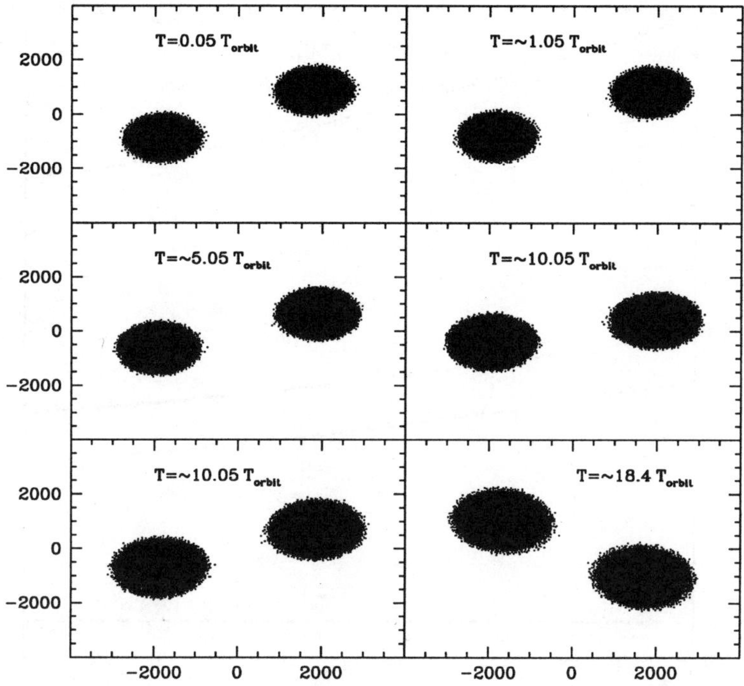

Figure 9. Time snapshots of two orbiting stars. The dots are physical SPH particles from a slice of the 3-dimensional simulations.

the mass. More important is the total angular momentum conservation. The angular momentum remains constant to the 0.01% level for over 15 orbits. Beyond this time, the close separation leads to too much viscous heating which ultimately degrades the orbit. Similar tests can be done with grid codes, but most tests in the literature are limited to fine-tuned codes for the orbit problem. The tricks used to do this fine-tuning can not be used in the supernova problem. It is important to determine just how well grid codes conserve angular momentum in a core collapse problem.

Even if the angular momentum is conserved, it can still be transported artificially. Fryer & Warren (2004) worried about just this effect. To test this, they lowered the artificial viscosity of their simulation by an order of magnitude. If numerical angular momentum transport were important, the angular momentum profile in this reduced viscosity simulation should

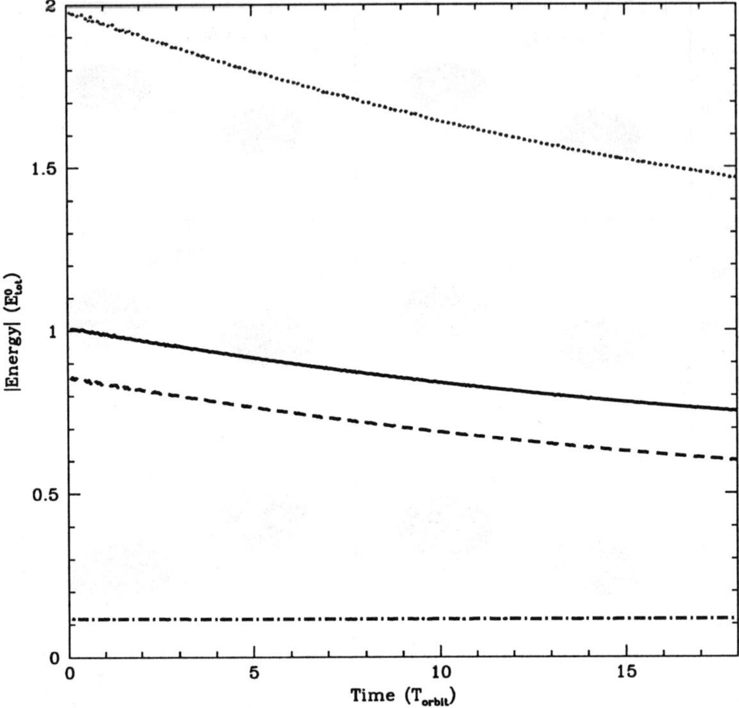

Figure 10. Components of the energy as a function of time in units of the total energy. The potential (dotted line) and, because the stars are bound, the total (solid) energies are negative. The magnitude of the total, potential, and thermal energy all decrease with time because of the expansion of the stars. The kinetic energy, the primary diagnostic of the orbits, remains relatively constant (which is a simple reflection of the conserved angular momentum and the rough conservation in the orbital radius).

be very different than the standard simulation. Instead, the total angular momentum of their collapsed core did not change by more than 10-20% (Figure 12). For SNSPH, numerical angular momentum transport is NOT an issue in the supernova problem.

If rotation is to be included in stellar collapse, both the total angular momentum conservation and the numerical angular momentum transport must be studied. Such tests are rarely done in stellar collapse, and must be done to trust this aspect of our computational models.

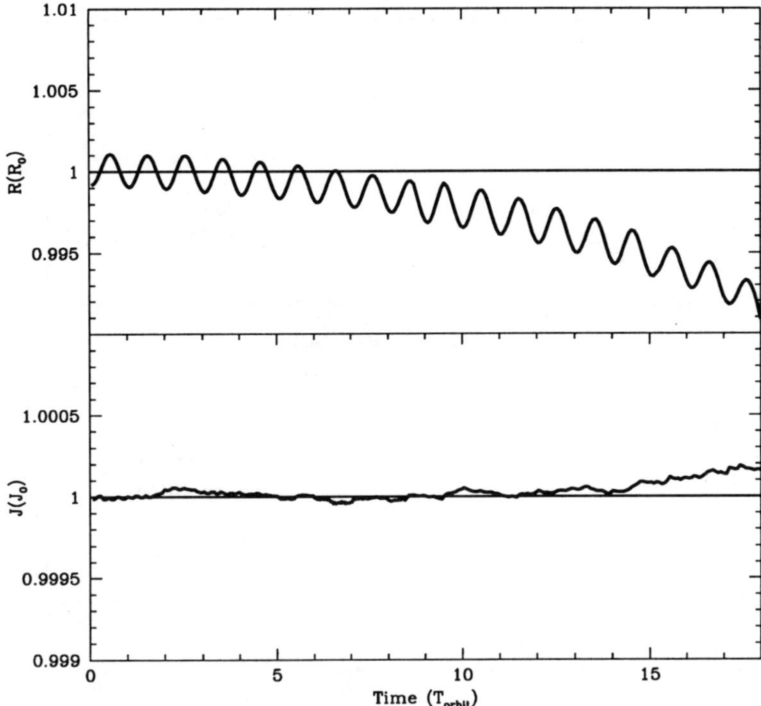

Figure 11. Orbital separation (top) and total angular momentum (bottom) versus time for our SNSPH simulation of a binary system.

5. Prospects for the Future

Understanding core-collapse supernovae is a very rich and complex problem. Physical and computational issues arise in all aspects of this complex problem and we have a great deal of work to be able to predict quantitatively accurate explosion energies for specific stars. In this proceedings, we discussed just two issues: the accuracy of our progenitors and our ability to model hydrodynamics. With progenitors, we found that the errors (determined by comparing two different stellar evolution codes) can be quite large. Densities beyon $1.5\,M_\odot$ can vary by more than an order of magnitude. In hydrodynamics, we listed a number of more detailed tests, both for the treatment of shocks and angular momentum, along with additional observables to study advection errors.

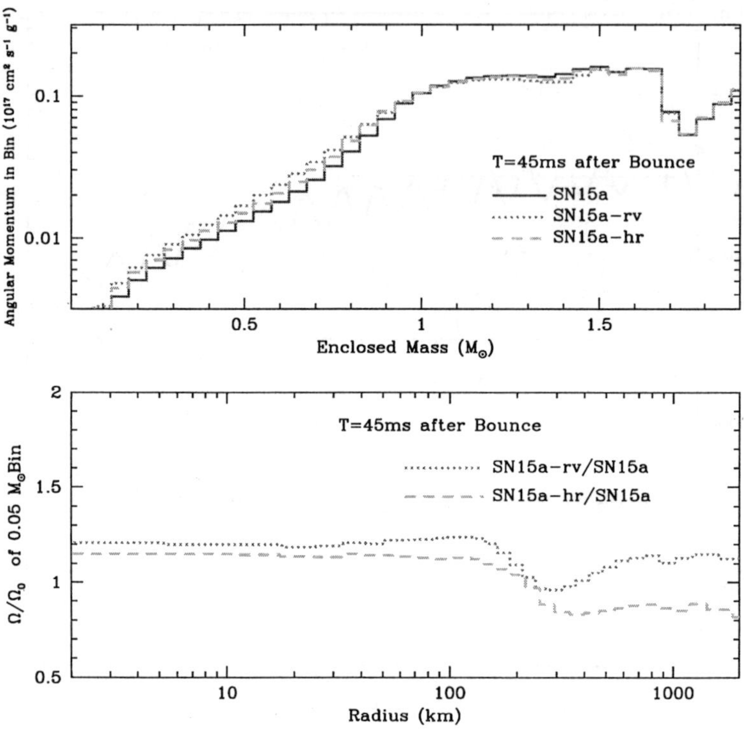

Figure 12. Angular momentum versus enclosed mass (top) for a rotating collapsing star (Fryer & Warren 2004; Fryer et al. 2005). The 3 lines represent the standard model (solid line), a model using an artificial viscosity reduced by a factor of 10 (dotted line), and a high resolution model (dashed line). The bottom slide shows the ratio of these angular velocities: reduced viscosity/standard (dotted line) and high resolution/standard (dashed). The answer does not change by more than 10-20% when reducing the artificial viscosity, and it is unlikely that we are off by more than that value for the angular momentum in our simulations.

The pessimistic view of the errors in these focused topics is that these errors alone make it impossible to calculate core-collapse with quantitative accuracy in the next decade. But this limitation really is just saying we can not say that a star with such-and-such a mass will produce this type of explosion and this mass remnant. However, we do have many of the tools to study trends. And we can learn a lot from these trends, especially when we compare these trends to the observations. So even though there remains much work in core-collapse supernovae before we can arrive at a final solution, there is considerable work that we can do better understand these objects.

What we need at this point for progenitors is to push harder on the progenitor modelers to real comparisons of their work and to start to think more carefully about how they might constrain their uncertainties. This will require moving from their current implicit codes to modeling some phases of the star's life hydrodynamically. Such work has already begun.

What we need in hydrodynamics is to insist upon further tests. Eulerian codes have a lot of advantages, but they also come with many disadvantages which, for the most part, have NOT been tested. Such work needs to be done before such codes can be trusted on the core-collapse supernova problem.

Acknowledgments

These simulations were run on LANL's Space Simulator and the ASC Q computer. Kim New, Rob Huckstadt, and Rob Coker provided welcome help on using the RAGE code. This work is supported by a DOE SciDAC grant number DE-FC02-01ER41176 and by NASA Grant SWIF03-0047-0037 under the auspices of the U.S. Dept. of Energy.

References

1. D. Arnett, *Supernova and Nucleosynthesis*, (Princeton: Princeton Univ. Press).
2. C. L. Fryer, W. Benz, M. Herant, and S. A. Colgate *ApJ*, **516**, 892 (1999).
3. C. L. Fryer and M. S. Warren *ApJ*, **601**, 391 (2004).
4. C. L. Fryer, G. Rockefeller, and M. S. Warren *submitted to ApJ*.
5. B. Fryxell, K. Olson, P. Ricker, F. X. Timmes, M. Zingale, D. Q. Lamb, P. MacNeice, R. Rosner, J. W. Truran, and H. Tufo, *ApJS*, **131**, 273 (2000).
6. M. Herant, W. Benz, W. R. Hix, C. L. Fryer, and S. A. Colgate, *ApJ*, **435**, 339 (1994).
7. T. Rauscher, A. Heger, R. D. Hoffman and S. E. Woosley, *Nuc. Phys. A*, **718**, 463 (2003).
8. G. Rockefeller, C. L. Fryer, F. Melia, and M. S. Warren, *ApJ*, **604**, 662 (1999).
9. B. Willems, M. Henninger, T. Levin, N. Ivanova, V. Kalogera, F.X. Timmes, C.L. Fryer *accepted by ApJ*
10. S. E. Woosley and T. A. Weaver, *ApJS*, **101**, 181 (1995).

MAGNETIC FIELDS IN CORE COLLAPSE SUPERNOVAE: POSSIBILITIES AND GAPS

J. CRAIG WHEELER AND SHIZUKA AKIYAMA

Department of Astronomy University of Texas
E-mail: wheel@astro.as.utexas.edu; shizuka@astro.as.utexas.edu

Spectropolarimetry of core collapse supernovae has shown that they are asymmetric and often, but not universally, bi-polar. The Type IIb SN 1993J and similar events showed large scatter in the Stokes parameter plane. SN 2002ap which showed very high photospheric velocities in early phases revealed that the dominant axes associated with hydrogen, with oxygen, and with calcium were all oriented substantially differently. Observational programs clearly have much more to teach us about the complexity of asymmetric supernovae and the physics involved in the asymmetry. Jet-induced supernova models give a typical jet/torus structure that is reminiscent of some objects like the Crab nebula, SN 1987A and perhaps Cas A. Jets, in turn, may arise from the intrinsic rotation and magnetic fields that are expected to accompany core collapse. We summarize the potential importance of the magneto-rotational instability for the core collapse problem and sketch some of the effects that large magnetic fields, $\sim 10^{15}$ G, may have on the physics of the supernova explosion. Open issues in the problem of multi-dimensional magnetic core collapse are summarized and a critique is given of some recent MHD collapse calculations. A crucial piece of information that can inform the discussion of potential MHD effects even in the absence of the explicit inclusion of magnetic fields is to give sufficient information from a rotating collapse to at least crudely estimate the time-dependent saturation field according to the prescription $v_a \sim r\Omega$. Many studies of rotating collapse produce such information, but fail to present it explicitly.

1. Introduction

Spectropolarimetry of supernovae has opened up a new window on these spectacular events and yielded remarkable new insights. A few rare, nearby supernovae and supernova remnants have revealed asymmetric images. Among these are the Crab nebula with its prominent jet/torus structure revealed by CXO, SN 1987A (Wang et al. 2002a) and Cas A (Fesen 2001; Hwang et al. 2004). It was not clear on this limited basis whether or not the strong asymmetries of these objects was important to the intrinsic process of the explosion. Spectropolarimetry has extended our knowledge of the composition-dependent geometry of core-collapse supernovae to numerous extragalactic supernovae. Spectropolarimetry of supernovae probes the geometrical structure of matter shed by a star before it explodes and the structure of the ejecta of the explosion with an effective spatial resolution far superior to any envisaged optical interferometry (Wang et al. 2002b). The structure revealed is closely related to the explosion mechanisms and to the progenitor systems.

Spectropolarimetry of supernovae continues to show that all core-collapse supernovae (those associated with young populations; Type II, Type Ib/c) are polarized and hence substantially asymmetric (Wang et al. 1996; Wang et al. 2001, 2002a,b, 2003a,b; Leonard et al. 2000; Leonard & Filippenko 2001; Leonard et al. 2001, 2002). The understanding that core-collapse supernovae are routinely asymmetric developed in parallel with the discovery that gamma-ray bursts are highly-collimated events. This supernova/gamma-ray burst connection was dramatically confirmed when SN 2003dh was revealed in the afterglow of GRB 030329 (Stanek et al. 2003; Hjorth et al. 2003; Kawabata et al. 2003).

Here we summarize some of the background on spectropolarimetry of core-collapse supernovae and the evidence that they are generically asymmetric. We discuss the importance of the magneto-rotational instability for the collapse problem and some of the attendant physics that may be expected. We outline some of the important issues involved in doing MHD collapse and give a summary and critique of some recent attempts to merge MHD physics with core collapse physics.

2. Results of Spectropolarimetry

The first qualitative insight of the "Texas" program of routine spectropolarimetry was that there is a distinct difference between Type Ia supernovae and core collapse events: Type II, Type IIn, Type IIb, Type Ib and Type Ic. The first systematic study (Wang et al. 1996) showed that core collapse supernovae are substantially polarized at the 1 % level, but that Type Ia were generally substantially less polarized. As more data were added, it became clear that the polarization of the core-collapse supernovae was deeply intrinsic to the explosion mechanism. The polarization grows as the photosphere recedes into the ejecta and tends to be higher for events with less thick blanketing hydrogen envelopes (Wang et al. 1996, 2001; Leonard et al. 2001), This implies that the basic machine that powers the explosion is asymmetric. Polarization of ~ 1 % implies an axis ratio of about 2 to 1 if interpreted in terms of ellipsoids of rotation (Höflich 1991).

The data often show a well-defined orientation suggesting that the explosion was substantially bi-polar. Figure 1 shows the data for the Type II plateau event SN 1999em. The data fall on the same line in the Stokes parameter plane as a function of time and of wavelength. This shows that there is a strongly favored axis to the geometry, hence that it is substantially bi-polar, a pattern repeated in several other events. We stress that there are exceptions. The Type IIb SN 1993J and the very similar event SN 1996cb showd large scatter in the Stokes parameter plane (Wang et al. 2001). SN 2002ap, a Type Ic that showed very high photospheric velocities in early phases revealed that the dominant axes associated with hydrogen, with oxygen, and with calcium were all oriented substantially differently (Wang et al. 2003b). Observational programs clearly have much more to teach us about the complexity of asymmetric supernovae and the physics involved in the asymmetry.

3. Asymmetric Core Collapse

We have learned that all core collapse supernovae are substantially asymmetric and often bi-polar. This alone does not prove that supernovae are exploded by jets, but numerical simulations (Khokhlov et al. 1999; Khokhlov & Höflich 2001; Höflich, Wang & Khokhlov 2001) have shown that bi-polar jets can, in principle, explode supernovae and produce these asymmetries with no aid from the classical powering process of neutrino deposition. The origin of any such jets remains a mystery. Rotation alone can induce asymmetric neutrino fluxes (Shimizu, Yamada, & Sato 1994; Fryer & Heger 2000), but rotation will inevitably lead to magnetic field amplification that can both produce MHD effects, including possibly jets (Wheeler et al. 2000, 2002; Akiyama et al. 2003), and affect neutrino transport (see §9 for a brief discussion). In practice, neutrino transport, probably itself asymmetric and bi-polar will remain an important ingredient in the phenomenon.

Asymmetries will also affect nucleosynthesis (Maeda et al. 2002; Nagataki et al. 2003). An important aspect of the jet-induced simulations is a characteristic feature of the chemical distribution. There will be a generic tendency for the iron-peak elements to be ejected along the jet direction with the traditional elements of bulk nucleosynthesis (oxygen, calcium) being ejected predominantly in the equatorial plane. There is evidence that SN 1987A shows just that sort of configuration (Wang et al. 2002a). An interesting challenge to this picture is the recent data from a long CXO

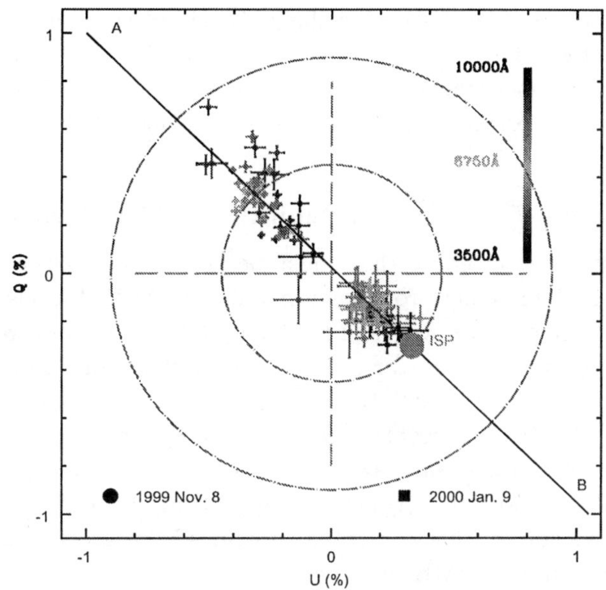

Figure 1.   Two epochs of spectropolarimetry on the Type II plateau supernovae 1999em showing the bi-polar nature of the ejecta that falls along a single locus in the plane of the Stokes vectors as a function of time and wavelength (Wang et al. 2001)

exposure of Cas A that clearly shows the jet and counter-jet structure long associated with Cas A, but predominantly in the element silicon, not, apparently, iron (Hwang et al. 2004). Another challenge is the displacement of the central compact object to the south of the center of expansion of the remnant, implying a "kick" to the compact object of about 330 km s^{-1} roughly normal to the locus of the jet (Thorstensen, Fesen & van den Bergh, 2001). It is possible that this complex dynamical structure is related to the multiple axes revealed in SN 2002ap (Wang et al. 2003b).

The tendency for collapse explosions to be bi-polar suggests that at the very least rotation is involved to provide a special, well-defined axis. There are strong arguments that rotation will naturally and unavoidably be attended with dynamo processes that generate and amplify magnetic fields. It is probably inconsistent to consider rotation in either the collapse process or the stellar evolution that precedes it without simultaneously and self-consistently considering the attendant magnetic field.

The ultimate problem of core collapse is one of three dimensions, rotation, magnetic fields, and neutrino transport. We have suspected this all along, but the polarization of supernovae and jets from GRBs demands that the issue of substantial asymmetries be met head on.

4. The Magneto-Rotational Instability and Core Collapse

An important physical effect that must be considered in the context of core collapse is the magneto-rotational instability (MRI; Balbus & Hawley 1991, 1998). Core collapse will lead to strong differential rotation near the surface of the proto-neutron star even for initial solid-body rotation of the iron core (Kotake, Yamada & Sato 2003; Ott et al. 2003). The criterion for instability to the MRI is a negative gradient in angular velocity, as opposed to a negative gradient in angular momentum for the Rayleigh dynamical instability. This condition is generally satisfied at the surface of a newly formed neutron star during core collapse and so the growth of magnetic field by the action of the MRI is inevitable.

More quantitatively, when the magnetic field is small and/or the wavelength is long (k $v_a < \Omega$) the instability condition can be written (Balbus & Hawley 1991, 1998):

$$N^2 + \frac{\partial \Omega^2}{\partial \ln r} < 0, \tag{1}$$

where N is the Brunt-Väisälä frequency. Convective stability will tend to stabilize the MRI, and convective instability to reinforce the MRI. The saturation field given by general considerations and simulations is approximately given by the condition: $v_a \sim \lambda\Omega$ where $\lambda \lesssim r$ or $B^2 \lesssim 4\pi\rho r^2\Omega^2$ where v_a is the Alfvén velocity.

Akiyama et al. (2003) have presented a proof-of-principle calculation that the physics of the MRI is inevitable in the context of the differentially-rotating environment of proto-neutron stars. The great power of the MRI to generate magnetic field is that while it works on the rotation time scale of Ω^{-1} (as does simple field-line wrapping), the strength of the field grows exponentially. This means that from a plausible seed field of 10^{10} to 10^{12} G that might result from field compression during collapse, only $\sim$ 7 - 12 e-folds are necessary to grow to a field of 10^{15} G. Akiyama et al. (2003) have shown that for rotation that is at all times sub-Keplerian, this instability will naturally grow any seed field exponentially rapidly to a saturation level of order 10^{15} to 10^{16} G in a few 10s of milliseconds, a timescale longer than the initial bounce timescale, but much less than popular late-time neutrino-heating mechanisms that work over hundreds of milliseconds.

Figure 2 shows the expected evolution of the angular velocity profile, the magnetic field and the associated MHD luminosity. The portion of the structure with decreasing angular velocity with radius, a generic feature at the boundary of the rotating proto-neutron star, represents structure that is unstable to the magneto-rotational instability. The predicted magnetic field is much larger than the quantum electrodynamic limit of $\sim 10^{13}$ G, but still smaller than the fields that would be directly dynamically important, of order 10^{17} to 10^{18} G. It remains to be seen whether this level of magnetic field will contribute substantially to asymmetries and jet formation in the explosions. The effects on the equation of state are estimated to be negligible near the PNS where the density is high (Duan 2004; Akiyama et al. 2004), but if a highly magnetic bubble is convected to a low density region, there could be important effects. There could also be effects on the neutrino cross sections as outlined briefly in §7. We note that these calculations have not yet considered the de-leptonization phase when the neutron star contracts and spins even faster, perhaps producing even larger fields on timescales of seconds.

The resulting characteristic MHD luminosity (cf. Blandford & Payne 1982) is:

$$L_{\mathrm{MHD}} \sim B^2 r^3 \Omega / 2 \sim 3 \times 10^{52} \text{ erg s}^{-1} B_{16}^2 R_{\mathrm{NS.6}}^3 \left(\frac{P_{\mathrm{NS}}}{10 \text{ ms}} \right)^{-1}. \qquad (2)$$

If this power can last for a significant fraction of a second, a supernova could result. The energy of rotation is approximately

$$E_{rot} \sim 1/2 I_{NS} \Omega_{NS}^2 \sim 1.6 \times 10^{50} \text{ erg } M_{NS} R_{\mathrm{NS.6}}^2 \left(\frac{P_{\mathrm{NS}}}{10 \text{ ms}} \right)^{-2}. \qquad (3)$$

A sufficiently fast rotation of the original iron core is needed to provide ample rotation energy. This will also promote a strong MHD luminosity.

For collapse to form a black hole, the velocities will be Keplerian and the associated, dynamo-driven, predominantly toroidal field will have a saturation strength, $B^2 \sim 4\pi \rho \lambda^2 \Omega^2$ with $\lambda \lesssim r$, of order $B \sim 10^{16}$G $\rho_{10}^{1/2}$ assuming motion, including the Alfvén speed, near the speed of light near the Schwarzschild radius and a characteristic density of order 10^{10} g cm^{-3} (MacFadyen & Woosley 1999). Fields this large could affect both the dynamics and the microphysics in the black hole-formation problem. Because of the nearly Keplerian motion in the black hole case, the fields generated will be much closer to pressure equipartition than in the neutron star case,

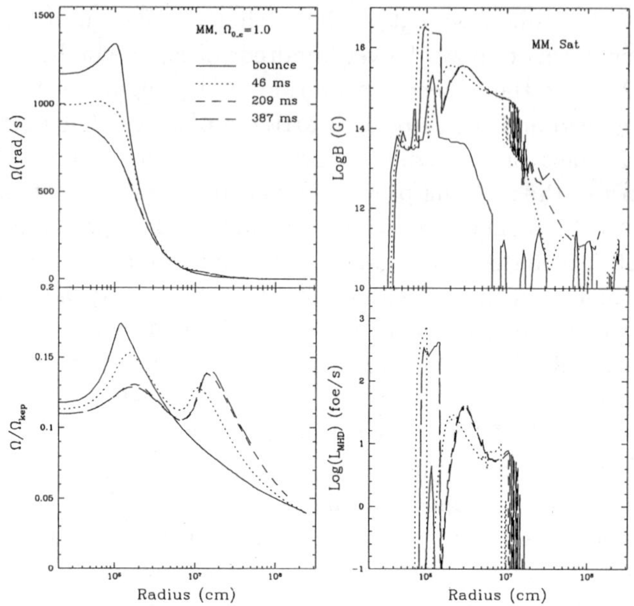

Figure 2.   Angular velocity, field strength and MHD luminosity(in units of 10^{51} erg s^{-1}) for a representative initial differential rotation of the iron core as a function of time from Akiyama et al. (2003)

and hence, perhaps, even more likely to have a direct dynamical effect. The associated MHD power in the black hole case would be roughly $10^{52} - 10^{53}$ erg s^{-1}.

The implication of this work is that the MRI is probably unavoidable in the differentially rotating ambience of core collapse for either "ordinary" supernovae and for those that produce gamma-ray bursts. Calculations that omit this physics are probably incorrect at some level. The magnetic field generated by the MRI should be included in any self-consistent calculation, but issues of how to capture this physics in numerical calculations are challenging. Balbus & Hawley (1998) summarize their work showing that the specific outcome of MRI calculations depends on the initial field configuration. In 2D, an initial magnetic field aligned with the rotation axis will give a streaming instability, whereas a configuration with a finite RMS field but with zero mean field will give a chaotic, turbulent field. We return to this point below (§8).

These implications need to be explored in much greater depth, but there is at least some possibility that the MRI may lead to strong MHD jets by the magneto-rotational (Meier, Koide & Uchida 2001) or other mechanisms. A key point is that the relevant dynamics will be dictated by strong, predominantly toroidal fields that are generated internally, and are not necessarily the product of twisting of external field lines that is the basis for so much work on MHD jet and wind mechanisms. Understanding the role of these internal toroidal fields in producing confining coronae (Hawley & Balbus 2002) or jets (Williams, 2003), in providing the ultimate dipole field strength for both ordinary pulsars and magnetars (Duncan & Thompson 1992), in setting the "initial" pulsar spin rate after the supernova dissipates (that is, the "final" spin rate from the supernova dynamicists point of view), and any connection to GRBs is in its infancy.

5. Open Issues

There are a large number of important open issues. Chief among them are whether or not the rotation and magnetic fields associated with core collapse lead to sufficiently strong MHD jets or other flow patterns to explode supernovae. This issue touches on many others:

- Magnetic effects in the rotating progenitor star
- Dynamos and saturation field strengths
- Effect of large fields on the equation of state
- Effect of large fields on the neutrino cross sections and transport
- Effect of large fields on structure and evolution of the neutron star
- Effect of large fields on jet formation
- Relevance of MRI and field generation to GRBs and "hypernovae"

6. Dynamo Theory and Saturation Fields

In traditional mean field dynamo theory, the turbulent velocity field that drives the "alpha" portion of the $\alpha - \Omega$ dynamo was specified and held fixed, but the turbulent velocity field cannot be constant. The buildup of small scale magnetic field tends to inhibit turbulence, cutting off the dynamo process for both small and large scale fields. Since the small scale field tended to grow faster than the large scale field, it appeared that the growth of the large scale field would be suppressed (Kulsrud & Anderson 1992; Gruzinov & Diamond 1994). In these theories, the magnetic field energy cascades to smaller length scales where it is ultimately dissipated at the resistive scale. Large scale fields tend to build up slowly, if at all.

A proposed solution to this problem has been the recognition (Blackman & Field 2000; Vishniac & Cho 2001; Field & Blackman 2002; Blackman & Brandenburg, 2002; Blackman & Field 2002; Kleeorin et al. 2002) that the magnetic helicity, $\mathbf{H} = \mathbf{A} \cdot \mathbf{B}$ is conserved in ideal MHD and that this conservation had not been treated explicitly in mean field dynamo theory. Incorporation of this principle leads to an "inverse cascade" of helical field energy to large scales that is simultaneous with the cascade of helical field energy from the driving scale to the dissipation scale. Basically, the large scale helical field and inverse cascade must exist with opposite magnetic helicity to that of the field cascading to small scale. The result (Blackman & Brandenburg 2002) is the rapid growth of large scale field in a kinematic phase (prior to significant back-reaction) to a strength where the field on both large and small scales is nearly in equipartition with the turbulent energy density. At that point, the back reaction sets in and there tends to be a slower growth to saturation at field strengths that can actually somewhat exceed the turbulent energy density. It may be that the early, fast, kinematic growth is the only phase that is important for astrophysical dynamos, especially in situations that have open boundaries so that field can escape (Brandenburg, Blackman & Sarson 2003; Blackman & Tan 2003) and that are very dynamic. The collapse ambience is clearly one of those situations.

A related insight is that the rapid kinematic phase of field growth can lead to magnetic helicity currents (Vishniac & Cho 2001). It is possible that these magnetic helicity currents can transport power out of the system in twisting, propagating magnetic fields. This is clearly reminiscent of jets or winds, but the physics is rather different than any that has been previously explored in driving jets or winds. This physics needs to be explored in the context of supernovae and gamma-ray bursts.

Vishniac & Cho (2001) argue that along with conservation of magnetic helicity, $\mathbf{H} = \mathbf{A} \cdot \mathbf{B}$, and the inverse cascade of magnetic field energy to large scales, one will get a current of magnetic helicity that can be crudely represented by

$$J_{\mathrm{H}} \sim B^2 \lambda \mathrm{v}, \tag{4}$$

where the characteristic length, λ, might be comparable to a pressure scale height, $\ell_{\mathrm{P}} = (\mathrm{d}\ln P/\mathrm{d}r)^{-1}$, and $\mathrm{v} \sim \mathrm{v_a} \sim \ell_{\mathrm{P}}\Omega$. The energy flux associated with this magnetic helicity current is $J_{\mathrm{H}}/\lambda \sim B^2 \mathrm{v_a}$, and so with $B^2 \sim \rho \ell_{\mathrm{P}}^2 \Omega^2$

the associated power is:

$$L = r^2 B^2 v_a \sim B^2 r^2 \ell_P \Omega \sim \rho r^5 \Omega^3 \left(\frac{\ell_P}{r}\right)^3. \tag{5}$$

Note that the next-to-last expression on the RHS is essentially just the characteristic Blandford-Payne luminosity; however, in this case the field is not externally given, but provided by the dynamo process so that the final expression on the RHS is given entirely in terms of local, internal quantities. The implication is that this amount of power is available in an axial, helical field without twisting an externally anchored field. Again, while this analysis has superficial resemblance to other jet mechanisms, it involves rather different physics and is self-contained. Whether this truly provides a jet remains to be seen.

Note that this process of creating a large scale field with an MRI-driven dynamo with its promise of naturally driving axial, helical flows does not require an equipartition field. As pointed out by Wheeler et al. (2002), the field does not have to have equipartition strength and hence to be directly dynamically important in order to be critical to the process of core collapse. The field only has to be significantly strong to catalyze the conversion of the free energy of differential rotation of the neutron star into jet energy. As long as this catalytic function is operative, the rotational energy should be pumped into axial flow energy until there is no more differential rotation. For the case of stellar collapse, this would seem to imply that, given enough rotational energy in the neutron star, this machine will work until there is a successful explosion. Even if the core collapses directly into a black hole, or does so after some fall-back delay, the basic physics outlined here, including magnetic helicity currents and their associated power should also pertain to black hole formation.

7. Neutrino Transport

Fields of order 10^{15} to 10^{16} G that will characterize both neutron star and black hole formation may affect neutrino transport. With a large magnetic field, direct $\nu - \gamma$ interaction is possible mediated by W and Z bosons. This would allow neutrino Cerenkov radiation, $\nu \to \nu + \gamma$, and would enhance plasmon decay, $\gamma \to \nu + \nu$ (Konar 1997).

In addition, processes like $\nu \to \nu + e^+ + e^-$ would no longer be kinematically forbidden. In that case, closed magnetic flux loops can trap pairs. The energy in pairs would grow exponentially to the point where annhilation cooling would balance pair creation. Thompson & Duncan (1993) estimated that an energy as much as $E_{pair} \sim 10^{50}$ erg could be trapped in this way. This is not enough energy to cause a robust explosion, but it is enough energy to drive the dynamics of core collapse in a substantially different way, perhaps by inducing anisotropic flow if the flux loops are themselves distributed anisotropically.

With substantial magnetic fields, the cross section for inverse beta decay, $\nu_e + n \to p + e^-$, would become dependent on neutrino momentum, especially for asymmetric field distributions, which would be the norm (Lai & Qian 1998; Bhattacharya & Pal 2003; Ando 2003; Duan & Qian 2004).

All these processes and more should be considered quantitatively in core collapse to form neutron stars and black holes.

8. Recent Work on Magnetic Core Collapse

In this section we will review, compare and contrast some recent work on rotating magnetic collapse and related issues that pertain to understanding asymmetric core collapse. Other relevant work that we do not discuss in detail is in Burrows & Hayes (1996) and Lai et al. (2001) and references therein that discuss the effects of neutrino flux asymmetry. A crucial piece of information that can inform the discussion of potential MHD effects even in the absence of the explicit inclusion of magnetic fields is to give sufficient information from a rotating collapse to at least crudely estimate the saturation field according to the prescription $v_a \sim r\Omega$ or $B \sim \sqrt{4\pi\rho}r\Omega$, that is, the angular velocity and the density profiles, or, even better, the product $\sqrt{\rho}\Omega$. Many studies of rotating collapse produce such information, but fail to present it explicitly. It would be very valuable if such information were presented explicitly as a function of time.

Akiyama et al. (2003) - As outlined above, Akiyama et al. did "shellular" rotating collapse calculations with multi-group flux limited diffusion of neutrinos, no angular momentum transport (although that possibility was discussed) and a heuristic treatment of the MRI. They concluded that core collapse is generically susceptible to the MRI and that the MRI could be important. They found fields of order 10^{15} to 10^{16} G could plausibly be generated in tens of milliseconds after bounce. Such fields are interestingly larger than the QED limit, but still not directly dynamically important. The magnetic field need not necessarily be dynamically important if the field can catalyze the dumping of the rotational energy of the neutron star into some useful, explosion-inducing form, jets or otherwise. This basic energy requirement puts a premium on rapid rotation of the progenitor and the proto-neutron star in order to have a sufficiently large energy reservoir on which to draw. Of course, the rotational energy may be abetted by the large neutrino flux.

Thompson, Quataert, & Burrows (2004) - Thompson et al. also did "shellular" rotating collapse calculations with a heuristic treatment of the MRI. Not surprisingly, given the similar assumptions and computations, they confirmed the field strength estimates of Akiyama et al. The new ingredient in this paper was to add viscous dissipation heating. Thompson et al. found that they could induce explosions for rapid enough rotation.

Fryer & Warren (2004) - In a series of works culminating, for now, in this paper, Fryer & Heger (2000) and Fryer & Warren (2002) have explored rotating core collapse. See also Fryer's contribution to these proceedings. This work has used an SPH code with Fryer & Heger and Fryer & Warren examining the 2D case and Fryer & Warren (2004) full 3D hydrodynamics. A feature that complicates the comparison of the results of this work with that from grid-based codes is that the SPH code yields prompt explosions in the basic non-rotating case, but no current grid-based code does so. The SPH code uses single energy, flux-limited diffusion. In the rotating calculations there are issues of angular momentum transport in SPH versus grid-based calculations. Fryer & Heger (2000) and Fryer & Warren (2002) found that rotation alone could induce bi-polar, asymmetric explosions with axis ratios of 2 to 1, but the calculations were not run into the free-expansion phase, so it is not clear that this large asymmetry will survive as required by the spectropolarimetry.

In their 3D calculation, Fryer & Warren take note of significant evolution in the angular momentum distribution. An important factor is the tendency for low angular momentum matter to flow in along the rotation axis while larger angular momentum matter tends to halt along the equator. This aspect of the dynamics cannot be captured in "shellular" calculations, but should be manifested in 2D calculations. Unfortunately, other papers have not commented specifically on this phenomenology which should be quite generic. It would be useful in making comparisons if others were to do so. Fryer & Warren (2004) do not include MHD, but use the heuristic prescription of the saturation field strength presented by Akiyama et al. to estimate the field strength for the angular velocity gradient and density they compute. Their estimates of the field strength are substantially less than found by Akiyama et al. even though, despite the very different dynamics (3D versus "shellular"), the resulting angular velocity gradients are rather similar. The difference seems to be that, with a prompt explosion, the density declines rapidly, thus decreasing the implied Alfvén speed and hence saturation field for a given angular velocity gradient.

Buras, Rampp, Janka, & Kifonidis (2003) - Buras et al. do a sophisticated rotating collapse with Boltzmann neutrino transport on radial rays. These calculations have no magnetic effects, but are of the sort that can establish the ambience in which MHD effects will occur. These calculations give bi-polar flow, but no explosion. The angular velocity profiles are not presented, so it is difficult to compare to other computations in that fundamental regard.

Ott, Burrows, Livne & Walder (2003) - Ott et al. used Livne's VULCAN/2D code to study rotating collapse. They include no neutrino transport, but do present useful information on angular velocity profiles at certain epochs. These calculations revealed the strong shear expected in core collapse and gave bi-polar flow patterns, but no explosion.

Kotake, Sawai, Yamada & Sato (2004) - Kotake et al. present 2D rotating, MHD collapse calculations using the ZEUS-2D code. They incorporate an approximate neutrino cooling with a leakage scheme. They assume the initial field prior to collapse is predominantly toroidal and explore the effect on anisotropic neutrino radiation. They find more effective neutrino heating near the axis in a way that affects the dynamics. These calculations assume rapid pre-collapse rotation and pre-collapse magnetic fields in the range $5 \times 10^9 - 10^{14}$ G. Such initial fields are probably unrealistically large. The calculations do produce phenomena that resemble MHD jets. The effects of field line wrapping are difficult to discriminate from the MRI, but Kotake et al. conclude that the MRI is likely to occur after bounce due to non-axisymmetric perturbations.

Yamada & Sawai (2004) - Yamada & Sawai also use ZEUS-2D but with a parametrized equation of state and no neutrinos. They assume rapid pre-collapse rotation and pre-collapse poloidal magnetic fields that are uniform, parallel to the rotation axis and with an amplitude of $\sim 10^{12}$ G. Again these large initial fields are probably unphysical. Yamada & Sawai find large fields "behind the shock" not in the core as for the pioneering calculation of LeBlanc & Wilson (1971). Once again it is difficult to see whether the growth of field strength is due to field line wrapping, especially with the initial axial field, or some aspect of the MRI, or both.

Madokoro, Shimizu, & Motizuki (2003) - Madokoro et al. (see also Shimizu et al. 1994) explore non-rotating models in which a prolate, anisotropic neutrino radiation field is imposed. They find that such an anisotropic neutrino flux gives a larger explosion energy for given neutrino luminosity.

Ardeljan, Bisnovatyi-Kogan, Kosmachevskii & Moiseenko (2004) - Ardeljan et al. (see also Ardeljan, Bisnovatyi-Kogan & Moiseenko 2004 and Moiseenko, Bisnovatyi-Kogan & Ardeljan 2004) present their own version of 2D MHD collapse and explosion for a collapsing bare white dwarf. They compute the collapse with rotation until the structure is nearly in hydrostatic equilibrium and then "turn on" a field that is subsequently amplified. They explore both dipole and quadrupole initial fields. The magnetic field subsequently grows to become comparable to the local pressure at which time an MHD shock is generated. The formation of the MHD shock may be related to the low density associated with the bare white dwarf collapse. They get some mass ejection with an energy of about 5×10^{50} ergs for a model in which the initial magnetic energy is a fraction 10^{-6} of the gravitational energy. If the gravitational energy corresponds to a neutron star with binding energy of order 10^{53} ergs this corresponds to an initial field of roughly 10^{15} G. Although the MRI is mentioned in Ardeljan, Bisnovatyi-Kogan & Moiseenko and by Moiseenko et al., few details are presented, so the mechanism of the field amplification is not clear. Unique among the calculations summarized here, this work follows the neutron star for ~ 10 s as it contracts and speeds up.

Hawley & Balbus (2002) - Hawley & Balbus performed the first MHD simulation of a collapse-related environment in which the MRI and jet formation were explicit ingredients. They use the ZEUS algorithms to solve the MHD equations. This was a 3D MHD simulation of the accretion of a torus of matter around a black hole. This is not the same as a true collapse calculation in the sense that there is no surrounding star, but it is still instructive. The torus accretes due to the turbulent stresses generated by the MRI. The resulting flow forms a hot, thick, nearly-Keplerian disk, a surrounding magnetized corona, and a jet up the axis. A key point is that their jet is not confined by hoop stress. It is held out by the centrifugal barrier and held in by the pressure of the highly magnetic ($\beta << 1$) corona. It is not clear how much this simulation would change if there were a surrounding, infalling star. Hawley & Balbus note that there was no significant dynamical difference between simulations that included or omitted resistive heating. Hawley & Balbus suggest that they get larger fields than in their closed box simulations, but do not discuss the reasons in any detail. They also do not explicitly discuss whether their fields are turbulent, changing sign on turbulent time scales, or are well-ordered and large scale.

Proga, MacFadyen, Armitage, & Begelman (2003) - Proga et al. also explored a somewhat different problem, the rotating magnetic core collapse onto a black hole rather than onto a proto-neutron star, but the similarities and differences are again instructive. Proga et al. also employ ZEUS-2D MHD to do calculations of accretion of a helium envelope around a pre-existing black hole. They adopt a particular initial angular momentum distribution that does not necessarily correspond to the other collapse models discussed here. They adopt an initial field that is purely radial, again a configuration unlike any of the other work mentioned here. They find that a thick, Keplerian torus forms with the subsequent development of a Poynting-flux dominated jet. They argue that the MRI is active in producing amplification of their initial field, but give few details of the operation of the MRI in their models. It is of particular interest that Proga et al. find their jet to be Poynting-flux dominated, since none of the canonical neutron star formation calculations seem to do so, even with rather large initial rotation and magnetic fields. This may be an artifact of their doing a calculation with a previously existing black hole onto which the helium envelope is dumped "impulsively." This may lead to a more evacuated, low density environment within the collapsing core and hence perhaps a tendency to be more nearly field dominated at a given field strength than if the collapse were followed *ab initio* from an iron core. There may also be some differences in the way field amplification evolves in this environment. There could also be differences in the intrinsic environment of proto-neutron stars versus black holes. It would be very interesting to explore in more detail the differences between the calculation of Proga et al. and that of, for instance, Kotake et al. (2004).

This summary raises a central practical issue. The grand goal is to do a full 3D, sufficiently resolved (whatever that means) simulation of rotational magnetic core collapse with anisotropic neutrino transport. Various groups, as just summarized, are taking the first steps in this direction. To what degree can current (or future) simulations be regarded as adequate to capture the physics? A case in point are simulations that may or may not reproduce the physics of the MRI.

A basic issue here is well-illustrated by Balbus & Hawley (1998) who point out that the Cowling dynamo theorem pertains to 2D MHD calculations. This theorem states that an axially-symmetric magnetic field cannot be produced by fluid flow (Cowling 1957). The implication, as pointed out by Balbus & Hawley (1998), is that sustained magnetic field amplification by axisymmetric turbulence is impossible in an isolated dissipative 2D system. Balbus & Hawley show, as pointed out earlier, that the behavior of

MRI simulations with initial vertical field and those with initial turbulent field ($<B> = 0$) are very different with the former yielding channel flow solutions and the latter turbulent solutions. In 2D simulations, however, both configurations yield a decaying field after an initial transient growth phase. Three-dimensional simulations, on the other hand, show a rapid growth to a saturation field. Balbus & Hawley argue that the results still retain some memory of the initial field configuration, but that the MRI works to grow the field exponentially rapidly even with $<B> = 0$ as long as the simulation is adequately resolved. The time to reach saturation depends on the initial value of $<B^2>^{1/2}$, but the saturation level is basically independent of the value of the initial mean field.

What, then, are we to make of the various simulations described above, especially those that are in 2D and may or may not see the effects of the MRI? It is possible that they are seeing the transient growth before the Cowling anti-dynamo effects set in. Is that adequate for a dynamical situation like a supernova where the field only needs to act long enough to produce an explosion? Perhaps, but that does not seem satisfactory when the goal is to get the physics right. In addition, the 2D simulations reproduced in Balbus & Hawley (1998) seem to show smaller ratios of peak fields to initial fields than the 3D calculation and that the field in 2D simulations decays on a timescale comparable to the rotation time, so there are indications already that one should not trust any 2D simulation to properly capture the physics of the MRI. We have also seen that various groups have made radically different assumptions about the initial field structure – axial, toroidal, radial – when the simulations indicate that the initial field configuration may affect the results, especially in 2D. There has been no systematic study of this possibility.

Another point to ponder is the stricture in the Cowling dynamo theorem that the system to which the theorem applies should be isolated and dissipative. A supernova is dissipative, but it is not isolated; the whole point is to eject matter through previously existing boundaries. The possibility of magnetic helicity currents as described by Vishniac & Cho (2001; §6) may also be relevant in this context. Is it possible that MHD calculations that produce jets and outflow are thereby not "isolated," and not subject to the anti-dynamo restrictions, and hence in some sense more trustworthy? Again this point has not been discussed at all in the literature.

This is not to deny that all the calculations described above have merit as this difficult regime is explored. The lesson is clear, however that the MRI is tricky to simulate and care must be taken in computing, presenting, and comparing results among various simulations.

9. Conclusions

Spectropolarimetric studies have shown that all core-collapse supernovae yet observed are significantly asymmetric, with geometries that can be predominantly bi-polar, but can also be more complex in ways that have yet to be sufficiently characterized or understood (see Wang et al. 2001, Wang et al. 2003b). This data, along with direct imaging of objects like SN 1987A, the Crab nebula and Cas A, means that the dynamics and the radiative transfer, both photons and neutrinos, are very likely to be significantly asymmetric. Account of this asymmetry must be made in the analysis of these events, including the derivation of such basic quantities as the ejecta mass and energy.

Core collapse is an intrinsically shearing environment and hence generically subject to dynamo-like instabilities such as the magneto-rotational instability. This means that both rotation and magnetic fields are intrinsic to the process of core collapse. This applies both to supernovae and to gamma-ray bursts, to the formation of neutron stars and of black holes.

Many of the points of this discussion have been made with the implicit assumption that the cause of the asymmetry of core collapse supernovae as revealed by the spectropolarimetry is in some way related to rotation which must be accompanied by some dynamo action and field growth. A last point concerns alternatives to this tacit assumption. Blondin, Mezzacappa, & DeMarino (2003) have shown that standing spherical accretion shocks such as those that arise in non-rotating, non-magnetic core-collapse supernova models are unstable to $\ell = 1$ and 2 modes of oscillation. See also Blondin in these proceedings, Foglizzo (2002), and the calculations of full collapse with neutrino transport illustrating this instability described by Janka (2004) and by Scheck et al. (2004). This instability and the rapid growth of turbulence behind the shock is driven by the injection of vorticity as the shock is perturbed from spherical. This interesting result means that one must, at least, be cautious about interpreting bi-polar symmetry as the result of rotation. On the other hand, sufficient rotation (never mind magnetic fields) may damp this particular instability (but see Scheck et al.). It is perhaps worth noting that this instability can only produce bi-polar structure whereas the observations show that strong deviations from this simple geometry are not uncommon. Of course, the simplest jet-induced models are also intrinsically axially symmetric and fail the same test. Clearly, this instability must be added to the zoo of possible phenomenology and treated self-consistently in future work.

The bottom line is that many new vistas have been opened by considering the intrinsic asymmetry of core-collapse supernovae and the ability of rotating, magnetic collapse to account for the observations. There is much work to be done to understand all the attendant physics.

Acknowledgments

The authors are grateful to Tony Mezzacappa for his invitation and encouragement to attend this workshop, to the Institute for Nuclear Theory for providing a rich and fruitful environment, and to John Hawley and Chris Fryer for especially rewarding conversations at the meeting. They also thank Dave Meier and Peter Williams for valuable on-going discussion of these topics. JCW is deeply grateful for the contributions of his collaborators Lifan Wang, Peter Höflich and Dietrich Baade and the excellent staff at the VLT, all of whom were critical to the success of our spectropolarimetry program. This work was supported in part by NASA Grants NAG5-10766 and NAG5-9302 and by NSF Grant AST-0098644.

References

1. Ando, S. 2003, Phy. Rev. D, 68, 63002
2. Akiyama, S. Wheeler, J. C., Meier, D. & Lichtenstadt, I, ApJ, 584, 954
3. Akiyama, S., Wheeler, J. C., Meier, D. L., & Duncan, R. C. 2004, in Cosmic Explosions in Three Dimensions, eds. P. Höflich, P. Kumar, & J. C. Wheeler (Cambridge: Cambridge University Press), in press
4. Ardeljan, N. V. Bisnovatyi-Kogan, G. S., & Moiseenko, S. G. 2004, MNRAS, in press (astro-ph/0410234)
5. Ardeljan, N. V. Bisnovatyi-Kogan, G. S., Kosmachevskii, K. V. & Moiseenko, S. G. 2004, Astronomy, 47, 37.
6. Balbus, S. A. & Hawley, J. F. 1991, ApJ 376, 214
7. Balbus, S. A. & Hawley, J. F. 1998, Review of Modern Physics, 70, 1
8. Bhattacharya, K. & Pal, P. B. 2003, hep-ph/0209053
9. Blackman E. G. & Brandenburg, A. 2002, ApJ, 579, 359
10. Blackman, E. G. & Field, G. B. 2000, ApJ, 534, 984
11. Blackman, E. G. & Field, G. B. 2002, Phys. Rev. Lett., 89, 265007
12. Blackman E. G. & Tan, J. 2003, in "Proceedings of the International Workshop on Magnetic Fields and Star Formation: Theory vs. Observation," in press (astro-ph/0306424)
13. Blandford, R. D. & Payne, D. G. 1982, MNRAS, 199, 833
14. Blondin, J. M., Mezzacappa, A., & DeMarino, C. 2003, ApJ, 584, 971
15. Brandenburg, A., Blackman E. G. & Sarson, G. R. 2003, Adv. Space Sci., 32, 1835

16. Buras, R., Rampp, M., Janka, H.-T., & Kifonidis, K. 2003, Phys. Rev. Lett., 90, 241101

17. Burrows, A., & Hayes, J., 1996, Phys. Rev. Lett., 76, 352

18. Cowling, T. G. 1957, Magnetohydrodynamics, (New York: Interscience)

19. Duan, H., & Qian, Y. -Z. 2004, Phys. Rev. D, 69, 123004

20. Duan, H. 2004, private communication

21. Duncan, R. C. & Thompson, C. 1992, ApJ, 392, L9

22. Fesen, R. A. 2001, ApJ Supp. , 133, 161

23. Field, G. B. & Blackman E. G. 2002, ApJ, 572, 68

24. Foglizzo, T. 2002, A&A, 392, 353

25. Fryer, C. L. & Heger, A. 2000, ApJ, 541, 1033

26. Fryer, C. L. & Warren, M. S. 2002, ApJ Lett., 574, L65

27. Fryer, C. L. & Warren, M. S. 2004, ApJ, 601, 391

28. Gruzinov, A. V. & Diamond, P. H. 1994, Phys. Rev. Lett., 72, 1651

29. Hawley, J. F. & Balbus, S. A. 2002, ApJ, 573, 738

30. Hjorth, J. et al. 2003, Nature, 423, 847

31. Höflich 1991, Astron & Astrophys.,, 246, 481;

32. Höflich, P., Khokhlov, A. & Wang, L. 2001, in Proc. of the 20th Texas Symposium on Relativistic Astrophysics, eds. J. C. Wheeler & H. Martel, (New York: AIP), 459

33. Howell, D. A., Höflich, P., Wang, L., & Wheeler, J. C. 2001, ApJ, 556, 302

34. Hwang, U. et al. 2004, ApJ Lett., 615, L117

35. Kawabata, et al. 2003, ApJ, 593, L19

36. Khokhlov, A. & Höflich, P. 2001, in Explosive Phenomena in Astrophysical Compact Objects, eds. H.-Y, Chang, C.-H. Lee & M. Rho, AIP Conf. Proc. No. 556, (New York: AIP), p. 301

37. Khokhlov A.M., Höflich P. A., Oran E. S., Wheeler J.C. Wang, L, & Chtchelkanova, A. Yu. 1999, ApJ, 524, L107

38. Kleeorin, N. I., Moss, D., Rogachevskii, I. & Sokoloff, D. 2002, A&A, **387**, 453

39. Konar, S. 1997, PhD. Thesis

40. Kotake, K., Yamada, S., & Sato, K. 2003, ApJ, 595, 304

41. Kotake, K., Sawai, H., Yamada, S., & Sato, K. 2004, ApJ, 608, 391

42. Kulsrud, R. M. & Anderson, S. W. 1992, ApJ, 396, 606

43. Lai, D. & Qian, Y.-Z. 1998, ApJ, 505, 844

44. Lai, D., Chernoff, D. F. & Cordes, J. M. 2001, ApJ, 549, 1111

45. Leonard, D. C., Filippenko, A. V., Barth, A. J., & Matheson, T. 2000, ApJ, 536, 239

46. Leonard, D. C. & Filippenko, A. V. 2001, PASP, 113, 920

47. Leonard, D. C., Filippenko, A. V., Ardila, D. R., & Brotherton, M. S. 2001, ApJ, 553, 861

48. Leonard, D. C., Filippenko, A. V., Chornock, R. & Foley, R. J. 2002, PASP, 114, 1333

49. MacFadyen, A. & Woosley, S. E. 1999, ApJ, **524**, 262

50. Maeda, K., Nakamura, T., Nomoto, K., Mazzali, P., Patat, F. & Hachisu, I. 2002, ApJ, 565, 405

51. Madokoro, H., Shimizu, T., & Motizuki, Y. 2003, ApJ, 592, 1035

52. Marietta, E., Burrows, A., & Fryxell, B. 2000, ApJS, 128, 615

53. Meier, D. L., Koide, S. & Uchida, Y. 2001, Science, 291, 84

54. Moiseenko, S. G., Bisnovatyi-Kogan, G. S. & Ardeljan, N. V. 2004, in "1604-2004 Supernovae as Cosmological Lighthouses," eds. M. Turatto et al. (San Francisco: ASP) astro-ph/0410330

55. Nagataki, S., Mizuta, A., Yamada, H., Takabe, H. & Sato, K. 2003, ApJ, ApJ, 596, 401

56. Ott, C. D., Burrows, A., Livne, E., & Walder, R. 2004, ApJ, 600, 834

57. Proga, D., MacFadyen, A. I., Armitage, P. J., & Begelman, M. C. 2003, ApJ Lett., 599, L5

58. Scheck, L., Plewa, T., Janka, H. Th., Kifonidis, K. & Müller, E. 2004, Phys, Rev. Lett., 92, 011103-1

59. Shimizu, T., Yamada, S., & Sato, K. 1994, ApJ. Lett., 432, L119

60. Stanek, K. Z. et al. 2003, ApJ, 591, L17

61. Thompson, C. & Duncan, R. C. 1993, ApJ, 408, 194

62. Thompson, T. A., Quataert, E., & Burrows, A. 2004, preprint (astro-ph/0403224)

63. Thorstensen, J. R., Fesen, R. A. & van den Bergh, S. 2001, ApJ, 122, 297

64. Vishniac, E. T. & Cho, J. 2001, ApJ, **550**, 752

65. Wang, L., Baade, D., Höflich, P. & Wheeler, J. C. 2002b, ESO Messenger, No. 109, 47

66. Wang, L., Howell, D. A., Höflich, P., & Wheeler, J. C. 2001, ApJ, 550, 1030

67. Wang, L., Wheeler, J. C., & Höflich, P. 1997, ApJ. Lett., 476, L27

68. Wang, L., Wheeler, J. C., Li, Z., & Clocchiatti, A. 1996, ApJ, 467, 435

69. Wang, L. et al. 2002a, ApJ, 579, 671

70. Wang, L. et al. 2003a, ApJ, 591, 1110

71. Wang, L. et al. 2003b, ApJ, 592, 457

72. Wang, L., Baade, D., Höflich, P., Wheeler, J. C., Kawabata, K., & Nomoto, K. 2004, ApJ Lett., 604, L53

73. Wheeler, J. C. 2004, in Cosmic Explosions in Three Dimensions: Asymmetries in Supernovae and Gamma-Ray Bursts, eds P. Höflich, P. Kumar & J. C. Wheeler (Cambridge: Cambridge University Press), astro-ph/0401323

74. Wheeler, J. C., Meier, D. L. & Wilson, J. R. 2002, ApJ, 568, 807

75. Wheeler, J. C., Yi, I., Höflich, P. & Wang, L. 2000, ApJ, 537, 810

76. Williams, P. T., 2003, IAOC Workshop *Galactic Star Formation Across the Stellar Mass Spectrum,*" *ASP Conference Series,*" ed. J. M. De Buizer, in press (astro-ph/0206230)

77. Yamada, S. & Sawai, H. 2004, ApJ, 608, 907

ADVANCES IN MULTI-DIMENSIONAL SIMULATION OF CORE-COLLAPSE SUPERNOVAE

F. DOUGLAS SWESTY AND ERIC S. MYRA *

*Dept. of Physics & Astronomy,
State University of New York at Stony Brook
Stony Brook, NY 11794–3800, USA
E-mail: dswesty@mail.astro.sunysb.edu
emyra@mail.astro.sunysb.edu*

We discuss recent advances in the radiative-hydrodynamic modeling of core collapse supernovae in multi-dimensions. A number of earlier attempts at fully radiation-hydrodynamic models utilized either the grey approximation to describe the neutrino distribution or utilized more sophisticated multigroup transport methods restricted to radial rays. In both cases these models have also neglected the $O(v/c)$ terms that couple the radiation and matter strongly in the optically thick regions of the collapsed core. In this paper we present some recent advances that resolve some shortcomings of earlier models.

1. Introduction: The Supernova "Problem"

For over a decade researchers have struggled to model the convective post-bounce epoch of core collapse supernovae. The radiative-hydrodynamic flows that occur in the region below the stalled prompt shock have held both promise and pitfall for the supernova modeler. The promise of this phenomenon is that it might explain the long sought-after mechanism that converts the core bounce into the observed explosion. The pitfalls are legion, mostly involving a complex convective flow structure that is three-dimensional in nature and couples neutrinos to matter strongly. For this reason there remain many open issues in modeling the convective epoch of core collapse supernovae.

The supernova "problem" persists, despite more than four decades of concentrated research. The problem is this: We have no convincing expla-

---

*Visiting Fellows, The Institute for Nuclear Theory, University of Washington, June, 2004.

nation as to how the core collapse, which ends the evolution of a massive star, rebounds in such a way as to generate the explosion we observe in nature. The most realistic supernova models collapse and rebound, but create a shock wave that doesn't eject matter, either on the hydrodynamic timescale (~ 10 ms), or the diffusive timescale of the escaping neutrino radiation ($\sim 1 - 10$ s), or on any other timescale we can model.

This is not a problem of overall energetics. The gravitational energy released during core collapse and the subsequent neutron-star cooling phase is several factors of 10^{53} erg. In contrast, the kinetic energy of the explosion required for consistency with observation is only 10^{51} erg. Instead of insufficient energy, the problem is one of energy conversion and transport—how a sufficient portion of the released gravitational energy is imparted to the material *ejectus*, giving it the requisite kinetic energy.

It has been understood for many years that neutrinos play a vital role in this process. In fact, essentially the entire remaining 99% of released energy (that which is not converted to kinetic energy of the matter) is radiated away as neutrinos. Thus, an accurate treatment of neutrino processes is a necessary component of any realistic model for a supernova.

In this article, we present what we currently regard as the most important issues in core-collapse supernova modeling. In Sec. 2, we discuss the major components that need to be part of any serious modeling endeavor. In Sec. 3, we present an outline of V2D, our new two-dimensional (2-D) supernova simulation code. Section 4 contains some preliminary results using this code. Our conclusions are in Sec. 5.

2. The Components of a Supernova Simulation

Broadly speaking, there are four main components to current supernova simulation models: (1) hydrodynamics, to track the collapse, rebound, and ejection of stellar material, (2) neutrino transport, to track the production of neutrino radiation and to follow its propagation and emission from the star, (3) nuclear microphysics, to describe the diverse states of matter encountered throughout a simulation, and (4) neutrino microphysics, to describe the reactions and interactions involving neutrinos and matter. It must be stressed that all of these components are tightly coupled to one another. Thus, the most effective models are designed with this coupling built in *ab initio*.

Hydrodynamics. For simplicity of implementation, it has been customary for supernova codes to employ explicit hydrodynamics, with either a Newtonian or a general relativistic formulation. Implicit algorithms have usually been avoided since they require the computationally expensive solution of large systems of non-linear equations.

Regardless of which approach is used, the hydrodynamic portion of the problem requires solution of some form the following equations, expressed here in Newtonian formalism:

$$\frac{\partial \rho}{\partial t} + \boldsymbol{\nabla} \cdot (\rho \mathbf{v}) = 0 \tag{1}$$

$$\frac{\partial (\rho Y_e)}{\partial t} + \boldsymbol{\nabla} \cdot (\rho Y_e \mathbf{v}) = -m_b \sum_f \int d\epsilon \left(\frac{\mathbb{S}_\epsilon}{\epsilon} - \frac{\bar{\mathbb{S}}_\epsilon}{\epsilon} \right) \tag{2}$$

$$\frac{\partial E}{\partial t} + \boldsymbol{\nabla} \cdot (E\mathbf{v}) + P\boldsymbol{\nabla} \cdot \mathbf{v} = -\sum_f \int d\epsilon \left(\mathbb{S}_\epsilon + \bar{\mathbb{S}}_\epsilon \right) \tag{3}$$

$$\frac{\partial (\rho \mathbf{v})}{\partial t} + \boldsymbol{\nabla} \cdot (\rho \mathbf{v}\mathbf{v}) + \boldsymbol{\nabla} P + \rho \boldsymbol{\nabla} \Phi + \boldsymbol{\nabla} \cdot \left\{ \sum_f \int d\epsilon \left(\mathbb{P}_\epsilon + \bar{\mathbb{P}}_\epsilon \right) \right\} = 0. \tag{4}$$

Equation (1) is the continuity equation for mass, where ρ is the mass density and $\mathbf{v}$ is the matter velocity, and where these quantities, and those in the following equations, are understood to be functions of position $\mathbf{x}$ and time t. Equation (2) expresses the evolution of electric charge, where Y_e is the ratio of the net number electrons over positrons to the total number of baryons. In the presence of weak interactions, the right hand side is non-zero to account for reactions where the number of electrons can change. Here, we express the net emissivity of a neutrino flavor (of energy ϵ) and its antineutrino by $\mathbb{S}_\epsilon$ and $\bar{\mathbb{S}}_\epsilon$, respectively. This expression is integrated over all neutrino energies and summed over all neutrino flavors f. The mean baryonic mass is given by m_b. Evolution of the internal energy of the matter is given by the gas-energy equation, Eq. (3), where E is the matter internal energy density and P is the matter pressure. Again, the right hand side of this equation is non-zero whenever energy is transferred between matter and neutrino radiation as a result of weak interactions. We note that it is also possible to substitute for Eq. (3) an expression for the evolution of the *total* matter energy (internal plus kinetic plus potential). Finally, Eq. (4) expresses gas-momentum conservation, where Φ is the gravitational potential, and P_ϵ and $\bar{P}_\epsilon$ are radiation-pressure tensors for each energy and flavor of neutrino and its anti-neutrino, respectively.

These equations must be discretized for solution within a computational framework. Traditionally, with one-dimensional models, it has been convenient to use Lagrangean methods, in which a computational mesh strictly co-moves with the mass elements of the fluid. With the advent of multi-dimensional models, however, it is common to use Eulerian hydrodynamics, where the mesh is fixed in an inertial frame of reference. This is because purely Lagrangean methods are difficult to implement in multi-dimensional schemes without the mesh suffering distortion and entanglement in convectively active regions.

For all the benefits of Eulerian meshes, they also present a number of thorny issues. This is especially true for spherical polar meshes, the most natural choice for supernova modeling. The most obvious issue is the coordinate singularity that exists when the polar angle, $\theta \to 0$. In addition, polar meshes exacerbate the problem of the timestep-restricting Courant-Friedrichs-Levy (CFL) condition at the center of the core. To deal with these issues, there are numerous resolutions and combinations of resolutions under active consideration. These include implicit methods, unstructured meshes, body-fitted meshes, and adaptive mesh refinement (AMR).

As mentioned above, it is also necessary to choose between a total energy and an internal energy formulation . For the supernova problem, an internal energy formulation, as given in Eq. (3), is preferred. This is because much of the energy is internal, as opposed to kinetic. Solving the gas-energy equation helps insure an accurate calculation of the entropy, which is critical in degenerate regimes where a small change in energy can lead to a large change in temperature.

The hydrodynamic algorithm must also have convergence properties that can deal with a realistic equation of state. This is particularly important in the regions of non-convex phase changes, such as the transition between nuclei and continuous nuclear matter.

Neutrino Transport. This component is the most difficult to implement in a supernova model and the most time-consuming computationally. This is because supernova neutrinos cannot, in general, be described by an equilibrium distribution function. A solution requires a complete phase-space description of each neutrino's position and momentum. To obtain such a solution, one must solve the six-dimensional Boltzmann Transport Equation or some reasonable approximation thereof. This extra dimensionality easily leads to the transport calculation completely dominating a simulation in terms of computer memory, execution time, and I/O requirements.

The Boltzmann Transport Equation (BTE) can be expressed in terms of the radiation intensity, $I = I(\epsilon, \mathbf{x}, \mathbf{\Omega}, t)$, where ϵ is the energy of a neutrino, $\mathbf{x}$ its position, and $\mathbf{\Omega}$ the solid angle into which the neutrino radiation is directed. In terms of I, the Newtonian BTE can be expressed as

$$\frac{1}{c}\frac{\partial I}{\partial t} + \mathbf{\Omega} \cdot \mathbf{\nabla} I + \sum_i a_i \frac{\partial I}{\partial p_i} = \left(\frac{\partial f}{\partial t}\right)_{\text{coll.}}, \tag{5}$$

where a_i is the ith component of the matter acceleration and p_i the ith component of the momentum of the neutrino. The right hand side of Eq. (5) lumps together the contributions from all interactions that a neutrino might experience and is collectively referred to as the collision integral.

A storm of issues faces one who implements a neutrino transport algorithm. Mezzacappa and Bruenn[18] have the only "full" solution to the BTE implemented in supernova simulations and then only with one-dimensional hydrodynamics. Upon moving to multi-dimensional models, the full solution of the BTE becomes yet more challenging to implement and more time-consuming to compute. However, this is the way that the field must ultimately go. (Livne and colleagues purport[17] to implement a two-dimensional S_n solution of the BTE. However, this solution omits critical matter-radiation coupling terms and no numerical details of the method have been disclosed.)

In the meantime, a number of approximate transport moment methods have emerged. The most successful of these is use of a finite series of angular moments of the BTE. When this approach is taken, a limiting scheme is then required to close the resulting equations. Breunn[2] implemented a P_1 scheme, with flux-limiting. Myra et $al.$[21] closed the zeroth angular moment of the BTE by implementing the Levermore and Pomraning flux limiter[16]. Bowers and Wilson[1] also used a flux-limiting scheme of their own device. More recently, Rampp and Janka[22] have implemented a variable-Eddington-factor approach to solve the first two angular moments of the BTE. In all the above cases, however, the implementation has been made in only one spatial dimension. (Janka et $al.$[12] are developing a two-dimensional Boltzmann transport code (MuDBaTH), but the numerical details are as yet unpublished.)

Since all the schemes noted so far derive monochromatic transport equations, yielding a separate equation for each neutrino energy, they are referred to as multi-group schemes. Those that combine multi-group and flux-limiting are known as multi-group flux-limited diffusion (MGFLD) schemes.

A much simpler alternative to multi-group schemes is so-call "grey" transport, which is derived by integrating the BTE over both neutrino energy and angle. To perform these integrals, one must assume a spectral shape for the neutrino distribution. Typically, this requires defining an arbitrary neutrino "temperature," and assuming that neutrinos can be parameterized by some kind of equilibrium Fermi-Dirac distribution. Among relatively recent models, this was first implemented by Cooperstein, van den Horn, and Baron,[7] and later by Swesty[29] and by Herant *et al.*[10] This approximation is still in active use by the latter group.[9]

Grey schemes have numerous shortcomings. First, work with multi-group schemes has shown that in areas where accurate neutrino transport is critical, neutrinos do not assume any kind of distribution that can parameterized once and for all as required by grey transport. Spectral distributions constantly evolve and, thus, a multi-group description is required to obtain even a qualitatively correct description. More troubling, Swesty[29] has shown that by adjusting the grey parameterization within very small bounds, it is possible to "dial" an explosion (or failed explosion) with the appropriate choices of these unknowable and unphysical tuning parameters. Hence, although grey codes have utility for making a sweeping exploration of parameter space, any scientific conclusions that rely on them should be viewed as highly suspicious, and not regarded as in any way definitive.

Regardless of which transport scheme is implemented, another critical issue that must be faced is matter-radiation coupling. Coupling occurs in Eqs. (2)-(4) for lepton number, energy, and momentum evolution. Coupling that occurs on the right-hand side of these equations has a conceptually simple analytic structure. However, momentum transfer is more troublesome in approximate schemes. This is because the radiation momentum equation is often truncated in a way that makes the accuracy of the calculated momentum transfer less certain.

Matter-radiation coupling also enters implicitly in the neutrino transport equation through spectral rearrangement terms and in the dynamic diffusion term. Both these terms are frequently and erroneously neglected in supernova models, even though the dynamic diffusion term is the leading order contribution in optically-thick regions.

Equation of State. Adequate modeling of stellar-core collapse requires an equation of state (EOS) that handles a density range of roughly $10^5 - 10^{15}$ g cm^{-3}, a temperature range of $0.1 - 25$ MeV, and an electron-fraction range of $0.0 - 0.5$. The EOS must also be able to handle different regimes of equilibrium states. Throughout most of the core, the material is in nuclear statistical equilibrium (NSE) and is usually modeled by one of the NSE equations of state. Although the gross features of nuclear matter are thought to be well-understood, there is still much open ground for investigation. Fertile regimes for such work include the EOS at supernuclear densities. In addition, little is known about the nature of nuclei at subnuclear densities when Y_e is small.

Matter in the silcon shell and beyond does not attain NSE until the bounce-shock wave passes through it. Dealing with the transition between NSE and non-NSE EOS's and with the network of nuclear reactions that joins them is a challenge that is only beginning to be addressed.[11]

Neutrino Microphysics. Since the energetics of a core-collapse supernova is primarily a neutrino phenomenon, it is necessary to have correct opacities and rates for the various neutrino processes that are important. The collection of reactions that are important, or possibly important, to the supernova problem is rich and has evolved through the years. Arguably the most important development came with the discovery of weak neutral currents, from which it could be inferred that the dominant contribution to neutrino opacity in a collapsing stellar core is from coherent elastic scattering of neutrinos from nuclei.[8]

The list of possible neutrino interactions is nearly endless, but those of demonstrated importance include the coherent scattering just mentioned, as well as conservative scattering from free nucleons. Also of undisputed importance are electron capture by protons (and protons bound in nuclei), neutrino production through electron-positron pair annihilation, and neutrino-electron scattering.

In recent years, with the experimental evidence pointing strongly to the existence of neutrino oscillations, it is also important to investigate the possible role of flavor-changing interactions to the supernova problem. Investigation into this has begun,[26] but has yet to be incorporated in a detailed simulation.

3. V2D: A New Code for Two-Dimensional Radiation Hydrodynamics

Our new radiation-hydrodynamic simulation code, V2D, is a two-dimensional, Newtonian, pure Eulerian, staggered-mesh code based on a modified version of the algorithm for ZEUS-2D by Stone and Norman.[23,24,25] Following Stone and Norman, it has been designed for use in a general orthogonal two-dimensional geometry, which makes its utility extend beyond the supernova problem.

V2D is an entirely new implementation, coded according to the Fortran 95 standard. It is a distributed-memory parallel code that uses calls to MPI-1 for message passing between processes. It has been designed for easy portability between computing platforms and currently runs on systems ranging from as small as a Linux-based laptop to as many as 2048 processors of an IBM SP. To aid in this portability, the input and output is formatted using parallel HDF5, which is built on the MPI-I/O portion of the MPI-2 standard.

One of the major design goals of V2D is componentization and, to adhere to this principle, we insist on completely separating microphysics from the numerical implementation of our radiation-hydrodynamics algorithm. This isolation of mathematics and computational science from physics has allowed significant contributions from applied mathematicians to enhancing the performance of our code.

The V2D algorithm relies on operator splitting, with advection steps split from source-term steps. Hydrodynamic and neutrino-transport source-term steps and coupling are interleaved. At the start of each simulation timestep, the gravitational mass interior to each point is calculated. Since the collapsed core is nearly spherically symmetric and highly condensed, we approximate the gravitational mass assuming that mass interior to the point of interest is in a spherically symmetric distribution. In future versions of our code, we will implement a more accurate Poisson solver to calculate the (slightly) non-spherical gravitational potential.

In a break from Stone and Norman's method, V2D next performs the advection sweeps in the radiation-hydrodynamic quantities (mass density, matter internal energy, velocities and momenta, electron fraction, and neutrino distributions). Following this, a neutrino transport step is performed for each flavor and each matter-radiation energy exchange is calculated.

The matter pressure is next updated, upon which the gravitational and matter- and radiation-pressure forces are applied to the matter. Artificial viscosity is calculated next and its contributions applied to the fluid. Finally, the gas-energy equation is solved.

This procedure is repeated for each timestep in a simulation, with the provision that advection sweeps are ordered alternately according to timestep (*i.e.*, x_1-direction first, followed by x_2, or *vice versa*).

3.1. *Neutrino Transport Implementation*

As an extension of earlier work by us,[21,30] we implement neutrino transport by taking the zeroth angular moment of the BTE to yield the following neutrino monochromatic energy equation in the co-moving frame:

$$\frac{\partial E_\epsilon}{\partial t} + \boldsymbol{\nabla} \cdot (E_\epsilon \mathbf{v}) + \boldsymbol{\nabla} \cdot \mathbf{F}_\epsilon - \epsilon \frac{\partial}{\partial \epsilon} (\mathsf{P}_\epsilon : \boldsymbol{\nabla} \mathbf{v}) = \mathbb{S}_\epsilon, \tag{6}$$

where E_ϵ is the neutrino energy density per unit energy interval at position $\mathbf{x}$ and time t, $\mathbf{F}_\epsilon$ is the neutrino energy flux per unit energy interval, and P_ϵ and $\mathbb{S}_\epsilon$ are as defined earlier. The expression $\mathsf{P}_\epsilon : \boldsymbol{\nabla}\mathbf{v}$ indicates contraction in both indices of the second-rank tensors P_ϵ and $\boldsymbol{\nabla}\mathbf{v}$. There is a corresponding equation to describe the antineutrinos. This pair of equations is repeated for each neutrino energy ϵ, and neutrino flavor. We currently track electronic, muonic, and tauonic neutrinos.

Equation (6) is closed using Levermore and Pomraning's prescription for flux-limited diffusion,[16] which allows us to express $\mathbf{F}_\epsilon$ as

$$\mathbf{F}_\epsilon = -D_\epsilon \boldsymbol{\nabla} E_\epsilon, \tag{7}$$

where D_ϵ is a "variable" diffusion coefficient that varies in such a way as to yield the correct fluxes for the diffusion and free streaming limits and an approximate solution in the intermediate regime. This prescription also provides the elements of the radiation-pressure tensor P_ϵ.

Presently, we employ the same prescriptions for electron capture and conservative scattering that we have used in the past.[21,2] The rates for these process are calculated on the fly within the course of a simulation. Our model also implements neutrino production via electron-positron pair annihilation, as in Yueh and Buchler[34] and Bruenn[2]. Neutrino-electron scattering has been implemented, but is not currently turned on in the preliminary results we present here. These latter two sets of processes use tables of precomputed rates, which are interpolated via a tri-linear interpolation scheme over neutrino energy ϵ, temperature T, and electron chemical potential μ_e.

Since the neutrino CFL restriction on a transport timestep is far too restrictive to permit an explicit solution, we use a purely implicit method to solve the transport. The equations comprising the description of each neutrino-antineutrino species are assembled in matrix form. We note that the second (advective) term in Eq. (6) is omitted from this process since it has been already treated during the operator splitting of the advective step described above. Blocking terms arising from Fermi-Dirac statistical restrictions on final neutrino states make this a system of non-linear equations. Fortunately, the system is sparse, which makes it amenable to solution by sparse iterative methods. A nested procedure is used, employing Newton-Krylov methods[30]. In the innermost loop, a linearized system is solved using preconditioned Krylov-subspace methods. The outer loop uses a Newton-Raphson iterative scheme to resolve the non-linearity of the system. Besides being an effective general procedure for sparse systems, our implementation of parallel preconditioners also insures that is amenable to large-scale solution on parallel architectures. This is the chief reason our code exhibits its high degree of scalability across many platforms.

3.2. *Equation of State*

V2D is designed to use an arbitrary equation of state and we use several in the course of testing the code. For production runs, however, we use the Lattimer-Swesty EOS[15,14] in tabular form. The thermodynamic quantities are tabulated in terms of independent variables, density, ρ, temperature, T, and electron fraction, Y_e. We have tabulated this EOS in a thermodynamically consistent way according to the prescription in Swesty[28]. (We refer to this combination collectively as LS-TCT.) We note that although the Lattimer-Swesty EOS is commonly used, and tabulations of it are also common, most tabulations are not constructed in such a way as to *guarantee* thermodynamic consistency. When non-thermodynamically-consistent tables are used, spurious entropy can be generated or lost. Such problems have been sometimes incorrectly attributed to the Lattimer-Swesty EOS, rather than erroneous tabulation of the otherwise consistent EOS.

To guarantee tabular thermodynamic consistency, LS-TCT uses biquintic Hermite interpolation in the Helmholtz free energy F, as a function of T and ρ. Functional dependence on Y_e changes slowly enough to permit linear interpolation. This procedure is required since satisfaction of the Maxwell relations requires consistency among the second derivatives of F. In addition, we desire fidelity of the interpolation and continuity of derivatives to the underlying tabular data for both F and its derivatives, $(\partial F/\partial T)_\rho$ and $(\partial F/\partial \rho)_T$.

Apart from thermodynamic considerations, we also want to insure that there are no discontinuities that might cause difficulties in the hydrodynamics. Hence, LS-TCT also insists that interpolations of the second derivatives of F approach the correct tabulated values. (This is the equivalent of saying that we require the derivatives of pressure and internal energy with respect to temperature and density be continuous.)

A final requirement is that we wish the interpolation function and its first and second derivatives be continuous across table-cell boundaries. This insures that nothing untoward happens as a fluid element migrates from one thermodynamic regime of interest to another.

3.3. *Software Validation and Verification*

When engaged in a major project, such as V2D, it is important for software developers to be constantly vigilant about the quality of software being produced. It is important that the software meet the requirements that it is intended to address (validation) and that the code yields correct answers (verification).

We take these issues seriously and, in an effort to address them, have implemented strict source-code control and testing procedures to ensure that our software meets our rigorous standards. One of the most important elements of our program has been the implementation of a suite of regression tests, which we currently run four times daily. This is not a static suite, but is constantly growing. Our eventual aim is to cover every major element of code in V2D. Currently our suite consists of about two dozen separate problems that include tests of the hydrodynamics, neutrino transport, parallel solvers, message passing, and parallel I/O. Wherever possible, we try to include problems with analytic or at least verifiable solutions.

Although implementation of these procedures is labor intensive and time-consuming, we feel it is a justifiable investment. With current regression tests, our procedures have already been effective in finding errors in our code. They have also served as an effective safeguard against introducing new errors as we continually enhance V2D's functionality.

4. Initial Results

We are using V2D to carry out our first 2-D multigroup models of the post-bounce epoch. To date, simulations have not reached a sufficient time that would allow us to be certain about whether an explosion is obtained. Nevertheless, the results warrant some discussion as they reveal important features of the post-bounce epoch.

4.1. *Initial Model*

For a progenitor model we employ the widely-used Woosley and Weaver[33] S15S7B2 $15M_\odot$ progenitor. Much previous work has focused on the evolution of this progenitor through collapse, core bounce, and convective phases. The Fe core, the Si shell, and a portion of the O shell area are zoned into a 256 radial-mass-zone mesh with zoning that is tuned (by trial and error) so as to yield a high spatial resolution grid in the proto-neutron star and the inner 200 km of the collapsed core at bounce. This tuned zoning sets up a radial grid that is compatible with subsequent 2-D Eulerian simulations. The neutrino-energy spectrum, ranging from 0–375 MeV, is discretized into 20 energy groups with group widths that increase geometrically with energy so as to resolve accurately the Fermi surface of the electrons and neutrinos in the proto-neutron star. The initial values for T, ρ, and Y_e, are interpolated from the original S15S7B2 data onto the Langragean mass grid and the initial radial coordinates of each mass shell are computed consistently with density. The neutrino energy densities E_ν are initialized to a small non-zero value that yield an initial neutrino luminosity that is many orders of magnitude below what the precollapse thermal pair-production luminosity would be. When the simulation is started in its pre-collapse quasi-static phase, the luminosity stabilizes within a light crossing time (5-10 ms).

4.2. *Lagrangean Collapse Calculations*

The initial model is collapsed using a 1-D Newtonian Lagrangean radiation hydrodynamics code RH1D that uses the mass and energy meshes described above. The evolution algorithm for each timestep utilizes operator splitting to first carry out a Lagrangean hydrodynamics step followed by Lagrangean neutrino evolution steps for each of the three neutrino flavors. After each Lgrangean neutrino evolution step the matter internal energy and electron fraction are corrected for any energy and lepton number exchange that has occurred.

The model contains neutrino microphysics as described in Bruenn[2] with two exceptions. The neutrino-nuclei scattering opacity has been modified to take into account the form-factor introduced by Burrows, Mazurek, and Lattimer[6] and we have neglected the effect of neutrino/anti-neutrino annihilation. Full neutrino-electron scattering as described in this paper is included in the code but is not turned on in the model described in this paper. Additional effects such as nucleon recoil, ion-ion correlations, etc. are being considered as follow-ons to this baseline model.

The EOS utilized is the TCT tabularized version of the nuclear EOS Lattimer-Swesty (LS-TCT) with the $k = 180$ MeV parameter set, with the exception of the electron EOS that has been updated to improve accuracy and the range of applicability. The original LS EOS chose a value for the alpha particle binding energy $B_\alpha = 28.3$ MeV that did not correctly account for the neutron-proton mass difference. This parameter has been subsequently corrected but in the model presented in this paper, the original value has been retained so as to allow comparison to other work.

Using RH1D with the microphysics and meshes described above the core collapses in approximately 233 ms. The central conditions at bounce are approximately $T \approx 10.3$ MeV, $\rho \approx 2.71 \times 10^{14}$ g cm^{-3}, $Y_e \approx 0.306$, and $Y_\ell \approx 0.37$. These conditions are in good agreement with those obtained in Lagrangean MGFLD and MGBT models [2,21,27,19].

4.3. *Eulerian 2-D Calculations*

Our 2-D models are carried out in spherical polar coordinates. The initial conditions for our 2-D simulations are taken from the 1-D Langrangean simulations as the central density of the core reaches the nuclear saturation density. The T, ρ, and Y_e profiles at this point are shown in Figure 1 and are taken at approximate time of 233.087 milliseconds. The choice of this epoch to begin our 2-D models was made for two reasons. First, the prompt shock propagates and stalls into a quasi-static equilibrium on the Eulerian mesh. We have found that if we attempt the transition from a Lagrangean algorithm to an Eulerian algorithm at the point where the shock has stalled, there will be noticeable differences in the quasi-static equilibrium caused by the presence of the nuclear "pasta" phase transition. These differences in the equilibrium point can set off spurious unphysical shocks. By choosing to let the prompt shock propagate a stall on an Eulerian mesh, we avoid this problem. The second reason for choosing the initial point for our 2-D models near the moment of bounce is that the radial coordinates of the zones in the outer part have achieved desirable values for an Eulerian simulation.

The radial zoning for the 2-D models is taken directly from the radial coordinates of the 1-D Langrangean model zones. In this way we avoid any need for remapping of data between radial zones. This eliminates the introduction of any spurious forces into the initial conditions for the 2-D models.

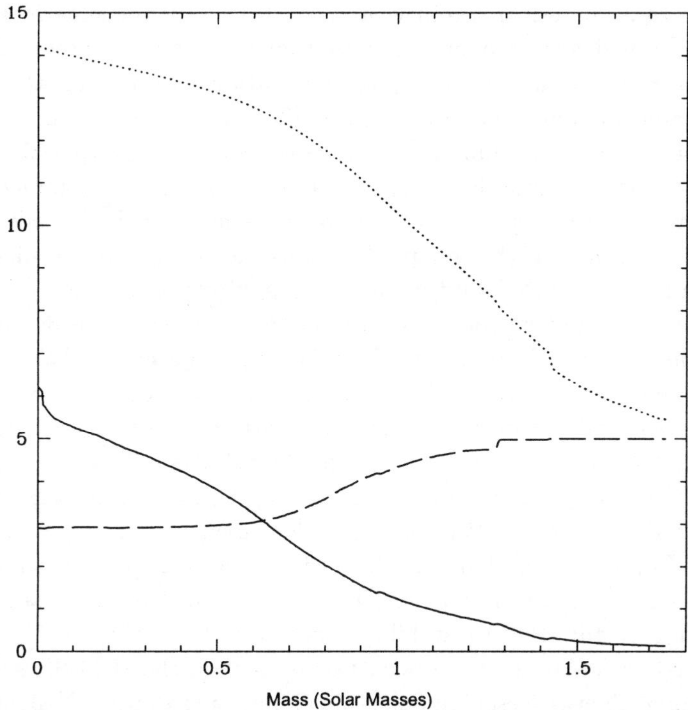

Figure 1. Initial radial profile for 2-D simulation. Solid line indicates temperature in units of MeV. Dashed line indicates $10 \times Y_e$. Dotted line indicates $Log10(\rho)$ in units of g/cm^3.

The initial data for T, ρ, Y_e, v_r, and the neutrino energy densities E_ν and $\bar{E}_\nu$ for each neutrino flavor are taken directly from the 1-D Langrangean model. The initial velocity in the θ direction is set to zero. The data are mapped in the polar-angular direction in a spherically symmetric fashion. The angular grid consists of 256 zones uniformly spaced in angle over the range of $0 \leq \theta \leq \pi$. We place a small sinusoidal perturbation in the electron fraction Y_e to seed convection in the region between 100 and 200 km of the form $(Y_e)_{\mathrm{perturb}} = (Y_e)_{\mathrm{Lagrangean}} + C_p \sin(4\theta)$ where $C_p = 10^{-6}$. The energy group structure is also left unchanged from the 1-D Langrangean runs so there is no need to remap the data in the energy dimension. In the remainder of this paper we shall refer to this model as Production Run 37 (PR37).

The use of spherical polar coordinates in combination with explicit hydrodynamic algorithms to model spatial domains that include the origin gives rise to a numerical stability problem. At $r = 0$ the spherical coordinate system is degenerate and all zones that include the origin as a vertex should be in instantaneous sonic communication with one another. However, standard explicit numerical finite-difference, finite-volume, or finite-element techniques are limited to nearest-neighbor type spatial coupling and do not include numerical coupling between all zones containing a given vertex. In order to circumvent this problem we numerically introduce "baffles" into the center of the collapsed core, as though it were a tank of fluid, to prevent movement of fluid in the angular direction inside a certain radius. Since no fluid movement occurs in the θ direction inside the baffle radius, there is no CFL restriction based on the zone size in the θ direction for zones inside that radius. Nevertheless, the zones on either side of the baffle are sonically connected as sound waves flow around the outer edge of each baffle. We strive to keep the baffle radius small, so that the flow in the θ direction remains unimpeded in any region where convective instabilities may develop. For the mode described in this paper the baffle radius is approximately 8.5 km which yields an average CFL timestep of about 5×10^{-7} s. As we will see, this baffle radius is well inside of any proto-neutron star (PNS) instability region.

The 2-D calculations have been carried out on the IBM-SP system at the National Energy Research Scientific Computing Center (NERSC). The models are run on 1024 processors with parallelism handled via message passing via calls to MPI libraries. Model PR37 required approximately 50,000 processor-hours of CPU time to reach a simulation time of 16 ms post-bounce.

4.4. *The Onset of Convection*

The 2-D models are carried out from the point of bounce. As expected the prompt shock weakens while propagating outward and finally stalls. In the 2-D models, the radius of the shock at 5 ms post-bounce is near 60 km and propagating outwards very slowly. This is in good agreement with our 1-D Lagrangean models.

One difference that we see from previous works including the grey models carried out by Swesty[29] is that convective instabilities are born much earlier when the 2-D models are initialized near the point of core bounce. By 10 ms after bounce, the model has developed two separate unstable layers. The outer layer, shown in Fig. (2), is the classic entropy driven Rayleigh-Taylor convection. The inner layer, shown in Fig. (3), seems to

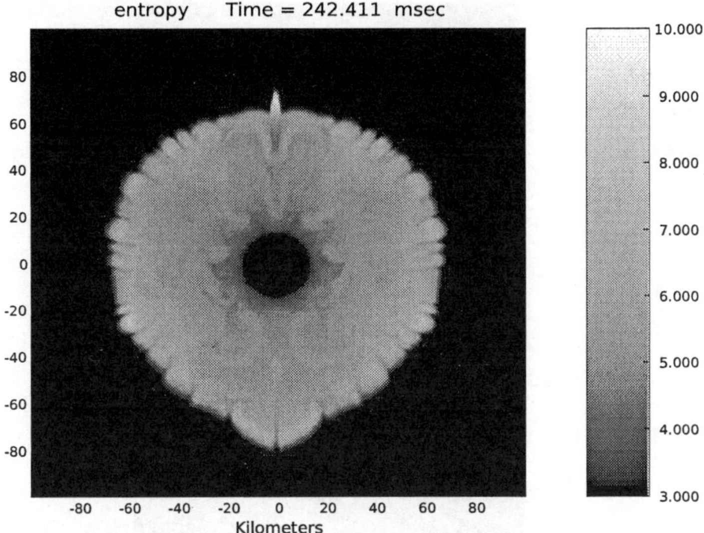

Figure 2.   The entropy per baryon for model PR37 at approximately 9 milliseconds after core bounce.

be an instability in the outer layers of the proto-neutron star similar to those seen in 2-D simulations carried out by by Keil *et al.*[13] and Mezzacappa *et al.*[20].

There is controversy about the existence of PNS instabilities. Originally, the work by Wilson and Mayle[32,31] claimed to find doubly-diffusive instabilities in the region below the neutrinosphere. Later work by Breunn and collaborators[3,4,5] cast doubt on the existence of such phenomena. The simulations of Keil,[13] which utilized the grey flux-limited diffusion approximation to transport neutrinos along radial rays, found a PNS instability that grew over the time period of approximately one second to encompass the entire proto-neutron star. In contrast, the subsequent work of Mezzacappa *et al.*[20] found a PNS instability that quickly damped out within a short time. The Mezzacappa *et al.* simulation utilized a multigroup flux-limited P_1 approximation to transport neutrinos along radial rays outward from an inner radius of r=20 km. The inner boundary conditions in this simulation were established as time-dependent data from the 1-D Langrangean code of Bruenn[2]. It is important to note that neither the simulations of Keil *et al.*[13] or Mezzacappa *et al.*[20] took into account the fully

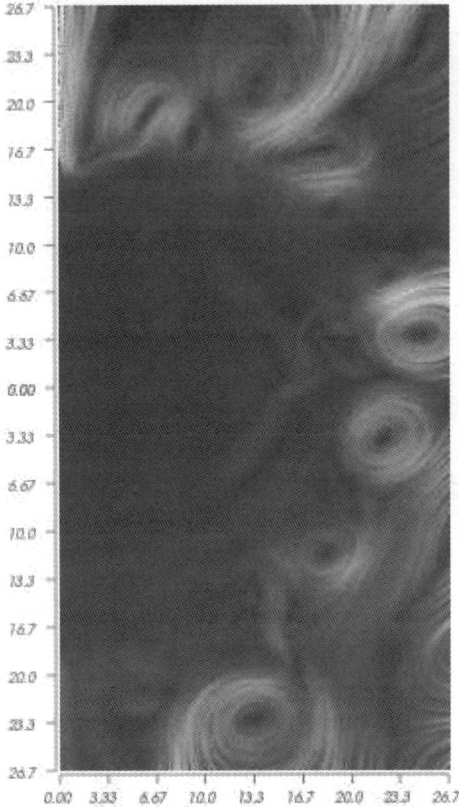

Figure 3. The velocity structure of the PNS instability layer in model PR37 at t=242.411 ms. The image depicts the instantaneous structure of the velocity field by means of a texture map visualization technique known as Lagrangean-Eulerian Advection (LEA).

radiation-hydrodynamic coupling via the compression and dynamic diffusion terms that are present in Eq. (6). Our simulations have confirmed the expected result that the effects of the dynamic diffusion term dominate the radiative diffusion term in the regions in which the optical depth is large.

Figure 3 shows the instantaneous structure of the velocity field in model PR37 at the same time as the data shown in Fig. (2). The velocity field is illustrated by means of a Lagrangean-Eulerian Advection (LEA) visualization technique that shows the direction of the vectors as streaks. One can clearly see eddies associated with the PNS instability layer at a radius of about 20–25 km. This is well outside the baffle radius of 8.5 km. In fact, there are approximately a minimum of 40 radial zones separating the innermost of the vortices and the outer edge of the baffles. We do not believe that the baffles in any way impede the dynamics of the vortices. Nevertheless other simulations are underway where the baffle radius is made substantially smaller to verify this claim. We are also developing an implicit hydrodynamic algorithm that will avoid the need for baffles altogether.

Whether the PNS instability will grow or diminish with time is as yet unclear since we have only evolved model PR37 to a time of approximately 16 ms at the time of this writing. The velocity structure at that time

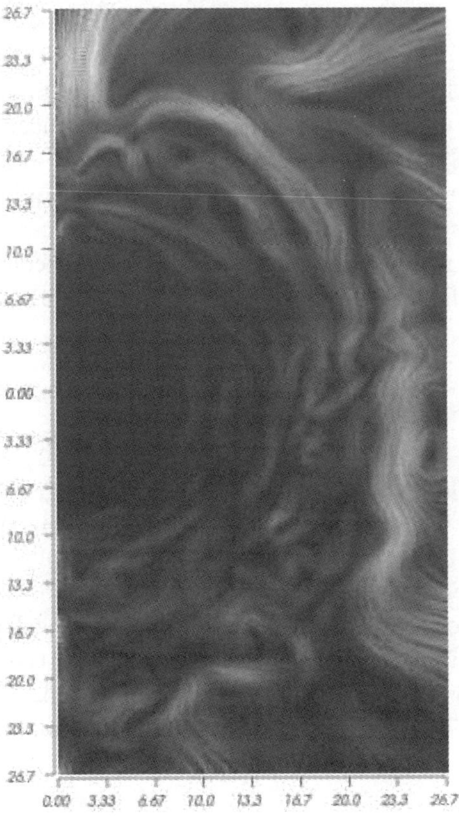

Figure 4. The velocity structure of the PNS instability layer in model PR37 at t=249.177 ms.

is shown in Fig. (4), which clearly reveals that much coherency has been lost in the vorticial structure of the PNS instability layer. This seems indicative of the decay of this PNS instability in a fashion similar to that described by Mezzacappa *et al.*[20]. However, it is necessary to evolve this simulation substantially farther in time before any definitive statements can be made about the long-term behavior of this sector of the proto-neutron star. During this relatively short timescale the outer convective zone seen in Fig. (2) does not exhibit significant growth. With evolution to the next 30 ms, we should be able to make comparative statements regarding model PR37 and earlier grey models carried out by Swesty[29].

5. Conclusions and Future Directions

The issues and work described in this paper fall far short of offering complete coverage of the active issues that remain in the area of the explosion mechanism of core collapse supernovae. Indeed, we have ignored many important issues such as magnetic fields, rotation, and neutrino flavor mixing. Clearly there is a large gulf of unexplored physics incorporated in those subjects.

Our own future efforts will be focused in two areas in the near future. The first of these is understanding the effects of numerous microphysics enchancements that will be added to the models. The second of these efforts involves extending the models to 3-D where a more realistic convective flow structure can arise.

Acknowledgments

The authors would like to thank Ed Bachta and Polly Baker of Indiana University at Indianapolis for their collaborative efforts in developing the visualization technology that went into Figures 3 and 4. We gratefully acknowledge the support of the U.S. Dept. of Energy, through SciDAC Award DE-FC02-01ER41185, by which this work was funded. We are also grateful to the National Energy Research Scientific Computing Center (NERSC) for computational support. Finally, we would like to thank the National Institute for Nuclear Theory at the University of Washington for its hospitality in hosting the workshop "Open Issues in Core-Collapse Supernovae," at which this work was presented in June 2004.

References

1. R. L. Bowers and J. R. Wilson, *Astrophys. J. Supp.*, **50**, 115 (1982).

2. S. W. Bruenn, *Astrophys. J. Supp.*, **58**, 771 (1985).

3. S. W. Bruenn and A. Mezzacappa, *Astrophys. J.*, **433**, L45 (1995).

4. S. W. Bruenn, A. Mezzacappa and T. Dineva, *Phys. Rep.*, **69**, 256 (1995).

5. S. W. Bruenn and T. Dineva, *Phys. Rep.*, **458**, L71 (1996).

6. A. Burrows, Mazurek, T. J. and J. M. Lattimer, *Astrophys. J.*, **251**, 325 (1981).

7. J. Cooperstein, L. J. van den Horn and E. Baron, *Astrophys. J.*, **309**, 653 (1986).

8. D. Z. Freedman, *Phys. Rev. D.*, **9**, 1389 (1974).

9. C. L. Fryer and M. S. Warren, *Astrophys. J.*, **601**, 391 (2004).

10. M. Herant, W. Benz, R. Hix, C. L. Fryer and S. A. Colgate, *Astrophys. J.*, **435**, 339 (1994).

11. R. Hix, personal communication, 2004.

12. H.-T. Janka, R. Buras, K. Kifonidis, T. Plewa and M. Rampp, in *From Twilight to Highlight: The Physics of Supernovae*, eds. W. Hillebrandt and B. Leibundgut, Springer, Berlin, 2003.

13. W. Keil, H.-T. Janka, and E. Muller, *Astrophys. J.*, **473**, L111 (1996).

14. Lattimer, J. M., C. J. Pethick, D. G. Ravenhall, and D. Q. Lamb, *Nucl. Phys. A*, **432**, 646 (1985).

15. Lattimer, J. M. and F. D. Swesty, *Nucl. Phys. A*, **535**, 331 (1991).

16. C. D. Levermore and G. C. Pomraning, *Astrophys. J.*, **248**, 321 (1981).

17. E. Livne, A. Burrows, R. Walder, I. Lichtenstadt and T. Thompson, *Astrophys. J.*, **609**, 277 (2004).

18. A. Mezzacappa and S. W. Bruenn, *Astrophys. J.*, **405**, 669 (1993).

19. A. Mezzacappa and S. W. Bruenn, *Astrophys. J.*, **405**, 685 (1993).

20. A. Mezzacappa, A. C. Calder, S. W. Bruenn, J. M. Blondin, M. W. Guidry, M. R. Strayer and A. S. Umar, *Astrophys. J.*, **493**, 848 (1998).

21. E. S. Myra, S. A. Bludman, Y. Hoffman, I. Lichtenstadt, N. Sack and K. A. Van Riper, *Astrophys. J.*, **318**, 744 (1987).

22. M. Ramp and H.-T. Janka, *Astron. & Astrophys.*, **396**, 361 (2002).

23. J. M. Stone and M. L. Norman, *Astrophys. J. Supp.*, **80**, 753 (1992).

24. J. M. Stone and M. L. Norman, *Astrophys. J. Supp.*, **80**, 791 (1992).

25. J. M. Stone, D. Mihalas and M. L. Norman, *Astrophys. J. Supp.*, **80**, 819 (1992).

26. P. Strack and A. Burrows, *Phys. Rev. D.*, **71**, 093004 (2005).

27. F. D. Swesty, J. M. Lattimer and E. S. Myra, *Astrophys. J.*, **425**, 195 (1994).

28. F. D. Swesty, *J. Comp. Phys.*, **127**, 118 (1996).

29. F. D. Swesty, in *Stellar Evolution, Stellar Explosions and Galactic Chemical Evolution*, ed. A. Mezzacappa, Institute of Physics Publishing, Bristol, p. 539, (1998).

30. F. D. Swesty, D. C. Smolarski and P. E. Saylor, *Astrophys. J. Supp.*, **153**, 369 (2004).

31. J. R. Wilson and R. W. Mayle, *Phys. Rep.*, **227**, 97 (1983).

32. J. R. Wilson and R. W. Mayle, *Phys. Rep.*, **163**, 63 (1988).

33. S. E. Woosley and T. A. Weaver, *Astrophys. J. Supp.*, **101**, 181 (1995).

34. W. R. Yueh and J. R. Buchler, *Astrophys. & Space Sci.*, **39**, 429 (1976).

THE LONG TERM: SIX-DIMENSIONAL CORE-COLLAPSE SUPERNOVA MODELS

C. Y. CARDALL[1,2], A. O. RAZOUMOV[1,2,3], E. ENDEVE[1,2,3], AND
A. MEZZACAPPA[1]

[1] *Physics Division,*
Oak Ridge National Laboratory,
Oak Ridge, TN 37831-6354, USA

[2] *Department of Physics and Astronomy, University of Tennessee, Knoxville,*
TN 37996-1200, USA

[3] *Joint Institute for Heavy Ion Research,*
Oak Ridge National Laboratory,
Oak Ridge, TN 37831-6374, USA

The computational difficulty of six-dimensional neutrino radiation hydrodynamics
has spawned a variety of approximations, provoking a long history of uncertainty in
the core-collapse supernova explosion mechanism. Under the auspices of the Terascale Supernova Initiative, we are honoring the physical complexity of supernovae
by meeting the computational challenge head-on, undertaking the development
of a new adaptive mesh refinement code for self-gravitating, six-dimensional neutrino radiation magnetohydrodynamics. This code—called *GenASiS*, for *Gen*eral
*A*strophysical *Si*mulation *S*ystem—is designed for modularity and extensibility of
the physics. Presently in use or under development are capabilities for Newtonian self-gravity, Newtonian and special relativistic magnetohydrodynamics (with
'realistic' equation of state), and special relativistic energy- and angle-dependent
neutrino transport—including full treatment of the energy and angle dependence
of scattering and pair interactions.

1. The Challenges of Core-collapse Supernovae

In taking stock of 'long-term' efforts to understand core-collapse supernovae, we reflect upon the fact that supernovae have been challenging us
for centuries. Their very existence helped overturn worldviews. Their explosion mechanisms and remnants involve all four fundamental forces, and
many (if not most) branches of physics. A cornucopia of electromagnetic radiation observables continues to provide intriguing puzzles, and yields some

clues regarding the violent proceedings of a massive star's death. But direct observational penetration of the secrets of neutron star birth and initiation of the explosive ejection of the stellar envelope demands extraordinary efforts aimed at the detection of gravitational waves and neutrinos, the only messengers carrying direct information from the extreme conditions of a newly-collapsed stellar core. And the associated theoretical penetration—required for both the prediction and interpretation of expected gravitational wave and neutrino signals—comprises algorithmic and computational issues that will challenge computational physicists and tax state-of-the-art super-computers for years to come.

In western civilization, supernovae played a role in changing prevailing notions of the universe in at least two eras. Remarkably, of the handful of supernovae in our Milky Way Galaxy recorded by humanity, two were observed by Tycho (1572) and Kepler (1604). Tycho's detailed observations established that the 'new and never previously seen star' of 1572—and also another transient celestial phenomenon, a comet of 1577—were beyond the moon's orbit, 'new phenomena in the ethereal world,' contributing to the overthrow of the Aristotelian worldview that included immutable heavens. In modern times, supernovae figured in the debate over whether the spiral nebulae were separate galaxies, each an 'island universe' comparable to our Milky Way. It was recognized that the 'novae' or 'new stars' seen in these nebulae would have to be much more luminous than typical novae occuring in our galaxy. The phrases "giant novae," novae of "impossibly great absolute magnitudes," "exceptional novae," and the German term "Hauptnovae" or "chief novae" were used during the 1920s.[1] In a review article, Zwicky explained that it was deduced that 'supernovae' were about a thousand times as luminous as 'common novae,' and claimed that "Baade and I first introduced the term 'supernovae' in seminars and in a lecture course on astrophysics at the California Institute of Technology in 1931."[2]

Supernovae are classified by astronomers into two broad classes based on their optical spectra.[3] These classes are 'Type I,' which have no hydrogen features, and 'Type II,' which have obvious hydrogen features. These types have further subcategories, depending on the presence or absence of silicon and helium features in Type I, and the presence or absence of narrow hydrogen features in the case of Type II. In particular, supernovae of Type Ia exhibit strong silicon lines, those of Type Ib have helium lines, and those of Type Ic do not have either of these. Astronomers have also identified a number of distinct characteristics in supernova light curves (total luminosity as a function of time).

There are two basic physical mechanisms for supernovae, but these do not line up cleanly with the observational categories of Type I and Type II. Type Ia supernovae are caused by a thermonuclear runaway that consumes an entire white dwarf, thought to be induced by accretion of matter from a companion star. Supernovae of Type Ib, Ic, and II are produced by a totally different mechanism: the catastropic collapse of the core of a massive star. The observational distinctions of presence or absence of hydrogen or helium turn out to be unrelated to the mechanism; they depend on whether the outer hydrogen and helium layers of the star—which have nothing to do with the collapsing core—have been lost to winds or accretion onto a binary companion during stellar evolution. Of the two physical mechanisms, core-collapse supernovae are the focus of the present discussion.

We now consider the core-collapse supernova process in more detail. Shortly after the discovery of the neutron in the early 1930s, Baade and Zwicky declared, "With all reserve we advance the view that supernovae represent the transitions from ordinary stars to *neutron stars,* which in their final stages consist of extremely closely packed neutrons."[4] This turned out to be true, at least for some 'core-collapse' supernovae (those of Type Ib/Ic/II); a black hole is another possible outcome. The dominant fleshing-out of the core collapse process in the last two decades[a] has been the delayed neutrino-driven explosion mechanism.[6,7]

A core-collapse supernova results from the evolution of a massive star. For most of their existence, stars burn hydrogen into helium. In stars at least eight times as massive as the Sun ($8\ M_\odot$), temperatures and densities become sufficiently high to burn through carbon to oxygen, neon, and magnesium; in stars of at least $\sim 10\ M_\odot$, burning continues through silicon to iron group elements. The iron group nuclei are the most tightly bound, and here burning in the core ceases.

The iron core—supported by electron degeneracy pressure instead of gas thermal pressure, because of cooling by neutrino emission from carbon burning onwards—eventually becomes unstable. Its inner portion undergoes homologous collapse (velocity proportional to radius), and the outer portion collapses supersonically. Electron capture on nuclei is one instability leading to collapse, and this process continues throughout collapse, producing neutrinos. These neutrinos escape freely until densities in the collapsing core become so high that even neutrinos are trapped.

Collapse is halted soon after the matter exceeds nuclear density; at

---

[a]See for example Ref. [5] for some information on earlier views of the mechanism.

this point ("bounce"), a shock wave forms at the boundary between the homologous and supersonically collapsing regions. The shock begins to move out, but after the shock passes some distance beyond the surface of the newly-born neutron star, it stalls as energy is lost to neutrino emission and endothermic dissociation of heavy nuclei falling through the shock.

It is natural to consider neutrino heating as a mechanism for shock revival, because neutrinos dominate the energetics of the post-bounce evolution. Initially, the nascent neutron star is a hot thermal bath of dense nuclear matter, electron/positron pairs, photons, and neutrinos, containing most of the gravitational potential energy released during core collapse. Neutrinos, having the weakest interactions, are the most efficient means of cooling; they diffuse outward on a time scale of seconds, and eventually escape with about 99% of the released gravitational energy.

Because neutrinos dominate the energetics of the system, a detailed understanding of their evolution will be integral to definitive accounts of the supernova process. If we want to understand the origin of the explosion with energy $\sim 10^{51}$ erg, we cannot afford to lose (or gain) more than this amount during the period covered by the simulation. This requires careful accounting of the neutrinos' much larger contribution to the system's energy budget. (For further discussion, and a review of work recognizing the importance of this point, see Ref. [8]).

What sort of computation is needed to follow the neutrinos' evolution? Deep inside the newly-born neutron star, the neutrinos and the fluid are tightly coupled (nearly in equilibrium); but as neutrinos are transported from inside the neutron star, they go from a nearly isotropic diffusive regime to strongly forward-peaked free-streaming. Heating behind the shock occurs precisely in this transition region, and modeling this process accurately requires tracking both the energy and angle dependence of the neutrino distribution functions at every point in space.

A full treatment of this six-dimensional neutrino radiation hydrodynamics problem is a major challenge, too costly for contemporary computational resources. While much has been learned over the years through simulation of model systems of reduced dimensionality, there is as yet no robust confirmation of the delayed neutrino-driven scenario described above (see Sec. 2).

Recent detections of a handful of unusually energetic Type Ib/c supernovae (often called 'hypernovae') in connection with gamma-ray bursts pose additional challenges to theory and observation. Prominent examples of this supernova/gamma-ray burst connection include SN1998bw /

GRB980425,[9,10] SN2002lt / GRB021211,[11] SN2003dh / GRB030329,[12,13] and SN2003lw / GRB031203;[14,15,16,17] there are probably many others (see, for example, Ref. [18]). Like many gamma-ray bursts without direct evidence for a supernova connection, GRB030329 has evidence of a jet; GRB980425 and GRB031203 do not, and are also underluminous gamma-ray bursts (but still unusually energetic Type Ib/c supernovae).[19,20] Determining the relative rates of jet-like hypernovae, non–jet-like hypernovae, and 'normal' supernovae—and the possible associated variety of mechanisms—are important challenges.

In summary, the details of how the stalled shock is revived sufficiently to continue plowing through the outer layers of the progenitor star are unclear. In normal supernovae, it may well be that some combination of neutrino heating of material behind the shock, convection, and instability of the spherical accretion shock leads to the explosion (see Sec. 2). It is tempting to think that rotation (for example, Refs. [21,22]) and magnetic fields (for example, Ref. [23]) in more massive progenitors may play a more significant role in the rare jet-like hypernovae, perhaps giving birth to 'magnetars,' the class of neutron stars with unusually large magnetic fields. This temptation appears to be sweetened by observational support.[24,25] (Observations of two nearby supernova remnants may suggest that rotation and magnetic fields also operate in normal supernovae, perhaps subdominantly.[26])

From the above discussion, several key aspects of physics that a core-collapse simulation must address can be identified; these are discussed in sections that follow, after a discussion of the history of approximate treatments of neutrino radiation transport and an overview of our new code.

2. History of Neutrino Radiation Hydrodynamics

While in general terms supernovae have been challenging us for centuries, the challenge of their simulation via computer modeling has 'only' been with us for a few decades—a 'short term' in comparison with centuries, but still a 'long term' in comparison with the time scales of individual academic careers.

Here we sketch the last two decades' progress on one critical aspect of core-collapse supernova simulations: the high dimensionality (three space and three momentum space dimensions—not to mention time dependence) of neutrino radiation hydrodynamics (see Table 1). The development of this aspect of the simulations is intertwined with important advances in the field, but of course does not represent every insight relevant to the explosion mechanism obtained via simulation or otherwise.

We pick up the story in 1982, when simulations showing the stalled shock reenergized by neutrino heating on a time scale of hundreds of milliseconds were first performed.[6] This was initially achieved in a simulation with a total of 2 dimensions (spherical symmetry, and energy-dependent neutrino transport). But these simulations required significant rezoning, possibly attended by nontrivial numerical error;[6] and further, with the introduction of full general relativity and a correction in an outer boundary condition,[34] it became clear that these models would not explode without a mock-up of a doubly-diffusive fluid instability in the newly-born neutron star that serves to boost neutrino luminosities[34,35,36,37]—a simulation of effective total dimensionality "2.5" (see Table 1). That the necessary conditions exist for this particular instability to operate has been disputed;[41,62,63] and though related phenomena may operate,[63] more recent simulations with energy-dependent neutrino transport and true two-dimensional fluid dynamics indicate that fluid motions are either suppressed by neutrino transport[42] or have little effect on neutrino luminosities and supernova dynamics.[59,61]

Recognizing that the profiles obtained in spherically symmetric simulations implied convective instabilities, and that observations of supernova 1987A also pointed to asphericities, several groups explored fluid motions in two spatial dimensions in the supernova environment in the 1990s. In two spatial dimensions, the computational limitations of that era required approximations that simplified the neutrino transport.

One class of simplifications allowed for neutrino transport in "1.5" or 2 spatial dimensions, but with neutrino energy and angle dependence integrated out, reducing a five dimensional problem to "1.75" or 2 effective total dimensions (see Table 1).[37,38,39] These simulations exhibited explosions, and elucidated an undeniably important physical effect: a negative entropy gradient behind the stalled shock results in convection that increases the efficiency of heating by neutrinos. However, in the scheme of Table 1, the inability to track the neutrino energy dependence in these simulations could be viewed as a minor step backwards in effective total dimensionality. The energy dependence of neutrino interactions has the important effect of enhancing core deleptonization, which makes explosions more difficult;[30,64,65] this raised the question of whether the exploding models of the early- and mid-1990s were too optimistic.

This concern about the lack of neutrino energy dependence received some support from a simulation in the late 1990s involving a different simplification of neutrino transport: the imposition of energy-dependent neutrino

Table 1.   Selected neutrino radiation hydrodynamics milestones in stellar collapse simulations studying the long-term fate of the shock.

Group	Year	Explosion	Total dimensions	Fluid space dimensions	ν space dimensions	ν momentum space dimensions
Lawrence Livermore[27,6,7]	1982	Yes*	2	1 (PN)	1	1 ($O(v/c)$)
Lawrence Livermore[28,29]	1985	Yes*	"2.25"	"1.5" NS (PN)	1	1 ($O(v/c)$)
Florida Atlantic[30,31,32,33]	1987	No	2	1 (GR)	1	1 ($O(v/c)$)
Lawrence Livermore[34,35,36]	1989	Yes*	"2.25"	"1.5" NS+HR (GR)	1	1 (GR)
Lawrence Livermore[37]	1992	Yes*	2	2 HR (N)	2	0 (N)
Los Alamos[38]	1993	Yes*	"1.75"	2 (N)	"1.5" thick/thin	0 (PN)
Arizona[39]	1994	Yes*	"1.75"	2 (N)	"1.5" ray-by-ray	0 (N)
Florida Atlantic[40,41]	1994	No	"2.25"	"1.5" NS (GR)	1	1 ($O(v/c)$)
Oak Ridge[42,43]	1996	No	"2.5"	2 (N)	1	1 ($O(v/c)$)
Max Planck[44,45,46,47]	2000	No, Yes* (ONeMg)	3	1 (N)	1	2 ($O(v/c)$)
Oak Ridge[48,49,50,51,52]	2000	No	3	1 (N)	1	2 ($O(v/c)$)
Arizona[53,54]	2002	No	3	1 (N)	1	2 ($O(v/c)$)
Oak Ridge[48,49,50,55,56,57]	2000	No	3	1 (GR)	1	2 (GR)
Los Alamos[38,58]	2002	Yes*	"2.5"	3 (N)	"2" thick/thin	0 (PN)
Max Planck[45,59,60,61,47]	2003	No, Yes* (180°)	"3.75"	2 (PN)	"1.5" ray-by-ray	2 ($O(v/c)$, PN)

Note: The "Yes" entries in the "Explosion" column are all marked with an asterisk as a reminder that questions about the simulations—described in the main text—have prevented a consensus about the explosion mechanism. "Total dimensions" is the average of "Fluid space dimensions" and "ν space dimensions," added to "ν momentum dimensions." The abbreviation "N" stands for 'Newtonian,' while "PN"—for 'Post-Newtonian'—stands for some attempt at inclusion of general relativistic effects, and "GR" denotes full relativity. A space dimensionality in quotes—like "1.5"—denotes an attempt at modeling higher dimensional effects within the context of a lower dimensional simulation. For the fluid, this is a mixing-length prescription in the neutron star ("NS") or the heating region ("HR") behind the stalled shock. For neutrino transport, it indicates one of two approaches: multidimensional diffusion in regions with strong radiation/fluid coupling, matched with a spherically symmetric 'light bulb' approximation in weakly coupled regions ("thick/thin"); or the (mostly) independent application of a spherically symmetric formalism/algorithm to separate spatial angle bins ("ray-by-ray").

distributions from spherically symmetric simulations onto fluid dynamics in two spatial dimensions.[43] Unlike the simulations discussed above, these did not explode, casting doubt upon claims that convection-aided neutrino heating constituted a robust explosion mechanism.

The nagging qualitative difference between spatially multidimensional simulations with different neutrino transport approximations motivated interest in the possible importance of even more complete neutrino transport: Might the retention of both the energy *and* angle dependence of the neutrino distributions improve the chances of explosion, as preliminary "snapshot" studies suggested?[66,53] Of necessity, the first such simulations were performed in spherical symmetry, which nevertheless represented an advance to a total dimensionality of 3 (see Table 1). Results from three different groups are in accord: Spherically symmetric models of iron core collapse do not explode, even with solid neutrino transport[44,52,54] and general relativity.[55,57] Recently, however, it has been shown that the more modest oxygen/neon/magnesium cores of the lightest stars to undergo core collapse (8-10 $M_\odot$) may explode in spherical symmetry.[46,47]

The current state of the art in neutrino transport in supernova simulations determining the long-term fate of the shock has been achieved by a group centered at the Max Planck Institute for Astrophysics in Garching, who deployed their spherically symmetric energy- and angle-dependent neutrino transport capability[45] along separate radial rays, with partial coupling between rays.[60] Initial results—from axisymmetric simulations with a restricted angular domain—were negative with regards to explosions (in spite of the salutary effects of convection, and also rotation),[59] apparently supporting the results of Ref. [43]. An explosion was seen in one simulation[67] in which certain terms in the neutrino transport equation corresponding to Doppler shifts and angular aberration due to fluid motion were dropped; this simulation also yielded a neutron star mass and nucleosynthetic consequences in better agreement with observations than the "successful" explosion simulations of the 1990s,[38,39] arguably because of more accurate neutrino transport in the case of both observables. The continuing lesson is that getting the details of the neutrino transport right makes a difference.

In addition to accurate neutrino transport, low-mode ($\ell = 1, 2$) instabilities that can develop only in simulations allowing the full range of polar angles may make a subtle but decisive difference, as in an explosion recently reported by the Garching group.[61,47] This achievement was presaged by earlier studies of the supernova context, which featured a demonstration of the tendency for convective cells to merge to the lowest order allowed by the spatial domain[68] and a newly-recognized spherical accretion shock instability[69] (discovered independently in a different context in Ref. [70]). These these global asymmetries may be sufficient to account for observed asphericities that have often been attributed to rotation and/or magnetic fields.

Surely every 'Yes' entry in the explosion column of Table 1 has been hailed in its time as 'the answer' (at least by some!), and as a community we cannot help hoping once again that these recent developments mark the turning of a corner; but important work remains to verify if this is the case. Several groups are committed to further efforts. For example, the Terascale Supernova Initiative (TSI, which includes authors of Ref. [69]) comprises efforts aimed at 'ray-by-ray' simulations[71] like those of the Garching group, as well as full spatially multdimensional neutrino transport, both with energy dependence only[72] and with energy *and* angle dependence (Sec. 6, and Ref. [73]). Delineation of the possible roles of rotation and magnetic fields are also being pursued by TSI. At least one other group is pursuing full spatially multidimensional neutrino transport.[74,75]

3. *GenASiS:* Philosophy and Basic Features

As discussed in Sec. 1, a core-collapse supernova involves a six-dimensional radiation hydrodynamics problem, making it a major computational challenge. Even three-dimensional pure hydrodynamics problems (with only space dimensions, no momentum space) have only become relatively common in the last few years, with manageable workflows on today's terascale machines ($\sim 10^{12}$ bytes of memory and flop/s). To begin to get a feel for the requirements of *radiation* hydrodynamics, consider just a five-dimensional problem, in which axisymmetry in the space dimensions is assumed. For example, supposing the numbers of spatial zones in spherical coordinates (r, θ) to be $(256, 128)$, and the numbers of momentum bins in energy and angle variables $(\epsilon, \vartheta, \varphi)$ to be $(64, 32, 16)$, of order 10^{10} bytes are required just to store one copy of one neutrino distribution function. While this gives rise to a taxing (but not necessarily insurmountable) workflow on terascale machines, it is apparent that the addition of the third spatial dimension—necessary for full exploration of the interacting effects of convection, rotation, and magnetic fields—will require petascale systems. But petascale systems will eventually be available (five to seven years is the current expectation); and given the long development time scales of sophisticated software, we believe it wise to develop our code with the full six-dimensional capability, even if it is only deployed in five dimensions in the near term.

The computational demands of radiation hydrodynamics can be ameliorated by "adapative mesh refinement" (AMR). The basic idea of AMR is to employ high resolution only where needed in order to conserve memory and computational effort. Our current expectation is to allow for refinement only in the space dimensions. This will help with the management of two difficulties: the large dynamic range in length scales associated with the density increase of six orders of magnitude that occurs during core collapse, and adequate resolution of particular features of the flow (the shock, for instance). With Eulerian codes in multiple space dimensions, these tasks require high resolution; and particularly for radiation hydrodynamics, the savings achievable by reducing the number of zones is considerable, since an entire three-dimensional momentum space is carried by each spatial zone.

Of the two basic types of AMR on structured grids—the block-structured and zone-by-zone varieties—we have chosen the zone-by-zone approach for use with neutrino radiation hydrodynamics. Block-structured AMR[76] involves the deployment of subgrids of a certain reasonable minimum size (e.g. eight or sixteen zones per side) at various levels of refinement. A basic solver routine is applied independently to each subgrid. Extra spatial zones (referred to as 'guard zones') are required on the edges of each subgrid, which carry information from neighboring subgrids; these become the boundary conditions applied by the solver routine. The strategy is designed for explicit solution algorithms, in which the functions describing time evolution need only be evaluated at the previous time step. However, the rapid time scales of neutrino interactions with the fluid require an *implicit* solution algorithm, in which the functions describing the evolution of the neutrino radiation field are evaluated at the *current* (that is, new) time step. Because of this mismatch with the intended purposes of block-structured AMR (implicit vs. explicit evolution), and the fact that popular block-structured AMR community packages did not seem readily amenable to handling momentum space variables in a natural way, we decided upon another flavor of AMR: the zone-by-zone refinement approach.[77] In this method, individual zones are refined (typically by bisection) and coarsened as needed. This provides more flexibility than the block-structured approach; the fine-grained control allows for maximum savings in the number of spatial zones deployed. A drawback for many users is that a single-grid explicit solver cannot be used "as is." Instead, new solution algorithms must be developed that address the entire hierarchical data structure (Ref.

[77] and our Sec. 4 are examples for explicit hydrodynamics); but we are required to develop such 'global solvers' for gravity (Sec. 5) and implicit neutrino transport (Sec. 6) anyway. And as a bonus, the need to carry memory-wasting 'guard zones' is obviated.

In implementing the zone-by-zone-refinement approach to the representation of spacetime, we have tried to follow object-oriented design principles to the extent allowed by Fortran 90/95. Figure 1 outlines the basic data structures we use to model the ideal of a continuous spacelike slice with a discretized approximation. (The hierarchy of structures, and our operations on them with well-controlled interfaces, are instances of the object-oriented principles of *inheritance* and *encapsulation*.) A region of a spacelike slice is represented by an object of `zoneArrayType`. Each such object contains an array of objects of `zoneType`, along with information about the coordinates of the zones and pointers to neighboring zone arrays. Each zone, an object of `zoneType`, contains various forms of stress-energy, each of which is a separate object. Figure 1 shows a perfect fluid and a radiation field; in the code we have an electromagnetic field as well. Each zone has a pointer to another object of `zoneArrayType`, whose allocation constitutes refinement of that zone; this structure can be extended to arbitrarily deep. A simple two-dimensional pure hydrodynamics test problem computed with our adaptive mesh code is shown in Fig. 2.

The word "General" that goes into the name of our code, *GenASiS*—for *Gen*eral *A*strophysical *Si*mulation *S*ystem—may give an initial impression of a messianic quest to create an impossibly all-purpose code for solving all conceivable problems in astrophysics and cosmology; but the code's 'generality' is, of course, considerably more modest: It refers to the use of Fortran 90's facility for function overloading. (This is an instance of the object-oriented principle of *polymorphism*.) This allows a generic function name to have several different implementations, providing for extensibility of the physics: Different equations of state, hydrodynamic flux methods, coordinate systems, gravity theories, and so forth can be employed by adding new implementations of generic function names, without having to go back and change basic parts of the code to implement new physics.

4. Magnetohydrodynamics

We employ a conservative formulation of the equations of magnetohydrodynamics. The Newtonian case will be described here. Conservation of

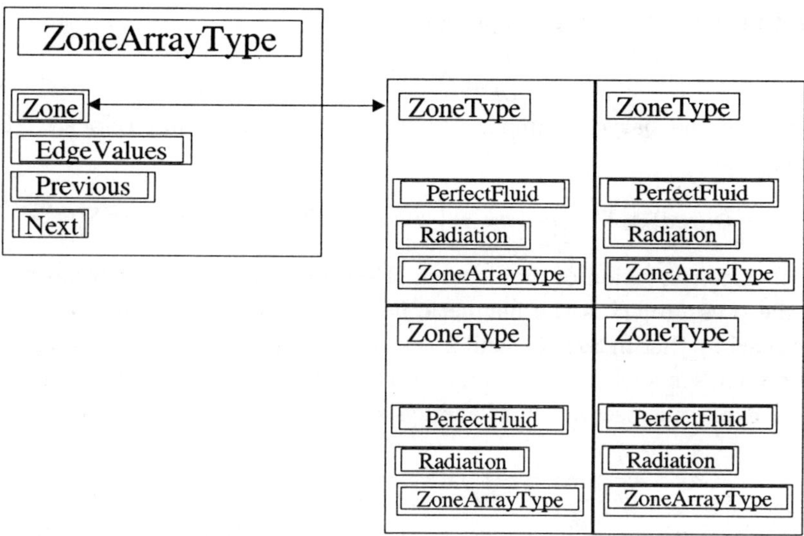

Figure 1. Data structures used in an adaptive mesh for radiation hydrodynamics.

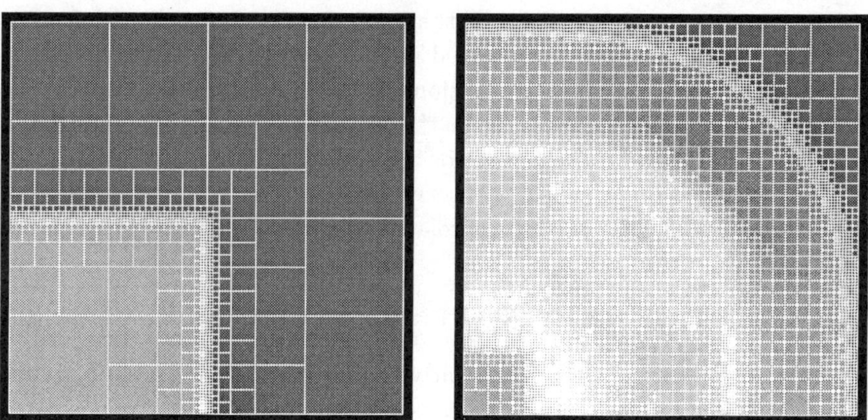

Figure 2. Density in a two-dimensional generalization of the shock tube. Red and blue indicate high and low density respectively. Left: Initial state. Right: Evolved state.

baryons is described by the equation

$$\frac{\partial n}{\partial t} + \boldsymbol{\nabla} \cdot (n\boldsymbol{v}) = 0, \tag{1}$$

where n is the baryon number density, and $\boldsymbol{v}$ is the fluid velocity. The equation

$$\frac{\partial}{\partial t}(mn\,\boldsymbol{v}) + \boldsymbol{\nabla} \cdot \left[mn\,\boldsymbol{vv} + \left(p + \frac{B^2}{2} \right) \mathbf{1} - \boldsymbol{BB} \right] = -mn\,\boldsymbol{\nabla}\Phi \tag{2}$$

describes conservation of momentum. Here m is the average baryon mass, p is the pressure, $\boldsymbol{B}$ is the magnetic field, Φ is the gravitational potential (discussed further in Sec. 5), and $\mathbf{1}$ is the unit tensor. Units of the magnetic field are chosen such that the vacuum magnetic permeability is unity. One way to express conservation of energy is

$$\frac{\partial}{\partial t}\left[e + \frac{mn}{2}\left(v^2 + \Phi \right) + \frac{B^2}{2} \right] +$$

$$\boldsymbol{\nabla} \cdot \left[(e + p + B^2)\boldsymbol{v} + \frac{mn}{2}\left(v^2 + \Phi \right)\boldsymbol{v} - \boldsymbol{B}(\boldsymbol{v} \cdot \boldsymbol{B}) \right] = -\frac{mn}{2}\left(\boldsymbol{\nabla} \cdot \boldsymbol{\Psi} + \boldsymbol{v} \cdot \boldsymbol{\nabla}\Phi \right) - n\left(\frac{\partial m}{\partial t} - \boldsymbol{v} \cdot \boldsymbol{\nabla}m \right), \tag{3}$$

where e is the internal energy density, and $\boldsymbol{\Psi}$ is a kind of "gravitational vector potential" (also discussed in Sec. 5). We have opted to employ the baryon number density, instead of the mass density, in explicit deference to the fact that mass is not conserved in the presence of nuclear reactions. The energy input from nuclear reactions has then resulted naturally in the derivation of Eq. (3), appearing in the last two terms of the right-hand side. We hope to add nuclear reaction networks to our code in the future.

This formulation is called "conservative" because volume integrals of the divergences in Eqs. (1)-(3) are related to surface integrals through the divergence theorem:

$$\int_V dV\,(\boldsymbol{\nabla} \cdot \boldsymbol{F}) = \oint_{\partial V} \boldsymbol{F} \cdot d\boldsymbol{A}. \tag{4}$$

The physical meaning of a conservative equation is that (modulo source terms) the time rate of change of a conserved quantity in a volume is equal to a flux $\boldsymbol{F}$ through the volume's enclosing surface. This meaning is built into the finite-difference representation of Eqs. (1)-(3); divergences are represented in discrete correspondence to their mathematical definition, using zone volumes V and face areas A:

$$\boldsymbol{\nabla} \cdot \boldsymbol{F} \to \frac{1}{V_{\leftrightarrow}} \sum_q \left[(A_q\,F^q)_{q\to} - (A_q\,F^q)_{\leftarrow q} \right]. \tag{5}$$

Here q runs over the three space dimensions. A double-headed arrow ($\leftrightarrow$) indicates evaluation at a zone center. Left-arrows ($\leftarrow q$) and right-arrows ($q \rightarrow$) denote evaluation at zone inner and outer faces respectively, in the q direction; dimensions other than q are evaluated at the zone centers. In this way the divergence theorem is replicated in every zone, ensuring global conservation to machine precision. Use of generalized zone volumes and areas in Eq. (5) enables the use of curvilinear coordinates ("ficticious forces" arising from curvilinear coordinates must also be included in the momentum equation).

The evolution of the magnetic field is described by Faraday's law:

$$\frac{\partial \boldsymbol{B}}{\partial t} = -\boldsymbol{\nabla} \times \boldsymbol{E}, \tag{6}$$

supplemented by the constraint

$$\boldsymbol{\nabla} \cdot \boldsymbol{B} = 0. \tag{7}$$

In Eq. (6), we take the electric field to be $\boldsymbol{E} = -\boldsymbol{v} \times \boldsymbol{B}$, in accordance with the usual astrophysical assumption of a perfectly conducting medium.

While Eq. (6) for the evolution of the magnetic field does not require conservation of the magnetic field, Eq. (7) requires the magnetic field to be divergence-free at all times. In the presence of discontinuous flow numerical solutions to Eqs. (1)-(6) can produce severe unphysical artifacts if this requirement is not met,[78] but it can be automatically enforced by the method of constrained transport.[79] Integrating Eq. (6) over a zone's enclosing surface, the left-hand side becomes the volume integral of $\boldsymbol{\nabla} \cdot \boldsymbol{B}$, via the divergence theorem. The right-hand side is a sum over area integrals over each zone face. With Stokes' theorem,

$$\int_A (\boldsymbol{\nabla} \times \boldsymbol{E}) \cdot d\boldsymbol{A} = \oint_{\partial A} \boldsymbol{E} \cdot d\boldsymbol{l}, \tag{8}$$

each of these surface integrals becomes a line integral around the zone face boundary. Summed over all faces, two line integrals in opposite directions cancel on every zone edge, enforcing $\boldsymbol{\nabla} \cdot \boldsymbol{B} = 0$ in the zone as desired. The method of constrained transport, then, is to evaluate $\boldsymbol{\nabla} \times \boldsymbol{E}$ on zone edges in discrete correspondence to the mathematical definition of the curl, using zone face areas A and edge lengths L:

$$(\boldsymbol{\nabla} \times \boldsymbol{E})_q \rightarrow \frac{1}{A_q} \sum_{r \neq q} [(L_s E^s)_{r \rightarrow} - (L_s E^s)_{\leftarrow r}]. \tag{9}$$

Here r runs over the two space dimensions orthogonal to a particular direction q, and s indicates the direction perpendicular to both q and r. Left-arrows ($\leftarrow r$) and right-arrows ($r \rightarrow$) denote evaluation along face inner and outer edges respectively, in the r direction. This ensures divergence-free evolution of the discrete representation of the area-averaged magnetic field, with components located on the appropriate zone faces, for all times to machine precision, provided the initial magnetic field satisfies Eq. (7).

Accurate computation of fluxes at zone faces and electric fields at zone edges is a key feature. So-called "central schemes" have been noted recently by astrophysicists for their ability to capture shocks with an accuracy comparable to Riemann solvers, but with much greater simplicity.[80] In particular, we employ so-called "HLL" versions of these schemes for both fluid conservation laws[81] and the magnetic induction equation.[82] We achieve second order in space by linear interpolation within zones (as usual, a slope limiter—deployed where necessary in order maintain discontinuities— reduces the treatment to first order).

While we have taken Khokhlov's zone-by-zone refinement approach[77] as our basic paradigm, we have made a novel extension to evolve the magnetic field, and use a different time-stepping scheme. As in Khokhlov's work, fluxes are computed at each zone interface only once, with the results used to update zones on both sides of the interface. At coarse/fine interfaces, fluxes are computed only on the faces of the refined zones. We have developed a similar approach for the induction equation: the electromotive forces on the zone edges are computed only once, and are used to update all zones sharing that edge. We use a second-order Runge-Kutta time stepping algorithm, made possible by the semi-discrete formulation of the central scheme. In doing so we evolve all levels of the mesh synchronously, unlike Khokhlov's approach of evolving refined levels with greater frequency. In addition to making it possible to take advantage of the semi-discrete formulation for time evolution, problems we encountered in self-gravitating systems with 'asynchronous' evolution a la Khokhlov were avoided.

Parallelization is achieved by giving each processor its share of spatial zones. Partitioning is accomplished by walking through all levels of the mesh in a recursive manner similar to a Morton space-filling curve; the result is a mapping of the multidimensional mesh to a one dimensional "string" of zones, which, when cut into pieces of uniform length, leaves each processor with roughly the same number of zones at each level of refinement.

An equation of state determines p, the quantity in Eqs. (1)-(3) whose determination has not yet been mentioned. So far, the code has two overloaded options. One is the familiar polytropic equation of state:

$$p = \kappa\, n^{\Gamma}, \tag{10}$$

$$e = (\Gamma - 1)^{-1} p, \tag{11}$$

where Γ is a specified parameter, and κ is updated in response to changes in e determined from Eq. (3). We have also implemented a "realistic" equation of state[83] suitable for problems involving nuclear matter. This equation of state takes as input the temperature, baryon number density, and electron fraction Y_e, defined by

$$Y_e = \frac{n_{e^-} - n_{e^+}}{n}, \tag{12}$$

where n_{e^-} and n_{e^+} are the number densities of electrons and positrons respectively. When using this "realistic" equation of state, an advection equation for Y_e must be added to the above list of conservation laws.

We have been working with a number of hydrodynanic and magnetohydrodynamic test problems, one of which is shown here. The rotor problem—which consists of a rapidly rotating dense fluid, initially cylindrical, threaded by an initially uniform magnetic field—was devised to test the onset and propagation of strong torsional Alfvén waves into the ambient fluid.[84] We have computed a version of the rotor problem with initial data identical to a so-called 'second rotor problem,'[85] and display the results in Fig. 3.

5. Newtonian Gravity

The Poisson equation for the Newtonian gravitational potential Φ is

$$\nabla^2 \Phi = 4\pi G\, mn, \tag{13}$$

where G is the gravitational constant. Our finite-differenced approach to Eq. (13) is based on the fact that the Laplacian is a divergence of a gradient:

$$\boldsymbol{\nabla} \cdot \boldsymbol{\nabla} \Phi = 4\pi G\, mn. \tag{14}$$

We discretize this equation in a manner similar to Eq. (5). There results a linear system for the values of Φ at the center of every leaf zone. In the matrix representation of this linear system, each row corresponds to the discrete version of Eq. (14) centered on a given "leaf zone" (those not having refined children). On a single-level grid, the resulting matrix would have three, five, and seven bands in one, two, and three dimensions respectively.

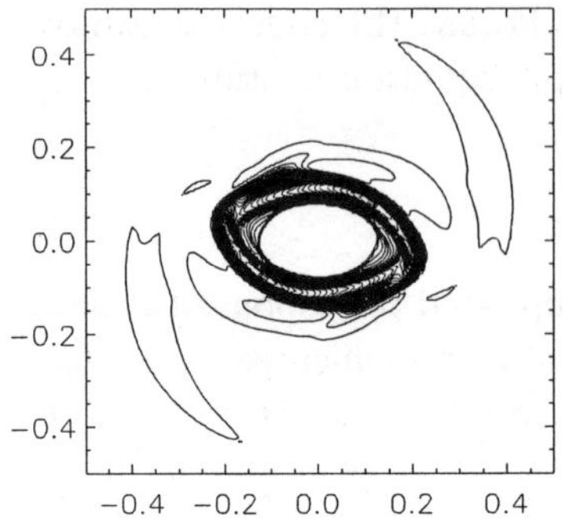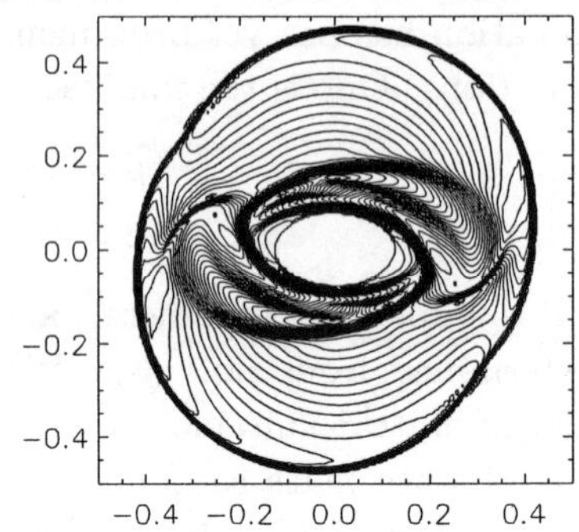

Figure 3. Density (left panel) and thermal pressure (right panel) at $t = 0.295$ for the second rotor problem given in Ref. [85]. 40 countours were used to produce the plots, with $0.512 \leq mn \leq 9.622$ and $0.010 \leq p \leq 0.776$. A 200 × 200 grid was used to produce the results.

With adaptive mesh refinement, the matrix structure becomes more diffuse, because several refined zones may contribute to the discrete representation of $\nabla\Phi$ at a zone face featuring a coarse/fine interface. Each processor fills in the portion of the matrix corresponding to its share of zones, and we rely on the PETSc library (`http://www-unix.mcs.anl.gov/petsc/petsc-2/`) to perform the distributed sparse matrix inversion.

We now discuss the contribution of gravity to the fluid energy evolution in Eq. (3). Without the gravitational potential energy included in the "total energy" in the time derivative, Eq. (3) appears in the more familiar form

$$\frac{\partial}{\partial t}\left[e + \frac{1}{2}mn\,v^2\right] + \nabla \cdot \left[(e + p + \frac{1}{2}mn\,v^2)v\right] = -mn\,v \cdot \nabla\Phi - $$
$$n\left(\frac{\partial m}{\partial t} - v \cdot \nabla m\right) (15)$$

Including the graviational potential energy density $(mn/2)\Phi$ in the time derivative, and making use of baryon conservation, we have

$$\frac{\partial}{\partial t}\left[e + \frac{mn}{2}\left(v^2 + \Phi\right)\right] +$$
$$\boldsymbol{\nabla}\cdot\left[(e+p)\boldsymbol{v} + \frac{mn}{2}\left(v^2 + \Phi\right)\boldsymbol{v}\right] = \frac{mn}{2}\left(\frac{\partial\Phi}{\partial t} - \boldsymbol{v}\cdot\boldsymbol{\nabla}\Phi\right) -$$
$$n\left(\frac{\partial m}{\partial t} - \boldsymbol{v}\cdot\boldsymbol{\nabla}m\right). \tag{16}$$

Using the formal solution for Φ,

$$\Phi(\boldsymbol{x},t) = -G\int\frac{m(\boldsymbol{x}',t)n(\boldsymbol{x}',t)\,d^3x'}{|\boldsymbol{x}-\boldsymbol{x}'|}, \tag{17}$$

together with baryon conservation (and neglecting time derivatives of m), one can show that

$$\frac{\partial\Phi}{\partial t} = -\boldsymbol{\nabla}\cdot\boldsymbol{\Psi}, \tag{18}$$

where $\boldsymbol{\Psi}$ satisfies a vector Poisson equation:

$$\nabla^2\boldsymbol{\Psi} = 4\pi G\,mn\,\boldsymbol{v}. \tag{19}$$

This may be solved in a manner similar to that used to solve for Φ (though the vector Poisson equation contains some additional terms in curvilinear coordinates). Conservation of total energy, including gravitational, is not local: the gravitational source terms (the first two terms on the right-hand side of Eq. (3)) vanish only upon integration over all space.

6. Neutrino Radiation Transport

Here we briefly describe our approach to the greatest computational challenge in supernova simulations: neutrino radiation transport. Neutrino distributions must be tracked in order to compute the transfer of lepton number and energy between the neutrinos and the fluid. There are three major challenges. One challenge is constructing a discretization that allows both energy and lepton number to be conserved to high precision. The two other challenges are associated with the limits of computational resources: the solution of a very large nonlinear system of equations, and neutrino interaction kernels of high dimensionality.

Energy conservation is an obvious measure of quality control, and care with the transport formalism and differencing can help achieve it. The importance of energy conservation is brought into focus by this question: How should we interpret the prediction of a $\sim 10^{51}$ erg explosion in a model where the total energy varies during the course of the simulation by $\sim 10^{51}$ erg or more? To achieve the required precision (say, global energy changes of less than $\sim 10^{50}$ erg), conservative formulations are a useful starting point, and relativistic treatments avoid quantitatively non-negligible conflicts at $O(v^2/c^2)$ between the number and energy transport equations.[86,57,87,8] Finite-differencing that simultaneously satisfies energy and lepton number conservation has been implemented in spherical symmetry,[57] and should be pursued in multiple spatial dimensions as well.

Our algorithm for solving the large nonlinear system has been described elsewhere.[88,73] The large system of equations requiring inversion (as opposed to explicit updates) results from the disparity between hydrodynamic and particle interaction time scales, which motivates implicit time evolution. The nonlinear solve is achieved with the Newton-Raphson method. A fixed-point method employing a preconditioner that splits the space and momentum space couplings is used for the linear solve required within each Newton-Raphson iteration.[88] An advantage of this linear solver method is that the dense blocks representing couplings in momentum space—which cannot all be stored at once—need only be constructed a few at a time, used in all steps required in a given fixed-point iteration, and discarded. In contrast, other linear solver algorithms seem to require dense blocks to be discarded and rebuilt multiple times in each iteration. We have sucessfully tested this solver on a two-dimensional problem in spherical coordinates with a static background and a simple emission/absorption interaction.

Because neutrino interactions are expensive to compute on-the-fly, we have implemented interpolation tables. Neutrino interactions depend on the neutrino momentum components and the state of the fluid with which the neutrinos interact. A grid of neutrino energies and angles is fixed, but the fluid density n, temperature T, and electron fraction Y_e vary throughout the simulation. Hence we employ tables that may be interpolated in n, T, and Y_e. Particularly for neutrino scattering and pair interactions—which depend on neutrino states before and after the collision—the interaction kernels are of high dimensionality, requiring a globally distributed table. On each processor a local table is constructed, which contains a copy of each n, T, and Y_e vertex required by the zones for which that processor is responsible. As n, T, and Y_e in a processor's zones evolve, the relevant vertices are pulled from the global table as needed. For each zone we construct a "cube" of pointers to the eight vertices surrounding the zone's values of n, T, and Y_e. This cube is then used for the necessary interpolations.

7. Outlook

We have made a promising start on *GenASiS*, a new code being developed to study the explosion mechanism of core-collapse supernovae. Our plan is to include all the relevant physics—including magnetohydrodynamics, gravity, and energy- and angle-dependent neutrino transport—in a code with adaptive mesh refinement in two and three spatial dimensions. Parallelization and implementation on the adaptive mesh are not yet complete, and the physics components have not yet been fully integrated; but steady progress and the successful completion of test problems give us confidence that we are well on our way towards a tool that will provide important insights into the supernova explosion mechanism.

Acknowledgments

We gratefully acknowledge S. W. Bruenn's contribution of subroutines for the computation of neutrino interaction kernels. E. J. Lentz collaborates on the development of a comprehensive neutrino radiation transport discretization scheme, still too immature to report here. We thank R. D. Budiardja and M. W. Guidry for discussions on the Poisson solver, and R. D. Budiardja for helping with that solver's interface to the PETSc library. This work was supported by Scientific Discovery Through Advanced Computing (SciDAC), a program of the Office of Science of the U.S. Department of Energy (DoE); and by Oak Ridge National Laboratory, managed by UT-Battelle, LLC, for the DoE under contract DE-AC05-00OR22725.

References

1. D. E. Osterbrock, *Bull. Am. Astron. Soc.* **33**, 1330 (2001).
2. F. Zwicky, *Rev. Mod. Phys.* **12**, 66 (1940).
3. A. V. Filippenko, *Annu. Rev. Astron. Astrophys.* **35**, 309 (1997).
4. W. Baade and F. Zwicky, *Phys. Rev.* **45**, 138 (1934).
5. A. Burrows, in *Supernovae*, edited by A. G. Petschek (Springer-Verlag, New York, 1990), pp. 143–181.
6. J. R. Wilson, in *Numerical Astrophysics: Proceedings of a Symposium in honor of James R. Wilson held at the University of Illinois in October, 1982*, edited by J. M. Centrella, J. M. LeBlanc, and R. L. Bowers (Jones and Bartlett, Boston, 1985), pp. 422–434.
7. H. A. Bethe and J. R. Wilson, *Astrophys. J.* **295**, 14 (1985).
8. C. Y. Cardall, E. J. Lentz, and A. Mezzacappa, submitted (2005).
9. T. J. Galama, P. M. Vreeswijk, J. van Paradijs, et al., *Nature* **395**, 670 (1998).
10. K. Iwamoto, P. A. Mazzali, K. Nomoto, et al., *Nature* **395**, 672 (1998).

11. M. Della Valle, D. Malesani, S. Benetti, et al., *Astron. Astrophys.* **406**, L33 (2003).

12. J. Hjorth, J. Sollerman, P. Møller, et al., *Nature* **423**, 847 (2003).

13. K. Z. Stanek, T. Matheson, P. M. Garnavich, et al., *Astrophys. J. Lett.* **591**, L17 (2003).

14. B. Thomsen, J. Hjorth, D. Watson, et al., *Astron. Astrophys.* **419**, L21 (2004).

15. B. E. Cobb, C. D. Bailyn, P. G. van Dokkum, et al., *Astrophys. J. Lett.* **608**, L93 (2004).

16. D. Malesani, G. Tagliaferri, G. Chincarini, et al., *Astrophys. J. Lett.* **609**, L5 (2004).

17. A. Gal-Yam, D.-S. Moon, D. B. Fox, et al., *Astrophys. J. Lett.* **609**, L59 (2004).

18. A. Zeh, S. Klose, and D. H. Hartmann, *Astrophys. J.* **609**, 952 (2004).

19. G. Ghirlanda, G. Ghisellini, and D. Lazzati, *Astrophys. J.* **616**, 331 (2004).

20. A. M. Soderberg, S. R. Kulkarni, E. Berger, et al., *Nature* **430**, 648 (2004).

21. C. L. Fryer and A. Heger, *Astrophys. J.* **541**, 1033 (2000).

22. T. A. Thompson, E. Quataert, and A. Burrows (2004), astro-ph/0403224.

23. J. C. Wheeler and S. Akiyama, INT workshop 'Open Issues in Understanding Core Collapse Supernovae,' Seattle, 2004 (2004), astro-ph/0412382.

24. B. M. Gaensler, N. M. McClure-Griffiths, M. S. Oey, et al., *Astrophys. J. Lett.* **620**, L95 (2005).

25. D. F. Figer, F. Najarro, T. R. Geballe, et al., *Astrophys. J. Lett.* **622**, L49 (2005).

26. A. Burrows, R. Walder, C. D. Ott, et al., Science Symposium on the Fate of the Most Massive Stars, Grand Teton National Park, Wyoming, 23-28 May 2004 (2004), astro-ph/0409035.

27. R. L. Bowers and J. R. Wilson, *Astrophys. J. Suppl. Ser.* **50**, 115 (1982).

28. R. W. Mayle, Ph.D. thesis, University of California, Berkeley (1985).

29. J. R. Wilson and R. W. Mayle, *Phys. Rep.* **163**, 63 (1988).

30. S. W. Bruenn, *Astrophys. J. Suppl. Ser.* **58**, 771 (1985).

31. S. W. Bruenn, *Phys. Rev. Lett.* **59**, 938 (1987).

32. S. W. Bruenn and W. C. Haxton, *Astrophys. J.* **376**, 678 (1991).

33. S. W. Bruenn, in *Nuclear Physics in the Universe: Proceedings of the First Symposium on Nuclear Physics in the Universe, held in Oak Ridge, Tennessee, USA, 24-26 September 1992*, edited by M. W. Guidry and M. R. Strayer (Institute of Physics Publishing, Bristol, 1993), pp. 31–50.

34. R. W. Mayle, in *Supernovae*, edited by A. G. Petschek (Springer-Verlag, New York, 1990), pp. 267–289.

35. R. W. Mayle and J. R. Wilson, in *Supernovae: The Tenth Santa Cruz Workshop in Astronomy and Astrophysics, July 9 to 21, 1989, Lick Observatory*, edited by S. E. Woosley (Springer-Verlag, New York, 1991), pp. 333–341.

36. J. R. Wilson and R. W. Mayle, *Phys. Rep.* **227**, 97 (1993).

37. D. S. Miller, J. R. Wilson, and R. W. Mayle, *Astrophys. J.* **415**, 278 (1993).

38. M. Herant, W. Benz, W. R. Hix, et al., *Astrophys. J.* **435**, 339 (1994).

39. A. Burrows, J. Hayes, and B. A. Fryxell, *Astrophys. J.* **450**, 830 (1995).

40. S. W. Bruenn and A. Mezzacappa, *Astrophys. J. Lett.* **433**, L45 (1994).

41. S. W. Bruenn, A. Mezzacappa, and T. Dineva, *Phys. Rep.* **256**, 69 (1995).

42. A. Mezzacappa, A. C. Calder, S. W. Bruenn, et al., *Astrophys. J.* **493**, 848 (1998a).

43. A. Mezzacappa, A. C. Calder, S. W. Bruenn, et al., *Astrophys. J.* **495**, 911 (1998b).

44. M. Rampp and H.-T. Janka, *Astrophys. J. Lett.* **539**, L33 (2000).

45. M. Rampp and H.-T. Janka, *Astron. Astrophys.* **396**, 361 (2002).

46. F. S. Kitaura Joyanes, Ph.D. thesis, Technical University, Munich (2003).

47. H.-T. Janka, R. Buras, F. S. Kitaura Joyanes, et al., 8th Symposium on Nuclei in the Cosmos, Vancouver, BC, Canada, 19-23 Jul 2004 pp. astro–ph/0411347 (2004a), `astro-ph/0411347`.

48. A. Mezzacappa and S. W. Bruenn, *Astrophys. J.* **405**, 669 (1993a).

49. A. Mezzacappa and S. W. Bruenn, *Astrophys. J.* **410**, 740 (1993b).

50. A. Mezzacappa and O. E. B. Messer, *J. Comput. Appl. Math.* **109**, 281 (1999).

51. M. Liebendörfer, Ph.D. thesis, University of Basel (2000).

52. A. Mezzacappa, M. Liebendörfer, O. E. Messer, et al., *Phys. Rev. Lett.* **86**, 1935 (2001).

53. A. Burrows, T. Young, P. Pinto, et al., *Astrophys. J.* **539**, 865 (2000).

54. T. A. Thompson, A. Burrows, and P. A. Pinto, *Astrophys. J.* **592**, 434 (2003).

55. M. Liebendörfer, A. Mezzacappa, F. Thielemann, et al., *Phys. Rev. D* **63**, 103004 (2001).

56. M. Liebendörfer, S. Rosswog, and F. Thielemann, *Astrophys. J. Suppl. Ser.* **141**, 229 (2002).

57. M. Liebendörfer, O. E. B. Messer, A. Mezzacappa, et al., *Astrophys. J. Suppl. Ser.* **150**, 263 (2004).

58. C. L. Fryer and M. S. Warren, *Astrophys. J. Lett.* **574**, L65 (2002).

59. R. Buras, M. Rampp, H.-T. Janka, et al., *Phys. Rev. Lett.* **90**, 241101 (2003).

60. H. T. Janka, R. Buras, K. Kifonidis, et al., in *Stellar Collapse*, edited by C. L. Fryer (Kluwer Academic Publishers, Dordrecht, 2004b), pp. 65–97, `astro-ph/0212314`.

61. H.-T. Janka, R. Buras, K. Kifonidis, et al., IAU Colloquium 192: Supernovae (10 Years after SN1993J), Valencia, Spain, 22-26 Apr 2003 (2004c), `astro-ph/0401461`.

62. S. W. Bruenn and T. Dineva, *Astrophys. J. Lett.* **458**, L71+ (1996).

63. S. W. Bruenn, E. A. Raley, and A. Mezzacappa, pp. astro–ph/0404099 (2004), `astro-ph/0404099`.

64. S. W. Bruenn, *Astrophys. J.* **340**, 955 (1989a).

65. S. W. Bruenn, *Astrophys. J.* **341**, 385 (1989b).

66. O. E. B. Messer, A. Mezzacappa, S. W. Bruenn, et al., *Astrophys. J.* **507**, 353 (1998).

67. H. T. Janka, R. Buras, and M. Rampp, *Nucl. Phys. A* **718**, 269 (2003), `astro-ph/0212317`.

68. M. Herant, W. Benz, and S. Colgate, *Astrophys. J.* **395**, 642 (1992).

69. J. M. Blondin, A. Mezzacappa, and C. DeMarino, *Astrophys. J.* **584**, 971

218

(2003).

70. T. Foglizzo, *Astron. Astrophys.* **392**, 353 (2002).

71. W. R. Hix, A. Mezzacappa, M. Liebendoerfer, et al., *Bull. Am. Astron. Soc.* **33**, 1445 (2001).

72. E. S. Myra and F. D. Swesty, *Am. Astron. Soc. Meet. Abs.* **205**, #172.04 (2004).

73. C. Y. Cardall, Workshop on Numerical Methods for Multidimensional Radiative Transfer Problems, Heidelberg, Germany, 24-26 Sep. 2003 (2004), `astro-ph/0404401`.

74. E. Livne, A. Burrows, R. Walder, et al., *Astrophys. J.* **609**, 277 (2004).

75. R. Walder, A. Burrows, C. D. Ott, et al. (2004), `astro-ph/0412187`.

76. M. J. Berger and P. Colella, *J. Comput. Phys.* **82**, 64 (1989).

77. A. M. Khokhlov, *J. Comput. Phys.* **143**, 519 (1998).

78. J. U. Brackbill and D. C. Barnes, *J. Comput. Phys.* **35**, 426 (1980).

79. C. R. Evans and J. F. Hawley, *Astrophys. J.* **332**, 659 (1988).

80. A. Lucas-Serrano, J. A. Font, J. M. Ibáñez, et al., *Astron. Astrophys.* **428**, 703 (2004).

81. L. Del Zanna and N. Bucciantini, *Astron. Astrophys.* **390**, 1177 (2002).

82. L. Del Zanna, N. Bucciantini, and P. Londrillo, *Astron. Astrophys.* **400**, 397 (2003).

83. J. M. Lattimer and F. Douglas Swesty, Nuclear Physics A **535**, 331 (1991).

84. D. S. Balsara and D. S. Spicer, *J. Comput. Phys.* **149**, 270 (1999).

85. G. Tóth, *J. Comput. Phys.* **161**, 605 (2000).

86. M. Liebendörfer, A. Mezzacappa, and F.-K. Thielemann, *Phys. Rev. D* **63**, 104003 (2001).

87. C. Y. Cardall and A. Mezzacappa, *Phys. Rev. D* **68**, 023006 (2003).

88. E. F. D'Azevedo, B. Messer, A. Mezzacappa, et al., SIAM J. Sci. Comput. **26**, 810 (2005).

Section 4
Neutrino Mixing

CONSEQUENCES OF NEUTRINO MASS AND FLAVOR MIXING FOR CORE COLLAPSE SUPERNOVAE

G. M. FULLER

Department of Physics,
University of California, San Diego,
La Jolla, CA 92093

It is now an experimental fact that neutrinos have mass and can transform their flavors. Though flavor-dependent neutrino interaction processes play pivotal roles in nearly every important aspect of the dynamics and heavy element nucleosynthesis associated with core collapse supernova events, neutrino flavor mixing has yet to be incorporated into supernova modeling in an organic fashion. This is worrisome, especially given that neutrino flavor inter-conversion can affect the flow of lepton number, entropy, and isospin throughout the regions of interest. Here I discuss how neutrinos might transform their flavors in the supernova environment and the possible implications of this for the explosion mechanism and for heavy element nucleosynthesis. Potentially even more revolutionary would be the existence of new kinds of neutrinos which mix with the known neutrinos. Such "sterile" neutrinos have not been revealed convincingly in the lab, though there are hints. The existence of light sterile neutrinos could completely alter the current core collapse supernova paradigm.

1. Introduction

The core collapse supernova problem is, in my view, fundamentally a weak interaction problem. Though nuclear physics and multi-dimensional hydrodynamics and convection are essential aspects of the core collapse supernova phenomenon, each of these ingredients plays out against a background of entropy, lepton number, and composition (*i.e.*, neutron-to-proton ratio) transport completely dominated by the neutrinos and the rest of the weakly interacting sector. When all is said and done, nearly all the energy in this problem resides in the seas of neutrinos generated during and after core collapse. This is an impressive $\sim 10^{53}$ ergs, the gravitational binding energy of the cold neutron star remnant, which is some 10% of the rest mass-energy of the core! The manner and the extent to which this energy is coupled to the supernova matter depends on the *flavor* states of the neutrinos. And therein lies the fundamental reason for why neutrino flavor mixing may play an important role in core collapse/explosion.

To make this case let us sketch out the present generally accepted paradigm for core collapse supernovae. In broad brush the stellar collapse/explosion event is thought to proceed as follows: (1) A Chandrasekhar mass core of iron peak composition in nuclear statistical equilibrium (NSE) at low entropy with a radius of order that of the earth collapses on a time scale of about one second to a configuration with a central density at or above nuclear saturation density and with a radius of about 40 km initially; (2) A shock is generated at the edge of the inner "piston" (or homologous core, whose mass is set by the integrated history of electron capture and neutrino lepton number/entropy loss) and begins to move out, losing energy to nuclear photo-dissociation and, in some 30 ms, evolves into a standing accretion shock; (3) The cherished hope is that the ν_e neutrinos (produced by electron capture) and the neutrino pairs of each flavor produced thermally by the passage of the shock but which are trapped in the core will diffuse out to the edge of proto-neutron star fast enough to effect efficient heating of the material behind the shock, principally through the charged current capture processes

$$\nu_e + n \rightarrow p + e^-, \tag{1}$$
$$\bar{\nu}_e + p \rightarrow n + e^+, \tag{2}$$

and so, possibly aided by convective energy transport, re-energize the shock, leading to an explosion with a total kinetic and optical energy of $\sim 10^{51}$ ergs; (4) Subsequently neutrinos diffuse out (diffusion time scale $\tau_{\text{diff}} \sim 2\,\text{s}$) and, over some tens of seconds, the proto-neutron star shrinks down to a radius $\sim 10\,\text{km}$, eventually losing its initial $\sim 10^{57}$ units of electron lepton number and, along the way, at perhaps time post-bounce $t_{\text{pb}} \sim 3\,\text{s}$ to $15\,\text{s}$, possibly producing the hot bubble, neutrino-driven wind conditions conducive to r-process nucleosynthesis. Given the pivotal role of neutrinos at every turn in this story, it may behoove us to know their fundamental properties!

Fortunately, the experimental neutrino physics revolution of the last few years has outstripped theory and has revealed many properties of the ghost-like neutrinos for the first time. We now know that the weak interaction (or flavor states) of the neutrinos are not coincident with the propagation energy eigenstates (or "mass" states) for these species and, as a consequence, we know that neutrinos have nonzero masses. The flavor states ν_α ($\alpha = e, \mu, \tau, ...$) are related to the mass states in vacuum ν_i ($i = 1, 2, 3, ...$)

through a unitary transformation $U_{\alpha i}$:

$$|\nu_\alpha\rangle = \sum_i U_{\alpha i}|\nu_i\rangle. \tag{3}$$

For the particular case of just the known active neutrinos, a 3×3 mixing problem, $U_{\alpha i}$ is parameterized by three mixing angles $(\theta_{12}, \theta_{23}, \theta_{13})$ and a CP-violating phase δ.

Neutrino oscillation experiments/observations involving the sun, atmospheric neutrinos, and reactor experiments have given us two of the four unitary transformation parameters for the 3×3 case and have placed a constraint on a third[1]. The atmospheric and solar neutrino data along with the KamLAND results imply that $\sin^2 2\theta_{23} \approx 1$ and $0.42 < \tan^2 \theta_{12} < 0.45$. The present data also restricts mixing between the electron flavor and the third mass state to be $|U_{e3}|^2 < 2.5\%$, or $\sin^2 2\theta_{13} < 0.1$. Long baseline neutrino oscillation experiments and/or the next generation reactor experiments could pin down θ_{13} and, if this is big enough, measure the CP-violating phase. It could be, however, that if θ_{13} is very small our best hope for measuring these last two quantities will lie with the detection of a Galactic supernova neutrino signal.

Note that if we take $\theta_{23} = \pi/4$ and $\delta = 0$, then we can reduce the dimensionality of the active-neutrino-only case to an effective 2×2 problem[2,3]. If we define

$$|\nu_\mu^*\rangle \equiv \frac{|\nu_\mu\rangle - |\nu_\tau\rangle}{\sqrt{2}}, \tag{4}$$

$$|\nu_\tau^*\rangle \equiv \frac{|\nu_\mu\rangle + |\nu_\tau\rangle}{\sqrt{2}}, \tag{5}$$

then only mixing in the channel $\nu_e \rightleftharpoons \nu_\tau^*$ is important because ν_μ^* is coincident with an energy eigenstate. The effective 2×2 vaccum mixing angle is $\theta \approx \theta_{13}$ when θ_{13} is small.

This trick could remain approximately valid throughout the supernova medium because the mu and tau neutrinos have nearly identical interactions, at least until (or if) large scale neutrino flavor conversion begins. Further, it simplifies the active neutrino flavor evolution problem in supernovae considerably because 2×2 neutrino flavor mixing is particularly simple. For this reason here I will adopt a 2×2 approach, though it must always be kept in mind that the supernova problem is a 3×3 mixing problem and approximating this as a 2×2 scheme may miss essential physics.

In vacuum we can represent the unitary transformation between the flavor and mass bases in this case by a simple rotation through a single "effective vacuum mixing angle" θ,

$$|\nu_\alpha\rangle = \cos\theta|\nu_1\rangle + \sin\theta|\nu_2\rangle, \tag{6}$$

$$|\nu_\beta\rangle = -\sin\theta|\nu_1\rangle + \cos\theta|\nu_2\rangle, \tag{7}$$

where $\alpha \neq \beta$ and where each index α and β take on values from e, μ^*, s.

Here ν_s represents a possible new neutrino, different from the three known active flavors, which must, on account of the Z^0 width limit, possess interactions which are much weaker than those of the Standard Model Weak Interaction. Hence, new flavors of neutrinos are given the inaccurate moniker "sterile." In fact, such a neutrino would not actually be sterile by virtue of its mixing with active neutrinos. The idea of sterile neutrinos is not particularly radical in particle physics and, indeed, right-handed neutrino states play a central role in models of neutrino mass. The natural mass scale for these species is a unification scale, well out of the range of any energy scale of interest in supernovae. It would be a surprise if there were *light* sterile neutrino states. Disturbingly, the LSND experiment results, if interpreted as vacuum mixing in the $\nu_\mu \rightleftharpoons \nu_e$ channel, suggests just this[4]. This result, crucial for our picture of core collapse supernovae, is currently being tested in the mini-BooNE experiment at FNAL.

Let us put aside the issue of sterile neutrinos for now. We do not know what the absolute neutrino masses are but we do know the mass-squared differences (the differences of the squares of the neutrino mass eigenvalues) from the experiments. If we take $m_3 > m_2 > m_1$, then $\delta m_{12}^2 \equiv m_2^2 - m_1^2 \approx 8 \times 10^{-5}\,\mathrm{eV}^2$ and $\delta m_{13}^2 \equiv m_3^2 - m_1^2 \approx 3 \times 10^{-3}\,\mathrm{eV}^2$. The tritium endpoint experiment constrains all the absolute active neutrino mass eigenvalues to be $< 2\,\mathrm{eV}$.

The absolute neutrino masses are themselves too small to be dynamically important in any aspect of the supernova problem, but the mass-squared differences may well be large enough to cause large scale neutrino and/or antineutrino flavor conversion in important epochs where the shock is being re-heated or where neutrino-heated nucleosynthesis takes place.

2. Matter-Enhanced Active-Active Neutrino Flavor Transformation

Neutrinos can have their effective masses and mixings modified by scattering processes in dense environments and, in turn, this can result in large scale conversion of neutrino and/or antineutrino flavors in the supernova environment. The seminal work of Wolfenstein and Mikheyev and Smirnov

(MSW)[5] explored this physics in the coherent limit, where the neutrino transport mean free paths are large compared to resonance widths. It has been the cosmology/early universe community that has examined this problem in the incoherent limit, where neutrino mean free paths are small compared to resonance widths, and where collisionally-induced de-coherence results in conversion of neutrino flavors. To date, however, essentially all of the work on incoherent neutrino flavor evolution pertaining to both the early universe and supernova cores has been directed at the active-sterile neutrino flavor conversion channel[6,7]. Let us consider each of these limiting cases in turn.

2.1. *Coherent Neutrino Flavor Conversion above the Neutrino Sphere: Coping with Macroscopic Coherence*

This limit is most appropriate for neutrinos propagating sufficiently far above the neutrino sphere (roughly the proto-neutron star surface) that neutrino and antineutrino inelastic scattering and absorption/emission are subdominant, though not necessarily unimportant, processes. Coherent neutrino propagation corresponds to the standard MSW regime, like neutrino flavor evolution in the sun, but with an added and crucial twist.

While forward scattering of neutrinos on electrons dominates the neutrino effective potential/mass terms (closely related to the neutrino refractive indices) in the sun and in particular regimes of the supernova environment, especially very close to the neutrino sphere, the huge fluxes of neutrinos of all kinds dictate that neutrino-neutrino forward scattering can dominate in key regimes above the neutrino sphere[8,9]. Neutrino-neutrino forward scattering-generated potentials render the neutrino flavor evolution problem nonlinear, at least in the sense that neutrino flavor states at a given time and location determine how neutrinos transform their flavors there.

And the problem is worse than that. Consider a coherently propagating neutrino of energy E_R whose world line above the neutrino sphere is directed radially outward. The flavor evolution history of this test neutrino could be as follows: (1) It experiences an MSW resonance at a location determined by E_R and the forward scattering-generated potentials; (2) Whether or not it transforms its flavor at this resonance depends on the flavor states of the "background" neutrinos flying through this location and upon which our test neutrino is forward scattering; (3) In turn, the flavor state of each of these background neutrinos with energy $E < E_R$ at this location is determined by what happened to them at earlier, deeper MSW resonances, closer to the neutrinos sphere; (4) Likewise, at these background MSW resonances neutrino flavor conversion was determined by the flavor states of the background neutrinos at these deeper positions. With this line of reasoning one sees that each neutrino-neutrino forward scattering event "entangles" these two neutrino states and that, essentially, the coherent flavor evolution history of one neutrino is geometrically and coherently coupled to, or entangled with the flavor evolution histories of every other neutrino[10,11]!

In other words, the neutrino flavor field sufficiently far above the neutron star is a classic, but rare example of macroscopic quantum coherence. So far as I know this is unique in astrophysics. It confronts the supernova community with an altogether new and vexing transport problem.

In fact, the entangled histories of neutrino flavor states beg the question of whether classic mean field Schroedinger approaches miss essential aspects of the physics[12]. The large number of scattering events, however, likely will suppress at least some of the non-mean field effects[13]. I will here adopt this line of argument and explore the evolution of the neutrino flavor field above the proto-neutron star with a Schroedinger mean field treatment. Furthermore, I will make another simplifying approximation: that all neutrinos of a given energy in the background, traveling on arbitrarily directed world lines (arbitrary angles relative to the radial direction) evolve with the same history as a radially directed neutrino of the same energy. In fact, as I will discuss later, this is *not* a good approximation, but it will suffice to make a qualitative argument about the evolution of the neutrino flavor field in the coherent regime.

For 2×2 neutrino mixing it is the *difference* in neutrino forward scattering potential between different flavor states that is physically significant. Since ν_e and $\bar{\nu}_e$ neutrinos can forward exchange scatter on electrons and positrons via the weak charged current coupling while mu and tau flavor neutrinos cannot, neutrinos feel an extra potential $\hat{H}_{e\nu} = A(t)|\nu_e\rangle\langle\nu_e|$ whenever they are in the electron flavor state[5]. Here t is any Affine parameter (*e.g.,* time, radius) along the neutrino's world line and $A(t) = \sqrt{2}G_{\mathrm{F}}(n_{e^-} - n_{e^+})$ and $n_{e^\pm}$ are the local electron (positron) number densities.

There is a neutrino-neutrino forward neutral current exchange analog of this process in supernovae[8]. The potential in this case is proportional to the difference in the asymmetries (numbers of neutrinos minus numbers of antineutrinos) of the two neutrino flavors. However, the neutrinos and antineutrinos, unlike electrons/positrons, will have anisotropic distribution functions in the regions above the neutrino sphere. The structure of the low energy weak current results in a $(1 - \cos\theta_{\mathbf{pq}})$ factor in the potential contribution for a neutrino of spatial momentum $\mathbf{p}$ forward scattering on a neutrino with spatial momentum $\mathbf{q}$, where $\theta_{\mathbf{pq}}$ is the intersection angle of their world lines[8]. In the limit where neutrino ensembles are isotropic this term averages to unity; whereas, two ultra relativistic and co-linear neutrinos contribute nothing to the potential because $\theta_{\mathbf{pq}} = 0$. (This makes sense because, in the limit of co-linear null world lines, these neutrinos never scatter!). In direct analogy to the neutrino-electron forward exchange case we can write the neutral current exchange potential seen by a neutrino with spatial momentum $\mathbf{p}$ as an sum over the background neutrinos:

$$\hat{H}_{\nu\nu} = \sqrt{2}G_{\mathrm{F}} \int (1 - \cos\theta_{\mathbf{pq}})\big[\hat{\rho}_{\mathbf{q}} - \hat{\bar{\rho}}_{\mathbf{q}}\big]d^3\mathbf{q}. \tag{8}$$

Here the density operators for neutrinos and antineutrinos at this location/time and for a pencil of directions/momenta $d^3\mathbf{q}$ are designated $\hat{\rho}_{\mathbf{q}}d^3\mathbf{q}$ and $\hat{\bar{\rho}}_{\mathbf{q}}d^3\mathbf{q}$, respectively. These just simply count neutrinos of all flavors in the pencil and, *e.g.*, for neutrinos

$$\hat{\rho}(t)_{\mathbf{q}}d^3\mathbf{q} \equiv \sum_{\alpha} dn_{\nu_\alpha}|\Psi_{\nu_\alpha}(t)\rangle\langle\Psi_{\nu_\alpha}(t)|, \tag{9}$$

where $|\Psi_{\nu_\alpha}(t)\rangle$ is the state vector for a neutrino *born* on the neutrino sphere in the α flavor eigenstate. (In the example we consider here α could be either e or τ^*, hereafter just "τ".)

The neutrino energy distribution functions above the neutrino sphere are only very roughly of Fermi-Dirac, thermal form. However, for argument's sake we will take all flavors to have this distribution function form. The Fermi-Dirac distribution is determined by two parameters: a "temperature" T_{ν_α} and a "degeneracy parameter" η_{ν_α}. If neutrinos are in strict thermal and chemical equilibrium with matter then T_{ν_α} is identical to the matter temperature and the degeneracy parameter is the ratio of the neutrino chemical potential and this temperature. Above the neutrino sphere, however, the neutrino energy spectra are generally harder (hotter) than for the local matter. With Fermi-Dirac form, the number density of ν_α neutrinos in a pencil of momenta/energies $d^3\mathbf{p}$ is

$$dn_{\nu_\alpha} \approx \frac{1}{2\pi^2}\left(\frac{d\Omega_\nu}{4\pi}\right)\frac{E_\nu^2 dE_\nu}{e^{E_\nu/T_{\nu_\alpha}-\eta_{\nu_\alpha}}+1} \tag{10}$$

$$\approx \frac{L_{\nu_\alpha}}{\pi R_\nu^2}\frac{1}{\langle E_{\nu_\alpha}\rangle}\left(\frac{d\Omega_\nu}{4\pi}\right)f_{\nu_\alpha}(E_\nu)\,dE_\nu \tag{11}$$

In Eq.s (10) & (11) the pencil of directions is given in terms of polar angle $\theta_{\mathbf{p}}$ and azimuthal angle ϕ by $d\Omega_\nu = \sin\theta_{\mathbf{p}} d\theta_{\mathbf{p}} d\phi$. In all of these expressions we assume that neutrinos have relativistic kinematics so that $d^3\mathbf{p} = p^2 dp\, d\Omega_\nu \approx E_\nu^2 dE_\nu d\Omega_\nu$, where $E_\nu = \sqrt{\mathbf{p}^2 + m_\nu^2}$ is the neutrino energy and we take $\hbar = c = 1$. Eq. (11) gives dn_{ν_α} in a form specialized for the spherically symmetric, sharp neutrino sphere (radius R_ν) case and is normalized to be consistent with the overall energy luminosity L_{ν_α} in neutrinos of this type.

The normalized thermal energy distribution function for neutrino species ν_α is then

$$f_{\nu_\alpha}(E_\nu) = \frac{1}{T_{\nu_\alpha}^3 F_2(\eta_{\nu_\alpha})} \frac{E_\nu^2}{e^{E_\nu/T_{\nu_\alpha} - \eta_{\nu_\alpha}} + 1}, \qquad (12)$$

where F_2 is a relativistic Fermi integral of order 2. The Fermi integrals are defined for general order k and argument η by

$$F_k(\eta) \equiv \int_0^\infty \frac{x^k dx}{e^{x-\eta} + 1}. \qquad (13)$$

The average neutrino energies are $\langle E_{\nu_\alpha} \rangle = T_{\nu_\alpha} F_3(\eta_{\nu_\alpha})/F_2(\eta_{\nu_\alpha})$. In the absence of neutrino mixing, the luminosities and/or the average energies for the mu and tau neutrinos and their antiparticles are expected to all be roughly the same at any epoch and at any location. This may not extend to the ν_e's and $\bar{\nu}_e$'s. In fact, especially at late times post-bounce, we might expect a generic hierarchy of average neutrino energies $\langle E_{\nu_\tau} \rangle \approx \langle E_{\bar{\nu}_\tau} \rangle \approx \langle E_{\nu_\mu} \rangle \approx \langle E_{\bar{\nu}_\mu} \rangle \geq \langle E_{\bar{\nu}_e} \rangle > \langle E_{\nu_e} \rangle$. Anytime there is a difference in these average energies in the spectra emergent from the neutrino sphere or anytime there are different luminosities for the different flavors, neutrino flavor transformation can be important both dynamically and in setting the local neutron-to-proton ratio[8,9,11].

Armed with these tools for counting neutrino states, and making all of the approximations outlined above, we can write down a flavor basis mean field Schroedinger-like equation for the evolution of the neutrino flavor amplitudes (a similar equation for the evolution of the antineutrino amplitudes follows in obvious fashion)

$$i\frac{\partial \Psi_f}{\partial t} \approx \frac{1}{2} \begin{pmatrix} A + B - \Delta\cos 2\theta & \Delta\sin 2\theta + B_{e\tau} \\ \Delta\sin 2\theta + B_{\tau e} & \Delta\cos 2\theta - A - B \end{pmatrix} \Psi_f, \qquad (14)$$

where we ignore terms proportional to the identity that only serve to give both flavor amplitudes a common overall phase, and define Ψ_f to be a column vector of flavor amplitudes

$$\Psi_f \equiv \begin{pmatrix} \langle \nu_e | \Psi_{\nu_\alpha}(t) \rangle \\ \langle \nu_\tau^* | \Psi_{\nu_\alpha}(t) \rangle \end{pmatrix}. \tag{15}$$

In the above I define the neutrino energy (E_ν) dependent quantity $\Delta(E_\nu) = \Delta \equiv \delta m^2/2E_\nu$. The neutrino-neutrino flavor diagonal potential is[8,9]

$$B = \sqrt{2}G_{\rm F} \int (1 - \cos\theta_{\bf pq}) \left(\left[\hat{\rho}_{\bf q} - \hat{\bar{\rho}}_{\bf q} \right]_{ee} - \left[\hat{\rho}_{\bf q} - \hat{\bar{\rho}}_{\bf q} \right]_{\tau\tau} \right) d^3{\bf q}; \tag{16}$$

while the corresponding flavor off-diagonal potential is[14,10,15]

$$B_{e\tau} = 2\sqrt{2}G_{\rm F} \int (1 - \cos\theta_{\bf pq}) \left[\hat{\rho}_{\bf q} - \hat{\bar{\rho}}_{\bf q} \right]_{e\tau} d^3{\bf q}. \tag{17}$$

Since the operator in Eq. (14) is Hermitian, B is real and the flavor off-diagonal potential need not be, but we must have $B_{\tau e} = B_{e\tau}^\dagger$. The flavor basis density operator matrix elements are

$$\left[\hat{\rho}_{\bf q}(t) - \hat{\bar{\rho}}_{\bf q}(t) \right]_{ee} d^3{\bf q} \equiv \langle \nu_e | \hat{\rho}_{\bf q}(t) d^3{\bf q} | \nu_e \rangle - \langle \bar{\nu}_e | \hat{\bar{\rho}}_{\bf q}(t) d^3{\bf q} | \bar{\nu}_e \rangle \tag{18}$$

$$\left[\hat{\rho}_{\bf q}(t) - \hat{\bar{\rho}}_{\bf q}(t) \right]_{\tau\tau} d^3{\bf q} \equiv \langle \nu_\tau | \hat{\rho}_{\bf q}(t) d^3{\bf q} | \nu_\tau \rangle - \langle \bar{\nu}_\tau | \hat{\bar{\rho}}_{\bf q}(t) d^3{\bf q} | \bar{\nu}_\tau \rangle \tag{19}$$

$$\left[\hat{\rho}_{\bf q}(t) - \hat{\bar{\rho}}_{\bf q}(t) \right]_{e\tau} d^3{\bf q} \equiv \langle \nu_e | \hat{\rho}_{\bf q}(t) d^3{\bf q} | \nu_\tau \rangle - \langle \bar{\nu}_e | \hat{\bar{\rho}}_{\bf q}(t) d^3{\bf q} | \bar{\nu}_\tau \rangle. \tag{20}$$

Note that, _e.g._, $\left[\hat{\rho}_{\bf q}(t) - \hat{\bar{\rho}}_{\bf q}(t) \right]_{ee} d^3{\bf q}$ gives the expectation value for the net number of ν_e's over $\bar{\nu}_e$'s in the pencil of momenta and directions $d^3{\bf q}$ at momentum ${\bf q}$. It is very important to recognize that the off-diagonal matrix elements of the density operators are nonzero and so contribute to $B_{e\tau}$ only if neutrinos do not remain in the flavor states in which they were born. That is, $B_{e\tau}$ will be zero until at least _some_ neutrinos mix appreciably.

As neutrinos propagate through the supernova medium the general relation between the flavor basis and the "instantaneous" mass (energy) states $|\nu_1(t)\rangle$ and $|\nu_2(t)\rangle$ can be written

$$|\nu_e\rangle = \cos\theta_M(t)|\nu_1(t)\rangle + e^{-i\beta(t)} \sin\theta_M(t)|\nu_2(t)\rangle \tag{21}$$

$$|\nu_\tau\rangle = -e^{i\beta(t)} \sin\theta_M(t)|\nu_1(t)\rangle + \cos\theta_M(t)|\nu_2(t)\rangle \tag{22}$$

where $\theta_M(t)$ is the effective matter mixing angle at location/time t, and $\beta(t)$ is a time and medium dependent phase. In general, as a neutrino propagates above the neutrino sphere in the presence of a neutrino background the flavor off-diagonal potential and at least some of the amplitudes in the instantaneous unitary transformation will be complex.

For now, however, in the interest of making a point, let us restrict our discussion to potentials and amplitudes that are strictly real. I will come back and relax this stricture later on. For real $B_{e\tau}$ we can write

$$\Delta_{\text{eff}} \cos 2\theta_M\,(t) \equiv \Delta \cos 2\theta - A - B \tag{23}$$

$$\Delta_{\text{eff}} \sin 2\theta_M\,(t) \equiv \Delta \sin 2\theta + B_{e\tau}. \tag{24}$$

From these relations it is clear that

$$\Delta_{\text{eff}} = \sqrt{(\Delta \cos 2\theta - A - B)^2 + (\Delta \sin 2\theta + B_{e\tau})^2}. \tag{25}$$

Likewise, for the antineutrinos the effective matter mixing angle $\bar{\theta}_M(t)$ is related to the potentials through,

$$\Delta_{\text{eff}} \cos 2\bar{\theta}_M\,(t) \equiv \Delta \cos 2\theta + A + B \tag{26}$$

$$\Delta_{\text{eff}} \sin 2\bar{\theta}_M\,(t) \equiv \Delta \sin 2\theta - B_{e\tau} \tag{27}$$

where it should be noted that antineutrinos experience potentials of opposite sign from those for the neutrinos.

An MSW resonance for a neutrino of energy E_{res} occurs at a time/location along this neutrino's world line where

$$\Delta\,(E_{\text{res}}) \cos 2\theta = \frac{\delta m^2 \cos 2\theta}{2 E_{\text{res}}} = A + B. \tag{28}$$

It is obvious that there will be maximal mixing $\theta_M(t_{\text{res}}) = \pi/4$ for neutrinos of energy E_{res} at the location t_{res} where the condition in Eq. (28) is met and where the total flavor diagonal potential $V = A + B$ is positive. If this potential is negative, there will be a resonance in the antineutrino channel at this location, $\bar{\theta}_M(t_{\text{res}}) = \pi/4$. Note that the effective matter mixing angles for neutrinos and antineutrinos will be vanishingly small wherever the magnitude of the total potential is large compared to its resonance value. Also note that the width, in total flavor-diagonal potential, of the resonance region is

$$\delta V \approx \Delta \sin 2\theta \left| 1 + \frac{2 E_\nu B_{e\tau}}{\delta m^2 \sin 2\theta} \right| \tag{29}$$

Wherever $B_{e\tau}$ is negligible we expect resonances to be narrow because, for our problem, the effective vacuum mixing angle will be small (*i.e.*, $\theta \approx \theta_{13}$). The extent of the resonance region in space/time is $\delta t \approx (dt/dV)\,|_{\text{res}}\delta V$, or $\delta t \approx H \tan 2\theta |1 + (2 E_\nu B_{e\tau})/(\delta m^2 \sin 2\theta)|$, where $H \equiv |V(dt/dV)|$ and will be comparable to t (*i.e.*, radius) wherever the neutrino-electron forward scattering potential A dominates the total potential. Mixing will be maximal or near maximal in the resonance region, but neutrino flavors can be regarded as efficiently transforming across this region if neutrino evolution is adiabatic.

Insight into the physical significance of an MSW resonance can be gleaned by examining the problem in the mass basis. With the above definitions and limits we can transform the flavor amplitude evolution equation, Eq. (14), to the instantaneous mass basis

$$i\frac{\partial \Psi_M}{\partial t} \approx \left[\frac{\Delta_{\rm eff}}{2}\begin{pmatrix} -1 & 0 \\ 0 & 1 \end{pmatrix} + \begin{pmatrix} 0 & -\dot{\theta}_M(t) \\ \dot{\theta}_M(t) & 0 \end{pmatrix} \right]\Psi_M, \tag{30}$$

where the amplitudes to be in the instantaneous mass states are arranged in a column vector

$$\Psi_M \equiv \begin{pmatrix} \langle \nu_1(t)|\Psi_{\nu_\alpha}(t)\rangle \\ \langle \nu_2(t)|\Psi_{\nu_\alpha}(t)\rangle \end{pmatrix}, \tag{31}$$

and where $\dot{\theta}_M(t)$ is the time derivative of the effective matter mixing angle at location/time t and where, again, we ignore terms proportional to the identity. When the ratio of the diagonal to off-diagonal terms in this equation, the adiabaticity parameter $\gamma = \Delta_{\rm eff}/2\dot{\theta}_M(t)$, is large, flavor amplitude evolution is adiabatic.

In the adiabatic limit, a neutrino born in an instantaneous mass state (or one close to it as $|\Psi_{\nu_\alpha}\rangle$ will be near the neutrino sphere) stays in this state. Given the asymptotic limits of (1) ultra high electron Fermi energy near the neutrino sphere and (2) vacuum outside the supernova, we see that adiabatic neutrino propagation through an MSW resonance corresponds to complete conversion of the neutrino from one flavor to another.

In fact the tracks of the squares of instantaneous effective mass eigenvalues $m_{1\rm eff}^2$ and $m_{2\rm eff}^2$ through an MSW resonance would cross at resonance except for a gap $\delta m_{\rm eff}^2 = m_{2\rm eff}^2 - m_{1\rm eff}^2$,

$$\delta m_{\rm eff}^2 \Big|_{\rm res} \approx \delta m^2 \sin 2\theta \left| 1 + \frac{2E_\nu B_{e\tau}}{\delta m^2 \sin 2\theta} \right|. \tag{32}$$

The larger this gap, the more adiabatic the neutrino's evolution and the more likely it will be to transform its flavor.

The Landau-Zener approximation is good for narrow resonances and suggests that the probability of conversion of a neutrino's flavor across the resonance width is $P \approx 1 - \exp(-\pi\gamma/2)^{16}$. The adiabaticity parameter γ is proportional to the ratio of the resonance width δt to the oscillation length at resonance $L_{\rm res} \approx 4\pi E_\nu/\delta m_{\rm eff}^2$,

$$\gamma \approx \frac{\delta m^2 H}{2E_\nu} \cdot \frac{\sin^2 2\theta}{\cos 2\theta} \cdot \left| 1 + \frac{2E_\nu B_{e\tau}}{\delta m^2 \sin 2\theta} \right|^2. \tag{33}$$

Let us set $B_{e\tau} = 0$. This corresponds to the classic MSW case: if $V = A + B > 0$ neutrinos in the supernova environment will experience a narrow MSW resonance and antineutrino mixing will be suppressed; and *vice versa* if $V < 0$. The neutrino (or antineutrino) energy which is resonant at a location/time where the total potential is $V = A + B$ is

$$E_{\rm res} = \frac{\delta m^2 \cos 2\theta}{A + B} \approx (0.02\,{\rm MeV}) \left(\frac{\delta m^2 \cos 2\theta}{3 \times 10^{-3}\,{\rm eV}^2} \right) \left(\frac{10^6\,{\rm g\,cm}^{-3}}{\rho(Y_e + Y_\nu)} \right), \quad (34)$$

where I have scaled to the atmospheric neutrino mass-squared difference and a density where, for example, we might be interested in shock re-heating or the neutron/proton ratio-altering effects of neutrino flavor transformation. Here Y_e is the net number of electrons per baryon and, similarly Y_ν is defined such that $B = \sqrt{2}G_{\rm F}\rho N_{\rm A} Y_\nu$. We can conclude that for typical conditions above the neutrino sphere, and with the *measured* values of neutrino mass-squared difference, we can expect only low energy neutrinos to transform unless $V = A + B \to 0$. (This can occur[10].) Put another way, because the neutrino mass-squared differences are so small, ordinary MSW evolution will produce negligible effects on the the shock re-heating process and likely subdominant effects in the later r-process epoch unless Y_ν is driven substantially negative.

The situation changes dramatically if $B_{e\tau}$ is significant. In this case numerical calculations suggest that neutrino *and* antineutrino mixing can be substantial and widespread in spatial extent. Recently Y.-Z. Qian and I have shown how this can come about[17]. We have found a simple limiting solution when the neutrino backgrounds dominate flavor evolution. This Background Dominant Solution (BDS) is valid (1) in the case where amplitudes and potentials are real, and (2) where the condition $|B_{e\tau}| \gg V$ is met. The former condition is unphysical but, we argue, adequate to show *where* $|B_{e\tau}|$-dominated mixing might occur and what it looks like qualitatively. The latter condition can be met in diverse circumstances in the supernova environment. In fact, when $|B_{e\tau}|$ is large, then $B \to 0$ and the condition to obtain the BDS becomes simply $|B_{e\tau}| \gg A$.

The essence of the BDS solution is as follows: neglect all potential terms compared to $B_{e\tau}$ and $|B_{e\tau}|$ in the Eqs. (23) through (27). In this limit note that the effective in-medium mixing angles become independent of neutrino/antineutrino energy over the range of energy for which $|B_{e\tau}| \gg A$

remains valid. These effective matter mixing angles are then

$$\theta_M \to \frac{\pi}{4}, \tag{35}$$

$$\bar{\theta}_M \to \frac{3\pi}{4}, \tag{36}$$

for $B_{e\tau}$ real and positive. If $B_{e\tau}$ is negative the limits become $\theta_M \to 3\pi/4$ and $\bar{\theta}_M \to \pi/4$. Either way this corresponds to *maximal* mixing, like in an MSW resonance, but now for both neutrinos *and* antineutrinos and nearly energy independent! Note that in the BDS, with maximal mixing for neutrinos and antineutrinos, $B \to 0$.

The BDS, if ever attained, presents us with a neutrino flavor mixing situation completely different from the classic MSW one: now both neutrinos and antineutrinos mix maximally and they can do so over a broad range of neutrino/antineutrino energy (perhaps corresponding to a broad spatial extent above the neutrino sphere). Furthermore, note that for large $|B_{e\tau}|$ resonance widths increase and, from Eq. (32), so does the effective mass-squared gap between the mass tracks. This clearly boosts the tendency for adiabatic flavor evolution and so increases the probability of asymptotic neutrino and antineutrino flavor conversion. (In the BDS the adiabaticity parameter becomes $\gamma \to H|B_{e\tau}|^2/A$.)

The BDS potential condition, $|B_{e\tau}| \gg A$, plausibly could be met during the shock break-out (neutronization burst) event during the shock re-heating epoch and more or less throughout the r-process/neutrino-driven wind epoch. This constraint, therefore, constitute *necessary* conditions for the BDS to exist in the supernova environment. Are these also *sufficient* to insure the attainment of the BDS?

The important question is then whether a treatment of the full neutrino and antineutrino evolution equations in a realistic supernova model ever finds the BDS or something like it. Numerical calculations give important hints that the answer may be yes. The Pastor and Raffelt calculation[18] finds "synchronization," a phenomenon which I argue is closely related to the BDS, but they do so with an un-physically large value of δm^2 and, as in all published numerical calculations to date, they neglect the effect of different flavor evolution histories on different trajectories (angles). A recent calculation by Yüksel and Balantekin[19], with the first of these short comings resolved but not the second, fails to find large scale synchronized mixing over restricted set of supernova conditions. It remains to be seen if they will find something like the BDS at any point in more general conditions and when the angles of neutrino trajectories are treated properly.

Both of these numerical calculations allow for complex amplitudes and potentials, and do so in a spin-1 representation of SU(2): the so-called spin polarization analogy. In this picture, which is crafted to be analogous to spin precession in a magnetic field, the flavor-basis matrix elements of the density operators (*e.g.*, Eq. 9) at location t are written as linear combinations of the identity and the Pauli matrices,

$$\begin{pmatrix} \left(\hat{\rho}_{\mathbf{q}}d^3\mathbf{q}\right)_{ee} & \left(\hat{\rho}_{\mathbf{q}}d^3\mathbf{q}\right)_{e\tau} \\ \left(\hat{\rho}_{\mathbf{q}}d^3\mathbf{q}\right)_{\tau e} & \left(\hat{\rho}_{\mathbf{q}}d^3\mathbf{q}\right)_{\tau\tau} \end{pmatrix} = \frac{1}{2}\begin{pmatrix} P_z + P_0 & P_x - iP_y \\ P_x + iP_y & P_0 - P_z \end{pmatrix} \tag{37}$$

and similarly for the antineutrinos. These, and their time derivatives, are evolved forward in time/location from the neutrino sphere with the Heisenberg representation analogs of Eq. (14). Here the "polarization vector" is $\mathbf{P} \Rightarrow \{P_x, P_y, P_z\}$. The flavor off-diagonal potentials $B_{e\tau}$ and $B_{\tau e}$ are related to $P_x - iP_y$ and $P_x + iP_y$, respectively, through, *e.g.*, Eq. (17). In general the vector $\mathbf{P}$ will precess around the z-axis at a frequency proportional to the total flavor diagonal potential V.

In the limit where the flavor-off diagonal potentials dominate (corresponding to the BDS), the off-diagonal terms in the above matrix equation will dominate, and so $\mathbf{P}$ will lie in the xy-plane, rotating in this plane with frequency A. This corresponds to maximal mixing. The Fuller & Qian BDS limit corresponds to the points in this cycle when $\mathbf{P}$ points along the positive x-axis (real, positive $B_{e\tau}$), or opposite to it (real, negative $B_{e\tau}$). Therefore, we can conclude that the necessary conditions ($|B_{e\tau}| \gg A$) for the BDS to obtain in the supernova environment derived above on the basis of assumed real potentials will also pick out the necessary conditions for synchronized maximal mixing when the full complex potentials are employed.

At this point we are awaiting a complete numerical treatment of the coherent neutrino flavor evolution problem above the neutrino sphere. This must include neutrino flavor transformation feedback on the electron fraction Y_e, through Eqs. (1) & (2), and also a correct and self consistent calculation of neutrino flavor history on differently-directed neutrino trajectories (different angles) from the neutrino sphere. This latter effect may be quite important. As argued above, neutrino and antineutrino mixing effects likely do not become dramatic until $|B_{e\tau}|$ becomes significant. In turn, $|B_{e\tau}|$ is zero until *some* neutrinos have swapped flavor labels. Low energy, high angle neutrinos, possibly at late times traveling on trajectories nearly tangential to the neutrino sphere, will see a large density scale height at MSW resonances and will therefore be the first neutrinos to transform. These will bring up $|B_{e\tau}|$ from zero and could then lead to a kind of avalanche, as neutrinos of higher energy transform in the higher $|B_{e\tau}|$ background, further increasing $|B_{e\tau}|$, etc.

Active-sterile coherent neutrino flavor evolution in the region above the neutron star in the supernova environment is much simpler than in the active-active channel. In the active-sterile neutrino mixing channel there are no flavor off-diagonal potentials. This is because there is no contribution to neutrino effective mass in forward scattering processes when the neutrino is in the sterile state.

As a result, coherent neutrino flavor evolution in the active-sterile channel is the classic MSW case discussed above. However, if the sterile neutrinos are like those suggested by the LSND experiment, then the mass-squared differences are large, $0.2\,\mathrm{eV}^2 \leq \delta m^2 \leq 25\,\mathrm{eV}^2$. The result is that conventional MSW mixing effects "kick in" in a dramatic fashion, relatively close to the neutrino sphere and so can affect shock re-heating and the r-process in an impressive manner as I will show below.

2.2. *Collision-Mediated Neutrino Flavor Conversion*

At very high matter density, *e.g.*, in the post-bounce supernova core, neutrino mean free paths between collisions will become very small (10's of centimeters to 10's of meters). These length scales may be considerably smaller than typical MSW resonance widths $\sim H \tan 2\theta < 0.3\,\mathrm{km}$. In this case neutrino flavor evolution through MSW resonances could not be regarded as coherent. We could not employ a mean field Schrödinger treatment to follow flavor evolution in this case, but instead must employ the quantum kinetic equations - the coherent limit of which is the Schrödinger equation and the completely incoherent limit of which is the Boltzmann equation. (See the recent work of Strack & Burrows[20] for a supernova formulation of the quantum kinetic equations.)

Most of the work done on this problem has been done in the active-sterile neutrino flavor conversion channel[6,7]. This is for three reasons: (1) there are no flavor off-diagonal potentials in the active-sterile channel; (2) wave function collapse in the active-sterile channel is well understood[21]; and (3) Dark Matter candidate sterile neutrinos have rest masses in the $\sim$ keV range, setting up MSW resonances deep in the supernova core and inviting speculation as to whether the supernova phenomenon can be used as a way to constrain sterile neutrino rest masses and their vacuum mixing with active species.

Extensions of this work to the active-active channel have hung up on an understanding of point (2) for the active-active regime and what may turn out to be a misconception. On this latter issue, active-active neutrino mass-squared differences are small and the densities in the core are large, so from Eqs. (23) through (27) it is obvious that effective in-medium mixing angles will be very very small, much smaller than vacuum mixings. However, given the small neutrino mean free paths, neutrinos will undergo a huge number of scattering events and one has to wonder whether this could trump the small flavor admixtures at each individual scattering event.

The overall net transport properties of , *e.g.*, electron neutrinos, might be altered by the fact that they are spending part of their time as muon/tau neutrinos or that the inter-conversion rate among these flavors is significant. Clearly this will be all the more dramatic in the active-sterile channel. The overall diffusion time for active neutrinos could be reduced dramatically if they are spending part of their time as sterile species which zip through the core at the speed of light!

The process of production of sterile neutrinos from an ensemble of active ones is a complicated quantum mechanical many-body problem. We can get a crude idea of what is going on with the neutrino fields if we describe what happens in a single particle framework. At each real scattering event the neutrino must be in an active flavor state. In this way scattering is like a measurement, forcing the neutrino state to "take a stand," and so inducing wave function collapse. At each collision then the neutrino's state is set to "active" but this state will evolve coherently with time until, at the next collision, there will be some coherent superposition of active and sterile components. At this point wave function collapse engendered by a collision could result in producing a sterile neutrino. This single particle picture inadequately describes what is inherently a many-body, second quantized process, but it serves to get across the basic idea.

An averaged Boltzmann approach has been applied to this process in the early universe and in supernova cores[6]. If $f_s(p,t)$ and $f_\alpha(p,t)$ are the sterile and active neutrino distribution functions with momentum p at time/location t, respectively, and $\hat{H}$ is a local hydrodynamic material expansion rate, then the averaged Boltzmann equation for the evolution of the sterile component is

$$\frac{\partial f_s(p,t)}{\partial t} - \hat{H}\frac{\partial f_s(p,t)}{\partial p} \approx \Gamma\left(\nu_\alpha \to \nu_s; p, t\right)\left[f_\alpha(p,t) - f_s(p,t)\right]. \tag{38}$$

Here the "averaging" comes about in this sense: the rate of sterile neutrino production is related to the total scattering rate $\Gamma_\alpha(p)$ for an active ν_α

neutrino with momentum p, the effective in-medium mixing angle θ_M at this location and energy $E_\nu = p$, and the local neutrino oscillation length $l_M \approx 2\pi/\Delta_{\text{eff}}$, by

$$\Gamma\left(\nu_\alpha \to \nu_s; p, t\right) \approx \frac{\frac{1}{2}\Gamma_\alpha\left(p\right)\sin^2 2\theta_M}{\left[1 + \left(\frac{1}{2}\Gamma_\alpha\left(p\right)l_M\right)^2\right]}. \tag{39}$$

With no off-diagonal potential in the active-sterile channel we can show that

$$\sin^2 2\theta_M \approx \frac{\tan 2\theta}{\sqrt{\left(1 \mp E_\nu/E_{\text{res}}\right)^2 + \tan^2 2\theta}}, \tag{40}$$

where the $E_{\text{res}} = \delta m^2 \cos 2\theta/2V$ is the resonance energy, the minus/plus sign is to be taken for the neutrino/antineutrino channel, θ and δm^2 are the appropriate active-sterile effective vacuum mixing angle and mass-squared difference, respectively, and the total active-sterile $\nu_\alpha \rightleftharpoons \nu_s$ channel potential V is, in terms of the number density of baryons n_b, electrons $n_e = n_b Y_e$, neutrons $n_n = n_b - n_e$ and number densities of active neutrino species n_{ν_α}, n_{ν_β}, n_{ν_γ} (with $\alpha \neq \beta \neq \gamma$ and α, β, γ drawn from e, μ, τ),

$$V = \sqrt{2}G_{\text{F}}\, n_M + \sqrt{2}G_{\text{F}}\left[2\left(n_{\nu_\alpha} - n_{\bar{\nu}_\alpha}\right) + \left(n_{\nu_\beta} - n_{\bar{\nu}_\beta}\right) + \left(n_{\nu_\gamma} - n_{\bar{\nu}_\gamma}\right)\right] \tag{41}$$

where $n_M = \left(n_e - \frac{1}{2}n_n\right)$ when $\alpha = e$, and $n_M = \left(-\frac{1}{2}n_n\right)$ otherwise.

Note the effect of the denominator in Eq. (39). Clearly, as the total ν_α scattering rate becomes very large this denominator suppresses the sterile neutrino production rate. This is the Quantum Zeno Effect. As the scattering rate increases there is little time for the coherently-developing neutrino to develop much phase (much sterile admixture) between scattering events ("measurements").

From the manner in which the sterile neutrino production rate is proportional to $\sin^2 2\theta_M$ in Eq. (39) we can conclude that in many circumstances the lion's share of neutrino conversion $\nu_\alpha \to \nu_s$ will take place at MSW resonances where, within the resonance width, $\sin^2 2\theta_M \approx 1$. Abazajian, Fuller, and Patel studied the evolution of these resonances and the overall active-sterile channel potentials (Eq. 41) in supernova cores[6]. They found a striking result: the potentials evolve to zero by conversion of active neutrinos or antineutrinos to steriles on time scales which, depending on parameters and equation of state issues, could be much faster than any other time scale in the problem, including the sound crossing time. In turn, this effect has led to the suggestion that the active-sterile admixture of Dark Matter sterile neutrinos along these lines could give rise to large pulsar kicks[22].

Net lepton numbers in any of the flavors at core bounce could make for nonzero potentials in active-active and active-sterile channels, with interesting effects on dynamics and the neutrino signal[23]. In the case of the active-sterile channel we know that these potentials will quickly evolve toward zero. It is an open issue whether this evolution toward zero potential would be accompanied by deleterious or even beneficial effects on the various problems associated with our models for core collapse supernova explosions and nucleosynthesis. The effect of the evolution of the active-active potentials in the core is relatively unexplored and is, at this point, a completely open issue.

3. Neutrino Flavor Transformation Effects in Shock Re-Heating and the r-Process

So far I have sketched out how neutrinos could transform their flavors both in active-active and in active-sterile channels in both the core and the near-free-streaming envelope of the supernova. This begs the question: How does neutrino flavor transformation affect models for core collapse supernovae and associated nucleosynthesis? Allow me to bring up two outstanding issues among the many open questions in this subject: shock re-heating; and r-process nucleosynthesis.

As first pointed out by Fuller, Mayle, Meyer, and Wilson[9] and subsequently examined by Bruenn and Mezzacappa[24], if there is a hierarchy in average neutrino energies $\langle E_{\nu_\tau} \rangle \approx \langle E_{\bar{\nu}_\tau} \rangle \approx \langle E_{\nu_\mu} \rangle \approx \langle E_{\bar{\nu}_\mu} \rangle \geq \langle E_{\bar{\nu}_e} \rangle > \langle E_{\nu_e} \rangle$ and/or similar hierarchies for luminosities, then matter-enhanced transformation in the channels $\nu_{\mu/\tau} \rightleftharpoons \nu_e$ or $\bar{\nu}_{\mu/\tau} \rightleftharpoons \bar{\nu}_e$ beneath the shock will make for enhanced heating through the reactions in Eqs. (1) & (2) . In principle this has a factor ~ 2 leverage in the total heating behind the shock and so could solve the explosion problem. However, there are two flies in the ointment: (1) we now know that neutrino mass-squared differences are too small to effect conventional MSW-enhanced heating under the shock - the MSW resonance regions are well out beyond the shock for relevant neutrino energies at the relevant times post-bounce; and (2) except for the brief neutronization burst associated with shock break-out through the neutrino sphere there is not likely to be a significant hierarchy in average neutrino energy or in luminosity.

However, as argued above, point (1) may no longer be an issue[17]. It is possible that the luminosity disparity between ν_e's and $\bar{\nu}_e$'s in the neutronization burst may be enough to "kick" the neutrino and antineutrino fields above the neutrino sphere into something like the BDS. Once mixing is significant it may tend to stay that way long enough to have some leverage on shock re-heating with, of course, the caveat at later times of point (2). Only detailed simulations which dynamically couple the neutrino field's flavor evolution with the heating and composition physics will be able to settle this issue.

There is, however, no doubt that sterile neutrinos could significantly alter the shock re-heating, delayed mechanism supernova explosion paradigm. As outlined above, active-sterile neutrino mixing could in principle allow active neutrinos to spend part of their time as sterile species, thereby increasing their effective mean free paths, decreasing neutrino diffusion times, and so boosting neutrino luminosities at the neutrino sphere. Again, heating under the shock will be enhanced and we might be able to solve the explosion problem in one-dimension. Note that a scenario along these lines would completely alter the way lepton number, energy, and entropy are transported in the core. Of course, the real issue here is whether sterile neutrinos with the relevant mass and mixing parameters exist. The mini-BooNE experiment at FNAL and the next generation X-Ray Observatories will either discover these particles or greatly narrow the available parameter space[25].

The r-process nucleosynthesis problem is another matter. Here there is growing evidence from meteoritic data and especially from observations of abundances on the surfaces of Ultra Metal-Poor halo stars that individual core collapse supernova events produce the r-process nuclides from nuclear mass $A = 100$ through the $A = 130$ and $A = 195$ abundance peaks in a *solar system abundance pattern*.[26] This is significant because in this abundance pattern the peaks are prominent and the total amount of material in the $A = 130$ is comparable to that in the $A = 195$ peak. This suggests that the r-process occurred in an environment that not only produced and preserved the abundance peaks but possibly also was neutron-rich enough to engender fission cycling that would tie together the total abundances of 130 and 195 peak material.

Our favorite candidate site for r-process nucleosynthesis is the hot bubble, neutrino-driven wind that may form at $t_{\rm pb} > 3\,{\rm s}$ - $20\,{\rm s}$. By this time the density is low and the entropy is high. High entropy dictates that freeze-out from nuclear statistical equilibrium here looks like the isospin mirror of Big Bang Nucleosynthesis, *i.e.*, alpha particles form aggressively, locking up all the free nucleons they can and thereby, for initially neutron-rich conditions, isolating free neutrons. One issue is how high this entropy could be, with the answer depending on the progress and tamper influence of the shock at late times. Another issue, almost completely dependent on neutrino flavor evolution, is how low Y_e could be. In the end, the material expansion rate, entropy per baryon, and Y_e in the hot bubble determine the all-important neutron-to-seed nucleus ratio. We need something like ~ 100 neutrons per baryon to get from iron peak seed nuclei to uranium.

That is a tall order for a slow expansion neutrino-driven wind. Fuller & Meyer & McLaughlin pointed out what we termed the "Alpha Effect."[27] It goes like this: (1) nucleons are gravitationally bound near the proto-neutron star by about $100\,\mathrm{MeV}$, while typical neutrino energies are $\sim 10\,\mathrm{MeV}$; (2) we must have high enough fluxes of ν_e's and $\bar{\nu}_e$'s to effect ejection of these nucleons to space, which implies that the fluxes of these neutrinos have to be high enough that each nucleon on average interacts some ~ 10 times with neutrinos; (3) this means, in turn, that when the alpha particles form and the free neutrons are isolated they will capture ν_e's, turn into protons, which are immediately incorporated into alpha particles. Each reaction $\nu_e + n \to p + e^-$, therefore, results in the loss of two free neutrons. In short order there may not be enough neutrons to make the required r-process abundance pattern.

Increased entropy, or a drastically faster material outflow rate could circumvent the Alpha Effect, but usually at the cost of some defect in the abundance yield and pattern, *e.g.*, no or reduced peaks in the fast outflow scenarios. Now here is the salient point: active-active neutrino mixing almost always makes this problem *worse* by tending to increase Y_e.

However, it has been pointed out that matter-enhanced active-sterile neutrino flavor conversion can engineer a neat solution to this conundrum[28,3]. In essence, the ν_e neutrinos have a high enough flux to effect baryon ejection in the regions nearer the neutron star where ejection is determined, but have disappeared, turned into sterile species, by the time/location where the alpha particles form, so there is no Alpha Effect.

And it is even better than that. Through ordinary MSW evolution in the channel $\nu_e \to \nu_s$, the ν_e flux falls, and since Y_e is determined by the competition[11,29] between the reactions in Eqs. (1) & (2), this quantity begins to fall and the material in the wind becomes more neutron-rich. Note that baryonic term in the total active-sterile potential in Eq. (41) is proportional to $\rho(Y_e - 1/3)$. Since this term dominates the potential in typical conditions, the total potential eventually will be driven negative. When this happens the $\bar{\nu}_e \to \bar{\nu}_s$ channel becomes matter-enhanced and $\bar{\nu}_e$'s are converted. This, however, does not result in a significant increase in Y_e

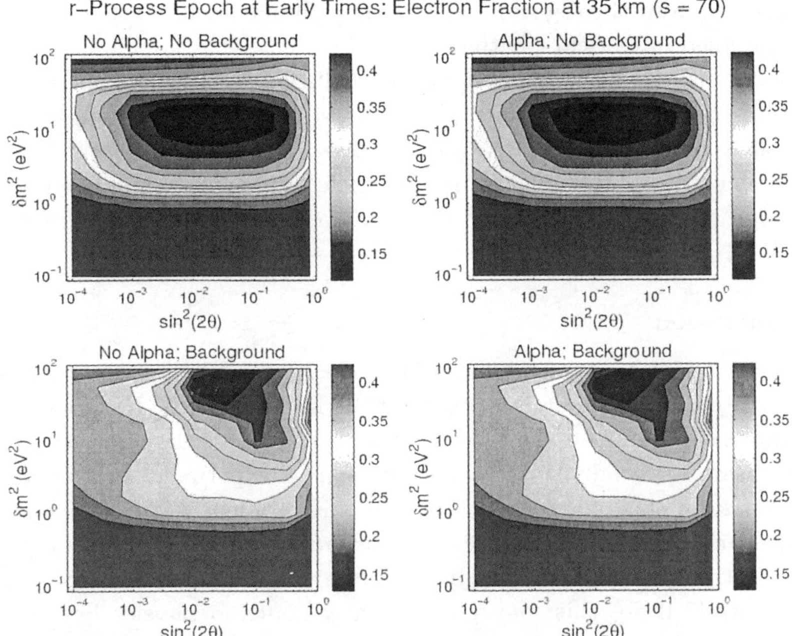

Figure 1. Contours of electron fraction Y_e for ranges of active-sterile channel mass-squared difference δm^2 and effective vacuum mixing $\sin^2 2\theta$ (see Y_e color code on right sides of figures) for an assumed entropy per baryon of 70 units of Boltzmann's constant per baryon. These results are based on a hydrostatic, adiabatic "wind" model. Results are given with and without the inclusion of the Alpha Effect as labled. Calculations by M. Patel.

because (1) the $\bar{\nu}_s$'s created undergo a second resonance at lower density and reconvert to $\bar{\nu}_e$'s, and (2) there is considerable time lag between the MSW-engineered reduction in the ν_e flux and the Y_e response on account of the processes in Eqs. (1) & (2) being *weak* interactions in a rapidly expanding background.

The net result can be a drastic reduction in Y_e with a greatly reduced alpha effect[28]. Wind model approximate numerical calculations by Mitesh Patel[30] illustrate this nicely and are shown in Fig. 1. There contours of Y_e are given as functions of active-sterile mass-squared splitting, δm^2, and effective vacuum mixing angle as parameterized by $\sin^2 2\theta$. These calculations employ a constant entropy and expansion rate in an adiabatic wind with entropy per baryon in units of Boltzmann's constant $S = 70$. It is obvious that, even with the Alpha Effect, very neutron-rich conditions (blue and deep blue in this figure) could result. Similar results are obtained in more detailed numerical treatments[28].

What matters in this effect is not really δm^2, but rather something more like $S^4 \delta m^2$, and the entropy per baryon S, as I argued above, is as yet poorly determined in supernova models. Therefore, a relatively small change in entropy could put the neutron-rich "sweet spot" right in the range of mixing parameters suggested by LSND and currently being probed by mini-BooNE.

4. Conclusion

We face a major task in our effort to properly and self consistently model the evolution of the neutrino and antineutrino flavor field histories in the supernova environment. However, I have argued above that this effort may be necessary, especially since we now know from experiment that neutrinos can transform their flavors. Solution of this problem in the active-active channels may lead to the unraveling of several outstanding problems in supernova dynamics and nucleosynthesis. On the other hand, it may make them worse! If so, this may point to new neutrino physics beyond the Standard Model or involving modest extensions of the Standard Model like active-sterile mixing channels. In any case, it would be an act of dangerous hubris to think that what we know now of neutrino physics is all there is, especially given the great sensitivity of the core collapse supernova phenomenon to flavor changing physics.

Acknowledgments

This work was supported in part by an NSF grant at UCSD and the TSI collaboration's DoE SciDAC grant at UCSD. I would like to thank K. Abazajian, P. Amanik, A. B. Balantekin, H. Duan, J. Hidaka, M. Patel, Y.-Z. Qian for much stimulating input , and I especially would like to thank A. Mezzacappa for useful discussions, inspiration, and a shared philosophy on the matter of uncertainties in the supernova problem, and Nancy Tate for her help in the Open Issues Workshop at the INT and both for great patience while this paper was prepared. Finally I would like to thank Wick Haxton and Linda Vilet and the INT at the University of Washington, where much of this work was performed, for loyal support and encouragement.

References

1. H. Back, *et al.*, arXiv:hep-ex/0412016.

2. A. B. Balantekin and G. M. Fuller, Phys. Lett. **B471**, 195 (2000).

3. D. O. Caldwell, G. M. Fuller, and Y.-Z. Qian, Phys. Rev. D **61**, 123005 (2000).

4. C. Athanassopoulos, *et al.*,, Phys. Rev. Lett. **75**, 2650 (1995); K. Eitel, New J. Phys. **2**, 1 (2000).

5. S. P. Mikheyev & A. Yu. Smirnov, Yad. Fiz. **42**, 1441 (1985) [Sov. J. Nucl. Phys. bf 42, 913 (1985)]; L. Wolfenstein, Phys. Rev. D **17**, 2369 (1978).

6. K. Abazajian, G. M. Fuller, and M. Patel, Phys. Rev. D **64**, 023501 (2001).

7. A. D. Dolgov, S. H. Hansen, G. Raffelt, and D. V. Semikoz, Nucl. Phys. **B590**, 562 (2000).

8. G. M. Fuller, R. Mayle, J. Wilson, and D. N. Schramm, Astrophys. J., **322**, 795 (1987).

9. G. M. Fuller, R. Mayle, B. Meyer, and J. Wilson, Astrophys. J., **389**, 517 (1992).

10. Y.-Z. Qian, and G. M. Fuller, Phys. Rev. D **51**, 1479 (1995).

11. Y.-Z. Qian, G. M. Fuller, G. J. Mathews, R. Mayle, J. R. Wilson, and S. E. Woosley, Phys. Rev. Lett. **71**, 1965 (1993).

12. N. Bell, A. Rawlinson, and R. F. Sawyer, Phys. Lett. **B573**, 86 (2003).

13. A. Friedland and C. Lunardini, Phys. Rev. D **68**, 013007 (2003); JHEP **43**, 0310 (2003).

14. J. Pantaleone, Phys. Lett. **B342**, 250 (1995).

15. G. Sigl and G. Raffelt, Nucl. Phys. **B406**, 423 (1993).

16. W. C. Haxton, Phys. Rev. D **36**, 2283 (1987).

17. G. M. Fuller and Y.-Z. Qian, astro-ph/0505240.

18. S. Pastor and G. Raffelt, Phys. Rev. Lett. **89**, 191101 (2002).

19. H. Yüksel and A. B. Balantekin, New J. Phys., **7**, 51 (2005).

20. P. Strack and A. S. Burrows, hep-ph/0504035.

21. L. Stodolsky, Phys. Rev. D **36**, 2273 (1987); B. McKellar and M. Thompson, Phys. Rev. D **49**, 2710 (1994); G. Raffelt, G. Sigl, and L. Stodolsky, Phys. Rev. Lett. **70**, 2363 (1993); K. Kainulainen, J. Maalampi, and J. T. Peltoniemi, Nucl. Phys. **B358**, 435 (1991); R. Foot, M. Thompson, and R. R. Volkas, Phys. Rev. D **53**, 5349 (1996); P. Di Bari, P. Lipari, and M. Lusignoli, Int. J. Mod. Phys. A **15**, 2289 (2000).

22. G. M. Fuller, A. Kusenko, I. Mocioiu, and S. Pascoli, Phys. Rev. D **68**, 103002 (2004).

23. J. T. Peltoniemi, hep-ph/9511323; H. Nunokawa, J. T. Peltoniemi, A. Rossi, and J. Valle, Phys. Rev. D **56**, 1204 (1997).

24. S. Bruenn and A. Mezzacappa, in "Sources and Detection of Dark Matter," World Scientific (1998).

25. K. N. Abazajian and G. M. Fuller, Phys. Rev. D **66**, 023526 (2002) [arXiv:astro-ph/0204293].

26. Y.-Z. Qian, P. Vogel, and G. J. Wasserburg, Astrophys. J. **554**, 578 (2001).

27. G. M. Fuller and B. S. Meyer, Astyrophys. J., **453**, 792 (1995); G. C. McLauhlin and G. M. Fuller, Astrophys. J. **455**, 202 (1995); B. S. Meyer, G. C. Mclaughlin, and G. M. Fuller, Phys Rev. C **58**, 3696 (1998).

28. G. C. McLaughlin, J. Fetter, A. B. Balantekin, and G. M. Fuller, Phys. Rev. C **59**, 2873 (1999); J. Fetter, G. C. McLaughlin, A. B. Balantekin, and

G. M. Fuller, Astropart. Phys. **18**, 433 (2002).

29. C. J. Horowitz and G. Li, Phys. Rev. Lett. **82**, 5198 (1999); C. J. Horowitz, Phys. Rev. D **65**, 043001 (2002).

30. M. Patel, private communication of UCSD thesis work..

Section 5
Neutrino Interactions

SHELL MODEL OF NUCLEI FOR STELLAR CORE COLLAPSE: CURRENT STATUS, FUTURE PROSPECTS

G. STOITCHEVA[1,2] AND D.J. DEAN[1]

[1]*Physics Division, Oak Ridge National Laboratory, P.O. Box 2008,*
Oak Ridge, Tennessee 37831, USA

[2]*Department of Physics, University of Tennessee,*
Knoxville, Tennessee 37831, USA

Electron capture on heavy nuclei plays a dominant role throughout core collapse. In this work, we discuss the nuclear input required to calculate electron capture rates in the supernova environment. This input is, in part, derived from understanding the nature of the Gamow-Teller resonance in medium-mass nuclei. Nuclear correlations affect the total strength, position, and width of the Gamow-Teller resonances. Therefore, an accurate description of electron capture on nuclei requires one to attack the nuclear quantum many-body problem. We outline the difficulties in the framework of the conventional shell-model diagonalization when approaching heavy systems ($A > 60$). Furthermore, we discuss the Auxiliary-field Shell Model Monte Carlo (SMMC) as our choice for reaching the properties of such nuclei of interest and the challenges in this approach. We demonstrate a promising development within the SMMC method that uses a shifted contour of integration to alleviate the Monte Carlo sign problem when performing calculations with realistic Hamiltonians.

1. Introduction

The quantum many-body problem is an interesting and challenging frontier of science. It governs the physical properties of atoms, molecules, nanophase systems, and nuclei. Distinguishing characteristics of nuclei include a complex force, which has yet to be completely determined, two different fermionic species (protons and neutrons), and the lack of an external confining potential. These three ingredients generate a range and diversity of behaviors that make the nucleus a truly unique quantum many-body system. Furthermore, nuclei often play important roles in the lives of stars and the generation of matter in the universe. For example, an important phenomenon that influences the way a type II supernova collapses involves electron capture on nuclei.[1] Once the interior of a star of roughly 10 to 25

solar masses has burned its nuclear fuel, its central region contains iron-like nuclei and their associated electrons. The star begins to collapse under its own gravitation but is stabilized by electron degeneracy pressure as long as the mass in the central region does not exceed the appropriate Chandrasekhar mass. If the core exceeds this mass, electrons are captured by nuclei.[2] This electron capture deleptonizes the iron core, and collapse continues. The capture of electrons on nuclei and protons plays a dominant role through collapse phase, significantly altering the electron fraction and entropy, thereby determining the strength and location of the initial supernova shock. Therefore, accurate calculation of electron capture rates will allow new improvements in understanding and treatment of the evolution of the supernovae.[3,4,5,6]

For many years, the electron capture rates by Fuller, Fowler & Newman[7] served as the standard in stellar evolution. The rates are based on the independent particle model (IPM), assuming a single Gamow-Teller (GT) transition (and in some cases a zero energy transition) and considering only single-particle states for proton and neutron numbers, $Z \leq 40$, $N \geq 40$. An important (and incorrect) consequence of this model is that the electron-capture rates on nuclei with $N \geqslant 40$ vanish since the GT transitions are Pauli-blocked.[7] This turns out to be an incorrect assumption.[8] As shell-model Monte Carlo (SMMC) calculations indicated, the nuclear correlations brought about by the effective nucleon-nucleon interaction can unblock the orbitals relevant for electron capture, and therefore the path to obtain electron capture on heavier nuclei is open.

Nuclear shell-model calculations use effective interactions between protons and neutrons (nucleons) within a nucleus. The bare nucleon-nucleon (NN) interactions are derived from NN scattering data.[9] One way to approach the quantum many-body problem is through diagonalization of the two-body nucleon-nucleon interaction in an appropriate many-body basis $\mid \Phi_i \rangle$. The free-space NN interactions are transformed (or 'down-folded') to an appropriately chosen model space through the techniques of many-body perturbation theory.[10] The down-folding procedure produces an effective two-body interaction for a given model (or basis) space. In the next step, one diagonalizes the effective interaction within the many-body model space to obtain solutions to the many-body problem. Nuclear physicists call this method the interacting shell model, while chemists call it a configuration-interaction method. This technique has been successfully applied to various regions of the periodic table and, provided that the nuclear Hamiltonian is sufficiently well determined, gives results that compare well to experiment.[11]

In the diagonalization approach, one distributes valence particles within a particular region of Hilbert space that approximates the single-particle levels just below and above the nuclear Fermi surface. For example, if one would like to study the properties of ^{54}Fe, one uses as a model space the $0f1p$-shell consisting of the forty single-particle orbitals above the ^{40}Ca closed core (20 orbitals for protons and 20 for neutrons). Mathematically, the shell model is a matrix-diagonalization problem in which one computes the matrix elements of the Hamiltonian of the system, H, $\langle \Phi_i | H | \Phi_j \rangle = H_{ij}$ between each of the many-body basis states. One then diagonalizes H_{ij} to find the eigenvalues (energy levels) and associated eigenvectors (many-body wave functions) of the matrix. Considerable effort has been devoted to studying nuclei within the shell model (developing effective interactions and operators, etc.), and impressive agreement between theory and experiment has been achieved. However, the standard diagonalization of the shell-model Hamiltonian matrix is limited by the combinatorial increase of the rank of the matrix with increasing model space and/or increasing numbers of valence particles within a given model space. Clearly, this brute force method will always reach a computational saturation limit beyond which it is impractical. Currently, this limit is on the order of 10^9 states, $A \sim 60$.[12] Therefore, as we move up to the region of heavy nuclei (from the pf-shell to the gds-shell and beyond), more computational problems must be resolved. In fact, the scaling of shell-model (it is an np-complete computational problem) problems with the number of single-particle orbitals is so dramatic that one should investigate other ways of calculating observables.

The auxiliary-field Monte Carlo shell-model method[13,14,15] (AFMC-SM, and also known as Shell Model Monte Carlo, SMMC) is an alternative approach which recasts the standard diagonalization problem into a problem of multidimensional integration. The SMMC method scales numerically far more gently with particle number and valence space size than standard diagonalization techniques. At present, this is the only method in the framework of the shell model that can reach huge many-body spaces (of dimension 10^{21} basis states and higher) and can provide exact results (to within statistical fluctuations) without truncations. The SMMC method has been applied to a number of interesting problems in nuclear structure including early investigations of nuclear properties in the iron region[16,17] and nuclear electron capture.[18] These calculations began with a realistic effective interaction in a given model space, such as the fp-shell interaction known as FPKB3.[19] Unfortunately, realistic shell-model two-body interactions generate a quantum Monte Carlo sign problem that makes calculations difficult due to fluctuations that occur in the sampling. This problem was treated by breaking the Hamiltonian into parts that have no sign problem, and the rest: $H_\lambda = H_G + \lambda H_B$, where H_G are the terms in the Hamiltonian that have no sign problem, and H_B are those terms which create a sign problem. One uses λ as a control parameter on the sign problem. Choosing $\lambda \leq 0$ generates a good Hamiltonian. Operationally, one computes observables for several λ's, and then extrapolates to $\lambda = 1$. This extrapolation method[20] was demonstrated to work well for many nuclei when compared to diagonalization results[21], but left the nagging question that there was an extrapolation from a region of known to unknown solutions.

One of our motivations in this work is to employ a different technology that has the potential to overcome the quantum Monte Carlo sign problem while still maintaining the applicability of the SMMC method. We introduce below a technique called the shifted-contour method that may allow us to push the Monte Carlo sign problem to low temperatures so that we may employ the SMMC method without reliance on extrapolations.

Our second motivation is an accurate evaluation of Gamow-Teller strength distributions for problems of interest to core-collapse supernova. Recently, Langanke et al.[6] performed calculation of electron capture rates based on a hybrid scheme. This scheme uses SMMC results with schematic Hamiltonians that do not involve a quantum Monte Carlo sign problem (such as the pairing+quadrupole Hamiltonians), but nevertheless capture the major nuclear structure information for a given nucleus. The thermally weighted occupation numbers calculated within the SMMC method are then used as input to Random Phase Approximation (RPA) calculations. While the approach is inconsistent (with different Hamiltonians used for the SMMC calculations and the RPA calculations), it gives a 'factor-of-two' result for the electron-capture rates in our region of interest. Using this approach, Langanke et al.[6] calculated electron capture rates for a sample of nuclei with $66 \leq A \leq 112$ for temperatures and densities appropriate for core collapse. They found that these rates are large enough so that electron capture in nuclei dominates over capture on free protons, in contrast to the previous IPM observations. These new rates made a significant impact on the core collapse trajectory and the properties of the core at bounce.[4] At present, only the hybrid SMMC+RPA method can provide the needed data. Verification of the hybrid rates should be undertaken within a consistent framework.

Ultimately we would like to use shifted-contour SMMC (SC-SMMC) methods to obtain realistic electron capture rates in a shell-model framework for nuclei heavier than iron. The first step in this process, and the principal discussion in this paper is a development of the SC-SMMC method. In the following sections we discuss the original SMMC method, outline our implementation of the shifted contour approach, and give some first results.

2. The Shell Model Monte Carlo method

The SMMC method relies on the ability to calculate observables using the imaginary-time many-body evolution operator, $\exp(-\beta H)$, where β is a real c-number and is interpreted as the inverse temperature of a system. Expectation values for any observable Ω may be calculated in a canonical formalism using

$$\langle \Omega \rangle = \frac{\mathrm{Tr}_A e^{-\beta H} \Omega}{\mathrm{Tr}_A e^{-\beta H}} , \tag{1}$$

where we use number-projected quantum-mechanical traces, Tr_A. Of course, direct computation of the many-body operators in this way would be prohibitively expensive, so we must resort to Monte Carlo methods to make calculations feasible.

In order to demonstrate how the AFMC methods are derived, we start with nuclear Hamiltonians, which are of the form

$$H = \sum_\alpha \tilde{\varepsilon}_\alpha a_\alpha^\dagger a_\alpha + \frac{1}{2} \sum_{\alpha\beta\gamma\delta} V_{\alpha\beta\gamma\delta} a_\alpha^\dagger a_\beta^\dagger a_\delta a_\gamma, \tag{2}$$

where $a_\alpha^\dagger$ and a_α are anti-commuting fermion creation and annihilation operators associated with the single-particle state α defined by the complete set of quantum numbers $(nljmt_z)$, and $\tilde{\varepsilon}_\alpha$ and $V_{\alpha\beta\gamma\delta}$ are the single-particle energies (defining a mean field) and the two-body matrix elements of the effective shell-model interaction, respectively. Nuclear Hamiltonians are rotationally invariant. Since protons do not behave exactly as neutrons, there is some possibility of isospin-symmetry breaking, although we do not consider this here.

It is always possible to bring the Hamiltonian in Eq. (2) into quadratic form

$$H = \sum_\alpha \varepsilon_\alpha \Theta_\alpha + \sum_\alpha \frac{1}{2} \lambda_\alpha \Theta_\alpha \Theta_\alpha , \tag{3}$$

where Θ_α is a density operator of the form $(a^\dagger a)_\alpha$, λ_α is the strength of the two-body interaction, and ε_α are modified single-particle energies. We

252

simplify the Hamiltonian in order to show how to proceed within the AFMC framework by setting $\alpha = 1$ to obtain

$$H = \varepsilon\Theta + \frac{1}{2}\lambda\Theta\Theta \, . \tag{4}$$

Then, by using the Hubbard-Stratonovich transformation[22] on the two-body part of the Hamiltonian, we recast the exponential operator into a calculable form:

$$e^{\frac{1}{2}\Lambda\Theta^2} = \sqrt{\frac{|\Lambda|}{2\pi}} \int d\sigma \, e^{-\frac{1}{2}\Lambda\sigma^2 + s\sigma\Lambda\Theta} \, , \tag{5}$$

where the phase, $s = \pm 1$ if $\Lambda \leq 0$ or $\pm i$ if $\Lambda > 0$. The imaginary-time evolution of the system becomes

$$e^{-\beta H} = \sqrt{\frac{\beta|\Lambda|}{2\pi}} \int d\sigma \, e^{-\frac{1}{2}\beta|\Lambda|\sigma^2} e^{-\beta h(\sigma)}, h(\sigma) = (\varepsilon + s\Lambda\sigma)\Theta. \tag{6}$$

The integral is a weighted sum over all possible fields σ. As a result, we reduce Hamiltonian H to the one-body Hamiltonian denoted by $h(\sigma)$. This reduction implies that the original many-body problem (generated by the exponential of the two-body operator) becomes an exponential of one-body operators, the trace of which can easily be computed. The price to pay is an integration in σ-fields.

In the general case, when $\alpha > 1$, we will have non-commuting operators, $[\Theta_\alpha, \Theta_\beta] \neq 0$ for $\alpha \neq \beta$, and we must split the exponential into N_t time-slices with $\beta = N_t\Delta\beta$,

$$e^{-\beta H} = \left[e^{-\Delta\beta H} \right]^{N_t} \, , \tag{7}$$

and for each time-slice, $n = 1, \cdots N_t$, we must perform a linearization similar to Eq. (6) using auxiliary fields $\sigma_{\alpha n}$. For the general Hamiltonian, the associated expectation value for a given observable is expressed by

$$\langle \Omega \rangle = \frac{\int D[\sigma] G(\sigma) \Phi(\sigma) \langle \Omega \rangle_\sigma}{\int D[\sigma] G(\sigma) \Phi(\sigma)}, \tag{8}$$

where $D[\sigma] = \prod_{n,\alpha} \left(\frac{\Delta\beta|V_\alpha|}{2\pi} \right)^{1/2} d\sigma_{\alpha n}$ is the volume element and $G(\sigma) = \exp\left[-\frac{1}{2}\Delta\beta \sum_{\alpha,n} |\lambda_\alpha| \sigma_{\alpha n}^2 \right]$ is the Gaussian factor, $U_\sigma = U_{N_t} \cdots U_2 U_1$, $U_n = e^{-\Delta\beta h(\sigma)}$, and

$$\langle \Omega \rangle_\sigma = \frac{\mathrm{Tr}_A U_\sigma \Omega}{\mathrm{Tr}_A U_\sigma} \, . \tag{9}$$

By using Eq. (8) in the limit of low temperature ($T \to 0$ or $\beta \to \infty$), the properties of the ground state can be extracted. Although this expression is exact, the fluctuations in $\Phi(\sigma) = \mathrm{Tr}_A U_\sigma / |\mathrm{Tr}_A U_\sigma|$, which is defined as the sign of the Monte Carlo weight function, determine the precision of the Monte Carlo evaluation. When the average sign is smaller than its uncertainty, this leads to the sign problem: the sign of the integrand fluctuates significantly and meaningful results become difficult, if not impossible, to obtain. When considering finite temperature using the canonical ensemble, the sign decreases exponentially when decreasing the temperature. It was shown that the sign problem is related to the time-reversal properties of the one-body Hamiltonian $h(\sigma)$ in Eq. (6).[23] In addition, for even-even and $N = Z$ nuclei, there is no sign problem if all $\lambda_\alpha \leq 0$. Such forces include interactions like pairing plus multipole interactions. However, for an arbitrary Hamiltonian, there is no guarantee that all $\lambda_\alpha \leq 0$. It was shown that all realistic Hamiltonians suffer form the sign problem and the direct application of the Monte Carlo method in this case fails.

The sign problem has been one of the most severe problems in the quantum Monte Carlo approach to nuclear structure calculations. It prohibits exact shell-model calculations in the framework of SMMC for nuclei when using realistic effective Hamiltonians. There have been many attempts trying to resolve this problem for nuclear structure. Unfortunately, so far all of them failed except for the approximate extrapolation technique described in the introduction. Before we demonstrate the application of shifted contours to SMMC, we will describe in more detail the nuclear Hamiltonian.

3. Nuclear Hamiltonians and Numerical Traces

We turn now to a more "nuclear" representation of the Hamiltonian. The goal is to rewrite the Hamiltonian in a form which is suitable for application of the Hubbard-Stratonovich transformation which requires square operators. In our representation, the two-body Hamiltonian we use can be written in terms of particle-hole angular-momentum coupled two-body matrix elements as:

$$H = \sum_{\alpha\beta,t=p,n} \varepsilon_{\alpha\beta}\rho_{\alpha\beta}^{00,t} + \frac{1}{2} \sum_{\alpha\beta\gamma\delta} \sum_{KM,T=0} (-1)^M E^{KT}(\alpha\gamma,\beta\delta)\rho_{\alpha\gamma}^{KMT}\rho_{\beta\delta}^{K-MT},$$

$$(10)$$

where

$$E^{KT=0}(\alpha\gamma,\beta\delta) = \frac{1}{2}(-1)^{j_b+j_c}\sum_J(-1)^J(2J+1)V_J^N(\alpha\beta,\gamma\delta)$$

$$\times\left\{\begin{array}{ccc} j_a & j_b & J \\ j_d & j_c & K \end{array}\right\}\sqrt{(1+\delta_{\alpha\beta})(1+\delta_{\gamma\delta})}, \qquad (11)$$

is the interaction after a Pandya transformation from the non-antisymmetrized particle-particle matrix elements V_{JT}^N (which are the angular-momentum coupled analogues of the $V_{\alpha\beta\gamma\delta}$ found in Eq. (2)). We note that only the $E^{KT=0}$ matrix elements are needed due to particular symmetries of the non-antisymmetrized matrix elements[15]. Note that $\{\alpha\gamma\beta\delta\}$ label the orbitals, $\{KMT\}$ are the particle-hole angular momentum, its projection and the isospin, while J denotes the particle-particle angular momentum. The one-body density operator is given by $\rho^{KMT} = \rho^{KM,p} + (-1)^T\rho^{KM,n}$, where

$$\rho^{KM,t}(a,b) = \sum_{m_a,m_b}(j_am_aj_bm_b\mid KM)\,a_{j_am_a,t}^\dagger\tilde{a}_{j_bm_b,t} \qquad (12)$$

in both the neutron and proton channels, and $\tilde{a}_{j_am_a} = (-1)^{j_a+m_a}a_{j_a-m_a}$. In diagonal form, the $E^{KT=0}(\alpha\gamma,\beta\delta)$ may be written in terms of its eigenvalues λ_K, and eigenvectors, $\vartheta_{\alpha K}$:

$$E^{KT=0}(\alpha\gamma,\beta\delta) = \sum_{kK}\vartheta_{\alpha\gamma,K}(k)\lambda_K(k)\vartheta_{\beta\delta,K}(k) \qquad (13)$$

and the two-body part of the H is:

$$H_2 = \frac{1}{2}\sum_{\alpha\beta\gamma\delta}\sum_{kKM,T=0}(-1)^M\vartheta_{\alpha\gamma,KT}(k)\lambda_{KT}(k)\vartheta_{\beta\delta,KT}(k)\rho_{\alpha\gamma}^{KMT}\rho_{\beta\delta}^{K-MT}.$$

$$(14)$$

The form of the two-body part of the Hamiltonian in Eq. (14) leads to the Lang's sign rule.[13] She showed that all "good" interactions satisfy the following rule: $\mathrm{Sign}(\lambda_K) = (-1)^{K+1}$. We still must transform to a completely diagonal representation in the operators. In order to do this, we define a new density ρ_k^{KMT} in terms of the original densities, and we introduce new operators Q and P. If we regroup various parts of this equation using a new density operator defined as

$$\rho_k^{KMT} = \sum_{\alpha\gamma}\theta_{\alpha\beta,KT}(k)\rho_{\alpha\beta}^{KMT}\ , \qquad (15)$$

we may then define operators P and Q in terms of the densities ρ_k^{KMT}:

$$Q_k^{KMT} = \begin{array}{ll} \frac{1}{2}\left(\rho_k^{KMT} + \rho_k^{K-MT}\right) & , M = 0 \\ \frac{1}{\sqrt{2}}\left(\rho_k^{KMT} + (-1)^M \rho_k^{K-MT}\right) & , M \neq 0 \end{array} , \qquad (16)$$

$$P_k^{KMT} = \begin{array}{ll} -\frac{i}{2}\left(\rho_k^{KMT} - \rho_k^{K-MT}\right) & , M = 0 \\ -\frac{i}{\sqrt{2}}\left(\rho_k^{KMT} - (-1)^M \rho_k^{K-MT}\right) & , M \neq 0 \end{array} , \qquad (17)$$

The Hamiltonian may then be written in diagonal form as

$$H = \sum_{\alpha\beta,t=p,n} \varepsilon_{\alpha\beta}\rho_{\alpha\beta}^{00,t} + \frac{1}{2}\sum_{k}\sum_{K,M\geqslant 0,T=0} \lambda_K(k)\left(\left(Q_k^{KMT}\right)^2 + \left(P_k^{KMT}\right)^2\right) . \qquad (18)$$

The form of Eq. (18) allows us to use the Hubbard-Stratonovich transformation, Eq. (5), such that the two-body part of the Hamiltonian is reduced to one-body representation of the form

$$h(\sigma) = \varepsilon + s\Lambda\sigma Q + s\Lambda\tau P. \qquad (19)$$

Here, σ and τ are the fluctuating fields that are coupled to P and Q operators. We have thus brought our Hamiltonian into an appropriate form suitable for Monte Carlo evaluation.

Another important ingredient in this discussion is the numerical representation of the many-body quantum-mechanical trace Tr_A. The number of valence particles treated in typical SMMC calculations ranges from 10-50, which is a small number of particles. The number fluctuations in the grand-canonical representation (where one would add to the Hamiltonian terms of the form $\mu_n N_n + \mu_p N_p$, where $\mu_{n,p}$ are chemical potentials for neutrons and protons, and $N_{n,p}$ are the associated number operators) become unacceptably large. The grand-canonical partition function is given by

$$Z = \mathrm{Tr}\, e^{\beta\mu N} U_\sigma = \det\left(1 + \mathbf{U}_\sigma\right) , \qquad (20)$$

where $\mathbf{U}_\sigma$ is the matrix representation of the operator U_σ, and μ is the chemical potential. The canonical partition function for A particles in the valence space is

$$Z_A = \mathrm{Tr}_A\, U_\sigma = \mathrm{Tr}\, P_A U_\sigma , \qquad (21)$$

where $P_A = \delta(A - N)$ is the number projector. We actually carry out this projection in the neutron and proton spaces separately. To perform the operation of P_A, we can write it as the exponential of the one-body operator N using the fact that N has integer eigenvalues $0, 1, \ldots, N_s$:

$$P_A = e^{-\beta\mu A} \int_0^{2\pi} \frac{d\phi}{2\pi} e^{-i\phi A} e^{(\beta\mu + i\phi)N} , \qquad (22)$$

where μ is an arbitrary c-number. The grand-canonical trace in Eq. (20) contains contributions from canonical ensembles of all A-particle partition functions within the valence space, and μ should be chosen to emphasize that contribution from the particular value of A desired. We do this in analogy with the usual thermodynamic treatment. In particular, if the eigenvalues of $\mathbf{U}_\sigma$ are ordered so that $\mathrm{Re}\varepsilon_1 \le \mathrm{Re}\varepsilon_2 \le \cdots \mathrm{Re}\varepsilon_{N_s}$ then a good choice for μ is $(\mathrm{Re}\varepsilon_A + \mathrm{Re}\varepsilon_{A+1})/2$.

We may evaluate few-body expectation values, such as the one found in Eq. 9 by considering the operator $U_\sigma(\varepsilon) = U_\sigma e^{\varepsilon\Omega}$ so that

$$\langle\Omega\rangle = \frac{d}{d\varepsilon}\ln\mathrm{Tr}U_\sigma(\varepsilon)\,|_\varepsilon = 0 \,. \tag{23}$$

Since $\mathrm{Tr}U_\sigma(\varepsilon) = \det\left(1 + \mathbf{U}_\sigma e^{\varepsilon\Omega}\right)$, and $\det(A + \varepsilon B)$ $(\det A)(1 + \varepsilon\mathrm{Tr}A^{-1}B)$ to linear order in ε, we have in a grand canonical ensemble,

$$\langle\Omega\rangle = \mathrm{tr}\frac{1}{1 + \mathbf{U}\Omega}\mathbf{U}. \tag{24}$$

For a number-projected one-body operator Ω , we can write

$$\mathrm{Tr}_A\, U_\sigma\Omega = e^{-\beta\mu A} \int_0^{2\pi} \frac{d\phi}{2\pi} e^{-i\phi A} \det(1 + e^{\beta\mu+i\phi}\mathbf{U}_\sigma)\mathrm{tr}\,\frac{1}{1 + e^{\beta\mu+i\phi}\mathbf{U}_\sigma}$$
$$\times e^{\beta\mu+i\phi}\mathbf{U}_\sigma\Omega \,. \tag{25}$$

Similar formulae hold for two-body density operators.[13]

4. Shifted-Contour Method for Shell Model Monte Carlo Calculations for nuclear structure

Our goal is to modify the Hubbard-Stratonovich transformation so that the formation of large-amplitude oscillations in the integrand are reduced. Let us consider once again the simplified Hamiltonian in Eq. (4). We will replace Θ in the two-body part of the Hamiltonian by $(\Theta - \langle\Theta\rangle)$ to obtain

$$H = (\varepsilon + \lambda\langle\Theta\rangle)\,\Theta + \frac{1}{2}\lambda\,(\Theta - \langle\Theta\rangle)\,(\Theta - \langle\Theta\rangle) - \frac{1}{2}\langle\Theta\rangle^2 \,. \tag{26}$$

Note that the manipulation has left the Hamiltonian invariant. It has shifted the single-particle energies, and has added an overall constant to the Hamiltonian. The next step is to perform the Hubbard-Stratonovich transformation on the term $(\Theta - \langle\Theta\rangle)^2$. This strategy is straightforward to bring over into the nuclear problem. It has the effect of reducing fluctuations in the numerical integrands.

We note that for the nuclear Hamiltonian, fluctuations may be reduced if we subtract the expectation value of the density from the operators. For this purpose, let us consider the two-body Hamiltonian, Eq. (10), in which the density operator, ρ, is shifted by an arbitrary density, $\tilde{\rho}$ (the densities are added and subtracted in order to maintain the same Hamiltonian):

$$
\begin{aligned}
H &= \sum_{\alpha\gamma} W_{\alpha\gamma}\rho_{\alpha\gamma} \\
&+ \frac{1}{2} \sum_{\alpha\beta\gamma\delta,KMT} E^K(\alpha\gamma,\beta\delta)\left(\rho_{\alpha\gamma}^{KMT} - \tilde{\rho}_{\alpha\gamma}^{KMT}\right)\left(\rho_{\beta\delta}^{K-MT} - \tilde{\rho}_{\beta\delta}^{K-MT}\right) \\
&- \frac{1}{2} \sum_{\alpha\beta\gamma\delta}\sum_{KMT} E^K(\alpha\gamma,\beta\delta)\tilde{\rho}_{\alpha\gamma}^{KMT}\tilde{\rho}_{\beta\delta}^{K-MT},
\end{aligned}
\tag{27}
$$

where $W_{\alpha\gamma} = \varepsilon_{\alpha\gamma}\sqrt{2j_\alpha + 1}\rho_{\alpha\gamma}^{00,T=0} + \sum_{KMT} E^K(\alpha\gamma,\beta\delta)\tilde{\rho}_{\beta\delta}^{KMT}\tilde{\rho}_{\beta\delta}^{K-MT}$ is the sum of the one-body part of H and the change from the two-body part of Eq. (18) is very convenient. Therefore, by following the AFMC formalism from the previous section, we introduce a set of density matrices, $\tilde{Q}_k$ and $\tilde{P}_k$:

$$
\tilde{Q}_k^{KMT} = \begin{cases} \frac{1}{2}\left(\tilde{\rho}_k^{KMT} + \tilde{\rho}_k^{K-MT}\right) & , M = 0 \\ \frac{1}{\sqrt{2}}\left(\tilde{\rho}_k^{KMT} + (-1)^M \tilde{\rho}_k^{K-MT}\right) & , M \neq 0 \end{cases},
\tag{28}
$$

$$
\tilde{P}_k^{KMT} = \begin{cases} -\frac{i}{2}\left(\tilde{\rho}_k^{KMT} - \tilde{\rho}_k^{K-MT}\right) & , M = 0 \\ -\frac{i}{\sqrt{2}}\left(\tilde{\rho}_k^{KMT} - (-1)^M \tilde{\rho}_k^{K-MT}\right) & , M \neq 0 \end{cases}.
\tag{29}
$$

This is similar to the set of operators introduced by Eqs. (16,17). Then, the shifted two-body part of the Hamiltonian in Eq. (27) in diagonal form can be written as:

$$
\begin{aligned}
H_2 &= \frac{1}{2}\sum_k \sum_{K,M\geqslant 0,T} \lambda_K(k)\left(\left(Q_k^{KMT} - \tilde{Q}_k^{KMT}\right)^2 + \left(P_k^{KMT} - \tilde{P}_k^{KMT}\right)^2\right) \\
&+ \sum_k \sum_{K,M\geqslant 0,T} \lambda_K(k)\left(Q_k^{KMT}\tilde{Q}_k^{KMT} + P_k^{KMT}\tilde{P}_k^{KMT}\right) \\
&- \frac{1}{2}\sum_k \sum_{K,M\geqslant 0,T} \lambda_K(k)\left(\tilde{Q}_k^{KMT}\tilde{Q}_k^{KMT} + \tilde{P}_k^{KMT}\tilde{P}_k^{KMT}\right),
\end{aligned}
\tag{30}
$$

where the second term is a one-body term and the third is a constant term, defined by the arbitrary densities. Next, we apply the Hubbard-Stratonovich transformation to Eq. (27), which lead us to the shifted one-body Hamiltonian, $h(\sigma)$

$$h(\sigma) = \sum_{\alpha\gamma} \varepsilon_{\alpha\gamma} \sqrt{2j_\alpha + 1}\, \rho^{00,T=0}_{\alpha\gamma} + \sum_k \sum_{K,M\geqslant 0,T} s\Lambda\sigma^{KMT}\left(Q_k^{KMT} - \tilde{Q}_k^{KMT}\right)$$

$$+ s\Lambda\tau^{KMT}\left(P_k^{KMT} - \tilde{P}_k^{KMT}\right). \tag{31}$$

We note that simple derivations show that the shift in H is equivalent to a shift of the auxiliary density σ to a line with the density, $\tilde{\rho}$. Formally, the contour of integration passes through a stationary point defined by the arbitrary density shift. The Hubbard-Stratonovich transformation applied to Eq. (31) becomes:

$$e^{\frac{1}{2}\Lambda\Theta^2} = const \int d\sigma\, e^{-\frac{1}{2}\Lambda\sigma^2 - \frac{1}{2}\Lambda\tau^2 + s(\sigma - s\tilde{Q})Q\Lambda + s(\tau - s\tilde{P})\Lambda P}, \tag{32}$$

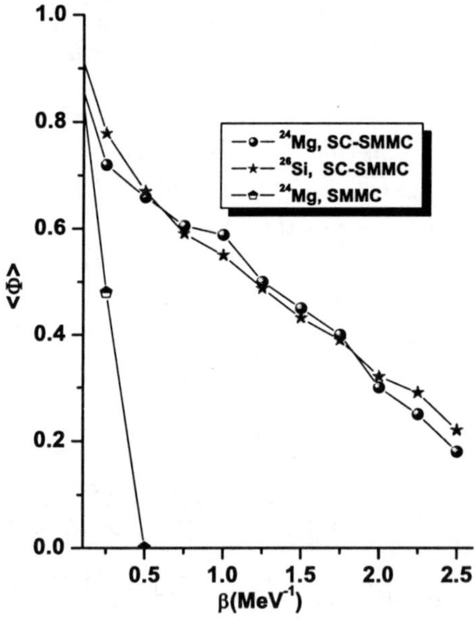

Figure 1.   The sign as a function of β for ^{24}Mg and ^{26}Si.

This technique was initially proposed and applied to problems in quantum chemistry[24,25], where the two-body interaction is the Coulomb force between two electrons. There, the authors investigated how such a shift in the Hamiltonian will affect the sign. In their case, the arbitrary density was chosen to be the density of Hartree-Fock. The calculations show a significant delay of the sign problem.

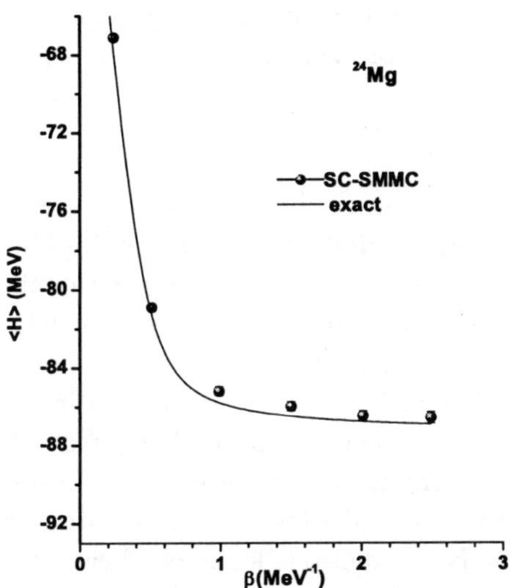

Figure 2. Plot of the energy as a function β for ^{24}Mg using the SC-SMMC method.

As in the case of electrons, for our calculations we also consider the arbitrary density, $\tilde{\rho}$, to be chosen as the expectation value of the Hartree-Fock density in the ground state. As a result, the oscillating part, $e^{s(\sigma - s\tilde{Q})\Lambda\Theta}$, is damped by the presence of $\tilde{Q}$ and the fluctuations are highly reduced. In addition, the choice of the density is very important for stabilizing the oscillations. This can be seen from Eq. (30) where the fluctuations as a result from the two-body interaction will be minimized if $\langle Q \rangle = \tilde{Q}$. In Figure 1, we plot the behavior of the sign versus different values of β. The change of the sign when increasing β based on the shifted-contour method is compared with the SMMC without applying the shift. The half-filled symbols (denoted by AFMC-SM in the figure) indicate the results without the shift. In this case, the sign clearly goes quickly to zero, and essentially renders the method useless. With the shifted fields, the sign is positive and manageable even out to fairly large values of β for $\triangle\beta = 1/64$ MeV^{-1} and 2500 decorrelated samples. Our results show a significant delay of the sign due to the mean-field shift. Stabilization is achieved and an overwhelming part of the sign problem is removed. The calculations are performed up to larger β; in these cases, $\beta \approx 2$MeV^{-1}. In Figure 2, we compute the thermal energy of the ^{24}Mg test case, and compare to the exact diagonalization. We find good agreement, and the promise is that the method may now be applicable to more general problems.

5. Conclusions

Further investigations of the applicability of the method and "the best" shifting density are needed and are under way.[26] The delay of the sign problem will be a major step in nuclear structure calculations. It will allow precision thermal calculations beyond the pf-shell where exact diagonalization can be carried out only for a few valence particles. As discussed above, the SMMC method is the only one in the framework of the shell-model approach which can calculate thermal and ground-state properties of heavier systems. Overcoming the sign problem will make such calculations possible. Furthermore, this will allow us to calculate electron capture rates using realistic two-body interactions for nuclei heavier than iron.

Acknowledgments

Supported by the U.S. Department of Energy under Contract Nos. DE-FG02-96ER40963 (University of Tennessee), DE-AC05-00OR22725 with UT-Battelle, LLC (Oak Ridge National Laboratory), and by the Scientific Discovery through Advanced Computing Terascale Supernova Initiative.

References

1. H. A. Bethe, G. E. Brown, J. Applegate, J. M. Lattimer, Nucl. Phys. A **324** 487 (1979).
2. H.A. Bethe, Rev. Mod. Phys. **62**, 801 (1990).
3. A. Heger, K. Langanke, G. Martinez-Pinedo, S. E. Woosley, Phys. Rev. Lett. **86**, 1678-1681 (2001).
4. W. R. Hix, O. E. B. Messer, A. Mezzacappa, M. Liebendoerfer, J. M. Sampaio, K. Langanke, D. J. Dean, G. Martinez-Pinedo, Phys. Rev. Lett. **91**, 201102 (2003).

5. W. R. Hix et. al., to be published.

6. K. Langanke, G. Martinez-Pinedo, J. M. Sampaio, D. J. Dean, W. R. Hix, O. E. B. Messer, A. Mezzacappa, M. Liebendoerfer, H.-T. Janka, M. Rampp, Phys. Rev. Lett. **90**, 241102 (2003).

7. G. M. Fuller, W. A. Fowler, M. J. Newman, Astrophys.J. Suppl. Ser. **42**, 447 (1980); Astrophys. J. Suppl. Ser. **48**, 279 (1982); Astrophys. J. **252**, 715 (1982), Astrophys. J. **293**, 116 (1985).

8. K. Langanke, E. Kolbe, and D. J. Dean, Phys. Rev. C **63**, 032801 (2001).

9. D.R. Emtem and R. Machleidt, Phys. Rev. C **68**, 041001 (2003); R. Machleidt, Phys. Rev. C **63**, 024001 (2001); R.B. Wiringa, V.G.J. Stoks, R. Schiavilla, Phys. Rev. C **51**, 38 (1995).

10. M. Hjorth-Jensen, T.T.S. Kuo, and E. Osnes, Phys. Rep. **261**, 125 (1995).

11. D.J. Dean, T. Engeland, M. Hjorth-Jensen, M.P. Kartamyshev, and E. Osnes, Prog. Part. Nucl. Phys. **53**, 419 (2004).

12. E. Caurier and F. Nowacki, Acta Physica Polonica **30**, 705 (1990).

13. C.W. Johnson, S.E. Koonin, G.H. Lang, and W.E. Ormand, Phys. Rev. Lett. **69**, 3157 (1992); G. H. Lang, C. W. Johnson, S. E. Koonin, and W. E. Ormand, Phys. Rev. C **48**, 1518 (1993).

14. W.E. Ormand, Prog. Theo. Phys. Supp. **124**, 37 (1996).

15. S. E. Koonin, D. J. Dean, and K. Langanke, Phys. Rep. **278**, 2 (1997).

16. K. Langanke, D.J. Dean, P.B. Radha, Y. Alhassid, and S.E. Koonin, Phys. Rev. C **52**, 718 (1995).

17. D.J. Dean, S.E. Koonin, K. Langanke, P.B. Radha, and Y. Alhassid, Phys. Rev. Lett. **74**, 2909 (1995).

18. D.J. Dean, K. Langanke, L. Chatterjee, P.B. Radha, and M.R. Strayer, Phys. Rev. C **58**, 536 (1998).

19. A. Poves and A. Zuker, Phys. Rep.**70**, 235 (1981).

20. Y. Alhassid, D.J. dean, S.E. Koonin, G. Lang, and W.E. Ormand, Phys. Rev. Lett. **72**, 613 (1994).

21. E. Caurier, G. Martinez-Pinedo, F. Nowacki, A. Poves, J. Retamosa, and A.P. Zuker, Phys. Rev. C **59**, 2033 (1999).

22. J. Hubbard, Phys. Rev. Lett. **3**, 77 (1959); R. L. Stratonovich, Dokl. Akad. Nauk. S.S.S.R. **115**, 1097 (1957).

23. C.W. Johnson and D.J. Dean, Phys. Rev. C **61**, 044327 (2000).

24. N. Rom, D.M. Charutz, and D. Neuhauser, Chem. Phys. Lett. **270**, 382 (1997).

25. S. Jacobi and R. Baer, Journal of Chem. Phys. **120**, 43 (2004).

26. G.S. Stoitcheva and D.J. Dean, to be published.

NEUTRINO-NUCLEUS INTERACTIONS IN CORE COLLAPSE SUPERNOVAE

W. R. HIX, A. MEZZACAPPA, D. J. DEAN

Physics Division, Oak Ridge National Laboratory, Oak Ridge, TN 37831, USA
E-mail: raph@ornl.gov

O.E.B. MESSER

Center for Computational Sciences, Oak Ridge National Laboratory, Oak Ridge,
TN 37831, USA and
Department of Astronomy and Astrophysics and Center for Astrophysical
Thermonuclear Flashes University of Chicago, Chicago, IL 60637, USA

K. LANGANKE, A. JUODOGALVIS

Gesellschaft für Schwerionenforschung, D-64291 Darmstadt, Germany

G. MARTÍNEZ-PINEDO

ICREA and Institut d'Estudis Espacials de Catalunya, Universitat Autònoma de
Barcelona, E-08193 Bellaterra, Spain

J. SAMPAIO

Centro de Fisica Nuclear da Universidade de Lisboa, 1649-003 Lisbon, Portugal

The most important nuclear interaction to the dynamics of stellar core collapse is electron capture. It has recently been re-discovered that electron captures on heavy nuclei (masses larger than 60) dominates electron capture on free protons. In prior simulations of core collapse, electron capture on these nuclei has been treated in a highly parameterized fashion, if not ignored. Advances in nuclear structure theory have allowed a more realistic treatment to be developed. With this new treatment of electron capture on heavy nuclei come significant changes in the hydrodynamics of core collapse and bounce. In this article we will review this recent work and the future prospects to better understand the importance of weak nuclear interactions to the mechanism of core collapse supernovae.

1. Introduction

Core collapse supernovae are among the most energetic events in the universe, emitting 10^{53} erg of energy, mostly in the form of neutrinos. Observationally categorized as Type II or Ib/c supernovae, these explosions mark the end of the life of a massive star and the formation of a neutron star or black hole. They play a preeminent role in the cosmic origin of the elements and serve as a principle heating mechanism for the interstellar medium.

With the formation of maximally-bound iron and neighboring nuclei in a massive star's core, thermonuclear fusion can no longer slow the inexorable contraction that results from the star's self gravity. Once this cold *iron core* grows too massive to be supported by the pressure of degenerate electrons, core collapse ensues. In the inner region of the core, this collapse is subsonic and homologous, while the outer regions collapse supersonically. When the inner core exceeds nuclear densities, it becomes incompressible. The suddenly stiffened inner core rebounds, colliding supersonically with the infalling outer core, producing a shock that initially drives these layers outward. However, this *bounce shock* is sapped of energy by the escape of neutrinos and nuclear dissociation and is generally thought to stall before it can drive off the envelope of the star[1] (though some recent investigations[2,3] have sought to resurrect this *prompt* mechanism, at least for the lowest mass supernovae).

The intense neutrino flux, which is carrying off the binding energy of the proto-neutron star (PNS), provides a tremendous source of heat to the matter between the neutrinospheres and the stalled shock. In the *neutrino reheating* paradigm, this heating reenergizes the shock, which drives off the concentric layers of successively lighter elements that lie above the iron core, producing the supernova. While this paradigm is widely believed, simulations of the neutrino reheating paradigm often fail to produce explosions. This includes recent spherically symmetric simulations implementing full multi-group Boltzmann neutrino transport[4-7] that fail to deliver sufficient heat to the envelope because of the stratification imposed by spherical symmetry. Models that break the assumption of spherical symmetry have achieved some success, either by enhancement of the neutrino luminosity due to fluid instabilities within the proto-neutron star,[8,9] or by enhanced efficiency of the neutrino heating due to large scale convection behind the shock[10-12]. The PNS instabilities are driven by lepton and entropy gradients, while convection behind the shock originates from gradients in entropy that result from the stalling of the shock and grow as the matter is heated from below. However, even with such convective enhancements, explosions are not guaranteed.[13–16] One potential cause of the failure to produce explosions in these numerical models is incomplete (or inaccurate) treatment of the wide variety of nuclear and weak interaction physics that is important to the supernova mechanism.

2. Electron capture in massive stars and supernovae

Once the supernova shock forms, emission and absorption of electron neutrinos and antineutrinos on the dissociation-liberated free nucleons are the dominant processes. However, neutrino interactions with nuclei can be important in unshocked regions, particularly the collapsing core. Bethe et al.[17] showed that the low entropy of the stellar core and resulting dominance of heavy nuclei over free nucleons causes electron capture processes on heavy nuclei to dominate the evolution of the electron fraction during the late stages of stellar evolution and the onset of stellar core collapse. Recent work[18,19] has shown that electron capture on nuclei plays a dominant role throughout collapse, significantly altering the electron fraction and entropy, thereby determining the strength and location of the initial supernova shock. These changes allow improvements in the treatment of electron capture to alter the initial conditions for the entire postbounce evolution of the supernova.

Calculation of the rate of electron capture during stellar evolution and in the collapsing core requires both knowledge of the isotopic composition and the appropriate electron capture reaction rates. The inclusion of electron capture within a multi-group neutrino transport simulation adds an additional requirement: information about the spectra of emitted neutrinos. In stellar evolution (and supernova nucleosynthesis) simulations, the nuclear composition is tracked in detail via a reaction network;[20,21] thus, specific reaction rates can be used. However in simulations of the supernova mechanism, the composition in the iron core is calculated by the equation of state assuming nuclear statistical equilibrium. Typically, the information on the nuclear composition provided by the equation of state is limited to the mass fractions of free neutrons and protons, α-particles, and the sum of all heavy nuclei as well as the identity of an average heavy nucleus, calculated in the liquid drop framework.[22–24] The advantage of such liquid drop equations of state is the smooth and consistent transition from matter dominated by nuclei to nuclear matter under the hot conditions found in a newly formed neutron star. Cooperstein and Wambach[25] did include approximations to the distribution of nuclei around this average in their single zone models; however, this improvement has not been utilized in more complete supernova simulations.

In the iron core, electron capture occurs predominantly via Gamow-Teller (GT) transitions changing protons in the $1f_{7/2}$ level of heavy nuclei into neutrons in the $1f_{5/2}$ level. For many years, the electron capture rates by Fuller et al.[26] (FFN) served as the standard in stellar evolution. Improved, shell model diagonalization calculations of weak interaction rates for electron/positron capture and β decays on the nuclei relevant for stellar evolution ($45 < A < 65$) have become available in recent years.[27,28] Stellar evolution calculations using these improved rates[29] show a marked increase in the electron fraction (Y_e) throughout the iron core before collapse. Because the final size of the homologous core, and therefore the shock formation radius, is proportional to the square of the mean trapped lepton fraction $< Y_l^2 >$ at core bounce,[30] the persistence of these initial differences in Y_e throughout collapse was predicted to have a discernible effect on supernova dynamics.

However, the electron/lepton fraction is greatly modified during the collapse of the stellar core. The increasing density, and concomitant increase in the electron chemical potential, accelerates the capture of electrons on heavy nuclei and free protons in the core, producing electron neutrinos that initially escape, *deleptonizing* the core. Thus the location at which the shock forms in the stellar core at bounce and the initial strength of the shock are largely set by the amount of deleptonization during collapse. Figure 1 summarizes the thermodynamic conditions throughout the core at bounce and displays the temperature, electron fraction (Y_e), electron chemical potential (μ_e), and mean electron neutrino energy (E_{ν_e}) in MeV as functions of the matter density. Also shown is the representative nuclear mass (A). The kinks near $3 \times 10^7 \, \mathrm{g\,cm^{-3}}$ mark the transition to the silicon shell. Deleptonization would be complete if electron capture continued without competition, but at densities of order $10^{11-12} \, \mathrm{g\,cm^{-3}}$, the electron neutrinos become "trapped" in the core, and the inverse neutrino capture reactions begin to compete with electron capture until the reactions are in weak equilibrium and net deleptonization of the core ceases on the core collapse time scale. The equilibration of electron neutrinos with matter occurs at densities between $10^{12-13} \, \mathrm{g\,cm^{-3}}$.

As the densities increase, the characteristic nuclei in the core increase in mass, owing to a competition between Coulomb contributions to the nuclear free energy and nuclear surface tension, until heavy nuclei are re-

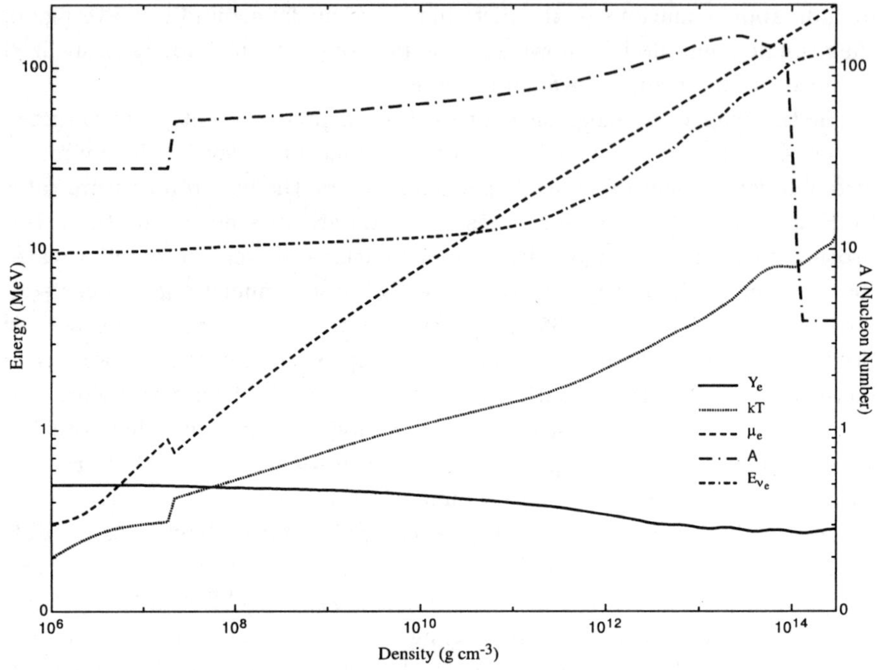

Figure 1. The energy scales and composition as a function of density in the collapsed stellar core at bounce for a 15 $M_\odot$ progenitor.

placed by nuclear matter for mass densities near that of the nucleons in the nucleus ($\sim 10^{14}\,\mathrm{g\,cm^{-3}}$). For densities of order $10^{13}\,\mathrm{g\,cm^{-3}}$, nuclei with mass > 100 dominate. Figure 2 demonstrates that the nuclear composition shows a wide spread in mass, with species with significant concentrations having masses that differ by 40, and that the abundances of nuclei with mass ~ 100 are significant as early as $10^{11}\,\mathrm{g\,cm^{-3}}$. Fuller[31] realized that electron capture on heavy nuclei would soon be quenched in the Bethe et al.[17] picture, as neutron numbers approach 40, filling the neutron $1f_{5/2}$ orbital. Independent particle model (IPM) calculations in the early eighties showed that neither thermal excitations nor forbidden transitions substantially alleviated this blocking,[25,31] leading to the belief that electron capture on protons dominated that on heavy nuclei during collapse. The dominance of captures on the small concentration of free protons results in a self-limiting response in the electron fraction. Unlike the concentration of heavy nuclei, changes of an order of magnitude in the free proton abundance can result from $\sim 10\%$ changes in the electron fraction. Thus a reduction in the electron fraction inhibits further electron capture by greatly reducing the free proton fraction and therefore the rate of electron capture.[32] If electron capture on protons is the dominant process, this self-limitation during the course of collapse has been shown[33–35] to erase differences in electron fraction like those demonstrated in recent stellar evolution simulations.

3. Re-examination of the role of heavy nuclei

It is well known that the residual nuclear interaction (beyond the IPM) mixes the fp and gds shells, for example, making the closed $g_{9/2}$ shell a magic number in stable nuclei ($N = 50$) rather than the closed fp shell ($N = 40$). This naturally calls into question whether the 20 year old perception that electron capture on protons dominates during core collapse is an artifact of the IPM. To examine the possibility that nuclear electron capture is not quenched at $N = 40$, cross sections for charged-current electron and electron-neutrino capture on many nuclei up to at least $A \sim 100$ are needed to accurately simulate core deleptonization. Full shell model diagonalization calculations remain impossible in this regime due to the large number of available levels in the combined $fp + gds$ system.[27,36] Langanke et al.[37] developed a "hybrid" scheme, employing Shell Model Monte Carlo (SMMC) calculations of the temperature-dependent occupation of the single-particle orbitals to serve as input to Random Phase Approximation (RPA) calculations for allowed and forbidden transitions to calculate the capture rate. With this approach, Langanke et al.[18] (LMS) calculated electron capture rates for a sample of nuclei with $A = 66 - 112$. As Figure 3 demonstrates, though the electron capture rates for heavy nuclei are individually smaller than that on protons, they are large enough that capture on the much more abundant heavy nuclei dominates the capture on protons throughout core collapse.

Hix et al.[19] used these LMS rates, along with the shell model diagonalization rates of Langanke and Martínez-Pinedo[27] (LMP) for lighter nuclei, to develop a greatly improved treatment of nuclear electron capture. To calculate the needed abundances of the heavy nuclei, a Saha-like NSE was used, including Coulomb corrections to the nuclear binding energy,[38,39] but neglecting the effects of degenerate nucleons.[40] Comparison between the long standard Bruenn[41] prescription (shown in Eq. 1) and this improved treatment of nuclear electron capture, which we will term the LMSH prescription, reveals two competing effects. In lower density regions, where the average nucleus is well below the $N = 40$ cutoff of electron capture on heavy

268

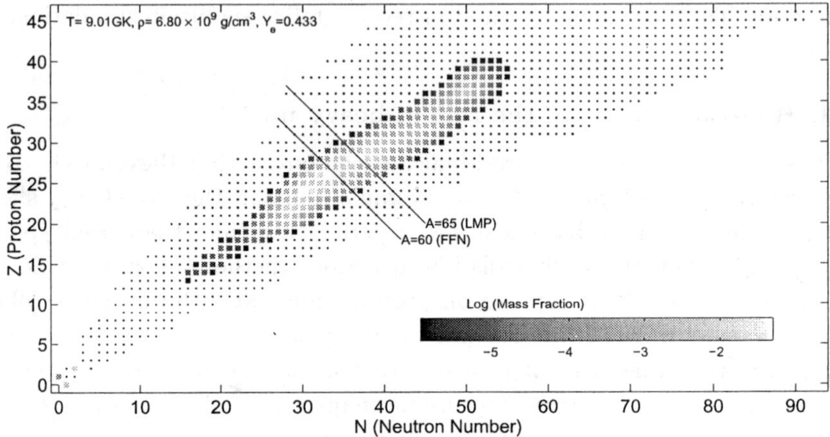

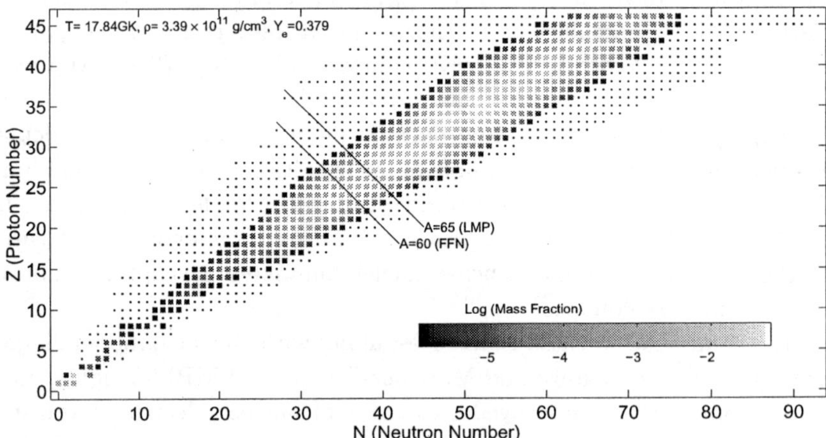

Figure 2. Details of the nuclear composition at two points during stellar core collapse. The diagonal lines at A=60 and A=65 denoted the upper limits to the electron capture rate tabulations of FFN and LMP, respectively.

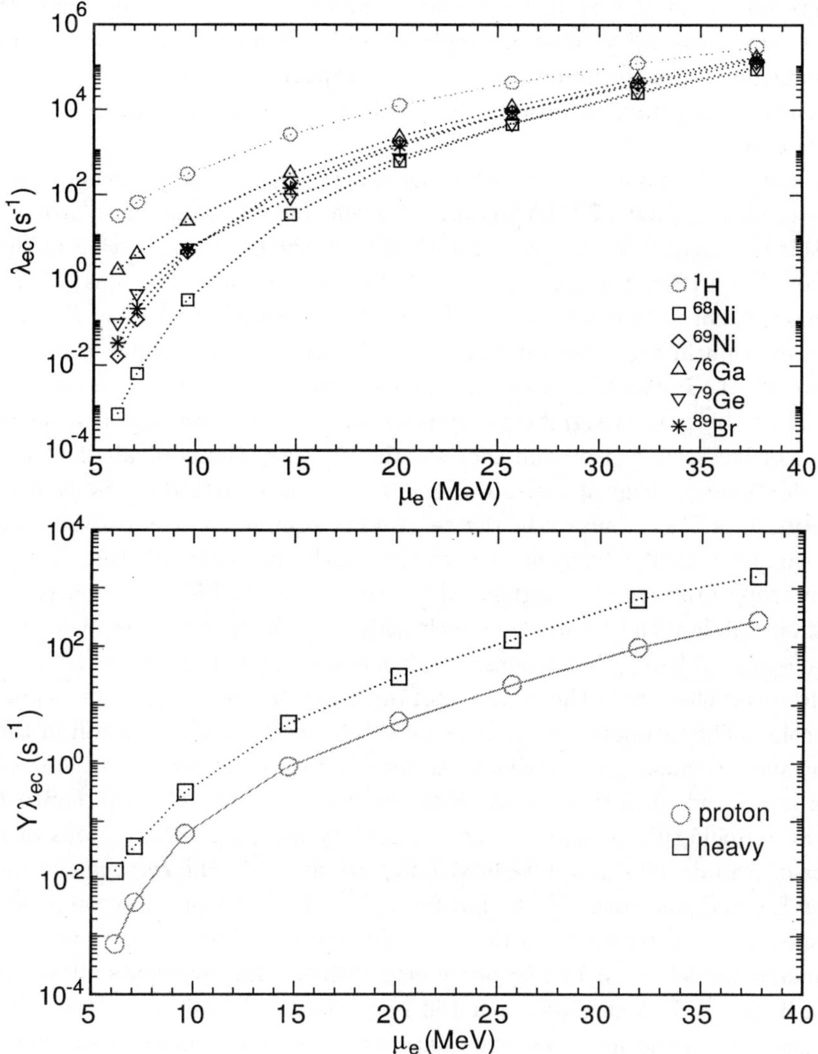

Figure 3. Comparison of the electron capture rates on heavy nuclei to that on protons as a function of the electron chemical potential over the range found in a collapsing stellar core. The upper panel shows the rates for individual species, the lower panel folds in the relative abundance of the protons and heavy nuclei.

nuclei, the Bruenn parameterization results in more electron capture than the LMSH case. This is similar to the reduction in the amount of electron capture seen in stellar evolution models[29] and thermonuclear supernova[42] models when the FFN rates are replaced by shell model calculations. In denser regions, the continuation of electron capture on heavy nuclei alongside electron capture on protons results in more electron capture in the LMSH case.

Hix et al.[19] demonstrated (using the spherically symmetric general relativistic AGILE-BOLTZTRAN code[43–46]) that for a 15 solar mass progenitor this produces a marked reduction (10%) in the electron fraction in the interior of the PNS, resulting in a nearly 20% reduction in the mass of the homologous core. As can be seen in Figure 4, this manifests itself at bounce as a reduction in the mass interior to the formation of the shock from .57 $M_\odot$ to .48 $M_\odot$ in the LMSH case for models employing General Relativity. A shift of this size is significant dynamically because the dissociation of .1 $M_\odot$ of heavy nuclei by the shock costs 10^{51} erg, the equivalent of the explosion energy. Figure 4 also evidences an 15% reduction in the central density and a 5% reduction in the central entropy at bounce, as well as a 15% smaller velocity difference across the shock and quite different lepton and entropy gradients throughout the core. Thus the LMSH prescription results in the launch of a weaker shock with more of the iron core overlying it, a change, in itself, that would inhibit a successful explosion.

However, changes in the behavior of the outer layers also play an important role in the ultimate fate of the shock. The lesser neutronization in the outer layers (resulting primarily from the LMP rates) slows the collapse of these layers, which further diminishes the growth of the electron capture rate by reducing the rate at which the density increases. Reductions of a factor of 5 in density and 40% in velocity are found in the regions outside of the homologous core. Such changes reduce the ram pressure opposing the shock, easing its outward progress. In general relativistic, spherically symmetric models for a 15 $M_\odot$ progenitor,[29] these improvements allow the shock in the LMSH case to reach 168 km, relative to 166 km in the fiducial case. Thus the lesser electron capture in the outer layers more than compensates for the greater mass overlying the shock when it was launched and the greater loss of energy to the neutrino burst. Additionally, changes in the electron capture rates also lead to changes in the core fluid gradients that may, in turn, drive fluid instabilities that are potentially important to the supernova mechanism. Within the inner 50 km, the entropy and lepton fraction gradients found in the LMSH model are considerably different from

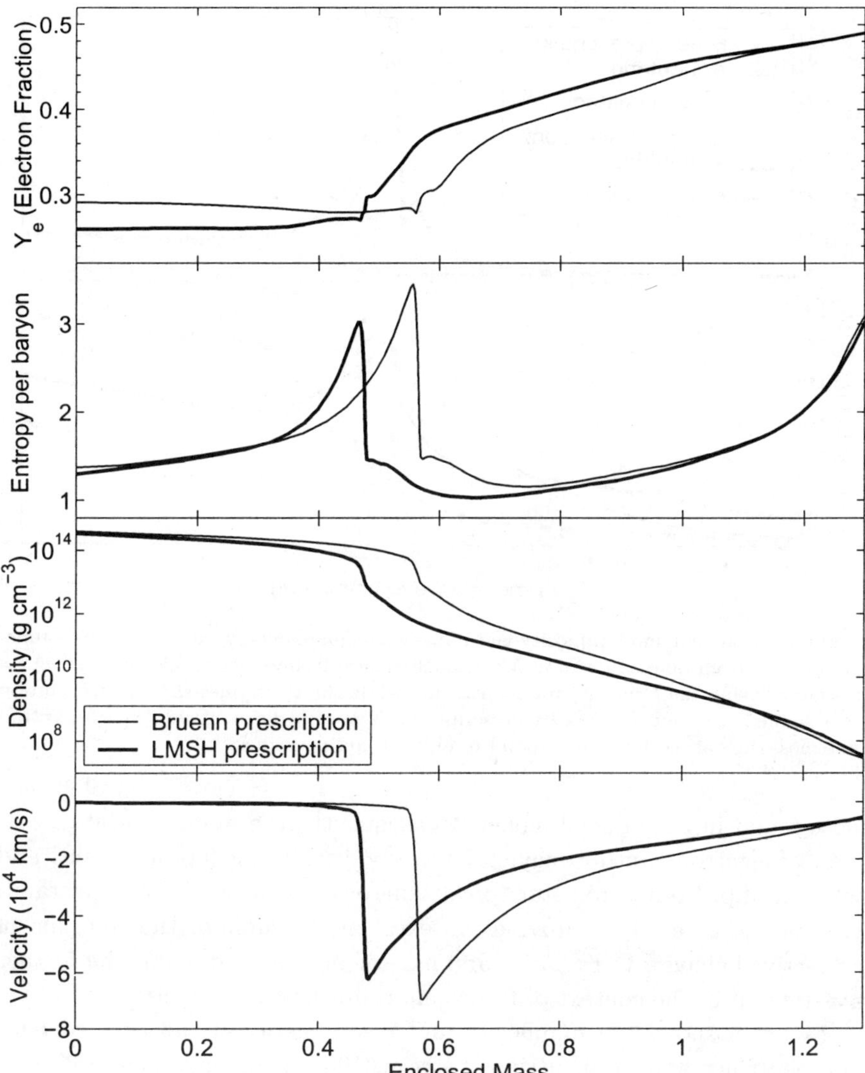

Figure 4. The electron fraction, entropy and velocity as functions of the enclosed mass at core bounce for a 15 $M_\odot$ model. The thin line is a simulation using the Bruenn parameterization while the thick line is for a simulation using the LMSH prescription. Both simulations implement General Relativity.

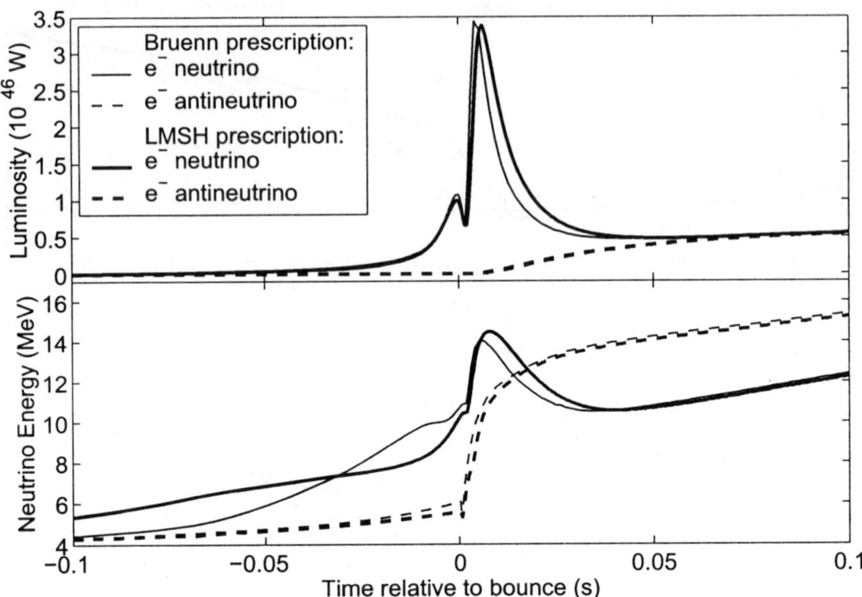

Figure 5. The neutrino luminosity and root-mean-square energy (at 500 km) as a function of time from bounce for a 15 $M_\odot$ model. The thin lines show this evolution for a simulation using the Bruenn parameterization while the thick lines show this evolution for a simulation using the LMSH prescription. The solid lines correspond to electron neutrinos, the dashed lines correspond to electron antineutrinos.

those found in the fiducial model. Consequently, the more accurate treatment of electron capture may significantly alter the location, extent, and strength of proto-neutron star convection, or other potential fluid instabilities, in the core. This provides an excellent example of the coupling of convective behavior to radiative and nuclear processes and must be further investigated in the context of future multidimensional models.

Figure 5 shows the luminosity and mean energy of the emitted electron neutrinos and antineutrinos between 100 milliseconds before bounce and 100 milliseconds after bounce. Clearly evident in the luminosity is a slight delay (2 ms) in the prominent "breakout" burst caused by the deeper launch of the shock in the LMSH case. Over the first 50 milliseconds after bounce, the LMSH model emits 15% more energy than the fiducial model, with a slightly lower luminosity at later times. This is largely the result of differences in the mean electron neutrino energy, which is as much as 1 MeV higher over the first 50 milliseconds in the LMSH case, but lower thereafter. This results from the neutrinospheres in the LMSH model occurring in deeper, hotter layers for the first 50 milliseconds, but cooler layers at later times. The differences in the neutrino spectrum during collapse, when electron capture on nuclei dominates, are larger than those described after bounce. For low densities, where capture on nuclei dominates in the Bruenn prescription as well, the approximate reaction Q-value derived from the free neutron and proton chemical potentials dramatically underestimates the Q-value, resulting in a much lower mean neutrino energy. As captures on protons begin to compete with captures on nuclei in the Bruenn prescription, the mean neutrino energy grows rapidly because of the higher Q value for capture on protons. It exceeds that found in the LMSH model by as much as 2 MeV in the 30 milliseconds just before bounce.

4. Open Issues

The work described in the previous section has reinvigorated interest in the role played by weak interactions for heavy nuclei in core collapse supernovae. However, it represents only the beginning of such studies. One outstanding issue in the treatment of neutrino interactions with nuclei is the continued neglect of inelastic scattering of neutrinos on heavy nuclei in virtually all supernova simulations. Bruenn and Haxton[47] pioneered the study of neutrino-nucleus reactions during core collapse. These authors used inelastic neutrino-nucleus scattering and neutrino-nucleus absorption rates for ^{56}Fe to approximate the effects on all heavy nuclei. In spherically symmetric supernova simulations, they found that the energy transfer due to neutrino-nucleus scattering was comparable to (but smaller than) that from the dominant neutrino-electron scattering during the collapse phase. At later times in the simulation, they found that neutrino-nucleus scattering dominated in the cooler iron-rich regions. However, ν_e capture on protons and neutrino scattering on ^{4}He were globally more important as dissociated matter occurred closer to the neutrinosphere. Bruenn & Haxton did not confirm an earlier suggestion[48] that neutrino-nucleus reactions can preheat the matter ahead of the shock during the early phase of the explosion, easing the passage of the shock, nor did their simulations reveal a significant contribution of neutrino-nucleus reactions to the revival of the stalled shock.

Juodagalvis et al.[49] have recently calculated neutral-current inelastic neutrino scattering rates for forty iron peak nuclei ($^{50-60}$Mn, $^{52-61}$Fe,

$^{54-63}$Co, $^{56-64}$Ni), using shell model diagonalization evaluations of the Gamow-Teller response and RPA estimations of the contributions from other multipoles. Unlike the work of Bruenn & Haxton, these rates include the effects of finite nuclear temperature, which enhance the low energy cross sections.[50] However, the total rates are dominated by higher energy neutrinos ($E_\nu \sim 15 - 30\,\mathrm{MeV}$) and thus similar on average to those used by Bruenn & Haxton. Therefore the conclusions of Bruenn & Haxton are likely to stand, although their re-examination is needed in light of recent improvements in supernova models. In addition to the microscopic rates for the interactions themselves, these conclusions also depend strongly on the other assumptions of their simulations, most notable the assumption of spherical symmetry, which causes thermal stratification and places iron far from the neutrinosphere, and the nuclear composition, which depends on the hydrodynamic state and neutronization. Therefore these conclusions must be revisited in light of recent improvements in supernova models. Of particular interest in this regard is the advent of multi-dimensional models and the impact of revisions in the treatment of nuclear electron capture that alter the thermodynamics and neutronization throughout the collapsing stellar core. Inclusion of the improved cross-sections for ^{4}He,[51] which sits closer to the neutinospheres, are also desirable. Finally, though the impact of neutrino-nucleus scattering on shock revival may be small, these scatterings could significantly impact the nucleosynthesis. In this case, rates for lighter nuclei are also desirable, since these nuclei experience the supernova's neutrino flux as well.

The other open issue is the continued pursuit of improvements in the treatment of electron capture on heavy nuclei. While the work of Hix et al.[19] and Langanke et al.[18] has established that core collapse is strongly affected when electron capture on heavy nuclei is not suppressed, this work does not demonstrate the impact that variations in the rates for nuclear electron capture may have. Messer et al.[35] have examined the sensitivity of the models to uncertainties in the nuclear electron capture rates with a parameter study taking (for simplicity and reproducibility by other groups) the Bruenn prescription[41] as a starting point. In this prescription, the emissivity from heavy nuclei is given by

$$j_{nuclear} = \frac{2}{7}\frac{(2\pi)^4 G_F^2}{\pi h^4 c^4}\, g_A^2\, \frac{\rho X_H}{m_B A}\, N_p(Z)N_h(N)(E + Q')^2 F_e(E + Q'), \quad (1)$$

where $F_e(E) = (1 - (M_e/E)^2)^{1/2}/(1 + \exp[(E - \mu_e)/k_b T])$. The functions

$N_p(Z)$ and $N_h(N)$ in equation (1), defined as

$$N_p(Z) = \begin{cases} 0 & Z < 20 \\ Z - 20 & 20 < Z < 28 \\ 8 & Z > 28 \end{cases} \quad \text{and} \quad N_h(N) = \begin{cases} 6 & N < 34 \\ 40 - N & 34 < N < 40 \\ 0 & N > 40 \end{cases},$$

respectively indicate the number of protons in the $1f_{7/2}$ level and the number of neutron holes in the $1f_{5/2}$ level of the average nucleus. It is the product $N_p N_h$ approaching zero as $N \to 40$ that allows electron capture on protons to dominate in this prescription. Instead of letting the product $N_p N_h$ in Eq. 1 vary as determined by the EOS, Messer et al.[35] set this product to several constant values in Newtonian collapse simulations. Figure 6 shows the effect of this variation on the velocity distribution at bounce, in comparison to the results of Newtonian models using the LMSH and Bruenn prescriptions. Clearly, a reduction in the total electron capture rate by a factor of 10 from those predicted by Langanke et al.[18] would erase the changes demonstrated by Hix et al.[19]. Likewise, a systematic increase by a factor of 10 would further reduce the initial PNS mass by at least 10%. Even changes intermediate to these would significantly alter location of shock formation, therefore further efforts to improve the treatment of electron capture on heavy nuclei are necessary.

Such improvements must address the approximations to all three components of the calculation of electron capture rates on heavy nuclei that were made.[18,19] To be used in core collapse simulations, electron capture rates must provide full coverage of electron capture on nuclei in the region $55 < A < 120$. These reaction rates must cover the full thermodynamic range of interest in supernovae (temperatures of $1 - 100\,\text{GK}$ and densities from $10^5 - 10^{13}\,\text{g}\,\text{cm}^{-3}$) and must also address the need for detailed spectra of the emitted neutrinos. At present, only the hybrid SMMC+RPA method and other, approximate methods[52] can provide the needed data, but ultimately better calculations will be needed.[36] In the interim, all such rates must vetted, either by direct measurements of neutrino capture or by experimental determinations of Gamow-Teller strength distributions. More detailed tracking of the nuclear composition is also necessary, in a form that retains the consistent transition to nuclear matter afforded by current equation of state schemes while allowing for accurate calculation of the rate of electron capture on heavy nuclei and, ultimately, for detailed nucleosynthesis. Only with such data, combining expertise from the nuclear and astrophysical communities, can we close these open issues in our understanding of the role that neutrino interactions with heavy nuclei play

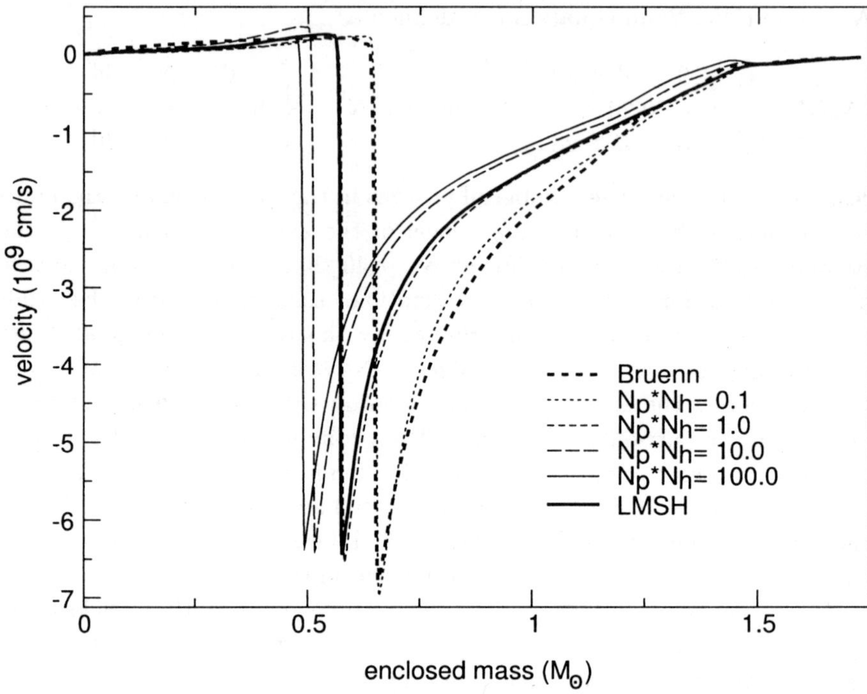

Figure 6.   Comparison of the velocity structure of the core of a 15 $M_\odot$ star at bounce. The thick dotted and solid lines indicate models using the Bruenn and LMSH prescriptions, respectively. The thin lines show results of models using a modification of the Bruenn prescription where the product of number of protons in the $1f_{7/2}$ level and the number of neutron holes in the $1f_{5/2}$ level is held constant for the course of the simulation.

in core collapse supernovae.

Acknowledgements The authors thank M. Liebendörfer, S.W. Bruenn and H.-Th. Janka for fruitful discussions. The work has been partly supported by the U.S. Department of Energy, through the Scientic Discovery through Advanced Computing Program of the Office of Science and through the Advanced Simulations & Computing Academic Strategic Alliances Program Program (Grant No. B523820), by the U.S. National Science Foundation under contract PHY-0244783, by the Danish Research Council, by the Portugese Foundation for Science and Technology and by the Spanish MCyT and European Union ERDF under contracts AYA2002-04094-C03-02 and AYA200306128. This research used resources of the Center for Computational Sciences at Oak Ridge National Laboratory. Oak Ridge National

Laboratory is managed by UT-Battelle, LLC, for the U.S. Department of Energy under contract DE-AC05-00OR22725.

References

1. A. Burrows and J. M. Lattimer, ApJ **299**, L19 (1985).
2. S. Wanajo, M. Tamamura, N. Itoh, K. Nomoto, Y. Ishimaru, T. C. Beers, and S. Nozawa, ApJ **593**, 968 (2003).
3. K. Sumiyoshi, H. Suzuki, S. Yamada, and H. Toki, Nucl. Phys. A **730**, 227 (2004).
4. M. Rampp and H.-Th. Janka, ApJ **539**, L33 (2000).
5. A. Mezzacappa, M. Liebendörfer, O. Messer, W. Hix, F.-K. Thielemann, and S. Bruenn, Phys. Rev. Lett. **86**, 1935 (2001).
6. M. Liebendörfer, A. Mezzacappa, F.-K. Thielemann, O. E. B. Messer, W. R. Hix, and S. W. Bruenn, Phys. Rev. D **63**, 103004 (2001).
7. T. A. Thompson, A. Burrows, and P. A. Pinto, ApJ **592**, 434 (2003).
8. J. R. Wilson and R. W. Mayle, Phys. Rep. **227**, 97 (1993).
9. W. Keil, H.-Th. Janka, and E. Müller, ApJ **473**, L111 (1996).
10. M. Herant, W. Benz, W. R. Hix, C. L. Fryer, and S. A. Colgate, ApJ **435**, 339 (1994).
11. A. Burrows, J. Hayes, and B. A. Fryxell, ApJ **450**, 830 (1995).
12. C. L. Fryer and M. S. Warren, ApJ **574**, L65 (2002).
13. H.-Th. Janka and E. Müller, A&A **306**, 167 (1996).
14. A. Mezzacappa, A. C. Calder, S. W. Bruenn, J. M. Blondin, M. W. Guidry, M. R. Strayer, and A. S. Umar, ApJ **493**, 848 (1998).
15. A. Mezzacappa, A. C. Calder, S. W. Bruenn, J. M. Blondin, M. W. Guidry, M. R. Strayer, and A. S. Umar, ApJ **495**, 911 (1998).
16. R. Buras, M. Rampp, H.-Th. Janka, and K. Kifonidis, Phys. Rev. Lett. **90**, 241101 (2003).
17. H. A. Bethe, G. E. Brown, J. Applegate, and J. M. Lattimer, Nucl. Phys. A **324**, 487 (1979).
18. K. Langanke, G. Martinez-Pinedo, J. M. Sampaio, D. J. Dean, W. R. Hix, O. E. B. Messer, A. Mezzacappa, M. Liebendoerfer, H.-Th. Janka, and M. Rampp, Phys. Rev. Lett. **90**, 241102 (2003).
19. W. R. Hix, O. E. B. Messer, A. Mezzacappa, M. Liebendoerfer, J. M. Sampaio, K. Langanke, D. J. Dean, and G. Martinez-Pinedo, Phys. Rev. Lett. **91**, 201102 (2003).
20. S. E. Woosley, in *16th Saas Fee Advanced Course, Nucleosynthesis and Chemical Evolution*, eds. B. Houck, A. Maeder, and G. Meynet (Geneva Obs., Sauverny, 1986), pp. 1–193.
21. W. R. Hix and B. S. Meyer, Nucl. Phys. A (2005), in press.
22. D. Q. Lamb, J. M. Lattimer, C. J. Pethick, and D. G. Ravenhall, Phys. Rev. Lett. **41**, 1623 (1978).
23. E. Baron, J. Cooperstein, and S. Kahana, Phys. Rev. Lett. **55**, 126 (1985).
24. J. Lattimer and F. D. Swesty, Nucl. Phys. A **535**, 331 (1991).
25. J. Cooperstein and J. Wambach, Nucl. Phys. A **420**, 591 (1984).

26. G. M. Fuller, W. A. Fowler, and M. J. Newman, ApJ **293**, 1 (1985).

27. K. Langanke and G. Martínez-Pinedo, Nucl. Phys. A **673**, 481 (2000).

28. T. Oda, M. Hino, K. Muto, M. Takahara, and K. Sato, Atomic Data and Nuclear Data Tables **56**, 231 (1994).

29. A. Heger, K. Langanke, G. Martínez-Pinedo, and S. E. Woosley, Phys. Rev. Lett. **86**, 1678 (2001).

30. A. Yahil, ApJ **265**, 1047 (1983).

31. G. M. Fuller, ApJ **252**, 741 (1982).

32. F. D. Swesty, J. M. Lattimer, and E. S. Myra, ApJ **425**, 195 (1994).

33. M. Liebendörfer, O. E. B. Messer, A. Mezzacappa, W. R. Hix, F.-K. Thielemann, and K. Langanke, in *Proceedings of the 11th Workshop on Nuclear Astrophysics, Ringberg Castle, Tegernsee, Germany, February 11-16, 2002*, eds. by W. Hillebrandt and E. Müller (2002), p. 126.

34. O. E. B. Messer, M. Liebendörfer, W. R. Hix, A. Mezzacappa, and S. W. Bruenn, in *Proc. of the ESO/MPA/MPE Workshop, From Twilight to Highlight: The Physics of Supernovae*, eds. by W. Hillebrandt and B. Leibundgut (Heidelberg: Springer, 2003), p. 70.

35. O. Messer, W. Hix, A. Mezzacappa, and M. Liebendörfer, ApJ (2005).

36. G. S. Stoitcheva and D. J. Dean, in *Open Issues in Core Collapse Supernovae*, edited by G. Fuller and A. Mezzacappa (World Scientific, Singapore, 2005).

37. K. Langanke, E. Kolbe, and D. J. Dean, Phys. Rev. C **63**, 32801 (2001).

38. W. R. Hix and F.-K. Thielemann, ApJ **460**, 869 (1996).

39. E. Bravo and D. García-Senz, MNRAS **307**, 984 (1999).

40. M. F. El Eid and W. Hillebrandt, A&AS **42**, 215 (1980).

41. S. W. Bruenn, ApJS **58**, 771 (1985).

42. F. Brachwitz, D. Dean, W. Hix, K. Iwamoto, K. Langanke, G. Martinez-Pinedo, K. Nomoto, M. R. Strayer, and F.-K. Thielemann, ApJ **536**, 934 (2000).

43. A. Mezzacappa and S. W. Bruenn, ApJ **405**, 669 (1993).

44. A. Mezzacappa and O. E. B. Messer, J. Comp. Appl. Math **109**, 281 (1999).

45. M. Liebendörfer, S. Rosswog, and F.-K. Thielemann, ApJS **141**, 229 (2002).

46. M. Liebendörfer, O. E. B. Messer, A. Mezzacappa, S. W. Bruenn, C. Y. Cardall, and F.-K. Thielemann, ApJS **150**, 263 (2004).

47. S. W. Bruenn and W. C. Haxton, ApJ **376**, 678 (1991).

48. W. Haxton, Phys. Rev. Lett. **60**, 1999 (1988).

49. A. Juodagalvis, K. Langanke, G. Martinez-Pinedo, W. R. Hix, D. J. Dean, and J. M. Sampaio, Nucl. Phys. A **747**, 87 (2005).

50. J. M. Sampaio, K. Langanke, G. Martínez-Pinedo, and D. J. Dean, Physics Letters B **529**, 19 (2002).

51. D. Gazit and N. Barnea, Phys. Rev. C **70**, 048801 (2004).

52. J. Pruet and G. M. Fuller, ApJS **149**, 189 (2003).

NEUTRINO PROCESSES IN HOT AND DENSE MATTER: CURRENT STATUS & OPEN ISSUES

SANJAY REDDY

Theoretical Division,
Los Alamos National Laboratory,
Los Alamos, NM 87545, USA
E-mail: reddy@lanl.gov

Production and propagation of neutrinos in hot and dense matter plays an important role in core-collapse supernova and thermal evolution of neutron stars. In this article we review the micro-physics that influences weak interaction rates in dense matter. We show that these rates depend sensitively on the strong and electromagnetic correlations between baryons. The current status of many body calculations of the medium response functions that are relevant for neutrino processes is discussed. Several physics issues have been identified but reliable quantitative predictions are still lacking - we present an overview of these open issues. Neutrino rates are also shown to be sensitive to the phase structure of matter at extreme density. We present a brief discussion of how these differences may affect core collapse supernova and the early evolution of a neutron star.

1. Introduction

Neutrinos play an important role in stellar evolution. By virtue of their weak interactions with matter neutrinos provide a mechanism for energy loss from the dense stellar interiors. In neutron stars, neutrino emission is the dominant cooling mechanism from their birth in a supernova explosion until several thousand years of subsequent evolution. The calculation of these rates are of current interest since several research groups are embarking on large scale numerical simulations of supernova and neutron star evolution [1]. Even moderate changes in the nuclear microphysics associated with the weak interaction rates at high density can impact macroscopic features that are observable. An understanding of the response of strongly interacting nuclear medium to neutrinos and its impact on neutron star evolution promises to provide a means to probe the properties of the dense medium itself.

In a recent article, Burrows, Reddy and Thompson reviewed the weak interaction processes in nuclear matter that are relevant in the supernova context [2]. Here we will recapitulate some of the subject covered in the aforementioned review and in addition discuss the possible role of phase transitions. In §2, we present a brief introduction to the macroscopic aspects of neutrino transport in a newly born neutron star. The discussion relating to neutrino-matter interactions is organized into two sections: (i) neutrino interactions in dense matter containing nuclei , nucleons and leptons (§3) and (ii) neutrino interactions in exotic new phases that are likely to occur in the dense inner core of the neutron star (§4).

2. Early evolution of the proto-neutron star

The illustration in Fig.1, shows the important stages of core-collapse supernova and the birth of a proto-neutron stars. Successive nuclear burning from lighter to heavier elements, which fuels stellar evolution, inevitably results in the formation of a iron core in massive stars ($M \gtrsim 8M_\odot$). Since iron is the most stable nucleus, further energy release through nuclear burning is not possible. The Fe-core is supported against gravitational collapse by the electron degeneracy pressure. When the mass of the Fe-core exceeds the Chandrasekhar mass ($M_{\rm ch} \simeq 1.4M_\odot$), it becomes unstable to gravitational collapse. Detailed numerical simulations indicate that the core collapses, from its initial radius $R_{\rm in} \simeq 1500$ km to a final radius $R_{\rm in} \simeq 100$ km, on a time-scale similar to the free-fall time-scale $\tau_{\rm free-fall} \simeq 100$ ms. Soon after the onset of collapse, the core density exceeds 10^{12} g/cm^3 and the matter temperature $T \simeq 5$ MeV. Under these conditions, thermal neutrinos become trapped on the dynamical time-scale of collapse. Consequently, collapse is nearly adiabatic. The enormous gravitational binding energy $B.E._{\rm Grav.} \simeq GM_{\rm NS}/R_{\rm NS} \simeq 3 \times 10^{53}$ ergs, is stored inside the star as internal thermal energy of the matter components, and thermal and degeneracy energy of neutrinos. The newly born neutron star looses this energy on a time-scale determined by the rate of diffusion of neutrinos [3,4]. Neutrinos emitted from the proto-neutron star can be detected in terrestrial detectors such as Super Kamiokande and SNO. Current estimates indicate that we should see $\sim 10,000$ events in Super Kamiokande and ~ 1000 events in SNO from a supernova at the center of our galaxy[5] (distance=8.5 kpc) . Understanding the micro and macro-physics that affects the spectral and temporal features of the neutrino emission is primarily motivated by this prospect. Supernova neutrinos are the only direct probes of both the dynamics of gravitational collapse and the properties of dense matter inside

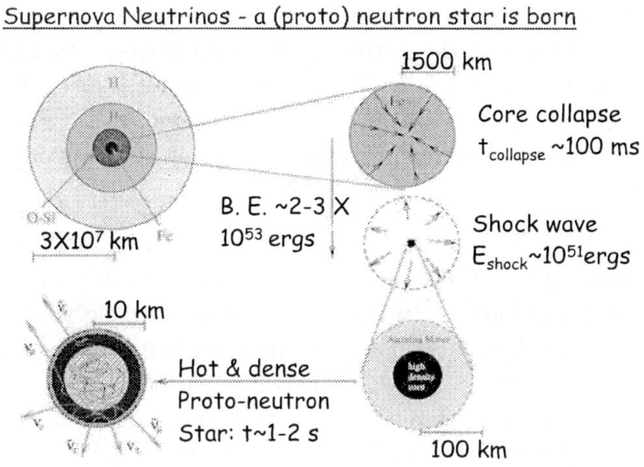

Figure 1. Schematic showing the various stages of a core-collapse supernova explosion.

the newly born neutron star. Since the neutrino emission time scale is set by neutrino diffusion, the duration over which we should expect to see neutrinos in terrestrial detectors is intimately connected with the neutrino opacity of matter inside the neutron star. We now turn to address micro-physical aspects of neutrino cross sections in dense matter.

3. Neutrino Interactions in Nucleon Matter

It was realized over a decade ago that the effects due to degeneracy and strong interactions significantly alter the neutrino mean free paths and neutrino emissivities in dense matter [6,7], it is only recently that detailed calculations have become available [8,9,10,11]. The scattering and absorption reactions that contribute to the neutrino opacity are

$$\nu_e + n \rightarrow e^- + p\,, \qquad \bar{\nu}_e + p \rightarrow e^+ + n\,,$$

$$\nu_X + A \rightarrow \nu_X + A\,, \qquad \nu_X + n(p) \rightarrow \nu_X + n(p)\,, \qquad \nu_X + e^- \rightarrow \nu_X + e^-\,,$$

where $n, p, e^{\pm}, A$ represent neutrons, protons, positrons, electrons and heavy Fe-like nuclei, respectively. At low temperature ($T \lesssim 3 - 5$ MeV) and relatively low density ($\rho \simeq 10^{12} - 10^{13}$ g/cm^3, heavy nuclei are present and dominate the neutrino opacity due to coherent scattering. When the density is higher, $\rho \simeq 10^{13} - 10^{14}$ g/cm^3, novel heterogeneous phases of matter, called the "pasta" phases have been predicted to occur, where nuclei become extended and deformed progressively from spherical to rod-like and slab-like configurations[12]. For densities greater than 10^{14} g/cm^3, matter is expected to be a homogeneous nuclear liquid. In what follows, we discuss the neutrino opacity in these different physical settings.

3.1. $\rho \simeq 10^{12}$ g/cm^3:

At low temperature ($T \lesssim 5$ MeV), matter at these densities comprises of heavy nuclei (fully ionized), nucleons and degenerate electrons. The typical inter-particle distance, $d \simeq 20 - 40$fm. At these large distances, the nuclear force is small and the correlations between particles is dominated by the coulomb interaction. Since nuclei carry a large charge ($Z \simeq 25$) , the coulomb force between nuclei $F_{\text{Coulomb}} \cong Z^2 e^2/d$ dominates the non-ideal behavior of the plasma. Further, for low energy neutrinos which couple coherently to the total weak charge Q_W $25 - 40$ of the nucleus, neutrino scattering off nuclei is far more important than processes involving free nucleons and electrons [13].

The elastic cross-section for low energy coherent scattering off a nucleus (A,Z) with weak charge $Q_W = A - Z + Z \sin^2 \theta_W$, where θ_W is the weak mixing angle, is given by[13]

$$\frac{d\sigma}{d\cos\theta} = \frac{1}{16\pi} \, G_{\text{F}}^2 \, Q_W^2 \, E_\nu^2 \, (1 + \cos\theta) \tag{1}$$

When neutrinos scattering off nuclei in a plasma we must properly account for the presence of other nuclei since scattering from these different sources can interfere. In the language of many-body theory, this screening is encoded in the density-density correlation function [6,7]. To make concrete, the relation to the density-density correlation function we begin by noting that the effective Lagrangian describing the neutral current interaction of low energy neutrinos with nuclei is given by

$$L_{\text{NC}} = \frac{G_{\text{F}}}{2\sqrt{2}} \, Q_W \, l_\mu \, j^\mu \tag{2}$$

where $l_\mu = \bar{\nu}\gamma_\mu(1 - \gamma_5)\nu$ is the neutrino neutral current. Nuclei are heavy, their thermal velocities are small, $v \ll c$, and it is good approximation to write the neutral current carried by the nuclei as $j^\mu = \psi^\dagger\psi \, \delta_0^\mu$. For simplicity, we will assume that nuclei are bosons characterized only by their charge and baryon number. Using Fermi's golden rule, we can calculate the neutrino cross sections from Eq.2. In matter, it is appropriate to define a cross section per unit volume rather than cross section off a single target particle since more than one target particle contributes to the response. We can write the differential cross section per unit volume in terms of the density operator in momentum space $\rho(\vec{q}, t) = \psi^\dagger \psi = \sum_{i=1 \cdots N} \exp(i\vec{q} \cdot \vec{r}_i(t))$, where $i = 1 \cdots N$ is the particle index. The cross section for scattering of a neutrino with energy transfer ω and momentum transfer $\vec{q}$ is given by

$$\frac{d\sigma}{V \, d\omega \, d\cos\theta} = \frac{G_{\rm F}^2}{16\pi} \, Q_W^2 \, (1 + \cos\theta) \, S(|\vec{q}|, \omega) \tag{3}$$

$$\text{where} \quad S(|\vec{q}|, \omega) = \frac{1}{2\pi \, N} \int dt \, \exp(i\omega t)\langle \rho(\vec{q}, t)\rho(-\vec{q}, 0)\rangle . \tag{4}$$

The function $S(|\vec{q}|, \omega)$ is called the dynamic structure function and embodies all spatial and temporal correlations between target particles arising from strong or electromagnetic interactions. To calculate the structure function we need to solve for the dynamics $(\vec{r}(t))$ of the ions as they move in the each others presence, interacting via the two body ion-ion interaction potential. The electrons are relativistic and degenerate, the time-scales associated with changes in their density distribution are rapid compared to the slow changes we expect in the density field of the heavy ions. The leading effect of the electrons is therefore to screen the Coulomb potential. Consequently, the ions interact through the potential $V(r) = Z^2 e^2 \exp(-r/\lambda_e)/(4\pi r)$, where λ_e is the electron Debye screening length. At 10^{12} g/cm^3, where the inter-ion distance $d \lesssim 30$ fm and $\lambda_e \gtrsim 80$ fm, the typical ion-ion interaction energy, $E_{\rm pot} \simeq Z^2 e^2/(4\pi d)$ is large. For temperature in the range $1 - 5$ MeV, the ratio of the potential energy to the kinetic energy, which is characterized by $\Gamma = Z^2 e^2/(4\pi d \, kT) \gg 1$. The dynamics of a strongly coupled plasma is not amenable to analytic methods of perturbation theory, nor are they tractable in approximate non-perturbative methods such as mean field theory or the Random Phase Approximation (RPA). For a classical system of point particles interacting via a 2-body potential it is possible to simulate the real system. Such numerical simulations confine N particles to a box with periodic boundary conditions and calculate the force on each particle at any time and evolve the particles by using their equations of motion ($Force = Mass \times Acceleration$). This method, which goes by the name molecular dynamics (MD), has been used extensively in condensed matter physics, plasma physics and chemistry [15]. For $T \gtrsim 1$ MeV, the De-Broglie wavelength $\Lambda_D \ll d$ so the ionic gas is classical and we can use MD to calculate $S(|\vec{q}|, \omega)$. Fig.2 , shows the results of such a calculation. We chose to simulate 54 ions (A=50,Z=25) in a box of length $L = 200$ fm. The background electron density was chosen so as to make the system electrically neutral. For the classical simulations, a single,

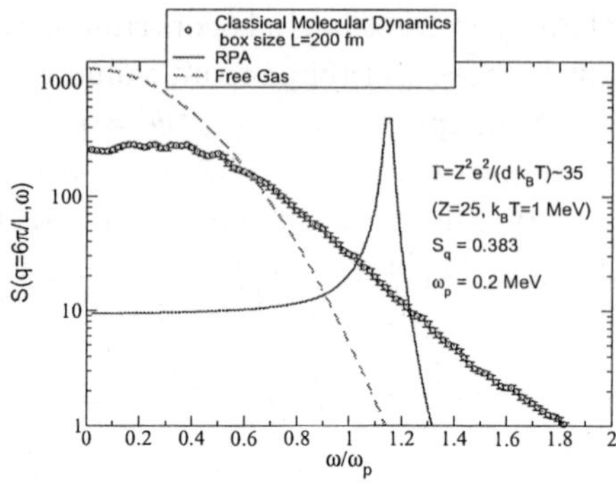

Figure 2. Dynamic structure function of a plasma of ions interacting as a function of energy transfer ω (measured in units of the plasma frequency $\omega_p = 0.2$ MeV) and fixed momentum transfer $|\vec{q}| = 6\pi/L = 18.6$ MeV.

dimensionless, quantity $\Gamma = Z^2 e^2/(4\pi d\, kT)$ characterizes the system. For $kT = 1$ MeV, $\Gamma \simeq 35$ for our system, corresponding to a strongly coupled plasma. The results in Fig.2 show a comparison between results obtained by MD simulations (dots with error bars) , free (Boltzmann) gas response (dashed-line) and response obtained using Random Phase Approximation (RPA). Details regarding the RPA will be discussed in more detail in the subsequent sections. For now, we simply note that in RPA, Coulomb interactions are accounted for by selective re-summation of bubble graphs involving free excitations which interact via the Coulomb interaction. In particular, RPA is able to correctly predict the presence of well defined collective mode, corresponding to a phonon in the weak coupling regime ($\Gamma \lesssim 1$) and is often expected to provide a fair description of the long-wavelength response of correlated systems. However, as the comparison in Fig.2 indicates, both RPA and the free gas response do poorly compared to MD. MD is exact in the classical limit and corrections to the classical evolution are small and expected to scale as λ_D/d. These results clearly illustrate that correlations can greatly affect the shape of the response and that approximate many-body methods such as RPA can fail when the coupling is strong, even at long-wavelengths.

3.2. $\rho \simeq 10^{13}\ g/cm^3$:

With increasing density, the nuclei get bigger and the inter-nuclear distance become smaller. Under these conditions, the nuclear surface and coulomb contributions to the Free energy of the system become important.

It becomes energetically favorable for nuclei to become very deformed and assume novel, non-spherical, shapes such as rods and slabs [12]. Further, the energy differences between these various shapes are small $\Delta E \simeq 10 - 100$ keV. The dynamics of such an exotic heterogeneous phase is a complex problem involving several scales and forces. For temperatures of interest, $T \lesssim 5$ MeV, the De-Broglie wavelength and the inter-particle distance are comparable and quantum effects cannot be neglected. Recently, there have been attempts to model the behavior of these pasta phases using quantum molecular dynamics[16] and also find rod and slab like configurations.

How does the heterogeneity and existence of several low energy excitations involving shape fluctuations influence the response of this phase to neutrinos ? In the simplest description, the structure size (r) and the inter-structure (R) distance characterize the system. We can expect that neutrinos with wavelength large compared to the structure size but small compared to the inter-structure distance can couple coherently to the total weak charge (excess) of the structure, much like the coherence we discussed in the previous section. The effects of this coherent enhancement in the neutrino cross-sections has recently been investigated [17]. In agreement with our naive expectation, this study finds that the neutrino cross sections are enhanced by as much as an order of magnitude for neutrinos with energy $1/r \gtrsim E_\nu \gtrsim 1/R$. In §4.1, we discuss similar effects in a higher density heterogeneous mixed phase composed of nuclear matter and novel high density phases such as quark matter.

3.3. $\rho \simeq 10^{14}$ g/cm^3:

With increasing density, the novel structures discussed previously merge to form a homogeneous liquid of neutrons, protons and electrons. The response of such a Fermi-liquid has been investigated by several authors[8,10,11]. The general expression for the differential cross section for the reaction $\nu_1 + 2 \rightarrow \nu_3 + 4$ is

$$\frac{1}{V}\frac{d^3\sigma}{d^2\Omega_3 dE_3} = -\frac{G_F^2}{128\pi^2}\frac{E_3}{E_1}\left[1 - \exp\left(\frac{-q_0 - (\mu_2 - \mu_4)}{T}\right)\right]^{-1}$$
$$\times\ (1 - f_3(E_3))\ \mathrm{Im}\ (L^{\alpha\beta}\Pi^R_{\alpha\beta}),\tag{5}$$

where the incoming neutrino energy is E_1 and the outgoing electron energy is E_3, 2 is the initial state of the target particle and 4 is its final state[8,10]. The factor $[1 - \exp((-q_0 - \mu_2 + \mu_4)/T)]^{-1}$ maintains detailed balance, for particles labeled '2' and '4' which are in thermal equilibrium at temperature T and in chemical equilibrium with chemical potentials μ_2 and μ_4, respectively. The final state blocking of the outgoing lepton is accounted for by the Pauli blocking factor $(1 - f_3(E_3))$. The lepton tensor $L_{\alpha\beta}$ and the target particle retarded polarization tensor $\Pi_{\alpha\beta}$ is given by[10]

$$\Pi_{\alpha\beta} = -i \int \frac{d^4 p}{(2\pi)^4} \mathrm{Tr} \, [T(G_2(p) J_\alpha G_4(p+q) J_\beta)] \,. \tag{6}$$

Above, k_μ is the incoming neutrino four-momentum and q_μ is the four-momentum transfer. The Greens' functions $G_i(p)$ (the index i labels particle species) describe the propagation of free baryons at finite density and temperature. J_μ is γ_μ for the vector current and $\gamma_\mu \gamma_5$ for the axial current.

To account for the effects of strong and electromagnetic correlations between target neutrons, protons and electrons we must find ways to improve $\Pi_{\alpha,\beta}$. This involves improving the Greens functions for the particles and the associated vertex corrections that modify the current operators. In strongly coupled systems, these improvements are notoriously difficult and no exact analytic methods exist. One usually resorts to using mean-field theory to improve the Greens functions. Dressing the single particle Greens functions must be accompanied by corresponding corrections to the neutrino – dressed-particle vertex function. The random-phase approximation (RPA) can be thought of as such a vertex correction. Within RPA, the polarization tensor[8,10]

$$\Pi^{RPA} = \Pi_{\mathrm{MF}} + \Pi^{RPA} D \Pi_{\mathrm{MF}} \,, \tag{7}$$

where D denotes the interaction matrix and Π_{MF} is the polarization tensor in Eq.6, but with the Green's functions computed in the mean-field approximation. Model calculations indicate that neutrino mean free paths computed in RPA tend to be a factor 2-3 times larger than in the uncorrelated system[10]. This is primarily because of repulsive forces in the spin-isospin channel, that suppress the axial response at low energies.

3.4. *Nuclear Response: Open issues*

The simple model, based on Fermi Liquid Theory and the Random Phase Approximation with central interactions, is rudimentary and relies on the poorly constrained determinations of the Landau parameters. While we can expect this simple analysis to capture qualitative aspects of many-body correlations it is not suitable for quantitative predictions. It has two important shortcomings, namely the neglect of multi-pair excitations and non-central interactions such as the tensor force (pion exchange), which is known to be important in nuclear systems. The discussions in §3.1 regarding the response of the ion-plasma clearly emphasized the need to incorporate finite propagation lifetimes for quasi-particles due to collisional damping. The role of collisional damping in the low-energy response is particularly well studied for the case of low-energy photon production from bremsstrahlung in many-particle systems and is called the Landau-Pomeranchuk-Migdal effect. Similar effects were first investigated in the context of the response of nuclear matter by [18]. They found that the rate of spin fluctuations $(1/\tau_{\mathrm{spin}})$ due to nucleon-nucleon collisions in the medium is rapid compared to the typical energy transfer ω in neutrino scattering. Consequently, the axial charge is dynamically screened for small $\omega \lesssim \tau_{\mathrm{spin}}^{-1}$, resulting in a suppression of the low-energy axial response.

The redistribution of response strength in energy is a generic feature of many-particle systems arising due to the finite lifetime of quasi-particles. However, the situation in nuclear systems is unique due to the presence of a strong tensor force. This has recently been clarified by [20]. The evolution of nucleon spin is dominated by tensor interactions, especially because the nucleon spin operator $\hat{\sigma}$ does not commute with the tensor operator in the nuclear Hamiltonian. The F-sum rule, discussed earlier in §3.1, for the spin response function does not vanish in the long-wavelength limit, since $[\mathcal{H}_{\mathrm{tensor}}, \hat{\sigma}] \neq 0$. Further, the relationship between the spin susceptibility and the Landau parameters is modified due to presence of the tensor interaction (see [20] for these revised relations). The preceding discussion indicates that interactions that do not commute with the spin operator may be especially important in determining the low-energy response. [20] estimate that as much as 60% of the low-energy axial response at long wavelength may reside in multi-particle excitations. This preliminary estimate warrants further investigation requiring both the inclusion of the tensor force and multi-pair excitations in the axial response.

When multi-pair excitations become important it is appropriate to work in terms of a correlated basis states rather than single particle bare neutron and proton states. The correlated basis states are expected to be close to the energy eigenstates of the system. Consequently, in this basis the residual interactions are weak. Further, the ground and excited states in the correlated basis states contain multi-particle hole states and the quasi-particles of the correlated basis are superpositions of neutron and proton states of both spins. This is particularly relevant for nuclear matter, where pion exchange can transform both spin and isospin of the bare nucleons. Recently, [21] have computed weak interaction matrix elements in the correlated basis obtained using a two-body cluster expansion. They find that spin and iso-spin correlations play an important role and result in quenching the weak interaction transition rates by $20 - 25\%$ at low energy.

4. Neutrino Interactions in Novel Phases at High Density

In this section we explore how phase transitions impact the weak interaction rates. Novel phases of baryonic matter are expected to occur at densities accessible in neutron stars. These new phases include pion condensates, kaon condensates, hyperons and quark matter. An understanding of how these phases might influence neutrino propagation and emission is necessary to inquire if these phase transitions occur inside neutron stars. We consider three specific examples of phase transitions: (i) generic first order transitions; (2) superconducting quark matter and (3) color-flavor locked superconducting quark matter to explore and illustrate the modification of neutrino rates in the novel high density phases of matter.

4.1. *Heterogeneous Phases: Effects of First Order Transitions*

First order phase transitions in neutron stars can result in the formation of a heterogeneous phase in which a positively charged nuclear phase coexists with a negatively charged novel phase which is favored at higher densities [22]. This is a generic possibility for first order transitions in matter with two conserved charges. In the neutron star context, these correspond to baryon number and electric charge. The pasta phase at sub-nuclear density, which was discussed earlier, where positively charged nuclei coexist with negatively charged neutron rich matter containing electrons is a familiar example of such a transition region. This mixed, heterogeneous, phase exists over a finite interval of pressure, unlike a mixed phase for a system with one conserved charge. Consequently, mixed phases in neutron stars occupy a finite spatial extent and understanding how neutrinos propagate through them becomes a relevant and interesting question.

Reddy, Bertsch and Prakash [23] have studied the effects of heterogeneous phases on ν-matter interactions. Based on simple estimates of the surface tension between nuclear matter and the exotic phase, typical droplet sizes range from $5 - 15$ fm, and inter-droplet spacings range up to several times larger. The propagation of neutrinos whose wavelength is greater than the typical droplet size and less than the inter-droplet spacing, i.e., 2 MeV $\leq E_\nu \leq 40$ MeV, will be greatly affected by the heterogeneity of the mixed phase, as a consequence of the coherent scattering of neutrinos from the matter in the droplet.

The Lagrangian that describes the neutral current coupling of neutrinos to the droplet is

$$\mathcal{L}_W = \frac{G_F}{2\sqrt{2}}\, Q_W\, \bar{\nu}\gamma_\mu(1-\gamma_5)\nu\, J_D^\mu\,, \tag{8}$$

where J_D^μ is the neutral current carried by the droplet and the total weak charge enclosed in a droplet of radius r_d is Q_W. For non-relativistic droplets, $J_D^\mu = \rho(x)\, \delta^{\mu 0}$ has only a time like component. Here, ρ is the density operator for the droplet and the form factor is $F(q) = (1/Q_W)\int_0^{r_d} d^3x\, \rho(x)\, \sin qx/qx$.
The differential cross section for neutrinos scattering from an isolated droplet is then

$$\frac{d\sigma}{d\cos\theta} = \frac{E_\nu^2}{16\pi}G_F^2 Q_W^2(1+\cos\theta)F^2(q)\,. \tag{9}$$

In the above equation, E_ν is the neutrino energy and θ is the scattering angle. To properly account for the presence of the other droplets in the medium we must calculate the droplet-droplet correlation function or $S(\vec{q},\omega)$ as was described in §3.1. However, such a calculation is yet to be done for this complex system. In what follows, we adopt a simple mean-field prescription to account for screening due to multiple droplet scattering[23]. This amounts to replacing the form factor as follows

$$F(q) \to \tilde{F}(q) = F(q) - 3\,\frac{\sin qR_W - (qR_W)\cos qR_W}{(qR_W)^3}\,. \tag{10}$$

The neutrino–droplet differential cross section per unit volume then follows:

$$\frac{1}{V}\frac{d\sigma}{d\cos\theta} = N_D\,\frac{E_\nu^2}{16\pi}G_F^2 Q_W^2(1+\cos\theta)\tilde{F}^2(q)\,. \tag{11}$$

Note that even for small droplet density N_D, the factor Q_W^2 acts to enhance the droplet scattering. Models of first order phase transitions in dense matter provide the weak charge and form factors of the droplets and permit the evaluation of ν–droplet scattering contributions to the opacity of the mixed phase. For the results shown in Fig.3, the quark droplets are characterized by $r_d \sim 5$ fm and inter-droplet spacing $2R_W \sim 22$ fm, and an enclosed weak charge $Q_W \sim 850$. For comparison, the neutrino mean free path in uniform neutron matter at the same n_b and T are also shown. It is apparent that there is a large coherent scattering-induced reduction in the mean free path for the typical energy $E_\nu \sim \pi T$. At much lower energies, the inter-droplet

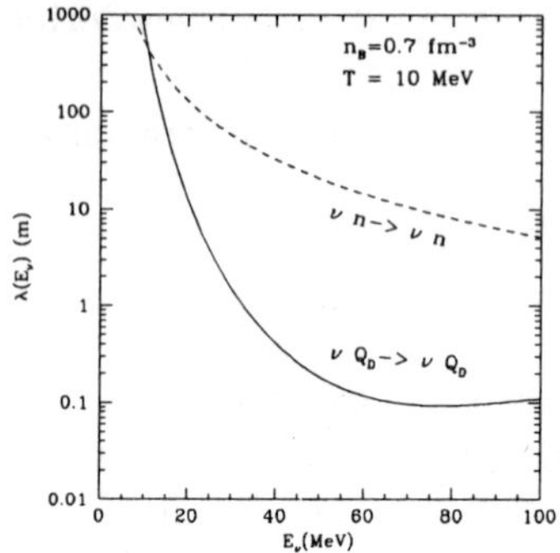

Figure 3.   Neutrino mean free path as a function of neutrino energy in a quark - hadron mixed phase. For comparison, the mean free path in uniform neutron matter at the same temperature and density are shown by dashed curve.

correlations tend to screen the weak charge of the droplet, and at higher energies the coherence is attenuated by the droplet form factor.

4.2. *Effects of Quark Superconductivity*

Although the idea of quark pairing in dense matter is not new [24], it is only recently that its role in the context of the phase diagram of QCD [25] has been appreciated. Model calculations, mostly based on four-quark effective interactions, predict an energy gap of $\Delta \sim 100$ MeV for a typical quark chemical potential of $\mu_q \sim 400$ MeV. In this section, we address how neutrinos propagate in superconducting and superfluid quark matter.

As discussed earlier, the main task will be to compute the equivalent of the current-current correlation function defined in Eq.6, but for the superconducting quark phase. This was addressed by Carter and Reddy [26]. The free quark propagators are naturally modified in a superconducting medium. As first pointed out by Bardeen, Cooper, and Schrieffer several decades ago, the quasi-particle dispersion relation is modified due to the presence of a gap in the excitation spectrum. More importantly quark propagators acquire off-diagonal components which are called anomalous propagators. This describes the process in which a pair is either created or destroyed from the superconducting ground state. These processes are naturally accounted for in the Nambu-Gorkov formalism described in Ref. [26]. In calculating these effects, we will consider the simplified case of QCD with two light quark flavors given that the light u and d quarks dominate low-energy phenomena. Furthermore we will assume that, through some unspecified effective interaction, which is attractive in the color antisymmetric channel, quarks pair in a manner analogous to the BCS mechanism. In this two flavor, spin zero superconductor (2SC) the anomalous (or Gorkov) propagator[25]

$$F(p)_{abfg} = -i\epsilon_{ab3}\epsilon_{fg}\Delta \left(\frac{\Lambda^+(p)}{p_o^2 - \xi_p^2} + \frac{\Lambda^-(p)}{p_o^2 - \bar{\xi}_p^2} \right) \gamma_5\, C\,. \tag{12}$$

Here, a, b are color indices, f, g are flavor indices, ϵ_{abc} is the usual antisymmetric tensor and we have conventionally chosen 3 to be the condensate color. This propagator is also antisymmetric in flavor and spin, with $C = -i\gamma_0\gamma_2$ being the charge conjugation operator. The normal quasi-particle propagators are given as

$$S(p)_{af}^{bg} = i\delta_a^b\delta_f^g \left(\frac{\Lambda^+(p)}{p_o^2 - \xi_p^2} + \frac{\Lambda^-(p)}{p_o^2 - \bar{\xi}_p^2} \right) (p_\mu\gamma^\mu - \mu\gamma_0)\,. \tag{13}$$

This is written in terms of the particle and anti-particle projection operators $\Lambda^+(p)$ and $\Lambda^-(p)$ respectively, where $\Lambda^\pm(p) = (1 \pm \gamma_0\vec{\gamma}\cdot\hat{p})/2$. The quasi-particle energy is $\xi_p = \sqrt{(|\vec{p}| - \mu)^2 + \Delta^2}$, and for the anti-particle $\bar{\xi}_p = \sqrt{(|\vec{p}| + \mu)^2 + \Delta^2}$. The appearance of an anomalous propagator in the

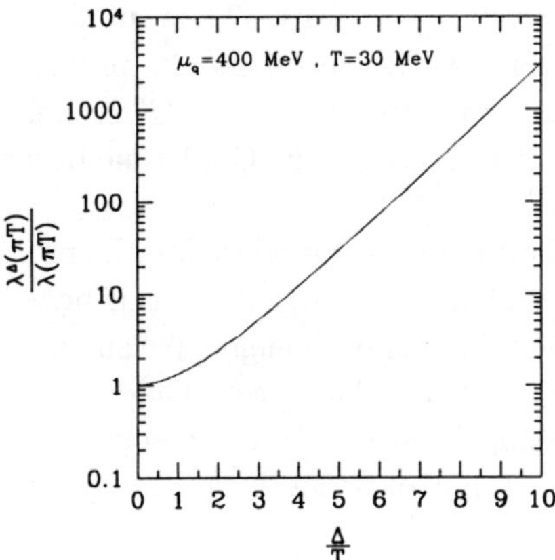

Figure 4. Ratio of the neutrino mean-free path in the superconducting phase to that in the normal phase as function of Δ/T.

superconducting phase indicates that the polarization tensor gets contributions from both normal quasi-particle propagators Eq.13 and anomalous propagator Eq.12 and is given by

$$\Pi_{\alpha\beta}(q) = -2i \int \frac{d^4p}{(2\pi)^4} \text{Tr} \left[S(p)\Gamma_\alpha S(p+q)\Gamma_\beta + F(p)\Gamma_\alpha \bar{F}(p+q)\Gamma_\beta \right]. \quad (14)$$

Fig.4 shows the results for neutrino mean free paths in the 2 flavor super-conducting quark phase. The ratio of mean free path in the superconducting phase to that in the normal phase is shown in the right panel. With increasing Δ, the gap in the quasi-particle excitation spectrum results in exponential attenuation of the scattering response and the observed exponential enhancement in the mean free paths.

4.3. *Neutrino Interactions with Goldstone bosons*

The discussion in the preceding section assumed that there were no low energy collective excitations to which the neutrinos could couple. This is true in the 2 flavor superconducting phase of quark matter. For three flavors and when the strange quark mass is negligible compared to the chemical potential the ground state is characterized by pairing that involves all nine quarks in a pattern that locks flavor and color [27]. Diquark condensation in the CFL phase breaks both baryon number and chiral symmetries. The Goldstone bosons that arise as consequence introduce a low lying collective excitations to the otherwise rigid state. Thus, unlike in the normal phase where quark excitations near the Fermi surface provide the dominant contribution to the weak interaction rates, in the CFL phase, it is the dynamics of the low energy collective states— the Goldstone bosons that are relevant [28,30,31].

The massless Goldstone boson associated with spontaneous breaking of $U(1)_B$ couples to the weak neutral current. This is because the weak isospin current contains a flavor singlet component. We should expect this mode to dominate the response because the psuedo-Goldstone modes arising from chiral symmetry breaking have non-zero masses. The amplitude for processes involving the $U(1)_B$ Goldstone boson H and the neutrino neutral current is given by

$$A_{H\nu\bar{\nu}} = \frac{4}{\sqrt{3}} G_F f_H \tilde{p}_\mu j_Z^\mu, \quad (15)$$

where $\tilde{p}_\mu = (E, v^2\vec{p})$ is the modified four momentum of the Goldstone boson. The decay constant for the $U(1)_B$ Goldstone boson has also been computed in earlier work and is given by $f_H^2 = 3\mu^2/(8\pi^2)$. Neutrinos of all energies can absorb a thermal meson and scatter into either a final state

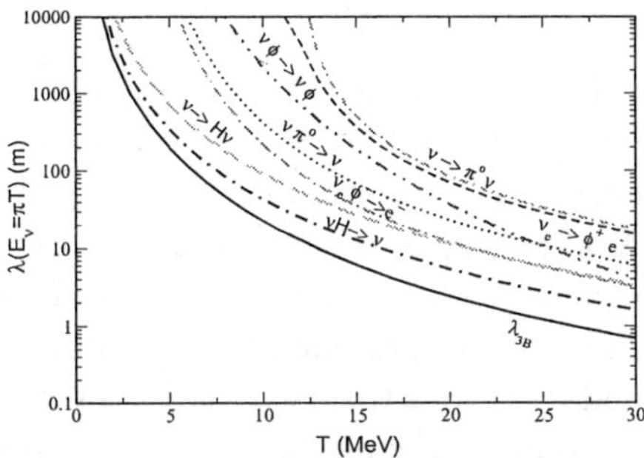

Figure 5. Neutrino mean free path in a CFL meson plasma as a function of temperature. The neutrino energy $E_\nu = \pi T$ and is characteristic of a thermal neutrino.

neutrino by neutral current processes like $\nu + H \rightarrow \nu$ and $\nu + \pi^0 \rightarrow \nu$ or via the charged current reaction into a final state electron by the process $\nu_e + \pi^- \rightarrow e^-$. They can also emit mesons through processes like $\nu \rightarrow H + \nu$ and $\nu \rightarrow \pi^0 + \nu$, since the Goldstone modes have space-like dispersion relations. Mean free path due to these processes, which we collectively refer to as Cerenkov processes can be computed[28] and are shown in Fig.5. In contrast to processes involving the emission or absorption of mesons by neutrinos, the usual scattering process involves the coupling of the neutrino current to two mesons is suppressed by the factor p/f_π where p is the meson momentum. The neutrino mean free paths in the CFL phase are very similar to those in the normal quark phase - both numerically and parametrically [28]. The low energy effective theory for the Goldstone modes is applicable only for temperatures that are small compared to the critical temperature. As one approaches T_c, the single-pair excitations that dominate the response in the normal phase become relevant. In this case a more microscopic description in terms of quarks becomes necessary. Such a description, based on the Nambu-Gorkov formalism described earlier in §4.2 but which also includes the collective Goldstone mode, can be found in Ref. [29]. In this work it was found that the single-pair (particle-hole) excitations play a role only in the vicinity of T_c. At lower temperature it is interesting to note that the existence of one massless mode compensates for the large gap in the particle-hole excitations spectrum. The contrast between the findings of the previous section, where we found a large enhancement in the mean free paths, is striking.

5. Discussion

In the simplest scenario, where rotation, magnetic fields and convection are ignored, the temporal structure of the neutrino emission is directly related to the supernova neutrino luminosity and energy spectrum depend solely on the weak interaction processes in the hot and dense protoneutron star. The detection of neutrinos from a galactic supernova in terrestrial detectors such as Super Kamiokande will provide detailed information regarding the temporal structure and motivates theory to address neutrino propagation in hot and dense matter. While there has been some progress in understanding the qualitative aspects of the response of dense matter to neutrinos, reliable quantitative calculations of neutrino mean free paths in dense nuclear and other novel phases of matter are yet to be performed. First principles calculations of the linear response of strongly correlated systems is a difficult problem, with limited success, and a long history in condensed matter, nuclear and particle physics. However, the prospect of probing the phase structure of matter at the most extreme densities through neutrinos from the next galactic supernova is compelling motivation to pursue these efforts. I hope this article has provided a glimpse of the promise and difficulties associated with this enterprise.

6. Acknowledgments

I would like to thank George Bertsch, Joe Carlson, Greg Carter, Chuck Horowitz, Jim Lattimer, Jose Pons, Madappa Prakash, Mariusz Sadzikowski, and Motoi Tachibana for enjoyable collaborations and/or useful discussions. This work is supported in part by funds provided by the U.S. Department of Energy (D.O.E.) under the D.O.E. contract W-7405-ENG-36.

References

1. R. Buras, et al., Phys. Rev, Lett. **90**, 241101 (2003); A. Mezzacappa et al., Phys. Rev. Lett. **86**, 1935 (2001); Burrows, et al., Astrophys. J. **539**, 865 (2000)
2. A. Burrows, S. Reddy and T. A. Thompson, arXiv:astro-ph/0404432.
3. A. Burrows, J.M. Lattimer: Astrophys. J. **307**, 178 (1986)
4. M. Prakash, J. M. Lattimer, J. A. Pons, A. W. Steiner and S. Reddy, Lect. Notes Phys. **578**, 364 (2001)
5. A. Burrows, D. Klein and R. Gandhi, Nucl. Phys. Proc. Suppl. **31**, 408 (1993).
6. R. F. Sawyer, Phys. Rev. D **11**, 2740 (1975).
7. N. Iwamoto and C. J. Pethick, Phys. Rev. D **25**, 313 (1982).

8. C.J. Horowitz, K. Wehrberger: Nucl. Phys. A **531**, (1991) 665 ; Phys. Rev. Lett. **66**, 272 (1991); Phys. Lett. B **226**, 236 (1992)

9. G. Raffelt, D. Seckel: Phys. Rev. D **52**, 1780 (1995)

10. S. Reddy, M. Prakash, J.M. Lattimer: Phys. Rev. D **58**, 013009 (1998); S. Reddy, M. Prakash, J.M. Lattimer, J.A. Pons: Phys. Rev. C **59**, 2888 (1999)

11. A. Burrows, R.F. Sawyer: Phys. Rev. C **58**, 554 (1998); A. Burrows, R.F. Sawyer: Phys. Rev. C **59**, 510 (1999)

12. D. G. Ravenhall, C. J. Pethick and J. R. Wilson, Phys. Rev. Lett. **50**, 2066 (1983).

13. D. Z. Freedman, Phys. Rev. D **9**, 1389 (1974).

14. L. B. Leinson, V. N. Oraevsky and V. B. Semikoz,

15. J.-P. Hansen, I. R. McDonald and E. L. Pollock, Phys. Rev. D **11**, 1025 (1975)

16. G. Watanabe, K. Sato, K. Yasuoka and T. Ebisuzaki, Phys. Rev. C **68**, 035806 (2003)

17. C. J. Horowitz, M. A. Perez-Garcia and J. Piekarewicz, arXiv:astro-ph/0401079.

18. Raffelt, G. & Seckel, D., Phys. Rev. D **52**, 1780 (1995)

19. Raffelt, G. & Seckel, D., Phys. Rev. Lett., **69**, 2605 (1998)

20. Olsson, E. & Pethick, C.J. , Phys. Rev. C **66**, 065803 (2002)

21. Cowell, S., & Pandharipande, V., Phys. Rev. C **67**, 035504 (2002)

22. N. K. Glendenning, Phys. Rev. D **46**, 1274 (1992).

23. S. Reddy, G. Bertsch and M. Prakash, Phys. Lett. B **475**, 1 (2000)

24. B.C. Barrois: Nucl. Phys. B **129**, 390 (1977)

25. M. Alford, K. Rajagopal, F. Wilczek: Phys. Lett. B **422**, 247 (1998) Nucl. Phys. B **357**, 443 (1999) *ibid.* **558**, 219 (1999) R. Rapp, T. Schäffer, E.V. Shuryak, M. Velkovsky: Phys. Rev. Lett. **81**, 53 (1998) Ann. Phys. **280**, 35 (2000)

26. G. W. Carter and S. Reddy, Phys. Rev. D **62**, 103002 (2000)

27. M. G. Alford, K. Rajagopal and F. Wilczek, Nucl. Phys. B **537**, 443 (1999)

28. S. Reddy, M. Sadzikowski and M. Tachibana, Nucl. Phys. A **714**, 337 (2003)

29. J. Kundu and S. Reddy, Phys. Rev. C **70**, 055803 (2004) [arXiv:nucl-th/0405055].

30. P. Jaikumar, M. Prakash and T. Schafer, Phys. Rev. D **66**, 063003 (2002)

31. S. Reddy, M. Sadzikowski and M. Tachibana, Phys. Rev. D **68**, 053010 (2003)

FLAVOR CHANGING NEUTRAL CURRENTS AND STELLAR COLLAPSE

P. S. AMANIK

*Department of Physics,
University of California, San Diego,
La Jolla, CA 92093*

We examine some of the effects on a core collapse supernova if neutrinos had the ability to change flavor. Flavor changing neutral currents (FCNC's) are predicted by some theories for beyond Standard Model physics. We will discuss how neutrino-quark FCNC's are "amplified" due to coherent elastic scattering with the nuclear matter present in the core collapse environment. We will show that this type of interaction has consequences for the supernova model even for ranges of coupling constant strength beyond that of current experimental limits.

1. Introduction

The current and ongoing effort to understand core collapse supernovae mostly involves physics that is already known and verified experimentally. The task of accounting for all known and understood phenomena of physics in supernova models is extremely challenging and very clever work is being done to make progress with it. However, it is also important to remember that we have not yet discovered all the laws of fundamental physics. In particular, we know the Standard Model (SM) is an incomplete description of particle physics and that some models which solve questions in the SM predict the existence of new particle interactions and even new particles. In this chapter we will investigate possible beyond SM neutrino physics in the context of our current description for core collapse supernovae. We will see that if such new physics exists in nature, even at levels beyond detection in current experiments, it would effect core collapse supernovae.

We could really identifiy the "open issue" of this chapter as any type of particle physics beyond the Standard Model. For example, new quark-quark interactions might effect the behavior of nuclear matter in the core during the various stages of collapse. This would add to the current open problem of determining the compostion of the nuclear matter in terms of known SM physics. Different types of new physics interactions will have different effects and consequences in supernovae. Neutrino physics is particularly interesting because of the important role neutrinos could play in the supernova mechanism.

Considering new physics in the supernova problem can lead to results which either favor or disfavor getting an explosion. Thus one might think either case would be evidence for or against the new physics. Of course, we do not yet know if our current models for the supernova mechanism are correct. Therefore, no definite conclusions can be drawn from the outcome of including new physics in the model. One overall goal of this chapter is to demonstrate, with a particular example, that particle physics does effect supernova models. Therefore, if new physics is discovered it must be included in the model. Likewise, detection of a supernova event could be used to probe new physics by comparing the output of simulations (which include the new physics) to the signal.

The neutrino physics we will consider is that of neutrino flavor changing neutral currents (FCNC's). In particular, we will look at interactions where neutrinos may change flavor by scattering with quarks. We will examine the effects that this type of neutrino FCNC will have in the supernova environment. We will do this analysis for the infall stage of the supernova and quantify ranges of interaction strength in which the new physics will still effect the model.

2. Core Collapse and Neutrinos

A star with mass of order $10\,M_\odot$ evolves in several stages over millions of years during which its core burns by thermonuclear fusion. Each stage of burning results in a core composed of heavier elements, the ashes of the last stage, which begin burning and producing a new core. The process ends when silicon burning produces a core of iron. Iron has the maximum binding energy per nucleon and therefore does not undergo fusion. This iron core is maintained against gravitational collapse by electron degeneracy pressure, but it continues to grow as Si burning adds more iron. When the core reaches its Chandrasekhar mass (about $1.4\,M_\odot$) it becomes unstable and begins to collapse.

The time scale for collapse is about one second. As the iron core collapses it increases in density. While electron degeneracy pressure is insufficient to prevent the core from collapsing, it does influence the structure of the core during collapse. Two seperate regions of the core form: an inner region in which the velocity of a fluid element is proportional to its radius, and an outer region in which fluid elements are falling in supersonically. The amount of core material in each of these regions is determined by electron pressure.

When the inner region of the core reaches nuclear density, or a few times nuclear density, its collapse halts. The inner region halts as a unit and causes the outer core to "bounce". More specifically, when the inner core halts a shockwave forms at the boundary (of the inner and outer regions) and begins to propogate outward. As the shock moves outward it dissociates nuclei in the outer core and the material beyond.

The shock looses energy as it breaks apart nuclei and simulations show that it eventually becomes a standing acretion shock and "stalls". It is believed that somehow the shock gets revived and continues to move outward and eventually explode the star. One of the goals in understanding core collapse supernovae is to explain the mechanism of shock revival. Neutrinos could in fact play an important part in this.[a]

2.1. *Neutrino's Participation During Infall*

As mentioned above, the time between the onset of collapse and bounce is about a second. We will focus on this infall regime and consider time intervals of order 100 milliseconds. At the onset of collapse, the iron core has a central density of $\rho \sim 10^{10} \, \mathrm{g\,cm^{-3}}$ and central temperature of $T \sim 1 \, \mathrm{MeV}$. The electron fraction is $Y_e \approx 0.42$, where

$$Y_f \equiv \frac{n_f - n_{\bar{f}}}{n_b}. \tag{1}$$

The entropy per baryon can be computed by considering all degress freedom and is found to be $s \approx 1\mathrm{k}$.

During infall, neutrinos are produced by two types of processes. Thermal emission mediated by the SM charged and neutral currents produces neutrino anti-neutrino pairs. Neutroniziation reactions produce electron neutrinos through electron capture on protons. These reactions are:

$$e^- + p \rightarrow n + \nu_e \tag{2}$$

$$e^- + A(Z, N) \rightarrow A(Z - 1, N + 1) + \nu_e. \tag{3}$$

The neutronization reactions are the dominant source of neutrino production. Electron capture obviously lowers the core electron fraction Y_e.

---

[a]See Refs. 1 and 2 for an introduction to the core collapse supernova model and review of the facts presented in the following subsections.

The entropy of the core at the onset of collapse is low ($s \sim 1k$), and remains low throughout collapse. Electron capture on nuclei increases the entropy of the core, and at the same time, neutrinos leaving the core lower its entropy. An electron capture reaction with a nucleus leaves the daughter nucleus in an excited state, thus heating the system and increasing the entropy. Neutrinos acount for some of the entropy in the core and carry entropy out as they escape.

The low entropy conditions favor nucleons to be bound in nuclei. As collapse proceeds, electron capture causes heavy neutron rich nuclei to form. The cross section for coherent elastic scattering of a neutrino on a nucleus via the weak neutral current is proportional to A^2, the nuclear mass squared.[b] The cross section for coherent elastic neutrino scattering on the heavy nuclei ($A \approx 100$) in the core is large enough that eventually the neutrino mean free path becomes smaller than the size of the core. The neutrinos scatter inside the core and cease to escape. This "trapping" of the neutrinos occurs at a density above $10^{11}\,\mathrm{g\,cm^{-3}}$ and a few hundred milliseconds after the onset of collapse. Core bounce occurs in a time of about 100 milliseconds after trapping.

After the neutrinos become trapped the entropy continues to stay low. The trapped neutrinos fill up a Fermi sea to the maximum level and electron capture then stops because the production of neutrinos is blocked. The entropy now does not change because the neutrinos are no longer escaping (and carrying entropy out of the core) and electron capture is no longer occuring (and increasing the entropy). The electron fraction at trapping is $Y_e^{\mathrm{trap}} \approx 0.35$ and, according to the current supernova model, does not significantly change before bounce.

2.2. *Importance of Particle Physics*

The iron core at the onset of collapse has a radius of about the size of the earth. In about one second it collapses to a neutron star with a radius of about 40 km and releases nearly 10^{52} ergs of gravitational binding energy. Approximately 10% of this energy, 10^{51} ergs, makes up the shock's initial energy. The remaining 90% of the energy is stored in the neutrino seas in the core. Observations of SN 1987A showed that the optical and kinetic energy of the explosion was about 10^{51} ergs. We know that the shock's initial energy of 10^{51} ergs is not enough to cause the star to explode becuase we

---

[b] $A = Z + N$ where Z and N are the number of protons and neutrons, respectively, in the nucleus.

300

know the shock looses some of this energy and eventually stalls. Clearly, the neutrinos have enough energy to be able to influence the explosion.

If it were not for the SM neutral current, neutrinos would release their energy on a much shorter time scale. In the absence of trapping, neutrinos could be emitted in as short as a collapse time scale. The energy stored in neutrinos is actually emitted on a diffusion time scale, the time it takes a neutrino to random walk out of the core. The random walk time is determined from the mean free path between nuclear scattering. This mean free path is small (and the diffusion time long) because of the A^2 amplification for coherent elastic scattering, which occurs only via the neutral current.

Neutrino scattering via the neutral current may seem less exciting than other particle interactions tested at accelerators today. However, it is very important in the context of core collapse supernovae. In fact, about 30 years ago, before the SM neutral current was discovered, explanations for the supernova mechanism took a different route. For example, it was once thought that the neutrino luminosity was high enough during collapse for the neutrinos to eject material beyond the core through momentum transfer by collisions with electrons. This theory was ruled out after the neutral current was discovered and included in the model — calculations showed the neutrino luminosity was too low because of trapping. This example illustrates how a complete treatment of particle physics is crucial to understanding core collapse supernovae.

2.3. *Effects of Neutrino Flavor Changing*

The initial energy of the shock at bounce is essentially the infall kinetic energy of the outer core, which gets converted to outgoing energy when the outer core bounces. The infall kinetic energy of the outer core can be estimated from the gravitational potential energy of the inner core which, obviously, depends on the inner core mass. As was discussed above, electron pressure determines the amount of material in the inner and outer regions of core. The amount of electrons in the core therefore determines the mass of the inner core and the initial shock energy. These can be quantified in terms of Y_e:[3]

$$M_{\text{inner}} \approx 5.8\, Y_e^2 M_\odot \tag{4}$$

$$E_{\text{shock}}^{\text{init}} \sim (Y_e)^{10/3}. \tag{5}$$

If the electron fraction were to change, the inner core mass and initial shock energy would change accordingly.

Lowering the electron fraction in the core disfavors getting an explosion. Clearly the initial shock energy is lowered. In addition, the inner region of the core is smaller. Thus, when the inner core bounces the shock will start deeper within the core and have more outer core material to pass through. The shock will then loose more energy in passing out of the core. A lower initial shock energy and more energy loss of the shock as it moves through the core both make the task of shock revival harder.

Any physics that allows neutrinos to change flavor can result in a lowered Y_e.[c] Recall that electron capture, e.g. Eq. (3), is blocked after neutrino trapping because the sea of ν_e's gets filled to the Fermi level. If by some process ν_e's could be converted to ν_μ's and/or ν_τ's, then holes would open in the ν_e sea and electron capture could procede. The neutrino flavor changing scattering with quarks that we will consider is just such a process. We will investigate the possible reduction in Y_e due to such interactions up to and beyond current constraints on their coupling constants.

The ability of neutrinos to change flavor through scattering has other consequences, though we will not study these in detail. For example, inelastic scattering is not blocked because ν_e's can scatter into a wide range of $\nu_{\mu,\tau}$ energy states. The neutrino transport problem is clearly altered. The entropy of the core can also increase because electron capture reactions are taking place while the neutrinos remain trapped. Finally, we expect the signal of a supernova event to be altered because more $\nu_{\mu,\tau}$'s are produced in the core on infall. These and other effects could be explored by including neutrino flavor changing reactions in supernova simulations.

3. Interactions and Cross Sections

There are different particle physics models which predict neutrino flavor changing through different interactions. For example, neutrinos could change flavor by scattering with other neutrinos, electrons or quarks by exchange of some other particle. The exchange particle, coupling constants, and specific vertices depend on the particular model. If the exchange particle is massive and the energy of the scattering particles is low, this type of interaction can be described by a low energy effective theory. A general low energy effective Lagrangian for such neutrino flavor changing scattering with a fermion f is

$$\mathcal{L} = \epsilon^f_{V_{ij}} G_F \bar{\nu}^i \gamma^\mu (1 - \gamma_5) \nu^j \bar{f} \gamma^\mu f + \epsilon^f_{A_{ij}} G_F \bar{\nu}^i \gamma^\mu (1 - \gamma_5) \nu^j \bar{f} \gamma^\mu \gamma_5 f. \qquad (6)$$

---

[c]See Ref. 4 for a discussion on the effects of neutrino flavor changing in core collapse.

The ϵ coefficients in front of each term are dimensionless constants whose values indicate how weak the process is compared to Standard Model interactions. In the context of the supernova environment, neutrino-quark flavor changing neutral currents (FCNC's) of the form in Eq. (6) have a larger effect than neutrino-electron or neutrino-neutrino FCNC's would. This is because coherent elastic scattering with the heavy nuclei in the core can occur via the neutrino-quark FCNC. The cross section for this type of scattering will have an A^2-like dependence, just as in the case of the Standard Model neutral current. This amplification to the cross section causes more neutrino flavor changing and thus greater reduction in Y_e than would result from neutrino FCNC's with other particles.

3.1. *Neutrino Nuclear Scattering*

A general zero momentum transfer cross section for coherent elastic neutrino flavor changing scattering with a spin-0 nucleus in vacuum is[5]

$$\sigma_{ij} \approx \frac{2}{\pi} |\epsilon^u_{V_{ij}} G_F(2Z + N) + \epsilon^d_{V_{ij}} G_F(Z + 2N)|^2 E_\nu^2. \tag{7}$$

The indices ij indicate initial and final neutrino states and E_ν is the neutrino energy. The approximate equals sign appears because terms of order E_ν/M, where M is the mass of the nucleus, have been neglected. The cross section does not depend on the axial vector coefficients because the matrix elements of the axial vector quark currents in Eq. (6) vanish between a spin-0 nuclear state. The quantities $(2Z + N)$ and $(Z + 2N)$ are the zero momentum transfer form factors for the matrix elements of the up and down quark vector currents. These form factors give the A^2-like dependence to the cross section.

Using the cross section of Eq. (7) to describe scattering in a supernova involves a number of approximations. First, we do not expect the vacuum cross section to apply in the hot dense environment of the core. It is possible there are corrections to the interactions from higher order scattering processes. Second, it is not known what composition the nuclear matter in the core takes as the core approaches nuclear density during the few hundred milliseconds between trapping and bounce. It is possible that some phase transitions take place and the matter could be in some "intermediate" state between nuclei and free nucleons. Finally, this cross section uses the zero momentum transfer limit for the form factors. The exact dependence on momentum transfer is not known.

All of these sources of uncertainty are present for the case of Standard

Model interactions as well. In making the approximation to ignore these uncertainties we are not treating our case any different from the way the SM interactions are studied in core collapse. Furthermore, on the issue of composition of nuclear matter in the core, coherent scattering can still occur in the various possible phases. Neutrinos could scatter coherently on groups of nucleons, or groups of quarks.[6] Therefore, using the cross section of Eq. (7) is a reasonable approximation to neutrino scattering in any of the phases of matter. As for the issue of the form factor, the variance from the value at zero momentum transfer should be negligible — the range of momentum transfer for elastic scattering of low energy neutrinos is small and therefore the form factor should vary little over this range. Work is being done to solve the problem of all these uncertainties for the SM interactions and when solutions are found they can be applied to new physics as well.

4. Quantifying the Effects Of Neutrino FCNC's

For current constraints on the FCNC's we are considering, the coefficients in Eq. (6) take values in the range of $10^{-1}G_F$ to $10^{-3}G_F$.[7] Given these values, the cross section for electron capture is larger than the cross section for flavor changing scattering on nuclei in the core.[d] The Fermi level of ν_e's in the core will then likely *not* change with the presence of FCNC's. This is because every time a hole is opened in the ν_e sea from a flavor changing event, it will be filled by a neutrino produced from electron capture. (It is also possible that sometimes a $\nu_{\mu,\tau}$ could have a flavor changing scattering to become a ν_e and fill the hole before an electron capture occurs.) Since the ν_e level will remain the same, the reduction in Y_e can be estimated as the increase in the fraction of ν_μ and ν_τ:

$$\Delta Y_e = -(\Delta Y_{\nu_\mu} + \Delta Y_{\nu_\tau}). \tag{8}$$

The FCNC channels $\nu_e \leftrightarrow \nu_\mu$, $\nu_e \leftrightarrow \nu_\tau$, and $\nu_\tau \leftrightarrow \nu_\mu$ are constrained at different levels (see Ref. 7) and so the reaction rates for these processes will be different. Y_{ν_μ} and Y_{ν_τ} could thus increase, for example, by net $\nu_e \to \nu_\tau$ and $\nu_\tau \to \nu_\mu$. In this example, the latter channel hinders the ν_τ Fermi sea from filling to the maximum level so that the former channel and electron capture can carry on and cause more reduction in Y_e than would result from only the first channel.

---

[d]Though, as discussed in Sec. 3.1, the extent of coherent amplifications for the nuclear matter in the core is not known and so this should really be regarded as a limiting case.

The number and rate of flavor changing scatterings for a neutrino in the core can be estimated from the mean free path for this scattering. The mean free path is computed from the number density of scattering targets and cross section for scattering on the targets. Such number and rate calculations have been done in Ref. 5 for the range of densities, in order of maginitude, the core passes through between trapping and bounce. There, the cross section $\sigma = (2/\pi)(\epsilon G_F)^2 A^2 E_\nu^2$ was used for the entire density range. As discussed earlier in Sec. 3.1, it is a reasonable approximation to use this cross section for such calculations. The calculations were done for values of $\epsilon = 10^{-1}, 10^{-2}, 10^{-3}$ and 10^{-4}. The case $\epsilon = 10^{-4}$ is an order of magnitude more stringent than the best current constraint.

To get the maximum reduction in Y_e, the net number of ν_e's which need to change flavor is twice the number of neutrinos that occupy up to the neutrino fermi level. The results from Ref. 5 indicate this situation can occur for values of ϵ within the range of current constraints because there would be many flavor changing scatterings for each neutrino before bounce. For the case $\epsilon = 10^{-4}$, Ref. 5 shows that each neutrino has of order one flavor changing scattering per millisecond just prior to bounce. Siginificant reduction of Y_e is possible even in this case! Note that the non-trivial form factors of the cross section in Eq. (7) were neglected in Ref. 5 where, for ease of calculation, a simple factor of $(\epsilon G_F A)^2$ was used. If included, the extra factors would increase the values for the rates and make maximum reduction of Y_e even more feasible. Of course, the calculations done in Ref. 5 and the discussion here did not take into account feedback of neutrino flavor changing on the core. A numerical simulation including FCNC reactions is required to determine an accurate value for the change in Y_e and quantify the other effects mentioned at the end of Sec. 2.3.

5. Conclusion

We have seen how new FCNC neutrino interactions can alter the current model for core collapse supernovae. The main effect we discussed was a lowered Y_e at bounce which, according to the current supernova model, results in a weaker shock. Depending on the channel, current constraints for the FCNC's we considered range from $10^{-1}G_F$ to $10^{-3}G_F$. We have seen that the undesirable effect of a lowered Y_e can occur for FCNC's as weak as $10^{-4}G_F$ for *all* channels. We cannot derive a constraint from this result though, because we do not know if our explanation for the supernova mechanism is correct. In addition, we have only studied the infall stage of collapse here. There could also be effects of neutrino flavor changing in the post-bounce stage that either favor or disfavor getting an explosion. Including FCNC's in numerical simulations in all stages of the model is necessary to "measure" all their effects.

Eventually physicists will discover and understand some piece of particle physics beyond the Standard Model. The new physics will most certainly have consequences in the supernova environment. It is possible that some new physics will be the key to solving the supernova puzzle. On the other hand, one day we may solve the supernova problem and then use our knowledge to study new physics not accessible in the lab. Either outcome will be very exciting for the supernova community, and the rest of the physics community as well!

Acknowledgments

This work was supported in part by NSF grant PHY-00-99499 and the TSI collaboration's DoE SciDAC grant at UCSD. I would like to thank A. Friedland, G. Fuller, B. Grinstein, B. Messer and A. Mezzacappa for useful discussions.

References

1. S. L. Shapiro and S. A. Teukolsky, Black Holes, White Dwarfs, and Neutron Stars, (John Wiley and Sons, Inc., New York, 1983).
2. H. A. Bethe, G. E. Brown, J. Applegate, and J. Lattimer, *Nucl. Phys.* **A324**, 487 (1979).
3. G. M. Fuller, Astrophys. J. **252**, 741 (1982).
4. G. M. Fuller, R. W. Mayle, J. R. Wilson and D. N. Schramm, *Astrophys. J.* **322**, 795 (1987).
5. P. S. Amanik, G. M. Fuller, and B. Grinstein, hep-ph/0407130.
6. S. Reddy, G. Bertsch and M. Prakash, *Phys. Lett.* **B475**, 1 (2000).
7. A. Friedland, C. Lunardini and C. Pena-Garay, Phys. Lett. **B594**, 347 (2004).

NEUTRINO PROCESSES IN STRONG MAGNETIC FIELDS

HUAIYU DUAN

Department of Physics
University of California, San Diego
La Jolla, CA 92093, USA
E-mail: hduan@ucsd.edu

YONG-ZHONG QIAN

School of Physics and Astronomy
University of Minnesota
Minneapolis, MN 55455, USA
E-mail: qian@physics.umn.edu

The processes $\nu_e + n \rightleftharpoons e^- + p$ and $\bar{\nu}_e + p \rightleftharpoons e^+ + n$ provide the dominant mechanisms for heating and cooling the material below the stalled shock in a core-collapse supernova. We summarize the major effects of strong magnetic fields on the rates of the above reactions and illustrate these effects with a simple supernova model. Due to parity violation of weak interaction the heating rates are asymmetric even for a uniform magnetic field. The cooling rates are also asymmetric for nonuniform fields. The most dramatic effect of strong magnetic fields of $\sim 10^{16}$ G is suppression of the cooling rates by changing the equations of state through the phase space of e^- and e^+.

1. Introduction

The neutrino processes

$$\nu_e + n \rightleftharpoons e^- + p, \tag{1}$$

$$\bar{\nu}_e + p \rightleftharpoons e^+ + n \tag{2}$$

play important roles in core-collapse supernovae. After the shock is stalled, neutrinos emitted from the protoneutron star exchange energy with the material below the shock mainly through these processes. The forward processes in Eqs. (1) and (2) heat the material through the absorption of ν_e and $\bar{\nu}_e$, while the reverse processes cool the material by emitting them. In the neutrino-driven supernova mechanism [1], the competition between heating and cooling of the material by these processes is expected to result in net energy gain for the stalled shock, which is then revived to make a successful supernova explosion. Unfortunately, the current consensus is that this mechanism does not work in spherically symmetric models [2, 3]. On the other hand, strong magnetic fields may be generated during the formation of protoneutron stars and in turn affect supernova dynamics. Observations have shown that some neutron stars possess magnetic fields as strong as $\sim 10^{15}$ G [4–6]. Although it is not clear how strong magnetic fields in supernovae could be, some calculations indicate that fields of 10^{15}–10^{16} G are not impossible [7]. While strong magnetic fields can affect supernova dynamics in many possible ways, here we consider their effects on the neutrino processes. Because the explosion energy is much smaller than the gravitational binding energy of the protoneutron star and nearly all of the latter is released in neutrinos, it is natural to expect that a small change in the neutrino physics input may have a large impact on the supernova mechanism.

The effects of strong magnetic fields on neutrino processes have been studied in various approximations [8–15]. In our recent work, we have calculated the effects of magnetic fields on the four processes in Eqs. (1) and (2) to the 0th [16] and 1st order [17] in E_ν/m_N, where E_ν is the neutrino energy and m_N is the nucleon mass. Here we summarize our results and list some issues that remain to be addressed.

2. Neutrino Processes in Strong Magnetic Fields

2.1. *General Effects of Magnetic Fields*

An obvious effect of the magnetic field is polarization of the spin of a nonrelativistic nucleon. When this effect is small, the polarization of the nucleon spin may be written as

$$\chi \simeq \frac{\mu B}{T} = 3.15 \times 10^{-2} \left(\frac{\mu}{\mu_N}\right) \left(\frac{B}{10^{16} \text{ G}}\right) \left(\frac{\text{MeV}}{T}\right), \qquad (3)$$

where μ is the nucleon magnetic moment, $\mu_N = e/2m_p$ is the nuclear magneton, B is the magnetic field strength, and T is the gas temperature. Due to parity violation of weak interaction polarization of the nucleon spin introduces a dependence on the angle Θ_ν between the directions of the neutrino momentum and the magnetic field for the cross sections of the forward processes in Eqs. (1) and (2) (see Sec. 2.2) and for the differential volume reaction rates of the reverse processes (see Sec. 2.3).

In addition, assuming a magnetic field in the positive z-direction, the motion of a proton in the xy-plane is quantized into Landau levels (see e.g. Ref. [18]) with kinetic energies

$$E_p(n_p, k_{pz}) = \frac{k_{pz}^2}{2m_p} + \left(n_p + \frac{1}{2}\right)\frac{eB}{m_p}, \qquad n_p = 0, 1, 2, \cdots, \tag{4}$$

where n_p is the quantum number of the proton Landau level and k_{pz} is the z-component of the proton momentum. We are interested in gas temperatures of $T \gtrsim 1$ MeV and magnetic fields of $B \sim 10^{16}$ G. As $eB/m_p = 63(B/10^{16}$ G$)$ keV, for such conditions a proton is able to occupy Landau levels with $n_p \gg 1$ and can be considered as classical.

For the conditions of interest here, e^- and e^+ are relativistic. Their Landau levels have energies

$$E_e(n_e, k_{ez}) = \sqrt{m_e^2 + k_{ez}^2 + 2n_e eB}, \tag{5}$$

where symbols are defined similarly to those for the proton. The above equation has taken spin into account. Note that the e^- or e^+ in the ground Landau level ($n_e = 0$) has only one spin state. This introduces an additional dependence on Θ_ν (independent of polarization of the nucleon spin) for the cross sections of the forward processes in Eqs. (1) and (2) (see Sec. 2.2) and for the differential volume reaction rates of the reverse processes (see Sec. 2.3). The effects of Landau levels are more prominent for e^- and e^+ than for nucleons. This can be seen from the quantum number for the highest Landau level occupied by e^- or e^+ with energy E_e,

$$(n_e)_{\max} = \left[\frac{E_e^2 - m_e^2}{2eB}\right]_{\mathrm{int}} = \left[8.45 \times 10^{-3}\left(\frac{E_e^2 - m_e^2}{\mathrm{MeV}^2}\right)\left(\frac{10^{16}\ \mathrm{G}}{B}\right)\right]_{\mathrm{int}}. \tag{6}$$

To account for the effects of Landau levels, part of the integration over the phase space of e^- or e^+ is changed to a summation over possible Landau levels, i.e.

$$2\int \frac{d^3 k_e}{(2\pi)^3} \rightarrow \frac{eB}{2\pi} \sum_{n_e=0}^{(n_e)_{\max}} g_{n_e} \int \frac{dk_{ez}}{2\pi}, \tag{7}$$

where g_{n_e} is the number of spin states for the n_eth Landau level ($g_{n_e} = 1$ for $n_e = 0$ and 2 for $n_e > 0$).

2.2. *Heating Processes*

To the 0th order in E_ν/m_N, the cross sections of the forward processes in Eqs. (1) and (2) are found to be [16]

$$\sigma_{\nu N}^{(0,B)} = \sigma_{B,1} \left[1 + 2\chi \frac{(f \pm g)g}{f^2 + 3g^2} \cos\Theta_\nu \right]$$

$$+ \sigma_{B,2} \left[\frac{f^2 - g^2}{f^2 + 3g^2} \cos\Theta_\nu + 2\chi \frac{(f \mp g)g}{f^2 + 3g^2} \right], \tag{8}$$

where the energy-dependent factors $\sigma_{B,1}$ and $\sigma_{B,2}$ are defined as

$$\sigma_{B,1} = \frac{G_{\mathrm{F}}^2 \cos^2\theta_{\mathrm{C}}}{2\pi} (f^2 + 3g^2) eB \sum_{n_e=0}^{(n_e)_{\max}} \frac{g_{n_e} E_e}{\sqrt{E_e^2 - m_e^2 - 2n_e eB}}, \tag{9}$$

$$\sigma_{B,2} = \frac{G_{\mathrm{F}}^2 \cos^2\theta_{\mathrm{C}}}{2\pi} (f^2 + 3g^2) eB \frac{E_e}{\sqrt{E_e^2 - m_e^2}}, \tag{10}$$

with $E_e = E_\nu \pm \Delta$ and Δ being the neutron-proton mass difference. In the above equations, the upper sign is for ν_e absorption on n and the lower sign for $\bar{\nu}_e$ absorption on p.

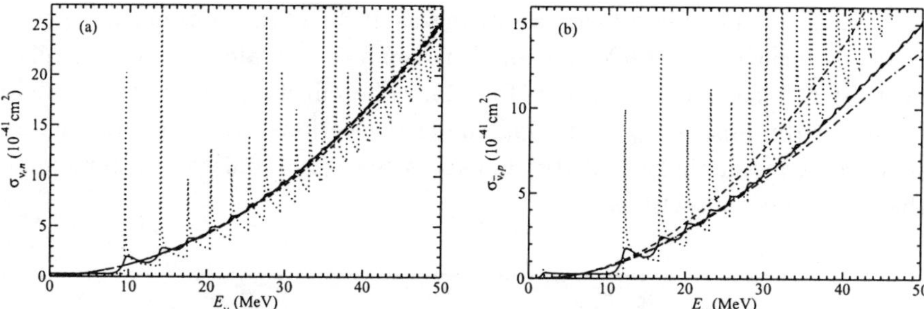

Figure 1. The cross sections of $\nu_e + n \rightarrow e^- + p$ (a) and $\bar{\nu}_e + p \rightarrow e^+ + n$ (b) as functions of neutrino energy E_ν. The angle Θ_ν between the directions of the neutrino momentum and the magnetic field is taken to be 0. The dotted and solid curves are the cross sections to the 0th and 1st order in E_ν/m_N, respectively. Both assume a magnetic field of $B = 10^{16}$ G. In addition, the solid curves assume a temperature $T = 2$ MeV for the nucleon gas. The short-dashed and dot-dashed curves are the cross sections to the 0th and 1st order in E_ν/m_N, respectively, but for $B = 0$. The long-dashed curve in (b) is for $B = 0$ and includes some corrections beyond the 1st order. The differences between the short-dashed, dot-dashed, and long-dashed curves in (b) at the high-energy end are mostly due to the combined effects of nucleon recoil and weak magnetism.

Comparing $\sigma_{B,1}$ with the well-known expression for the 0th-order cross section in the absence of magnetic fields,

$$\sigma_{\nu N}^{(0)} = \frac{G_F^2 \cos^2 \theta_C}{\pi}(f^2 + 3g^2)k_e E_e, \tag{11}$$

one can see that the only difference is the change in phase space [see Eq. (7)]. To illustrate the effects of strong magnetic fields, we plot the cross sections of neutrino absorption on nucleons for $B = 10^{16}$ G and 0 in Fig. 1 (see Ref. [17] for more details). An immediate observation is that the cross sections are enhanced (dotted curves in Fig. 1) if the energy of the outgoing e^- or e^+ satisfies the condition

$$E_e = \sqrt{m_e^2 + 2n_e eB} \tag{12}$$

for $n_e > 0$. This is because a new Landau level opens up when Eq. (12) is satisfied. Just as discrete energy levels of atoms lead to absorption lines in the light spectra, ideally the presence of strong magnetic fields would produce sharp dips in the neutrino energy spectra where Eq. (12) holds. However, the nucleons absorbing neutrinos have thermal motion, which will smear out these sharp dips. We have included the thermal motion of nucleons and calculated the cross sections to the 1st order in E_ν/m_N [17]. Even for magnetic fields as strong as 10^{16} G, the thermal motion of nucleons with $T \sim 2$ MeV is enough to smooth out almost all the spikes in $\sigma_{\nu N}(E_\nu)$ (see solid curves in Fig. 1). The effect of the magnetic field is further diminished by averaging the cross sections over neutrino energy spectra. A magnetic field of 10^{16} G causes changes of only a few percent to the average cross sections.

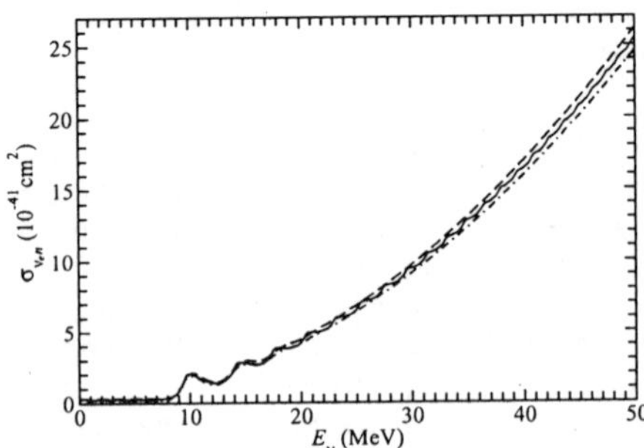

Figure 2. The cross section of $\nu_e + n \rightarrow e^- + p$ to the 1st order in E_ν/m_N for $B = 10^{16}$ G and $T = 2$ MeV. The angle Θ_ν between the directions of ν_e and the magnetic field is taken to be 0 (dot-dashed curve), $\pi/2$ (solid curve), and π (dashed curve), respectively.

The term proportional to $\sigma_{B,1}$ in Eq. (8) depends on the direction of the incoming neutrino with respect to the magnetic field. This is due to parity violation of weak interaction. In fact, as long as the target nucleon has a polarization χ, the lowest-order expression of the cross section in the absence of magnetic fields can be written as

$$\sigma_{\nu N}^{(0)}(\chi) = \sigma_{\nu N}^{(0)}\left[1 + 2\chi\frac{(f \pm g)g}{f^2 + 3g^2}\cos\Theta_\nu\right],\qquad(13)$$

which has exactly the same angular dependence as the $\sigma_{B,1}$ term in Eq. (8). The appearance of the term proportional to $\sigma_{B,2}$ in this equation is due to the fact that there is only one spin state for the ground Landau level of e^- or e^+. This term has an angular dependence even if the nucleon polarization $\chi = 0$. Nevertheless, this dependence is again due to parity violation as the e^- or e^+ in the ground Landau level is polarized. For crude estimates of the angular dependence, we use Eq. (13). As the nucleon form factors $f = 1$ and $g = 1.26$ are close in numerical value, $\sigma_{\bar{\nu}_e p}^{(0)}(\chi_p)$ has little dependence on Θ_ν. On the other hand, the Θ_ν-dependent term for $\sigma_{\nu_e n}^{(0)}(\chi_n)$ is $\sim \chi_n \cos\Theta_\nu$. The angular dependence of $\sigma_{\nu_e n}$ for a strong magnetic field of $B = 10^{16}$ G and a gas temperature of $T = 2$ MeV is shown in Fig. 2.

2.3. *Cooling Processes*

Because e^- and e^+ do not have definite velocities [18], we define a volume reaction rate Γ_{eN}, which gives the rate of e.g., e^+ capture per neutron when multiplied by the e^+ number density n_{e+}. The differential volume reaction rates to the 0th order in E_ν/m_N are found to be [16]

$$\frac{d\Gamma_{eN}^{(0,B)}}{d\cos\Theta_\nu} = \frac{\Gamma_{eN}^{(0)}}{2}\left[1 + 2\chi\frac{(f \pm g)g}{f^2 + 3g^2}\cos\Theta_\nu\right]$$

$$+\delta_{n_e,0}\frac{\Gamma_{eN}^{(0)}}{2}\left[\frac{f^2 - g^2}{f^2 + 3g^2}\cos\Theta_\nu + 2\chi\frac{(f \mp g)g}{f^2 + 3g^2}\right],\qquad(14)$$

where

$$\Gamma_{eN}^{(0)} = \frac{G_F^2\cos^2\theta_C}{2\pi}(f^2 + 3g^2)E_\nu^2\qquad(15)$$

with $E_\nu = E_e \pm \Delta$ is the volume reaction rate in the absence of magnetic fields. In the above equations, the upper sign is for e^+ capture on n and the lower sign for e^- capture on p. The angular dependence for neutrino emission in Eq. (14) is the same as that for neutrino absorption in Eq. (8) and is due to parity violation of weak interaction as explained in Sec. 2.2.

However, the volume reaction rates for the cooling processes obtained by integrating the differential rates in Eq. (14) over Θ_ν are isotropic for a uniform magnetic field. This is in contrast to the cross sections in Eq. (8) for the heating processes.

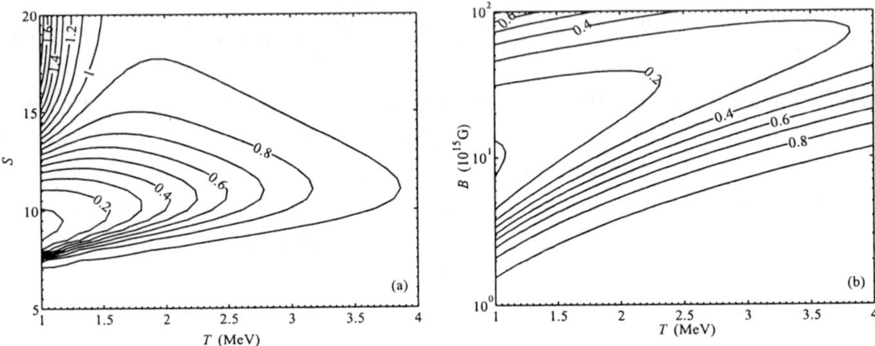

Figure 3. Contours of the ratio of the total cooling rate with magnetic fields to that without. A magnetic field of 10^{16} G is assumed for (a) and a total entropy per nucleon $S = 10$ is assumed for (b). A constant electron fraction $Y_e = 0.5$ is assumed for both.

Like the cross sections for the heating processes, the volume reaction rates of the cooling processes are not much affected even for magnetic fields as strong as 10^{16} G. However, the cooling rates could be severely suppressed by such strong magnetic fields due to changes in the equations of state through the phase space of e^- and e^+ [16]. Given the electron fraction Y_e, the total entropy per nucleon S, and the gas temperature T, one can solve the equations of state

$$\frac{\rho Y_e}{m_N} = \mathfrak{n}_{e^-} - \mathfrak{n}_{e^+}, \tag{16}$$

$$S = S_N + S_\gamma + S_{e^-} + S_{e^+}, \tag{17}$$

to obtain the baryon mass density ρ and the electron degeneracy parameter η_e for cases of strong and no magnetic fields. The cooling rates (per nucleon) in each case can then be calculated by integrating the volume reaction rates over the energy-differential number densities of e^- and e^+. The suppression of the total cooling rate by strong magnetic fields is shown in Fig. 3. The reason for this suppression is that compared with the case of no magnetic fields, there are more low-energy e^- and e^+ in magnetic fields of $\sim 10^{16}$ G as most of the e^- and e^+ reside in the ground Landau level ($n_e = 0$).

3. Application to Core-Collapse Supernovae

To illustrate the effects of strong magnetic fields on supernova dynamics, we consider a simple supernova model. All neutrinos are assumed to be emitted at the same radius $R_\nu = 50$ km. The shock is stalled at a radius $R_s = 200$ km. The electron fraction and the total entropy per nucleon are taken to be $Y_e = 0.5$ and $S = 10$, respectively, and held constant between R_ν and R_s. We adopt the temperature profile

$$T(r) = T(R_\nu)\frac{R_\nu}{r} = 4\left(\frac{50\text{ km}}{r}\right)\text{ MeV}. \tag{18}$$

With the above assumptions, we calculate the total heating and cooling rates as functions of radius r for cases of strong and no magnetic fields [the equations of state (16) and (17) are solved to obtain ρ and η_e for calculating the total cooling rate in both cases]. In the absence of magnetic fields, the total heating rate is found to be equal to the total cooling rate at a gain radius $R_g = 137$ km, above which heating dominates cooling.

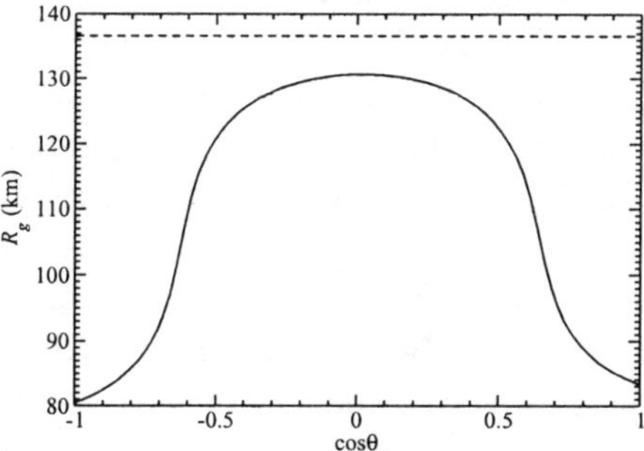

Figure 4. The gain radius R_g as a function of $\cos\theta$ (solid curve) for a dipole magnetic field. Compared with the case for $B = 0$ (dashed curve), the gain radius is substantially reduced at the north and south poles where the magnetic fields are the strongest.

Consider a magnetic field of dipole configuration in spherical coordinates (r, θ, ϕ),

$$\mathbf{B} = B_0\left(\frac{R_\nu}{r}\right)^3(2\cos\theta\,\hat{\mathbf{r}} + \sin\theta\,\hat{\boldsymbol{\theta}}). \tag{19}$$

We calculate R_g for $B_0 = 5 \times 10^{16}$ G and plot it as a function of $\cos\theta$ in Fig. 4. Due to suppression of the cooling rates by strong magnetic fields, R_g becomes smaller in the presence of strong magnetic fields compared with the case of no magnetic fields. As a result, there is more region of net heating below the stalled shock and the neutrino-driven supernova mechanism may work more efficiently. The cooling rates are most suppressed at north and south poles where the magnetic fields are the strongest. The stalled shock is likely to be revived earlier and more energetically in these directions. It is also worth mentioning that there is a small difference in the gain radius between the north and south poles (see Fig. 4). This is due to the angular dependence of the heating processes discussed in Sec. 2.2. This small asymmetry may result in a kick to the protoneutron star that could explain the space velocities observed for pulsars.

4. Open Issues

Although we have demonstrated that strong magnetic fields have important effects on the dynamics of core-collapse supernovae, our results depend on how strong the magnetic fields in supernovae could be. This is the biggest open issue. Alternatively, our results can be used to gauge whether magnetic fields would affect supernova dynamics by changing the rates of neutrino processes. In this regard, we find that magnetic fields weaker than 10^{15} G would have negligible effects on the neutrino processes, while fields of $\sim 10^{16}$ G would dramatically change supernova dynamics through neutrino physics. Of course, magnetic fields weaker than 10^{15} G may already have important hydrodynamic effects in supernovae. This is not considered here but should be investigated by future studies. Another open issue is how to model supernova explosions by including both hydrodynamic effects and changes in the neutrino processes due to strong magnetic fields of $\sim 10^{16}$ G. The processes $\nu_e + n \to e^- + p$ and $\bar{\nu}_e + p \to e^+ + n$ not only provide the dominant mechanisms for heating the material below the stalled shock, but also are the main opacity sources for determining the thermal decoupling of ν_e and $\bar{\nu}_e$ from the protoneutron star, and hence, their emission energy spectra. An interesting issue is whether strong magnetic fields in supernovae could leave detectable imprints on the neutrino energy spectra.

Acknowledgements

HD is very grateful to the hospitality of Tony Mezzacappa, George M. Fuller, and the Institute for Nuclear Theory during the workshop. He also wants to thank Arkady Vainshtein for helpful discussions. This work was supported in part by DOE grant DE-FG02-87ER40328.

References

1. H. A. Bethe and J. R. Wilson, *Astrophys. J.* **295**, 14 (1985).
2. M. Rampp and H.-T. Janka, *Astrophys. J.* **539**, L33 (2000), astro-ph/0005438.
3. M. Liebendörfer, A. Mezzacappa, F.-K. Thielemann, O. E. Messer, W. R. Hix, and S. W. Bruenn, *Phys. Rev.* **D63**, 103004 (2001), astro-ph/0006418.
4. C. Kouveliotou, T. Strohmayer, K. Hurley, J. van Paradijs, M. H. Finger, S. Dieters, P. Woods, C. Thompson, and R. C. Duncan, *Astrophys. J.* **510**, L115 (1999), astro-ph/9809140.
5. E. V. Gotthelf, G. Vasisht, and T. Dotani, *Astrophys. J.* **522**, L49 (1999), astro-ph/9906122.
6. A. I. Ibrahim, J. H. Swank, and W. Parke, *Astrophys. J.* **584**, L17 (2003), astro-ph/0210515.
7. S. Akiyama, J. C. Wheeler, D. L. Meier, and I. Lichtenstadt, *Astrophys. J.* **584**, 954 (2003), astro-ph/0208128.
8. E. Roulet, *JHEP* **01**, 013 (1998), hep-ph/9711206.
9. L. B. Leinson and A. Pérez, *JHEP* **9809**, 020 (1998), astro-ph/9711216.
10. D. Lai and Y.-Z. Qian, *Astrophys. J.* **505**, 844 (1998), astro-ph/9802345.
11. P. Arras and D. Lai, *Phys. Rev.* **D60**, 043001 (1999), astro-ph/9811371.
12. A. A. Gvozdev and I. S. Ognev, *JETP Lett* **69**, 365 (1999), astro-ph/9909154.
13. D. Chandra, A. Goyal, and K. Goswami, *Phys. Rev.* **D65**, 053003 (2002), hep-ph/0109057.
14. C. J. Horowitz, *Phys. Rev.* **D65**, 043001 (2002), astro-ph/0109209.
15. K. Bhattacharya and P. B. Pal, *Pramana* **62**, 1041 (2004), hep-ph/0209053.
16. H. Duan and Y.-Z. Qian, *Phys. Rev.* **D69**, 123004 (2004), astro-ph/0401634.
17. H. Duan and Y.-Z. Qian (2005) (to be submitted to Phys. Rev. D).
18. L. D. Landau and E. M. Lifshitz, *Quantum Mechanics: non-relativistic theory* (Pergamon, Oxford, 1977), 3rd ed.

Section 6
The Equation of State

THE EQUATION OF STATE FOR BARYONIC MATTER

J. RIKOVSKA STONE

Department of Physics, Oxford University,
Parks Road,
Oxford, OX1 3PU, UK
and
Physics Division
Oak Ridge National Laboratory
Oak Ridge, TN 37831, USA
E-mail: j.stone@physics.ox.ac.uk

The current status of the Equation of State (EOS) used in strongly interacting matter and models of the supernova phenomenon and neutron stars is reviewed. The key role of understanding the nucleon-nucleon interaction and interactions in exotic matter (strange baryons, meson condensates and deconfined quarks) is emphasised. A brief survey of experimental and observational quantities which may help to characterize these interactions is given. Example of such characterization is given for the phenomenological Skyrme interaction. Preliminary results for the EOS based on a nucleon-nucleon interaction derived from a non-relativistic one-gluon-exchange-quark model are shown for pure neutron and symmetric nuclear matter.

This contribution surveys the present state of our knowledge of the Equation of State of strongly interacting baryonic matter over a wide range of density. Recent developments and future objectives are discussed. Sec. 1 gives basic definitions and formulae. Models of the nucleon-nucleon interaction, as used in the construction of EOS, are reviewed in Sec. 2, grouped by their underlying physics. Sec. 2 also gives illustration of some applications of the EOS to various phases of baryonic matter. Possible constraints on the choice of a realistic EOS, provided by observational and experimental data, are described in Sec. 3. Two new developments are outlined in Sec. 4: (i) the first fully selfconsistent EOS obtained using 3D Skyrme-Hartree-Fock model at finite temperatures and (ii) the first EOS based on a one-gluon-exchange quark model. Sec. 5 concludes the discussion.

1. Nuclear Equation of State

1.1. *Strongly Interacting Matter*

Nuclear forces play an important role in many stellar environments, acting in concert with gravitational forces to form compact objects. For example, the observed maximum mass of cold neutron stars cannot be explained without considering the pressure, due to strong nuclear forces, to oppose the gravitational collapse. It follows that strongly interacting matter is one of the fundamental systems that we have to study to understand stellar phenomena as well as properties of finite nuclei.

It is useful to distinguish two forms of strongly interacting matter, homogeneous and inhomogeneous. Homogeneous matter is of constant density, usually idealised as infinite nuclear matter with no surface effects and zero net electric charge; this is a very convenient system for testing models of forces acting between particles constituting the matter. At densities between $\sim$1–3$\times$nuclear saturation density of 0.16 fm^{-3} homogeneous matter is considered to be composed of light baryons (nucleons). Such matter has very similar properties to that existing in cold neutron stars between the inner crust and upper layer of the core. At higher densities it contains a variety of heavy baryons (hyperons), possibly meson condensates and, speculatively, deconfined quarks. Such a matter exists in the core of neutron stars. Inhomogeneous matter consists of finite nuclei (or regions having a high density of nucleons) coexisting with free neutron and/or electron gas and forms e.g. the crust of neutrons stars and pre-supernovae during the initial stages of collapse, when the density of the collapsing star is lower than nuclear saturation density.

Composition and make-up of both these forms of matter is strongly density dependent. As discussed later in Sec. 2.2 in more detail, the ground state of the matter evolves through a variety of compositions of inhomogenous phase at density region from about 5×10^{-15} fm^{-3} to about 0.1 fm^{-3} and makes a transition to a homogeneous phase at higher densities. The challenge for the theories of strongly interacting matter is to provide a consistent description of all the forms of matter at all relevant densities.

1.2. *Equation of State at T=0*

The theoretical framework for description of properties of strongly interacting matter is the formalism of Equation of State that relates the pressure and total energy density of the system. This EOS is derived from very general considerations. Let us assume that the expectation value of the energy of a many body system is expressed as

$$E = < \Phi|(T+U)|\Phi >, \tag{1}$$

where $T(U)$ is the kinetic(potential) energy of the system, and Φ is the total wave function. The description of the interaction between particles in the system is contained in the potential U. A fundamental difficulty in theoretical nuclear physics at present is that the precise form of this internucleon potential, (and of the potential acting between nucleons and other strongly interacting particles present in the matter), is not known. Consideration of many-body interactions in phenomenological models at present is largely restricted to the mean-field approximation, in which individual nucleons move independently in the field generated by the other particles in the system. Particle-particle correlations are typically limited to treatment of two- and three-body effects. In models derived from bare nucleon-nucleon scattering data again only two- and three- body forces are taken into account; higher order effects are assumed to be small. All current models depend on many adjustable parameters, which are determined by fitting to experimental data. It follows that theories of nuclear media are model dependent and results obtained for the EOS are subject to the approximations of the particular nuclear model.

Having chosen a model for U in (1), the expression for E can be found in terms of baryon number density n. For asymmetric nuclear matter, consisting of a fixed number A nucleons but variable numbers of protons Z and of neutrons $N = A - Z$, an asymmetry $I = (N - Z)/(N + Z)$ is assumed. Then the energy per particle can be calculated

$$E/A = \mathcal{E}(n, I). \tag{2}$$

This equation is the basic relation that is the starting point for derivation of all other properties of strongly interacting matter. It is easy to see that it can be generalized for matter containing other particles than nucleons. The potential U then includes terms describing interactions amongst all particles in the system and the *asymmetry I* takes a more general form dependent on relative numbers of all components present. The expression (2) for energy per particle as a function of particle number density is sometimes called the *equation of state.*

The ground state of the matter at each n and I is found by minimization of $\mathcal{E}$. Other properties are then calculated using derivatives of $\mathcal{E}$ with respect to n or I at the minimum. The *pressure P* is given by[1]:

$$P(n, I) = n^2 \frac{\partial \mathcal{E}}{\partial n} = n \frac{\partial \epsilon}{\partial n} - \epsilon, \tag{3}$$

322

where $\epsilon = n(\mathcal{E} + m_0 c^2)$ is the total energy density (including the rest mass energy), and m_0 is the nucleon rest mass. This relationship between the pressure P and the total energy density ϵ, as a function of baryon number density n, can easily be cast into a form that is conventinally known as the *equation of state* $P = P(V, T)$ considering $n = A/V$.

There are several more observables that impose conditions on a realistic EOS. The *incompressibility* K is related[2] to $\mathcal{E}$ and P

$$K(n, I) = 9n^2 \frac{\partial^2 \mathcal{E}}{\partial n^2} + 18 \frac{P(n, I)}{n} \tag{4}$$

and the *speed of sound* v_s in the medium, an important quantity in models of supernovae collapse, is expressed as

$$\frac{v_s}{c} = \sqrt{\frac{K}{9(mc^2 + \mathcal{E} + P/n)}}. \tag{5}$$

Clearly, the EOS derived from (2) will be dependent on unknown parameters of the models for the potential U in (1). The usual way of determination of some (or all) of these parameters is to fit them to experimental data on finite nuclei and empirical values of some observable determined for special extreme cases of infinite nuclear matter. These are the symmetric nuclear matter (SMN, $I = 0$), the lowest energy state of nuclear matter, and pure neutron matter (PNM, $I = 1$), which have fundamentally different properties. The energy per particle $\mathcal{E}$ in SNM reaches a negative minimum value (i.e. it saturates) at a *saturation density* n_0, and this then corresponds to the ground state of SNM. The value of $\mathcal{E}$ at saturation, $\mathcal{E}_0$, is usually taken to be the coefficient of the volume term a_v in the liquid-drop model based semiempirical mass formula[3], obtained by fitting the binding energies of a large number of nuclei. This procedure gives $\mathcal{E}_0 = -(16.0 \pm 0.2)$ MeV. The density n_0 of SNM at saturation is taken as[2] $n_0 = (0.160 \pm 0.005)$ fm^{-3} based on calculating the charge distribution in heavy nuclei. $\mathcal{E}$ in PNM is always positive, i.e. PNM does not exist in a bound state. If it did, we would observe nuclei made of neutron only. Similarly, neutron stars are not made of only neutrons but represent an electrically neutral system containing a small proportion of a rich variety of species (protons, heavy baryons, mesons and leptons) which reaches generalised β-equilibrium with time (neutron star matter). Obviously PNM is not a ground state of nuclear matter as some of the neutrons will β-decay to a lower energy state until β-equilibrium is reached.

The empirical properties of SNM and PNM place serious constraints on the parameters of the nucleon-nucleon interaction U. In particular, although the exact density dependence of $\mathcal{E}$ in PNM is not known experimentally, it is the main factor determining the EOS of neutron star matter, at least at densities up to 2-3$\times n_0$ as will be discussed later (Sec. 3.1). The very existence of neutron stars implies that $\mathcal{E}$ should monotoneously increase with density, generating enough baryon pressure to oppose gravitational collapse of the neutron star.

The value of the incompressibility K_∞ at saturation density of SNM provides a weaker constraint on models of nuclear matter as it is a derived quantity and its 'best' value is model dependent. The estimates range from 200–300 MeV. Non-relativistic models imply a lower value range of 220–240 MeV; somewhat higher values of 250–270 MeV are predicted by recent relativistic analysis[4]. The speed of sound (5) also provides a serious constraint. It depends on the density of the medium but cannot exceed the speed of light at any density.

One very important variable for discussions of properties of nuclear matter is the *symmetry energy* $\mathcal{S}$, defined as the difference in energy between symmetric and pure neutron matter

$$\mathcal{S}(n) = \mathcal{E}(n, I = 0) - \mathcal{E}(n, I = 1). \tag{6}$$

$\mathcal{S}$(n) can be expanded about the value of the energy for symmetric nuclear matter[1], with the second-order term being related to the *asymmetry coefficient* a_s in the semi-empirical mass formula[3],

$$a_s = \frac{1}{2} \left. \frac{\partial^2 \mathcal{E}}{\partial I^2} \right|_{I=0}. \tag{7}$$

The accepted (model dependent) range of values of a_s at nuclear saturation density n_0 is 27–36 MeV with the higher values again coming from relativistic models[4,5]. The density dependence of a_s is not known experimentally and is a matter of considerable interest for both the nuclear and heavy-ion collision communities[6,7]. These effects are discussed in Sec. 3 where the density dependence of the symmetry energy is found to be a useful fingerprint for identification of properties of the phenomenological Skyrme potentials used in construction of the EOS.

For completeness we mention here another specific state of nuclear matter, a system of neutrons, protons, electrons and muons in chemical equilibrium with respect to β-decay (BEM) that can provide certain constraints on parameters of potentials used in construction of the EOS. The equilibrium state is reached over a longer time scale than the rapid processes that occur during supernova collapse and the subsequent explosion. However,

BEM is believed to form the bottom layers of the inner crust and the top layers of the core of cold neutron stars. Equilibrium with respect to

$$n \leftrightarrow p + e^- \leftrightarrow p + \mu^-$$

implies that the chemical potentials satisfy the conditions:

$$\mu_n = \mu_p + \mu_e, \qquad \mu_\mu = \mu_e, \tag{8}$$

with each μ being defined by

$$\mu_j = \frac{\partial \epsilon}{\partial n_j}, \tag{9}$$

where ϵ is the total energy density (including the rest masses of the particles involved) and the n_j's are the number densities of each type j of particles. The latter are used to calculate particle fractions with respect to the total baryon number density $n_b = n_n + n_p$:

$$y_j = \frac{n_j}{n_b}. \tag{10}$$

The requirement of charge neutrality implies $n_p = n_e + n_\mu$. BEM is described by the total energy density written as the sum of nucleon and lepton contributions[2]:

$$\epsilon(n_p, n_n) = \epsilon_N(n_p, n_n) + n_n m_n c^2 + n_p m_p c^2 + \epsilon_e(n_e) + \epsilon_\mu(n_\mu), \tag{11}$$

where $\epsilon_N = n_b \mathcal{E}$. The EOS is then given as

$$\rho(n_b) = \epsilon(n_b)/c^2 \qquad P(n_b) = n_b^2 \frac{d\epsilon/n_b}{dn_b}, \tag{12}$$

where ρ is the mass density of matter.

The above expressions for nucleon+lepton BEM can easily be generalised to any baryon+lepton system in its ground state[8]. If the matter consists of ν_j components that are conserved on time-scales longer than the lifetime of the system, the chemical potentials μ_j of all these components must satisfy the condition

$$\nu_j \mu_j = 0. \tag{13}$$

The equilibrium condition (13) can be easily evaluated by noting that the chemical potential μ of any particle is a linear combination of the chemical potentials of the neutron μ_n and electron μ_e, weighted by the baryon number b and electric charge q carried by the particle[8],

$$\mu = b\mu_n - q\mu_e. \tag{14}$$

1.3. *Equation of State at T≠0*

The discussion in this section has so far been for cold (T=0) systems. At finite (nonzero) temperatures, it is conventional to characterise a system by the Helmholtz free energy density

$$\mathcal{F} = \mathcal{E} - TS, \tag{15}$$

where S is the entropy density of the system. All the properties of nuclear matter defined above can be calculated analoguously to the Eqs. (1)–(14) with the energy density $\mathcal{E}$ replaced by the free energy density $\mathcal{F}$. For example, for matter consisting of protons, neutrons and electrons in exact thermodynamical equilibrium, the minimization of the free energy will give[9,10]

$$P = n^2 \partial(\mathcal{F}/n)/\partial n|_{T,Y_e} \tag{16}$$

$$S = -\partial(\mathcal{F}/n)/\partial T|_{n,Y_e} \tag{17}$$

$$\mu_n - \mu_p - \mu_e = -\partial(\mathcal{F}/n)\partial Y_e|_{n,T}, \tag{18}$$

where Y_e is the electron fraction in the matter (equal to the proton fraction y_p due to the requirement of charge neutrality of the matter) and μ_j are chemical potentials (8). A complete set of thermodynamical relations required to determine the EOS theory at finite temperature may be found, for example in Ref. [9,10].

2. Nucleon-Nucleon Interaction and the EOS

2.1. *Models of the Nucleon-Nucleon Potentials*

Various approaches have been used to model the nucleon-nucleon potential[1,11,12] for astrophysical applications. This is true also of low-energy nuclear structure physics, in which many very different models of nuclear potentials are used without there being any clear preference for a specific model. It is necessary in astrophysics not only to make assumptions about the nature of the hadron-hadron potentials, but particularly as regards their behaviour as a function of the baryon number density of the system under consideration.

2.1.1. *Empirical Models*

The simplest models of bulk nuclear properties that do not attempt to describe details of internucleon interactions are based on the compressible liquid drop model (semiempirical mass formula[3]), used on different levels of sophistication, even including microscopic elements[13]. Alternatively, the nucleon-nucleon interaction is parametrized purely analytically[14,15]. Free parameters of the models are fitted to saturation properties of symmetric nuclear matter and/or ground state binding energies of finite nuclei. These models have been rather successful in modelling of nuclear matter[9,10,14,16] as well as basic properties of finite nuclei[13]. They however offer only a limited prospect for better understanding of fundamental nuclear processes occuring in stars owing to the a wide region of densities and varying composition of stellar baryon matter.

2.1.2. *Non-Relativistic Density Dependent Effective Interactions*

More elaborate models are based on non-relativistic density dependendent effective nucleon-nucleon interactions. These interactions are not derived from 'realistic' nuclear potentials (see below) but are directly parametrized at the starting point of the model calculation. About 10-15 variable parameters, determined by fitting to several hundred of pieces of experimental data on ground states of doubly-closed-shell even-even nuclei and saturation properties of infinite nuclear matter, are needed. The most well-known representatives of such interactions are the zero-range Skyrme[17,18], the finite-range Gogny[19,20] and indefinite-range SMO[21] potentials. Using these effective potentials, the ground state properties of finite nuclei and infinite matter are calculated in the mean-field Hartree-Fock (HF) model[22] or Extended-Thomas-Fermi[23] with Strutinski Integral[24] (ETFSI) method[25]. The latter is a good approximation to the full HF method in most cases with less demand on computational time. The advantage of these approaches is that the models are very simple to calculate and yet they are rather successful in global description of finite nuclei and nuclear matter. The disadvantage is that they do not allow unique determination of the highly correlated parameters for any of the potentials. This implies that, for each potential, an infinite number of combinations of possible values of these parameters could be found which would yield fits of comparable quality to experimental data. Further, the parameters are fitted to properties of stable nuclei and have unknown isospin and density dependence. This limits severely the predictive power of these models for finite nuclei far from stability and for asymmetric nuclear matter at densities significantly different from nuclear saturation density.

2.1.3. *Meson-Exchange Between Point-Like Nucleons*

The starting point of all models based on meson-exchange theory of nuclear force is the construction of a two-body potential between bare nucleons. The potential is spin and isospin dependent and consists of a short-range ($r<r_c \simeq 0.4$ fm) repulsive part and intermediate and long range terms containing both attractive and repulsive components. The long range part ($r>\sim 2$ fm) is usually well described by the asymptotic form of the Yukawa one-pion-exchange (OPE) potential[26]. This potential is uniquely determined by the properties of the pion and its coupling strength to the nucleonic field. To model the intermediate range part of the potential, two-pion and ρ and ω (and possibly some other heavier meson) exchange have to be introduced. The short range part has to modelled phenomenologically ('hard' [27] or 'soft' [28] core). It turns out that to obtain high-precision agreement with experimental data using such a potential, additional phenomenological terms (central, tensor, spin-orbit and some arbitrary radial functions[22]) have to be added. The total number of independent variable parameters in these 'realistic' potentials is typically 40–60. They are determined by fit to a data base of several thousand data on bare nucleon-nucleon (np, nn, pp) scattering cross-sections and phase shifts and the properties of the deuteron and ^{3}He (if three-body terms are included in the potential). Frequently used parametrizations of the non-relativistic 'realistic' potentials include Paris[29], Reid-93, Nijmegen-I and II[30], Argonne $v14$[31] and $v18$ (A18) [32] and A18+δv+UIX*[34].

In the next step, in order to use the 'realistic' potentials in models of nuclear systems, they have to be replaced by an effective potential, including the effect of the nuclear medium on the interaction between bare nucleons. The reason is that the presence of other nucleons fundamentally changes the bare interaction in a way which is not fully understood to date. Such an effective interaction can be theoretically constructed by an infinite sum of scattering process of two nucleons in the nuclear medium. This sum would include all the nucleon correlations contributing to the energy of the system. The first term of the sum represents the Born approximation and it is hoped that the higher order terms may be treated in perturbation theory. However, the infinite sum may not converge and various approximations have to be applied to obtain a reasonable estimate of the energy and wavefunctions of the system. One of the best known effective interactions used in this context is the Brueckner G-matrix[35] which, for the case of two nucleons in a medium, is an equivalent of the scattering matrix of two nucleons

328

in free space. The G-matrix can be used in the standard Hartree-Fock approximation (Brueckner-Hartree-Fock), density dependent Hartree-Fock (DDHF)[36] or Brueckner-Bethe solution of coupled cluster equations[37]. To avoid complications of finding the effective 'realistic' potential in order to calculate the total ground state energy, a variational approach[38,39] has been adopted. Using this method, the upper bound E_v on the expectation value of the ground state energy of a nuclear system $< E >$ is found. $< E >$ is minimized with respect to a trial wavefunction, dependent on the parameters of the 'realistic' potential[37] and E_v is then evaluated in a diagrammatic cluster expansion[33,34,40].

The potentials and many-body theories discussed above do not include relativistic effects. However, relativistic corrections grow in importance with increasing density of the nuclear medium and thus are relevant for the EOS at high densities for neutron stars and supernova models.

The starting point of relativistic models is the Lagrangian involving the interaction between Dirac nucleons with, in principle, all participating meson fields, usually scalar, vector and isovector mesons. The nucleon-meson coupling constants can be algebraically related to properties of nuclear matter. The Lagrangian is then used to determine amplitudes of relativistic bare two-nucleon scattering which in turn are used to construct the long-range part of a relativistic 'realistic' potential instead of the Yukawa long-range potential in non-relativistic models[42]. Other terms and parameters in the fully relativistic 'realistic' potentials are equivalent to the non-relativistic ones. All parameters are fitted to a high precision in the standard way to bare nucleon-nucleon scattering data[41]. These potentials include the Bonn A,B,C [41], CD-Bonn[42] and Groningen[43] parametrizations. The Lagrangians can be easily extended to include hyperons and leptons[44]. Coupling constants involving hyperon fields can be determined on the bases of experimental data on hypernuclei[45]. Similar to the non-relativistic case, many-body techniques have to be used to apply relativistic 'realistic' potentials for nuclear systems. Relativistic Dirac-Brueckner-Hartree-Fock [46], Dirac-Brueckner[46,47] and density dependent relativistic hadron field theory DDRH[48] are frequently used.

Relativistic field theory is also widely used in a similar way to the mean-field models with non-relativistic effective density dependent interactions, discussed above. The relativistic mean-field (RMF) approach originates from the $\sigma - \omega$ model of Walecka[49] with a standard Lagrangian and its extension to the non-linear Walecka model[50,51,52] involving non-linear self-interaction $U(\sigma)$. The five coupling constants between the meson and nucleon fields are replaced by mean values in a static approximation and fitted to saturation properties of symmetric nuclear matter and chosen values of the symmetry energy coefficient and the incompressibility[8]. Extended Lagrangians designed to describe finite nuclei as well as nuclear matter use fitting of the eight parameters (σ, ω, ρ meson masses, corresponding coupling constants and two parameters of $U(\sigma)$) to binding energies, charge and neutron radii of nuclear ground-states[53]. A number of parameter sets of the Lagrangians exist similar to those of the Skyrme or Gogny non-relativistic effective interaction. Density dependence of σ, ω, ρ meson coupling constants has been investigated[54] as well as coupling of mesons to derivatives of nucleon fields[55]. Finally, based on the effective field theory, various non-linearities involving σ, ω, ρ fields have been explored[56] with application to high density EOS.

Some doubts ahve been expressed about the validity of the mean-field approximation for the meson fields used in RMF theory in the density range relevant for neutron stars[34]. In particular, for the RMF approximation to be valid requires $\mu r \ll 1$, where μ is the inverse Compton wavelength of the meson involved and r is the mean interparticle spacing. Over the 1–5 n_0 density range, r estimated using a b.c.c. lattice ranges from 2 to 1.2 fm. Thus μr is between 1.4 to 0.8 for the pion and between 7.8 to 4.7 for vector mesons. These values are obviously not much less then unity. More general comment on the validity of meson exchange models of the nuclear force in general can be found in the contribution by T. Barnes to this volume.

2.1.4. *Meson Exchange Between Quarks*

The role of quark degrees of freedom in models of nuclei and nuclear matter is still a question of debate. It is admittedly true that many-body theories based on effective potentials between point-like nuclei are reasonably successful. However, the desire for more a fundamental description of the nucler force has motivated a considerable effort in exploration of quark effects in nuclei. Chiral effective Lagrangians[57] and quark-meson coupling models[58] were reported but did not yield strikingly different results from models based on nucleon-meson coupling. Very recently a new interesting application of the quark-meson-coupling (QMC) model to description of the quark structure of the nucleon immersed in a nuclear environment[59] has appeared. The interactions of the quarks with the nuclear medium are represented by the exchange of mesons between quarks of different nucleons, with coupling constants treated as free parameters fitted to properties of nuclear matter. Effective-Lagrangian-based models at the mean-field level, combined with a bag model for quarks have been also used to to examine properties of the mixed baryon-quark phase in high density neutron star matter [8,34,60,61].

2.1.5. *One-Gluon-Exchange Between Quarks*

As the latest development, discussed separately later in Sec. 4, the baryon-baryon potential based on one-gluon-exchange in the framework of the non-relativistic quark model has been applied to nuclear matter in the mean-field approximation. The model has 5-7 well defined parameters and yields results for symmetric and pure neutron matter comparable to those of A18 and A18+δv+UIX* potentials[34].

2.2. *Application of the EOS to Astrophysical Problems*

To understand the supernova phenomenon and the physics of neutron stars, a detailed knowledge of the baryon EOS at the relevant densities and temperatures is required. This in turn depends upon an appropriate choice of baryon interaction. The subnuclear density region relevant for supernova models spans[63] about 4×10^{-8} to about 0.1 fm^{-3}. The nuclear matter that is formed in the high density stage of the collapse of the progenitor core ranges from $\sim$0.1 fm^{-3} to about 0.6 fm^{-3} [62]. Neutron star models involve even wider density range starting from $\sim$6$\times10^{-15}$ fm^{-3} (estimated density at the surface of neutron stars) to about 0.6–1.0 fm^{-3} (expected in the center of neutron stars). Subnuclear (inhomogeneous matter) and supernuclear (homogeneous matter) differ considerably in composition and physical phenomena occuring in the matter. The fundamental difference between the two phases is that finite nuclei are present only in the former. The latter consist of unbound baryons, leptons and possibly mesons and quarks. If follows that EOSs suitable for the inhomogeneous phase must be based on nucleon-nucleon potentials (Sec. 2.1) that are suitable for description of heavy nuclei. This excludes all the meson-exchange 'realistic' potentials that have been used so far to calculate only nuclei up to A=10 [64]. Most of the currently used EOSs in both the subnuclear and supernuclear density regions do not cover the whole region but are composed of several EOSs reflecting the evolution of the character of matter with changing density in smaller intervals. This situation and its consequences are described in more detail in the next sections.

2.2.1. *Inhomogeneous Phase*

At densities below $\sim 6 \times 10^{-11}$ fm^{-3} the matter in this ground state consists of ^{56}Fe nuclei arranged in a lattice to minimize their Coulomb energy. A fraction of electrons are bound to the ^{56}Fe nuclei. The EOS of Feynman, Metropolis and Teller[65], with the Thomas-Fermi model to calculate electronic energies, is usually used in this density region. In the density range $\sim 6 \times 10^{-11}$ fm$^{-3} < n < 2.5 \times 10^{-4}$ fm^{-3} the equilibrium nuclei, now immersed in a relativistic electron gas but still distributed in a lattice, become more neutron-rich as electron capture and inverse β-decay take place but the direct β-decay is inhibited by the lack of unoccupied electron states. The EOS of Baym, Pethick and Suntherland[66] is appropriate in this regime. At about 2.5×10^{-4} fm^{-3} (the neutron drip-line) finite nuclei cannot support the neutron excess any longer and start to emit neutrons to continuum neutron states. From this density up to about 0.1fm^{-3}, where the transition to homogeneous nuclear matter occurs, it is customary to use the Baym, Bethe and Pethick[67] EOS. However, many other more modern possibilites exist (see e.g. Ref. [71,72]) as discussed in Sec. 2.2.3. This region of still bound nucleons, coexisting with free neutron and electron gas, is the most difficult one to deal with as the equilibrium arises as a result of a delicate balance between nuclear surface energy (favouring large nuclei) and the Coulomb energy, the sum of the positive nuclear self energy and negative lattice energy, which favours small nuclei. The current understanding is that the nuclei form exotic phases ('the pasta phase') like rods, slabs, tubes, inside-out bubbles, spaghetti or lasagna, stabilized by the Coulomb interaction and characterized by various crystal lattice structures.

2.2.2. *Homogeneous Phase*

Above baryon number density ~ 0.1 fm^{-3}, nuclear matter makes a transition to a homogenous phase where bound nuclei are no longer present. Most of the models of the nucleon-nucleon interaction, both 'realistic' and 'empirical', discussed in the previous paragraphs, are used in this regime in construction of the EOS. The matter, consisting of neutrons, protons and electrons and muons, develops β-equilibrium (see Sec. 1.2). With increasing density, creation of heavy baryons (replacing nucleons), meson condensates [73] and possibly deconfined quarks becomes energetically favourable. There is great uncertainty in modelling this exotic matter. The fundamental problem arises from the fact that the interactions between nucleons, mesons and hyperons are not reliably known. This affects the calculation of thresholds for appearance of a different heavy particle in nuclear matter. In the Fermi-gas (non-interacting) approximation usually adopted, the baryon chemical potentials μ_B are taken equal to the vacuum rest mass of the particle. However, in reality, μ_B depends also on the interaction strength of species B in the medium and may seriously differ from the simple estimate $\mu_B = m_B c^2$. This in turn affects the matter composition and calculated properties of stars. One of the most compelling questions concerning high density composite matter is whether or not the presence of other than nucleon components in the matter affects the slope of the EOS with increasing density. Most of the existing models predict softening (decrease in slope) of the EOS. However, there is some suggestion that this is not always the case and the effect depends strongly on the nature of the hyperon-hyperon and hyperon-nucleon potentials[48,74].

There is extensive literature on composite nuclear matter above the nuclear saturation density. Since the pioneering work of Pandharipande[75], Bethe and Johnson[76], Mozskowski[77] and Glendening[78], various methods of construction of EOS have been applied (see e.g. Ref. [48,79,80] and references therein). Ultimately, speculations about quark matter, alone or in coexistence with hadrons, are considered[8,34,44,58,61]. Some authors ignore the possibility of such exotic matter completely and extrapolate instead the EOS used for n+p+e+μ matter to higher densities[2,18,34,40]. Justification of such a process might be found if predictions of the maximum mass and radius of cold neutron stars would match observation. Direct experimental data on high density nuclear matter are scarce. The objects in nature expected to reach these densities are mainly interiors of neutron stars. Alternatively, observables resulting from terrestrial experiments with heavy ion collisions can yield information on exotic matter, but there the identification of a particular matter composition is much more complicated than for neutron stars[6,7].

2.2.3. *EOS for Supernova and Neutron-Star Models at $T \neq 0$*

Most of the EOS discussed above in this section do not take into account the fact that during supernova collapse and explosion the temperature of the matter rises as high as 10-20 MeV. In such a situation the Helmholtz free energy is used for construction of EOS in the way outlined in Sec. 1. Empirical EOS, tailored to simulation of supernovae during gravitational collapse and bounce, was developed by Bower and Wilson[81]. The EOS included relativisitic electrons, He nuclei, heavy nuclei of varying atomic weight, non-relativistic nucleons, photons, positrons and neutrinos and their antiparticles at all relevant densities and temperatures and exhibited smooth behaviour at phase transitions throughout the whole density range. Empirical EOS derived by Lattimer et al.[9], based on the compressible liquid drop model, has been later simplified[10] to the Lattimer-Swesty EOS that has become standard in supernovae simulations during the last decade. This EOS describes inhomogeneous matter as a mixture of electrons, positrons, photons, free neutrons and protons, alpha particles and a single species of heavy nuclei and uses Maxwell construction at phase boundaries. These empirical models have however some limitations, mainly in modelling the 'pasta' phase in spherical Wigner-Seitz approximation without microscopic treatment of more subtle effects like shell corrections and pairing, discontinuities at phase boundaries and simplified approach to high density homogeneous matter not including exotic components. Some of these deficiencies can be removed in a selfconsistent approach using Hartree-Fock method with effective interaction discussed in the next paragraph.

Self-consistent non-relativistic EOS's use HF method to determine the ground state energy of matter as a function of density, temperature and proton fraction by minimization of the HF Hamiltonian in coordinate space. In this way the equilibrium energy per particle can be calculated as a continuous function of density and the composition of nuclear matter is determined naturally without *a priori* assumptions about particular physical processes in certain density regions. One of the first EOS of this kind was developed by Bonche and Vautherin[82] for inhomogeneous matter assuming spherical symmetry (1D space) using the SkM* Skyrme interaction. Later it was adopted by Hillebrandt and Wolff[68] with RATP, Ska and SkM Skyrme parametrizations for tabulation of the EOS in a dense grid of energy density, temperature (or entropy) and electron fraction for use in hydrodynamical codes. Main interest of these calculations focussed on subnuclear densities, in particular the 'pasta' phase. The Gogny finite range interaction D1 has been also used to construct finite temperature EOS [19,20] based on application of the variational principle to the thermodynamical potential with special attention to the liquid-gas phase transition. The ETFSI method with RATP, SkM* and SkSC4-10 Skyrme interaction was also applied to derive EOS at finite temperatures for inhomogeneous [69] and homogeneous[70] nuclear matter under conditions appropriate to a collapsing star assuming spherical symmetry. The EOS based on the RMF theory with the TM1 parameter set for the RMF Lagrangian[83] was used in modelling of supernova physics. Again both homogeneous and inhomogeneous matter are treated within the model. However, the inhomogeneous matter is not calculated fully selfconsistently but is approximated as a b.c.c lattice of spherical nuclei and the free energy is minimized with respect to cell volume, proton and neutron radii and surface diffuseness of the nuclei and densities of the outside neutron and proton gases.

The major limitation of the models discussed so far, i.e. the assumption of spherical symmetry for inhomogeneous matter, has been removed in the recent work of Magierski and Heenen[71] who treated nuclear matter in the density region of the 'pasta' phase for the first time fully self-consistently in 3D coordinate space (allowing for triaxial shapes) using HF method with SLy4 Skyrme interaction at T=0. The Hartree-Fock equations are solved in a rectangular box with periodic boundary conditions and includes naturally shell effects in neutron gas and pairing. In calculation of the Coulomb energy, the model goes beyond the Wigner-Seitz approximation of spherical cells and includes both interactions between nuclei in different lattice sites and protons and electrons. Extention[72] of this work to homogeneous matter and finite temperature is discused fully in Sec. 4.1.

3. Constraints on the Equation of State

In face of such a wide variety of models and calculation methods it is vital to use all possible evidence to narrow down the range of possible theories by setting constraints on specific properties of the models. As the main constituents of stellar matter are nucleons, the natural fingerprints of the true EOS are sought in nuclear matter as it occurs in compact objects, mainly supernovae and neutron stars at all phases of their life. Some observables in finite nuclei, like neutron skin, may also be important. There is no systematic investigation available to date that would involve all representative EOS based on different nuclear and particle theories tested on a full set of experimental and observational data. Furthermore, the sensitivity of theories, underlying EOSs, to different types of data has not been comprehensively studied. It is beyond the scope of this paper to give an exhaustive review of the search for EOS constraints. Thus only a few examples are given of how such tests could be carried out.

3.1. *The Skyrme Models*

A comprehensive investigation of performance of phenomenological Skyrme parametrisations which have been applied by different studies in homogeneous nuclear matter has been recently reported[18] by Rikovska Stone et al.

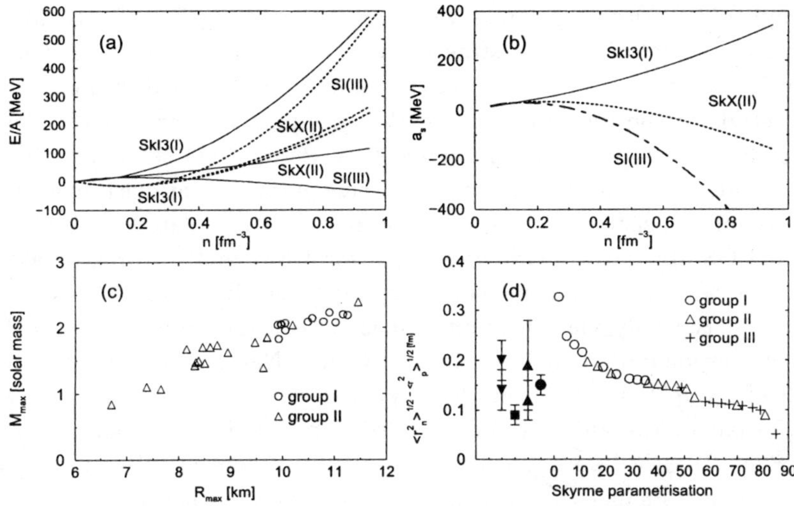

Figure 1. (a) The energy per particle for SNM (solid line) and PNM (dotted line) plotted as a function of the baryon number density n for the SkI3, SkX, and SI Skyrme interactions (typical representatives of group I, II and III)[18]. (b) The asymmetry coefficient a_s is plotted as a function of the baryon number density n for the SkI3, SkX and SI Skyrme interactions. (c) The maximum gravitational masses of T=0 non-rotational neutron-star models plotted against the corresponding radius calculated for BEM nucleon+lepton matter for Skyrme interactions of group I and group II. Skyrme interactions of group III do not produce realistic neutron-star models. (d) Neutron skin of ^{208}Pb, calculated for representative Skyrme parametrisations from each group (arbitrary identification number of each Skyrme model used is displayed on the x-axis). Experimental data from antiprotonic atoms (full circle), GDR (full tringle up) and from older elastic proton scattering experiments (full triangle down) are taken from Ref. [85]. Recent elastic proton scattering data (full square) are from Ref. [86].

Energy per particle was calculated as a function of baryon number density in two extreme cases, pure neutron and symmetric matter (Fig. 1, panel (a)). The results fell clearly into three groups (I-III), with each group formed by around 30 different sets of parameters of the Skyrme interaction. The first group (I) predictes increase of E/A with density for both SNM and PNM at comparable rate, with the SNM remaining the ground state of the matter at all densities. The same result is obtained[34] with the Av18+δv+UIX* 'realistic' potential. In group II the rate of growth of E/A with density calculated for PNM is slower than that in SNM. At some density the two curves cross and PNM becomes the ground state which is, as discussed in Sec. 1.2, unphysical. Parametrizations of group III yield a very slow increase of E/A for PNM at low densities followed by decrease at higher densities in contrast with the fast growing E/A for SNM. The transition to PNM ground state occurs systematically at lower densities than for parametrizations of group II, again an unphysical result.

Details of the interpretation of this grouping in terms of the density dependence of the *symmetry energy* (Eqs. (6) and (7) see panel (b)) are given in Ref. [18]. This division of the Skyrme parametrizations into groups is remarkably consistent with other physical results, calculated using Skyrme-based EOS. Thus the EOSs of group I for BEM matter, extrapolated to densities $\sim 3 \times n_0$ and complemented by BBP and BPS EOS at subnuclear densities, typically yield maximum mass of cold neutron stars over $2M\odot$ and corresponding radii between 10-13 km. Neutron star models using EOS of group II predict considerably smaller stars, with maximum mass lower than $1.8M\odot$ and radii between 8-11 km (see Fig. 1 panel (c)). EOSs of group III either do not yield stable neutron stars at all or the objects produced are smaller than permitted by any conventional EOS[84].

This correlation can be easily understood considering that the baryon pressure (Eq. 3) in the star, opposing the effect of gravity, is given by the derivative of E/A with respect to density. Clearly, growing pressure with density (group I) will sustain a relatively large star. For the other two groups the pressure is either lower than for group I or decreasing with increasing density which leads to smaller stars or objects that undergo a gravitational collapse. When observational data on the maximum mass-radius correlation in neutron stars become more definitive, it would serve as a sensitive selection criterion of Skyrme parametrisations.

It is very interesting to notice that this general correlation between the density dependence of the symmetry energy and the size of a neutron star persists to finite nuclei. The correlation between the calculated size of the neutron skin $< r_n^2 >^{1/2} - < r_p^2 >^{1/2}$ in ^{208}Pb and the type of the Skyrme parametrisation in shown in Fig. 1, panel (d). The calculation in finite nuclei includes surface and Coulomb effects, not present in calculation of the symmetry energy in nuclear matter and yet the distinction between parametrisations on the basis of the density dependence of the symmetry energy manifests itself. Parametrisations of group I predict the thickest neutron skin as compared to the thinnest, predicted by parametrisations of group III. The current experimental data, extracted in a model dependent way from measured cross-sections of various nuclear reactions and giant dipole resonances (GDR) do not have reguired precision as yet[85,86] to give preference to any parametrization group. However, an experiment has been proposed in JLAB for measurement of the neutron skin in ^{208}Pb to

1.5% precision[87] via atomic parity violation effects in electron scattering. If this experiment is successful, it will provide a very powerful constraint on Skyrme parametrisations. Prediction of the observed neutron skin in very neutron-rich nuclei is thus linked to modelling nuclear matter at the transitional region between subnuclear and supernuclear densities and has consequence for supernova simulations.

3.2. *Other Constraints*

Improved astronomical data are urgently needed to assist development of theories yielding EOS close to reality. Masses and radii of compact objects, in correlation and individually, cooling rates of neutron stars and moments of inertia and periods of rotating neutron stars are amongst the most valuable pieces of information. Lattimer and Prakash [84] investigated constraints on EOS and showed that the neutron star radius, primarily determined by the behaviour of pressure of matter in the vicinity of nuclear matter equilibrium density, measured with precision better than 1 km, may provide a useful constraint of EOS. Based on observation of a pulsar – white dwarf binary[88] PSR J0751+1807 , limits on the mass of the millisecond pulsar were deduced to be 1.6 and 2.8 $M_\odot$ reported[89] to narrow down to 2.2±0.2 $M_\odot$. This new value constitutes a very serious constraint on EOS, strongly preferring non-relativistic potential models. Observation[90] of an isolated neutron star RXJ1856-3754 with mass of 1.5 $M_\odot$ led to determination of its radius of 13.7±0.6 km. This result was used to constrain the EOS. The test, involving twelve EOSs described in Ref. [84], resulted in preference for EOSs stiff at nuclear density. It is interesting to note that most EOSs based on 'realistic' potentials included in the test, gave radii outside the limit provided by the experiment. Refined models of the cooling mechanism and data on thermal emission for selected neutron stars have been used for testing of EOS [91]. The minimal cooling scenario was proposed as a constraint for EOS at supranuclear densities. High sensitivity to choice of EOS of the moment of inertia of a slowly rotating neutron star was recently reported[92], using data on the binary pulsar J0737-3039A with M=1.337 $M_\odot$ and spin frequency Ω=276.8 s^{-1} as an example. Representative EOS were chosen based on non-relativistic 'realistic' potential models, relativistic mean-field approximation, and strange quark matter. Although no conclusion is drawn in this work, it is clear that measurement of the moment of inertia to about 10% accuracy would serve as a sensitive constraint. Differential rotation of the J0737-3039A pulsar has been also

338

used to constrain the EOS [93]. Very recently, the very low, precisely measured, mass of 1.250 ± 0.005 $M_\odot$ of Pulsar B in the double pulsar system J0737-3039 was used to constrain EOS of neutron star matter *under the assumption* that it was formed in an electron-capture supernova [94].

The most sought after constraint would be based on success in modelling a type II core-collapse supernova. There is no doubt that the supernova phenomenon exists in nature and is accompanied by an explosion. Current models fail to reproduce this explosion. One of the possible solutions of this problem is improvement of the EOS since the EOS at subnuclear densities controls the rate of collapse, the amount of deleptonization and thus the size of the collapsing core and the bounce density. An interesting comparison has been made recently by Janka et al.[95] who followed the development of the shock radius and electron neutrino luminosity over the first 200 ms after core collapse using three different EOSs, Lattimer and Swesty[10], Shen et al.[83] and Hillebrandt and Wolff[68]; this simulation found certain differences in the results. It is however surprising that for EOSs based on such different physics as Lattimer and Swesty[10] versus Shen et al.,[83] these differences were not more noticeable.

Evolution of the supernova core from the beginning of gravitational collapse of a 15 $M_\odot$ star up to 1 second after core bounce was modelled to compare the Lattimer-Swesty[10] and Shen et al.[83] EOSs. Many aspects of the simulation were compared including the radial positions of shock waves as a function of time after bounce, mass and lepton fractions and nuclear species in the supernova core, velocity profiles as function of baryon mass coordinates and heating rates and temperature profiles as a function or radius. Although some differences were found they were not significant enough to explain the lack of explosion[96]. Thus further search for aspects of supernova explosions that would show strong sensitivity to the physics of the EOS is very important.

As the supernova simulation models often involve different computational techniques, it is also important to compare different codes using the same EOS. This has been done successfully for the codes AGILE-BOLTZTRAN of Oak Ridge-Basel group and VERTEX of the Garching group for Lattimer-Swesty EOS for modelling core collapse and post-bounce evolution [97] yielding very consistent results.

4. New Developments

4.1. *Selfconsistent 3D Hartree-Fock EOS at $T \neq 0$*

As a new development, Newton and Rikovska Stone[72] have extented the work of Magierski and Heenen[71] (Sec. 2.2.3) to a more general general application for systems at finite temperature and to densities up to $3 \times n_0$ for use in supernova simulation models. The preliminary results are in particular interesting in the density region corresponding to the 'pasta' phase. A wide range of exotic nuclear shapes immersed in a free neutron and electron gas is observed, including, but not confined to, the 'spaghetti' and 'lasagne' phases previously identified. This picture smoothly evolves both ways, to homogeneous nucleon distribution with increasing density and to finite nuclei in a lattice, surrounded by free neutrons and electrons, with decreasing density. These results have serious consequence, for example, for calculation of neutrino scattering cross-sections on the 'pasta' phase. It is clear that these cross-sections cannot be just taken as contributions from finite isolated nuclei and free neutron and electron gas, but the complex density distribution of particles will have to be taken into account.

4.2. *EOS based on One-Gluon-Exchange Quark Model*

The nucleon-nucleon potentials discussed thus far have not taken the structure of the nucleons themselves into account . As a very recent development in this section we present preliminary results obtained for the EOS, based on the nonstrange baryon-baryon scattering amplitudes of Barnes et al.,[98] derived in the nonrelativistic quark model using the 'quark Born diagram' formalism. The model is described in more detail in the contribution by T. Barnes to this volume. These quark model calculations give a baryon-baryon potential in a given isospin, spin channel (I,S) (we note that in this section I stands for isospin quantum number and not the *asymmetry* as in Sec. 1.1) which is the sum of the contributions of four independent quark scattering diagrams

$$V_{BB}^{I,S}(r) = \frac{8\alpha_s}{3\sqrt{(\pi)}m_q^2} \sum_{n=1}^{4} V_n^{I,S} \tag{19}$$

where for harmonic oscillator baryon wavefuctions the individual potentials are

$$V_n^{I,S} = \frac{w_n^{I,S}\eta_n^{I,S}}{(A_{I,s} + B_{I,s})^{3/2}} \exp\left(-\frac{(r/\hbar c)^2}{A_{I,s} + B_{I,s}}\right) \tag{20}$$

340

All the symbols used in Eqs. (19) and (20) are defined in Ref. [98]. The potential, given as a sum of single gaussian functions, is nonsingular and depends only on two adjustable parameters, α_s/m_q^2 (the spin-spin hyperfine interaction strength) and α (the baryon wavefunction length scale). There is a relatively narrow range of plausible values for these quark-model parameters. These potentials are used to derive the EOS for the proton-neutron nuclear matter in the independendent Fermi gas model[99]. Translational in-

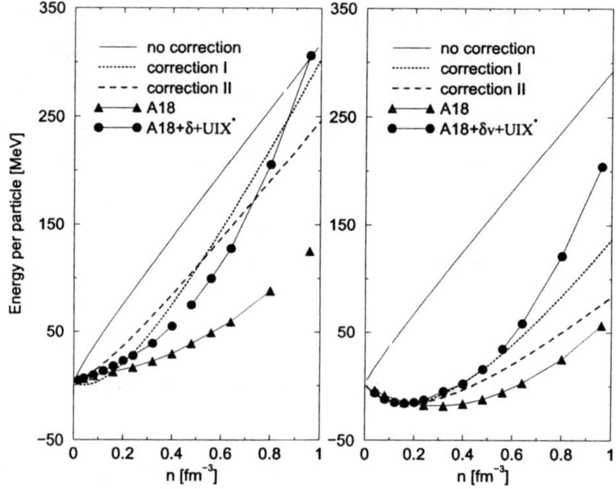

Figure 2. EOS for pure neutron and symmetric nuclear matter. The EOSs constructed using the A18+δv+UIX* and A18 potentials [34] are added for comparison.

variance of a uniform medium implies that the single-particle wave functions are plane waves, which are already solutions to the Hartree-Fock equations in this case. The expectation value of the Hamiltonian in a non-interacting Fermi system (each particle moves independently in the mean field generated by all the other particles present) gives the first approximation to the ground state energy of nuclear matter,

$$E_0 + E_1 = <F|H|F>, \qquad (21)$$

where F is the total wave function of the system. This expectation value is equivalent to the minimum energy obtained in a variational calculation of the system for the case of uniform matter. Therefore to construct the EOS for uniform proton-neutron matter, the sum

$$E_0 + E_1 = 4\sum_{k}^{k_F}\frac{\hbar^2 k^2}{2m} + \frac{1}{2}\sum_{k\lambda\rho k'\lambda'\rho'}^{k_F}\sum^{k_F} [< k\lambda\rho k'\lambda'\rho'|V_{BB}|k\lambda\rho k'\lambda'\rho' > \quad (22)$$

$$- < k\lambda\rho k'\lambda'\rho'|V_{BB}|k'\lambda'\rho'k\lambda\rho >]$$

is evaluated, where the summation runs over all occupied single-particle states $\phi_k(\mathrm{x})\eta_\lambda\xi_\rho$ up to the Fermi level (ϕ, η and ξ are the spatial, spin and isospin wave-fuctions, respectively). Results for $(E_0 + E_1)/A$ are shown in Fig. 2 for two extreme cases, pure neutron matter (left panel) and symmetric nuclear matter (right panel). EOSs obtained with the A18 and A18+δv+UIX* potentials[34] are also shown for comparison. The simplest quark-model EOS (no correction) is rather stiff for pure neutron matter, and does not show saturation in symmetric nuclear matter. This is expected, as the chosen V_{BB} potential is due to the contact spin-spin interaction alone and is purely repulsive. The calculated EOS is however at a reasonable energy scale and, if an empirical intermediate-range interaction were added, would clearly lead to a modified EOS that gives realistic results. This can be demonstrated empirically by the addition of a weak gaussian potential to the total V_{BB}^{IS} in each IS channel,

$$V'_{I,S} = V_{BB}^{I,S} + v_0^{I,S} \exp\left[-\frac{1}{2}\left(\frac{r}{R_0^{I,S}}\right)^2 \right] \quad (23)$$

Strength $v_0^{I,S}$ and length scale $R_0^{I,S}$ parameters have to be determined by fitting to properties of nuclear matter. Fig. 2 shows results based on two trial sets of these parameters. Assuming amplitudes of equal magnitude 0.5 GeV in each I,S channel, positive (repulsive) for $I = S$ and negative (attractive) for $I \neq S$ for simplicity (correction I), $R_0^{1,1}$=0.57 fm, $R_0^{1,0}$=0.71 fm, $R_0^{0,1}$=0.512 fm, $R_0^{0,0}$=0.75 fm are obtained. More microscopically, the additional potential (23) may arise from one-pion-exchange. The ratio of relative amplitudes in this case is $v_0^{1,1}:v_0^{1,0}:v_0^{0,1}:v_0^{0,0}$ as $(1/9):(-1/3):(-1/3):1$ (correction II). Given those ratios only a single strength parameter $v_0 = v_0^{0,0}$ is required. For this illustration v_0=0.8 GeV is assumed. A reasonable description of the saturation properties of symmetric nuclear matter and the density dependence of energy per particle of pure neutron matter was achieved with $R_0^{1,1}=R_0^{1,0}=R_0^{0,1}=R_0=0.78$ fm and $R_0^{0,0}=R_0-r_0$, with r_0=-0.06 fm, which introduces two length scale parameters R_0 and r_0. The total number of adjustable parameters of the model with correction I increases to seven and with correction II to five. Examination of Fig. 2

shows a surprising result; this five (seven)-parameter model, based on one-gluon-exchange complemented by a phenomenological intermediate-range potential can produce results for PNM and SNM of the same quality as the A18+δv+UIX* potential, which has over 40 free parameters. Work on the extension of this model to the description of neutron star matter, including strange baryons, as well as on understanding the microscopic origin of the phenomenological intermediate-range potential, is in progress.

5. Conclusions

The currently unsatisfactory state of nuclear EOS calculations highlights the problems of understanding atomic nuclei and the physics of processes, such as stellar evolution, in which nuclei and nucleons play a fundamental role. To date there have been many attempts to describe the nuclear EOS using a wide range of models. Although a reasonably convincing description of nuclear and neutron-star matter exists at moderate densities, the spectrum of observables predicted in these systems is very wide. With increasing experimental precision and better observational data the situation should improve through the establishment of certain constraints upon acceptable EOS theory. Recently it has been realised that potential constraints may link very different phenomena. However, even if some EOSs are eliminated by application of experimental constraints, the question remains whether these constraints can be used to identify the physics underlying the 'allowed' EOSs. In other words, it is imperative not only to allow or eliminate certain EOSs, but also to understand the physics the EOS is based on. It is likely that many of the parameters of the existing EOSs are insensitive to the underlying physics, or are important only in correlation with other parameters.

As regards the importance of the EOS for supernova simulations, it is not clear how important the details of the EOS are in modelling the explosion after core-collapse. In particular, the newly developed selfconsistent EOS, which has thus far been applied to only one class of nuclear potentials but can easily be adaptable to any suitable potential, may shed some light on these problems. The quark model based EOS brings into focus new degrees of freedom which have not previously been considered in this context and which may offer new insights into high density physics. This may have profound implications in our understanding of neutron star and supernova physics.

Acknowledgement

Discussions with T. Barnes, C. R. Downum, W. R. Hix, A. Mezzacappa, J. C. Miller, W. G. Newton, N. J. Stone and E. S. Swanson are gratefully acknowledged. This work has been supported from DOE Scientific Discovery through Advanced Computing Grant, INT Seattle and US DOE grant DE-FG02-94ER40834.

References

1. H. Heiselberg and M. Hjorth-Jensen, *Phys.Rep.* **328**, 237 (2000)
2. E. Chabanat, P. Bonche, P. Hansel, J. Meyer and R. Schaeffer, *Nucl.Phys.* **A627**, 710 (1997)
3. C. F. von Weizsacker, *Z.Phys.* **96**, 431 (1935)
4. G. Colo and N. Van Giai, *Nucl.Phys.* **A731**, 15 (2004)
5. A. E. L. Diepering, Y. Dewulf et al., *Phys.Rev.* **C68**, 064307 (2003)
6. Bao-An Li, *Phys.Rev.Lett.* **88**, 192701 (2002)
7. P. Danielewicz, R. Lacey, W. G. Lynch, *Science* **298**, 1592 (2002)
8. N. K. Glendenning, *Compact Stars*, Springer, 2nd edition (2000)
9. J. M. Lattimer, C. J. Pethick, D. G. Ravenhall and D. Q. Lamb, *Nucl.Phys.* **A432**, 646 (1985)
10. J. M. Lattimer and F. D. Swesty, *Nucl.Phys.* **A535**,331 (1991)
11. J. M. Lattimer and M. Prakash, *Phys.Rep.* **334**, 121 (2000)
12. H. Heiselberg and V. Pandharipande, *Annu.Rev.Nucl.Part.Sci.* **50**, 481 (2000)
13. P. Moller and J. R. Nix, *Nucl.Phys.* **A361**, 117 (1981)
14. M. Prakash, I. Bombaci et al., *Phys.Rep.* **280**, 1 (1997)
15. C. Horowitz, M. A. Péres-Garcia and J. Piekarewicz, *Phys.Rev.* **C69**, 045804 (2004)
16. S. Balberg and A. Gal, *Nucl.Phys.* **A625**, 435 (1997)
17. D. Vautherin and D. M. Brink, *Phys.Rev.* **C5**, 626 (1972)
18. J. Rikovska Stone, J. C. Miller, R. Koncewicz, P. D. Stevenson and M. R. Strayer, *Phys.Rev.* **C68**, 034324 (2003)
19. J. Heyer, T. T. S. Kuo, J. P. Shen and S. S. Wu, *Phys.Lett.* **B202**, 465 (1988)
20. J. Ventura, A. Polls et al.,*Nucl.Phys.* **A545**, 247 (1992)
21. J. Rikovska Stone, P. D. Stevenson, J. C. Miller and M. R. Strayer, *Phys.Rev.* **C65**, 064312 (2002)
22. P. Ring and P. Schuck, *The Nuclear Many-Body Problem*, Springer Verlag, (1980), Chapts. 4,5
23. B. Grammaticos and A. Voros, *Ann.Phys.(N.Y.)* **123**, 359 (1979) and **129**, 153 (1980)
24. A. K. Dutta, J. P. Arcoragi et al., *Nucl.Phys.* **A458**, 77 (1986)
25. M. Brack, C. Guet and H. B. Häkansson, *Phys.Rep.* **123**, 275 (1985)
26. H. Yukawa, *Proc.Phys.-Math.Soc.Japan* **17**, 48 (1935) and **19**, 1082 (1937)
27. T. Hamada and I. D. Johnson, *Nucl.Phys.* **34**, 382 (1962)
28. P. V. Reid, *Ann.Phys.(N.Y.)* **50**, 411 (1968)

29. M. Lacombe et al., *Phys.Rev.* **C21**, 861 (1980)

30. V. G. J. Stoks, R. A. M. Klomp, C. P. F. Terheggen and J. J. de Swart, *Phys.Rev.* **C49**, 2950 (1994)

31. R. B. Wiringa, R. A. Smith and T. L. Ainsworth, *Phys.Rev.* **C29**, 1207 (1984)

32. R. B. Wiringa, V. G. J. Stoks and R. Schiavilla, *Phys.Rev.* **C51**, 38 (1995)

33. A. Akmal and V. R. Pandharipande, *Phys.Rev.* **C56**, 2261 (1997)

34. A. Akmal, V. R. Pandharipande and D. G. Ravenhall, *Phys.Rev.* **C58**, 1804 (1998)

35. K. A. Brueckner, *Phys.Rev.* **97**, 1353 (1955)

36. J. W. Negele, *Phys.Rev.* **C1**, 1260 (1970)

37. B. D. Day and R. B. Wiringa, *Phys.Rev.* **C32**, 1057 (1985)

38. V. R. Pandharipande and R. B. Wiringa, *Rev.Mod.Phys.* **51**, 821 (1979)

39. E. Lagaris and V. R. Pandharipande, *Nucl.Phys.* **A359**, 331 (1981) and **A359**, 349 (1981)

40. R. B. Wiringa, V. Fiks and A. Fabrocini, **Phys.Rev. C38**, 1010 (1988)

41. R. Machleidt, *Adv.Nucl.Phys.* **19**, 189 (1989)

42. R. Machleidt, F. Sammarruca and Y. Song, *Phys.Rev.* **C53**, R1483 (1996)

43. F. de Jong and H. Lenske, *Phys.Rev.* **C57**, 3099 (1998) and **C58**, 890 (1998)

44. G. F. Burgio, M. Baldo, P. K. Sahu and H.-J. Schulze, *Phys.Rev.* **C66**, 025802 (2002)

45. C. M. Keil, F. Hofmann and H.Lenske, *Phys.Rev.* **C61**, 064309 (2000)

46. H. Muether, M. Prakash and T. L. Ainsworth, *Phys.Lett.* **B199**, 469 (1987)

47. L. Engvik, M. Hjorth-Jensen et al., *Phys.Rev.Lett.* **73**, 2650 (1994)

48. F. Hofmann, C. M. Keil and H. Lenske, *Phys.Rev.* **C64**, 025804 (2001)

49. J. D. Walecka, *Ann.Phys.(N.Y.)* **83**, 491 (1974)

50. J. Boguta and A. R. Bodmer, *Nucl.Phys.* **A292**, 413 (1977)

51. B. D. Serot, *Rep.Prog.Phys.* **55**, 1855 (1992)

52. N. K. Glendening, F. Weber and S. A. Moszkowski, *Phys.Rev.* **C45**, 844 (1992)

53. G. A. Lalazissis, J. König and P. Ring, *Phys.Rev.* **C55**, 540 (1997)

54. S. Typel and H. H. Wolter, *Nucl.Phys.* **A656**, 331 (1999)

55. S. Typel, T. v. Chossy and H. H. Wolter, *Phys.Rev.* **C67**, 034002 (2003)

56. H. Müller and B. D. Serot, *Nucl.Phys.* **A505**, 508 (1996)

57. H. Hanauske, D. Zschiesche et al., *Astrophys.J.* **537**, 958 (2000)

58. S. Pal, M. Hanauske et al., *Phys.Rev.* **C60**, 015802 (1999)

59. P. A. M. Guichon and A. W. Thomas, *Phys.Rev.Lett.* **93**, 132502 (2004)

60. M. Prakash, J. R. Cooke and J. M. Lattimer, *Phys.Rev.* **D52**, 661 (1995)

61. P. K. Panda, D. P. Menezes and C. Providencia, *Phys.Rev.* **C69**, 058801 (2004)

62. I. Bombaci, T. T. S. Kuo and U. Lombardo, *Phys.Rep.* **242**, 165 (1994)

63. S. W. Bruenn, *Astrophys.J.Supplement* **58**, 771 (1985)

64. S. C. Pieper, K. Varga and R. B. Wiringa, *Phys.Rev.* **C66**, 044310 (2002).

65. R. P. Feynman, N. Metropolis and E. Teller, *Phys.Rev.* **75**,1561 (1947)

66. G. Baym, C. Pethick and P. Sunderland, *Astrophys.J.* **170**, 299 (1971)

67. G. Baym, H. A. Bethe and C. Pethick, Nucl.Phys. A175, 225 (1971)

68. W. Hillebrandt and R. G. Wolff, in *Nucleosynthesis:Challenges and New*

Developments, eds. D. Arnett and J. W. Truran, Univ.Chicago Press, 1985
69. M. Onsi, H. Przysiezniak and J. M.Pearson, *Phys.Rev.* **C55**, 3139 (1997)
70. M. Onsi, H. Przysiezniak and J. M.Pearson, *Phys.Rev.* **C50**, 460 (1994)
71. P. Magierski and P.-H. Heenen, *Phys.Rev.* **C65**, 045804 (2002)
72. W. G. Newton, *DPhil Thesis*, Oxford (2005)
and W. G. Newton and J. Rikovska Stone, to be published
73. R. Knorren, M. Prakash and P. J. Ellis, *Phys.Rev.***C52**, 3470 (1995)
74. M. Yonge, *Research Project*, Oxford University, Oxford (2005) (unpublished)
75. V. R. Pandharipande, *Nucl.Phys.* **A178**, 123 (1971)
76. H. A. Bethe and M. Johnson, *Nucl.Phys.* **A230**, 1 (1974)
77. S. Mozskowski, *Phys.Rev.* **D9**, 1613 (1974)
78. N. K. Glendening, *Phys.Letts.* **114B**, 392 (1982)
79. M. Baldo, G. F. Burgio and H.-J. Schulze, *Phys.Rev.* **C61**, 055801 (2000)
80. I. Vidana, A. Polls et al., *Phys.Rev.* **C62**, 035801 (2000)
81. R. Bowers and J. R. Wilson, *Astrophys.J.* **263**, 366 (1982),
and *Astrophys.J.Supplement* **50**, 115 (1982)
82. P. Bonche and D. Vautherin, *Nucl.Phys.* **A372**, 496 (1981)
and *Astron. Astrophys.* **112**, 268 (1982)
83. H. Shen, H. Toki, K. Oyamatsu and K. Sumioshi,
Nucl.Phys. **A637**, 435 (1998)
84. J. M. Lattimer and M. Prakash, *Astrophys.J.* **550**, 426 (2001)
85. A. Krasznahorkay et al., *Nucl.Phys.* **A731**, 224 (2004)
86. B. C. Clark, L. J. Kerr and S. Hama, *Phys.Rev.* **C67**, 054605 (2003)
87. C. J.Horowitz, S. J. Pollock, S. J. Souder and R. Michaels,
Phys.Rev. **C63**, 025501 (2001)
88. D. J. Nice, E. M. Splaver and I. H. Stairs, arXiv:astro-ph/0311296 (2003)
89. J. M. Lattimer and M. Prakash, *Phys.Rev.Lett.* **94**, 111101 (2005)
90. T. M. Braje and R. W. Romani, *Astrophys.J.* **580**, 1043 (2002)
91. D. Page, J. M. Lattimer, M. Prakash and A. W. Steiner,
Astrophys. J. Supplement **155**, 623 (2004)
92. I. A. Morrison, T. W. Baumgarte, L. S. Shapiro and V. R. Pandharipande,
Astrophys.J. **617**, L135 (2004)
93. I. A. Morrison, T. W. Baumgarte, L. S. Shapiro, *Astrophys.J.* **610**, 941 (2004)
94. Ph. Podsiadlowski, J. D. Dewi, P. Lesaffre, J. C. Miller, W. Newton and
J. Rikovska Stone, submitted to *Mon.Not.R.Astron.Soc.*
95. H.-T. Janka, R. Buras, K.Kifonidis, A. Marek and M. Rampp,
Proceedings IAU Coll. **192**, eds. J. M. Marcaide and K. W. Weiler, Springer
Verlag, (2003)
96. K. Sumioshi et al., accepted for publication in *Astrophys.J.*
97. M. Liebendörfer, M. Ramp, H.-Th. Janka and T. Mezzacappa,
Astrophys.J. **620**, 840 (2005)
98. T. Barnes, S. Capstick, M. D. Kovarik and E. S. Swanson,
Phys.Rev. **C48**, 539 (1993)
99. A. L. Fetter and J. D. Walecka, *Quantum Theory of Many-Particle Systems*,
McGraw-Hill Book Company, 1971
100. E. S. Swanson, private communication (2005)

QCD AND SUPERNOVAS

T.BARNES

Department of Physics and Astronomy
University of Tennessee
Knoxville, TN 37996, USA
and
Physics Division
Oak Ridge National Laboratory
Oak Ridge, TN 37831, USA
E-mail: tbarnes@utk.edu

In this contribution we briefly summarize aspects of the physics of QCD which are relevant to the supernova problem. The topic of greatest importance is the equation of state (EOS) of nuclear and strongly-interacting matter, which is required to describe the physics of the proto-neutron star (PNS) and the neutron star remnant (NSR) formed during a supernova event. Evaluation of the EOS in the regime of relevance for these systems, especially the NSR, requires detailed knowledge of the spectrum and strong interactions of hadrons of the accessible hadronic species, as well as other possible phases of strongly interacting matter, such as the quark-gluon plasma (QGP). The forces between pairs of baryons (both nonstrange and strange) are especially important in determining the EOS at NSR densities. Predictions for these forces are unfortunately rather model dependent where not constrained by data, and there are several suggestions for the QCD mechanism underlying these short-range hadronic interactions. The models most often employed for determining these strong interactions are broadly of two types, 1) meson exchange models (usually assumed in the existing neutron star and supernova literature), and 2) quark-gluon models (mainly encountered in the hadron, nuclear and heavy-ion literature). Here we will discuss the assumptions made in these models, and discuss how they are applied to the determination of hadronic forces that are relevant to the supernova problem.

1. Introduction: QCD in the Supernova Problem

The evolution of a supernova depends on the nature of a compact object known as a proto-neutron star (PNS) formed during the stellar collapse. In particular, the supernova may itself be launched by the intense flux of neutrinos emitted by the PNS. Since the PNS is composed of strongly interacting matter (primarily nucleons) at moderately high densities, its physical properties such as mass, composition profile, and interactions with neutrinos depend on the nature of the strong interaction between nucleons. In the core of the PNS and the subsequent neutron star remnant (NSR), the high nucleon density and energy scales suggest that the formation of other hadronic species, notably strange baryons, may be an important effect. For this reason an understanding of the interactions between strange baryons may prove to be important in determining the properties of the PNS and NSR.

Five simple questions about the nature of the interactions of strongly interacting particles (especially baryons) which can supply the information needed to establish the physics of the PNS and NSR are as follows:

(1) What hadrons, and more generally what phases of strongly interacting matter, should be included in PNS and NSR models?
(2) How do the relevant hadrons interact?
(3) What is the baryon pair interaction $V_{BB'}(r)$?
(4) What is the EOS for strongly interacting matter?
(5) How do the PSR and NSR interact with neutrinos?

QCD in principal tells us everything about the nature of strongly interacting particles, and is implicitly able to answer the first four questions. Unfortunately, the dynamical equations of QCD are nonlinear, and the quark-gluon and gluon-gluon couplings are very strong; for this reason much of the physics of hadrons is nonperturbative, and must be studied using models or specialized techniques such as lattice QCD (LQCD).

In the following discussion we will first review the basic physics of QCD, following which we will discuss the first three basic questions listed above in more detail. The problem of determining the EOS for strongly interacting natter from pair interactions (and more generally) is discussed by J.Stone,[1] and the interactions of the PSR and NSR with neutrinos is discussed by S.Reddy[2] (both in these proceedings).

2. QCD basics

QCD, quantum chromodynamics, is now accepted as the correct theory of the strong interaction. It is summarized by the QCD lagrangian,

$$L_{QCD} = \sum_q \left\{ \bar{\psi}_q^i \left((\gamma_\mu p_\mu - m_q)\delta_{ij} - g\frac{\lambda_{ij}^a}{2}\gamma_\mu A_\mu^a \right) \psi_q^j \right\} - \frac{1}{4} G_{\mu\nu}^a G_{\mu\nu}^a \ . \quad (1)$$

This lagrangian specifies the basic interactions between the six types "flavors" of quarks ($q = u, d, s, c, b, t$) and the gluons (g). These include a

q-q-g interaction (a "vertex" in Feynman diagram terminology), which is the trilinear term $-g\bar{\psi}_q(\lambda^a/2)\gamma_\mu A^a_\mu\psi_q$, analogous to the electron-photon vertex of QED. This term gives a q-q-g vertex because there are three quantum fields present, ψ_q, $\bar{\psi}_q$ and A^a_μ, each of which creates or annihilates one quark, one antiquark, or one gluon. In addition there are g-g-g and g-g-g-g three- and four-gluon vertices implicit in the pure gluon term $-G^a_{\mu\nu}G^a_{\mu\nu}/4$, since these field strength tensors are actually linear and quadratic in the gluon field A^a_μ (and hence in creation and annihilation operators), $G^a_{\mu\nu} = \partial_\nu A^a_\mu - \partial_\mu A^a_\nu + gf^{abc}A^b_\mu A^c_\nu$. The f^{abc} are group theoretic coefficients, which are specified by the internal QCD "color" symmetry group SU(3).

The spin-1/2 quarks are described by Dirac fields ψ^i_q (for each quark flavor q), and the massless, transverse spin-1 gluons are described by vector fields A^a_μ. These are basically identical in form to the Dirac electron field ψ and photon field A_μ of QED, except for the additional "color" degree of freedom. Color is an internal state label for quark and gluon states and fields, for example ψ^i_q for quarks ($i = 1\ldots3$) and A^a_μ for gluons ($a = 1\ldots8$), and arises from the requirement that QCD possesses a non-Abelian (more than 1 component) gauge invariance. Gauge invariance is actually required for renormalizability of a quantum field theory with vector fields, so it is not surprising that QCD is a gauge theory.

The original experimental evidence that the strong interactions were described by a *non-Abelian* gauge theory was the observation that the interactions grew weaker with increasing energy and momentum scales. In 1973, Gross and Wilczek [3] and Politzer [4] noted that this property of "asymptotic freedom" was satisfied only by non-Abelian gauge theories, of which QCD is a special case. The specific numbers of colors in QCD itself, 3 for quarks and $8 = 3^2 - 1$ for gluons, follow from the fact that the gauge symmetry group of QCD happens to be SU(3); the reason that nature has chosen this particular symmetry group for the strong interaction is not known.

3. Five Questions for Supernova Simulations

3.1. *What hadrons exist; more generally, what phases of strongly interacting matter exist in nature?*

This is a central issue in QCD in general, and although some basic rules regarding physically allowed hadron states are well known, the general question of what hadrons exist is still an open one.

As QCD is a theory of quarks and gluons, one would naively expect the quarks and gluons to be observed as free particles, just as are their QED analogues (electrons and photons). In a perturbative description of QCD this is what one assumes; exchange of a single gluon (one gluon exchange, "OGE") between two quarks, two antiquarks or a quark-antiquark pair gives a $1/r$ "color Coulomb" potential, with a strength proportional to the q-q-g coupling constant squared, $\alpha_s = g^2/4\pi$. Numerically, $\alpha_s \approx 0.6$ for light hadrons, although the actual value "runs" with momentum transfer, which is sometimes treated in models as an r-dependent α_s). In QCD, this α_s/r potential is multiplied by a numerical matrix that depends on the color state of the interacting qq or $q\bar{q}$ pair; the most attractive case is the color-neutral "color-singlet" quark-antiquark state,

$$|(q\bar{q})_1\rangle = \sum_{i=1}^{3} \frac{1}{\sqrt{3}} |(q_i \bar{q}_i)\rangle \tag{2}$$

for which $V_{q\bar{q}}(r) = -(4/3) \cdot \alpha_s/r$.

Although one can indeed see evidence of this attractive color Coulomb force in the spectroscopy of heavier $q\bar{q}$ mesons such as charmonium $c\bar{c}$ and bottonium $b\bar{b}$, one surprising result in QCD is that a much stronger *non-perturbative* effect, known as confinement, is dominant in determining the physical spectrum of states. (Nonperturbative here refers to physics that results from the interactions of a divergent number of gluons, which cannot be treated using the perturbative Feynman diagram expansion.)

Theoretical studies of confinement require nonperturbative methods such as lattice QCD (LQCD), which can be used to evaluate certain ground state matrix elements using the full gluonic lagrangian (the $-G^a_{\mu\nu} G^a_{\mu\nu}/4$ of Eq.1). LQCD simulations of the response of the gluon field to static quark and antiquark sources show that a gluonic "flux-tube" forms between the sources, as is shown by Bali *et al.*[5] in our Fig.1. (This is actually the action density for the simpler color gauge group SU(2); the physical SU(3) case gives qualitatively similar results.)

A determination of the corresponding potential energy $V_{q\bar{q}}(r)$ between static color sources as a function of their separation is shown in Fig.2. (As in Fig.1 this is for the unphysical SU(2) color gauge group, but very similar results are found in the physical case of SU(3).) This approximately linear potential makes it impossible to separate a quark from its partner antiquark in a $q\bar{q}$ meson, unless the flux tube is allowed to break through the formation of a new, intermediate $q\bar{q}$ pair. (This is actually a strong decay process, which leads to two separate $q\bar{q}$ mesons.) A similar confining interaction is present between the constituents of a baryon, which in the

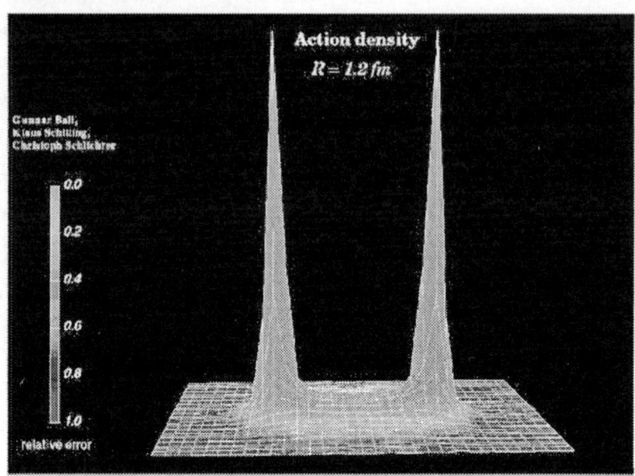

Figure 1.   A lattice simulation of QCD with static quark and antiquark sources, showing the formation of a gluonic flux tube.

naive quark model is a color-singlet bound state of three quarks. (In color SU(N), the simplest "quark model" color singlet states are $q\bar{q}$ mesons and q^N baryons.)

The lowest-lying color-singlet mesons and baryons, which are of greatest interest in the context of the QCD EOS for the supernova and neutron star problems, are given in Tables 1,2. These results are abstracted from the 2004 Review of Particle Properties[6]. The mesons are primarily of interest as contributors to the nuclear force, although there have been suggestions that a significant kaon population might exist in the interior of a neutron star[7]. The baryons are mainly of interest as possibly important hadronic species in the interior of neutron stars, in conditions of high density that result in a nucleon Fermi energy above the thresholds for production of these additional baryons.

Almost all of the known hadrons can be classified as these "minimal" $q\bar{q}$ meson and qqq baryon (and $\bar{q}\bar{q}\bar{q}$ antibaryon) color singlet bound states. More complicated color-singlet combination of quarks and gluons *a priori* might exist as physical hadrons, and if such resonances do exist this will affect the predicted QCD EOS. Considerable effort has been expended in particular in experimental searches for multiquark states, such as $q^2\bar{q}^2$ "baryonia" and recently on $q^4\bar{q}$ "pentaquarks". (The controversial $n^4\bar{s}$ pentaquark candidate $\theta(1540)$[8,9] has been the subject of intense interest in recent years.) Despite predictions of many such states in the context of specific models, there is little strong experimental evidence for these multiquarks; it appears that binding into the minimal color singlets $q\bar{q}$ and q^3 is so strong that most light multiquark systems spontaneously dissociate into these conventional quark model states. Whether this is always the case in light quark systems remains an open queston for research. There *is* general agreement that color-singlet gluonic hadrons, known as "glueballs" (for example g^2) and "hybrids" (for example $q\bar{q}g$ and q^3g) exist in the spectrum of physical hadron states. Numerical studies of the spectrum of these states however suggest that they first appear at rather large masses, ca. 1.6 GeV for the lightest glueball[10] and ca. 2.0 GeV for the lightest hybrid meson.[11] There are several experimental candidates for these states, which are rather controversial at present. These large mass scales in any case suggest that multiquarks and gluonic hadrons are of much less importance for the QCD EOS than the much lighter conventional meson and baryon resonances.

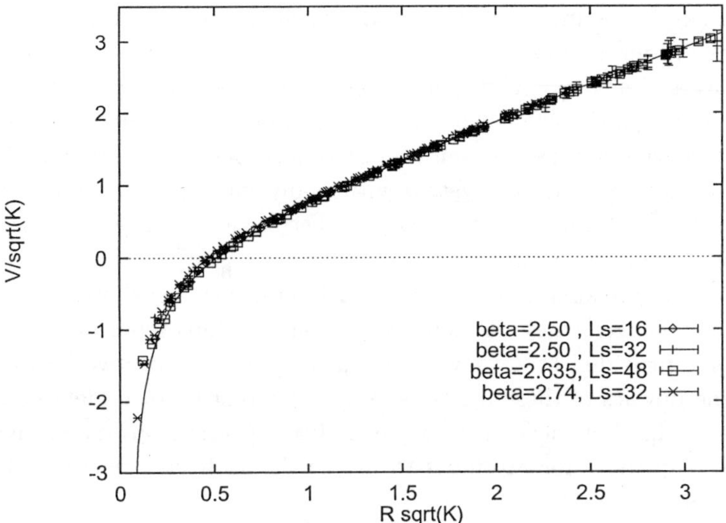

Figure 2.   The potential energy between a static quark-antiquark pair in lattice QCD. Note the approximately linear behavior at large separation (confinement). The string tension K (usually called b in the literature) provides length and energy scales for the figure. The actual string tension between a quark and antiquark in a $q\bar{q}$ meson is $b \approx$ 0.9 GeV/fm, which in English units is a very large *16 tons*.

Determination of the spectrum (masses and quantum numbers) of hadrons is of course a crucial step in determining the EOS, since the evaluation of thermodynamic properties requires thermal averages over all states of the system. For example, the partition function is given by

$$Z(\beta) = \sum_i d_i \, e^{-\beta E_i} \tag{3}$$

where the sum is over all energy eigenstates. For an approximate evaluation of the EOS, for example in the proto-neutron star, it is presumably a good lowest-order approximation to treat the system as a simple low-temperature Fermi gas of nucleons.[1] Corrections to this first approximation arise from interactions between nucleons and from the presence of other hadronic species; these may include strange and excited nonstrange baryons, strange and nonstrange mesons, and possibly other phases of strongly interacting matter, such as a quark-gluon plasma (QGP).

Although the spectrum of low-lying "quark model" mesons ($q\bar{q}$) and baryons (qqq) with $q = u, d, s$ is reasonably well established to approximately 1.5 GeV, it is remarkable that the answer to the simple question of what other types of hadrons exist at low mass scales is still open. This is a crucial issue for estimates of the EOS, which involves a sum over all excited states. A notable example is the topic of multiquark states, such as the controversial $n^4\bar{s}$ ($n = u, d$) pentaquark mentioned previously; if one such narrow multiquark resonance exists, there may be a large family of such states, and even macroscopic multiquark objects such as "quark stars". Clearly, the EOS of a hadronic system with many multiquark resonances will be very different from an essentially pure Fermi gas of nucleons with a dominant repulsive interaction.

Although there are simple reasons why such multiquark systems should not exist as resonances (they can simply "fall apart" into separate quark model hadrons, provided that they are at energies above the two-hadron threshold), the physics of these multiquark systems is still not understood, and future experimental developments in hadron spectroscopy may have important consequences for attempts to model the hadronic matter EOS at high densities.

3.2. *How do hadrons interact?*

3.2.1. *Defining the problem*

In view of the present uncertainties in the spectrum of hadrons at higher masses, it appears appropriate to initially take a rather conservative approach, and assume that the hadronic matter EOS required to describe the PNS and NSR is a gas of the known, well established mesons and baryons. At these energy and density scales the dominant hadron species should be nucleons, although significant populations of light mesons, strange baryons, and excited nonstrange baryons may also are expected. (Nucleons because the matter collapsing into the PNS is largely nucleons; light mesons and excited nonstrange baryons simply because the meson-baryon coupling constants such as $g_{NN\pi}$ and $g_{N\Delta\pi}$ are quite large, so that mesons and excited nonstrange baryons will inevitably be created in virtual hadronic transitions. Similarly, strange mesons and baryons are formed virtually through strong transitions that create $s\bar{s}$ pairs from the vacuum, since the corresponding couplings such as $g_{N\Lambda K}$ and $g_{N\Sigma^* K}$ are comparable in strength to the nonstrange couplings. In addition, at the higher densities of the NSR core, Pauli blocking implies a relatively high Fermi energy for the nucleons, so that it may be energetically favorable to have a significant core population of strange and excited nonstrange baryons.

The strong interaction between nucleons leads to corrections to the free Fermi gas equation of state, and incorporation of these strong forces is presumably the most important correction to the low density free fermion EOS. (As a related and rather trivial observation, the existence of the condensed NSR itself is a result of the weak intermediate-range attraction between nucleons.) Once again, this aspect of the strong interaction, as well as the forces between any pair of interacting hadrons, is in principle determined by the equations of QCD. In practice, however, there is at present no consensus between the nuclear and hadronic communities as to what mechanism within QCD is responsible for the dominant forces between hadrons. Two apparently quite distinct mechanisms for the strong forces have been discussed in the literature and elaborated in models, which are 1) meson exchange, and 2) quark-gluon forces. The differences between these models, and how one uses them to predict hadronic interactions, is the principal topic of the remainder of this contribution.

3.2.2. *Meson exchange models*

Historically, the strong force that binds the nucleons into a nucleus was assumed first assumed (by Yukawa[12]) to be due to the exchange of a massive meson in t-channel. (This was by analogy with the electromagnetic interaction being due to t-channel photon exchange.) The static nucleon limit of meson exchange gives an internucleon "Yukawa potential" which is exponentially damped in r-space by the mass m_m of the meson,

$$V_{NN}(r) \propto \frac{g^2_{NNm}}{4\pi} \frac{e^{-m_m cr/\hbar}}{r}. \tag{4}$$

Note that the characteristic range of the Yukawa force is

$$R = \frac{\hbar}{m_m c} \, . \tag{5}$$

Since the strong force presumably has a range specified by the nuclear length scale of ~ 1 fm, this suggested a mass for Yukawa's hypothetical strongly interacting meson of

$$m_m \approx \frac{\hbar}{c \cdot 1 \, \text{fm}} \approx 200 \text{ MeV}. \tag{6}$$

The discovery of the pion in cosmic ray emulsions in 1947 by Powell and collaborators,[13] with a mass of $\approx$ 135-140 MeV (see Table 2), confirmed Yukawa's expectation that strongly interacting particles with roughly this mass existed. The subsequently determined very strong low-energy pion-nucleon coupling constant, $g_{NN\pi} \approx 13$, showed that exchange of this meson could be responsible for the longest-ranged part of the internucleon force. One pion exchange is now known to give a good description of the higher-L partial waves in NN scattering; these are dominated by exchange of the lightest (longest-ranged) strongly interacting system, which is a single pion.

The success of this pion exchange picture motivated the suggestion that other aspects of the nucleon-nucleon interaction, observed at higher energies and larger momentum transfers (and hence at smaller internucleon separations) might also be due to t-channel meson exchange. The strongest of these forces is the short-range repulsive "core" interaction, which is partly responsible for the observed "saturation" of nuclear forces. (In large nuclei, binding energies per nucleon and nucleon densities approach roughly constant limits, as would be expected for a combined intermediate-range attraction and hard-core repulsion.) Since the repulsive core and intermediate-range attractions are present with comparable strengths in both NN S-wave isospin channels, the obvious assumption in the meson exchange picture is that they are due to the exchange of I=0 mesons.

Since the pion itself has I=1, one pion exchange is not a good candidate for these NN forces. (Two pion exchange does have an I=0 component, and has been suggested as a possible source of the intermediate-range attraction. Another problem with single pion exchange is that the pion is a pseudoscalar meson (it has spin-parity $J^P = 0^-$), which gives a spin-spin force rather than a spin-independent potential as the dominant interaction. Theoretical studies of different meson exchanges in the 1950s showed that the exchange of an I=0 vector meson ($J^P = 1^-$) of sufficient mass would give the required short-range repulsion in both NN isospin channels, and the subsequent discovery of the ω meson (see Table 2), with a mass near 780 MeV (and hence a Yukawa range of $R = \hbar/m_\omega c \approx 0.25$ fm, comparable to the estimated length scale of the core repulsion) appeared to confirm vector meson exchange as the origin of the short-range repulsive core.

It was also noted that the intermediate range attraction could be explained by the exchange of a somewhat lighter I=0 scalar meson, known as the σ. Unfortunately, evidence for this light scalar meson is controversial, with believers and nonbelievers in the hadron community divided into opposing camps. One problem with resolving the question of the existence of a light I=0 scalar σ meson is that it should appear in I=0 $\pi\pi$ S-wave elastic scattering; the data shows that if a light meson (significantly below 1 GeV) *is* produced in this channel, it much be extremely broad, and hence very hard to distinguish from nonresonant interactions.

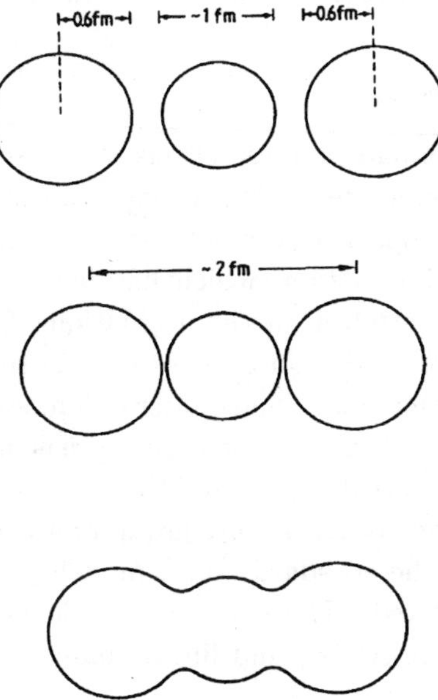

Figure 3. A simple geometrical argument (due to Maltman and Isgur) showing that meson exchange models of nuclear forces are difficult to justify for NN distances smaller than about 2 fm (corresponding to exchanged meson masses greater than about 100 MeV).

A fundamental problem with the meson exchange description of the NN interaction at intermediate and short ranges was noted by Maltman and

Isgur in 1983, in two quark model papers that derived the NN interaction at the level of quarks and gluons.[14,15] They noted that the length scales of nucleons and mesons determined experimentally (for example from electromagnetic form factors) correspond to diameters of about 1 fm, so the picture of a meson with free space properties being exchanged between two baryons is unrealistic at internucleon separations much below 2 fm (see Fig.3). Although pion exchange appears plausible, as this force has a range of $R = \hbar/m_\pi c \approx 1.4$ fm, exchanges of higher-mass mesons such as the ω and the hypothetical σ appear absurd geometrically, since these would have ranges of only $R \approx 0.25$ fm. As these very small distances correspond to overlapping quark wavefunctions rather than physically distinct hadrons, a physically realistic description of hadronic interactions in this regime presumably requires modeling at the quark-gluon level, with the scattering amplitudes described as matrix elements of the quark-quark interaction between explicit quark model hadron wavefunctions.

3.2.3. *Quark-gluon models*

To determine hadron-hadron interactions at distances significantly smaller than 2 fm, we must evaluate matrix elements of the quark-quark interaction between overlapping quark wavefunctions. In principle one should include the full interquark interaction, which has been established in analytical and numerical QCD studies, in the evaluation of hadron-hadron scattering amplitudes. One might anticipate that the scattering amplitudes are nonperturbative in the quark-quark interaction, since this is a strong coupling problem. In response to these *a priori* reasonable considerations, the earliest hadron-hadron scattering amplitude calculations used nonperturbative methods, specifically resonating group or variational approaches, and treated as much of the standard quark model quark-quark interaction as could be accommodated. The standard quark model interaction consists of one gluon exchange (OGE) and linear confinement terms, and is of the form

$$H_{q_i q_j} = (V(r) + H_{so} + H_{hyp}) \, (\lambda^a/2)_i \cdot (\lambda^a/2)_j \tag{7}$$

where $V(r)$ consists of Color Coulomb and linear confinement potentials, and H_{so} and H_{hyp} are the spin-orbit and OGE hyperfine interactions respectively. The spin-orbit term receives contributions from both OGE and the confining interaction. The OGE hyperfine interactions consist of a tensor term and a spin-spin contact hyperfine term.

Many studies of hadron interactions using this constituent approach have the NN system, due to the importance of the NN interaction specialized to the NN sector, due to its importance to nuclear physics. The first such study was due to Liberman in 1977;[16], and subsequent work (to 1993) on deriving the NN interaction from the quark model is reviewed by Barnes *et al.*[17]

Most of these studies assumed a nonrelativistic quark model formalism, and used nonperturbative techniques to iterate the quark-quark interaction. The resonating group method has been the most frequently used approach, since it gives scattering amplitudes directly. Maltman and Isgur[14,15] in contrast used a variational method with a rather general six-quark wavefunction to extract effective NN potentials. These two approaches led to rather similar conclusions.

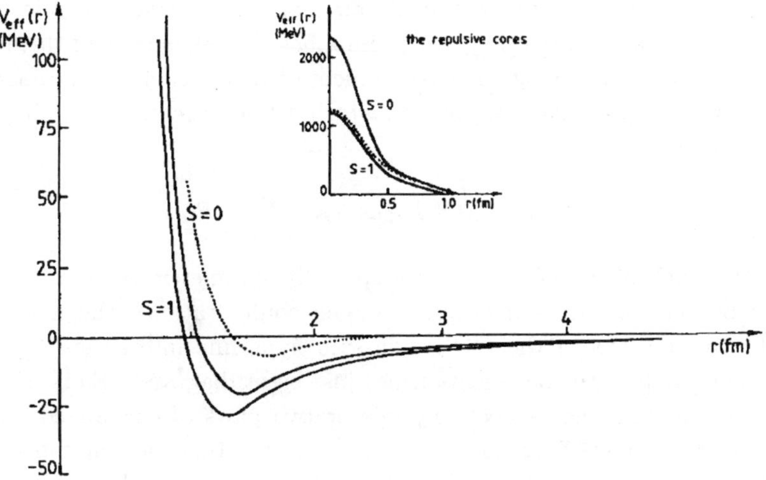

Figure 4.   Quark model results for the potentials between two nucleons in S-wave found by Maltman and Isgur. The two different (I,S) states (0,1) ("S=0") and (1,0) ("S=1") are shown as solid lines. The dashed lines are an initial, simplified calculation.

These calculations were successful in generating a short-range repulsive core in the NN interaction. Rather remarkably, it was found that this core was dominated by a single term in the quark-quark interaction, the OGE

358

spin-spin contact "hyperfine" term,

$$H_{ss} = \sum_{a,i<j} \left[-\frac{8\pi\alpha_s}{3m_i m_j}\, \delta(\vec{r}_{ij}) \right] \left[\vec{S}_i \cdot \vec{S}_j \right] \left[(\lambda^a/2)_i \cdot (\lambda^a/2)_j \right] . \tag{8}$$

In contrast, the very important intermediate-range attraction, which is responsible for the existence of nuclei, did not have such a simple origin in quark-gluon forces. Maltman and Isgur[14,15] did find a realistic intermediate-range attraction in their NN variational calculation, but this was due to a rather complicated "color van der Waals" spatial distortion of the overlapping quark wavefunctions. Finally, the large spin-orbit interaction observed in NN scattering in higher partial waves, which is crucial for understanding nuclear structure, has not yet been understood in terms of quark-gluon forces. This unsolved spin-orbit problem has been referred to by Isgur as the "holy grail" of quark-gluon calculations of the NN force.

The observation that the short-range repulsive core is reasonably well described by a Born-order matrix element of the relatively simple OGE spin-spin hyperfine interaction vastly simplifies the calculation of this force. In our work we calculate the matrix element of this scattering Hamiltonian between two-baryon scattering states, which defines a reduced Hamiltonian matrix element h_{fi};

$$_f\langle BB'|H_{scat}|BB'\rangle_i \equiv h_{fi}\, \delta(\vec{P}_f - \vec{P}_i) . \tag{9}$$

Since the nucleon wavefunctions are spatially symmetric (assuming dominance by the lowest, S-wave quark model configuration), the many different contributions of the OGE spin-spin hyperfine interaction between quark pairs reduce to coefficients times just eight diagrams. Pairs of these eight diagrams having respectively one or two pairs of exchanged quarks are related by (CD) final baryon interchange, so there are actually only four independent diagrams in the NN scattering process; these are shown below.

diagram $D_1 =$

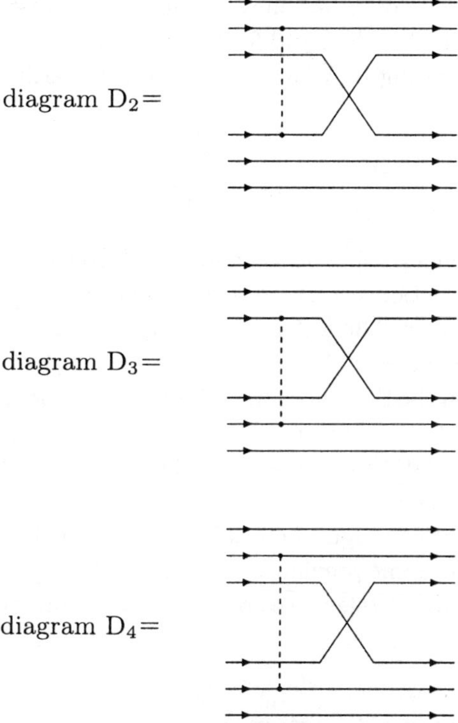

In a given baryon-baryon channel the spatial overlap integrals associated with each diagram are multiplied by channel-dependent numerical coefficients ($w_1 \ldots w_4$), which are group theoretic numbers that are determined by spin, flavor and color matrix elements of the $\lambda \cdot \lambda \, \vec{S}_i \cdot \vec{S}_j$ OGE hyperfine operator. The resulting reduced Hamiltonian matrix element h_{fi} in a specified channel is then given by the channel-dependent weight factors times diagram-dependent spatial overlap integrals,

$$h_{fi} = \sum_{n=1}^{8} w_n \, I_n(D_n) \ . \tag{10}$$

Evaluation of these quantities is discussed in detail elsewhere.[17]

This h_{fi} matrix element may be expressed as an equivalent low energy baryon-baryon potential near threshold, which only involves the first four diagrams and their weight factors (the contribution of the final four quark diagrams is generated automatically by the potential scattering formalism for identical particles, since these are related to the first four diagrams by crossing). These equivalent potentials may be determined as Fourier transforms of the low energy limit of h_{fi}, and for SHO ground-state baryon wavefunctions (assuming scattering only by the spin-spin hyperfine term) are given by

$$V_{BB}(r) = \frac{8\alpha_s}{3\sqrt{\pi}m_q^2} \sum_{n=1}^{4} \frac{w_n \eta_n}{(A_n + B_n)^{3/2}} \exp\left\{ -\frac{(r/\hbar c)^2}{(A_n + B_n)} \right\}. \qquad (11)$$

The diagram-dependent parameters A_n and B_n and overall constants η_n are determined by spatial wavefunction overlap integrals, and are explicitly[17] $A_1 = B_1 = 2/3\alpha^2$, $A_2 = A_3 = 20/33\alpha^2$, $B_2 = B_3 = 4/11\alpha^2$, $A_4 = 1/3\alpha^2$ and $B_4 = 0$; $\eta_1 = 1$, $\eta_2 = \eta_3 = (12/11)^{3/2}$ and $\eta_4 = (3/4)^{3/2}$. The parameters A_n and B_n contain the SHO baryon wavefunction scale α, which we take to be $\alpha = 0.4$ GeV. The strength of the hyperfine interaction we assume is $\alpha_3/m_q^2 = 0.6/0.33^2$ GeV^{-2}; these are standard parameters in the nonrelativistic constituent quark model.

The group-theoretic spin-color weight factors $(w_1 \ldots w_4)$ for S-wave NN scattering are given in Table 3. Some previously unpublished coefficients for strange baryon pairs, which are of relevance to the neutron star problem, are also given.

3.3. *What is the baryon pair interaction $V_{BB'}(r)$?*

Here we will evaluate examples of these potentials assuming the quark-gluon model of the previous section. Although the meson exchange models do give a good description of the NN interaction, they assume exchange of many different mesons with many free parameters, and as such are effectively parametrizations of the experimentally well determined scattering amplitudes rather than predictions. In contrast there are only two parameters in the quark-gluon calculation, the baryon scale α and the OGE hyperfine interaction strength α_s/m_q^2, and these are already determined by fits to the spectrum of individual hadrons.

First we consider the most important case, NN interactions. Since individual nucleons have isospin and spin I=1/2 and S=1/2, and $1/2 \otimes 1/2 = 1 \oplus 0$ the two-nucleon system can in general have four sets of quantum numbers, (I,S)= $(1,1), (1,0), (0,1)$ and $(0,0)$. The dominant interactions at these low energies are in the S-waves (L=0, no NN orbital angular momentum). Since the Pauli exclusion principle only allows antisymmetric states of identical fermions, the only allowed NN S-wave channels are (I,S)= $(1,0)$ and $(0,1)$. We evaluate the NN potentials in these two S-wave channels using Eq.11, with the weight factors of Table 3. The results are shown in Fig.5.

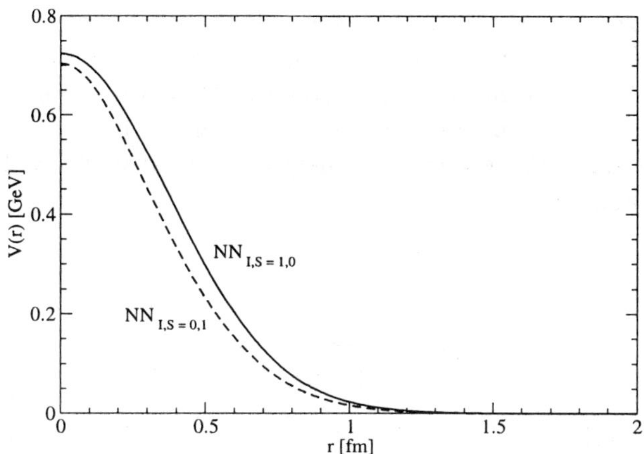

Figure 5.   Quark model prediction of the core potential between two nucleons in S-wave. The two different (I,S) states are shown; these are accidentally quite similar functions.

It is clear by inspection that the basic features of the NN repulsive core interaction, a sub-fermi range, a maximum in the 1 GeV range and similar potentials for the two (I,S) channels, are indeed reproduced by this calculation. A more useful exercise is to calculate the NN S-wave phase shifts that result from these potentials (see Fig.2 of Barnes *et al*;[17]); this shows that at higher energies (away from the effects of the intermediate range attraction) these potentials are indeed of approximately the right magnitude and range required by experimental NN S-wave scattering. This agreement with experiment and the similarity of the two NN potentials is quite nontrivial, since other BB' channels have very different weight factors. There is even experimental evidence that the prediction of a slightly stronger (I,S)= $(1, 0)$ potential than $= (0, 1)$ is correct.

This quark-gluon description of baryon-baryon interactions is especially useful for the supernova problem. An understanding of the internal structure of the SNR requires predictions of the interactions between strange baryons, for which there is little data. Since the q-g model has only two parameters, which require little modification for application to strange baryons (specifically one should replace m_q by m_s), we can generate BB' core potentials for all the pairs of strange baryons likely to be of importance in the SNR. Here we will give two examples of strange baryon potentials; a systematic study of these is a principal future goal of this research project.

Since the lightest strange baryon is the Λ, Λ-pair interactions will certainly be of interest. There is only one S-wave $\Lambda\Lambda$ channel, (I,S)= $(0,0)$. For a first "zeroth-order" calculation we use the same parameters as in the NN problem, and only change the group-theoretic diagram weight factors of Table 3. The resulting $V_{\Lambda\Lambda}(r_{\Lambda\Lambda})$ potential is shown in Fig.6. Evidently it is also a simple repulsive core, rather similar to the two NN cores but somewhat weaker. (This was also evident in the relative NN and $\Lambda\Lambda$ weight factors in Table 3.)

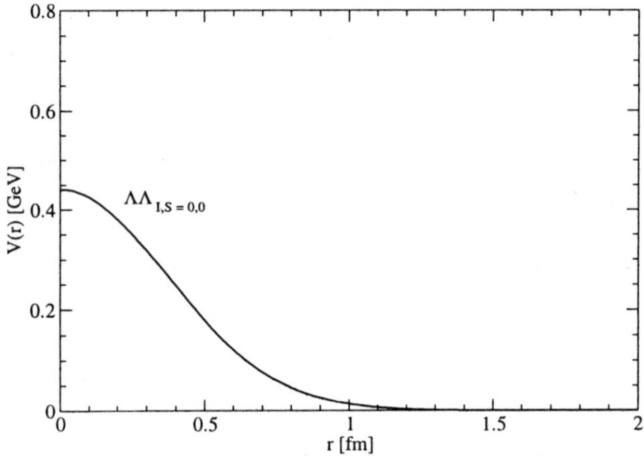

Figure 6. Quark model prediction of the core potential between two Λ baryons in S-wave.

The next lightest strange baryon is the Σ, and since this baryon has I=1 there are three accessible S-wave $\Sigma\Sigma$ channels, which have (I,S) = (2,0), (1,1) and (0,0). The diagram weights for these channels are also given in Table 3. The first two $\Sigma\Sigma$ channels have repulsive cores, as we found for NN and $\Lambda\Lambda$. The third $\Sigma\Sigma$ channel is unusual, however, as the diagram weights are all negative; this implies an attractive core potential. (These $\Sigma\Sigma$ potentials are shown in Fig.7.) This unusual short-range *attraction* may significantly enhance the population of Σ baryons in the core of the SNR over the naive expectation for noninteracting strange baryons.

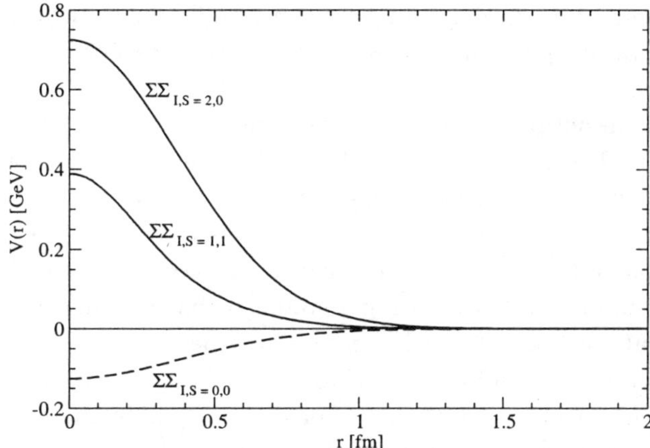

Figure 7.   Quark model prediction for the core potentials between two Σ baryons in the three allowed S-wave channels. Note that the (I,S)=(0,0) $\Sigma\Sigma$ channel has an attractive core.

3.4. *Future*

Future research on QCD in supernovas might most usefully address several topics in the physics of the PNS and NSR and their interactions with neutrinos. These topics range from straightforward calculations of short-range baryon interactions to rather profound questions, which may require considerable effort as well as new theoretical developments.

In the near term, the work discussed above on the short-range core forces between baryon pairs should be extended to the full range of relevant baryons, including at a minimum all the nonstrange and strange baryons in the ground-state octet and decimet. This work is straightforward and uses well established quark model techniques. These core potentials should then be incorporated in EOS calculations for nuclear matter including all these baryon species, and the effect of using this EOS on the structure and neutrino interactions of the NSR should be evaluated.

The origin of the intermediate-range attraction is a very important but rather more difficult question. This force is of course responsible for the existence of condensed nuclear matter, and has been attributed to several mechanisms in QCD, but these have never been convincingly tested. The existence of bulk strange nuclear matter will depend on the the strength and range of this intermediate-range force between strange baryons, which cannot be determined theoretically without an understanding of the QCD interactions responsible for this force. Of course a convincing determination of origin of this attraction is a long-standing and difficult problem, but as it is fundamental to much of nuclear physics it is correspondingly important to address.

The determination of the EOS from these hadron pair interactions at present uses a Fermi gas model, and includes the interactions as perturbative corrections. Since these forces are quite large and short-ranged, the applicability of a perturbative approach is open to question; thus it will be very important to evaluate higher-order corrections. Ideally, in future the EOS should be evaluated using a nonperturbative numerical technique, provided that such an approach can be developed.

The topic of the importance of forms of hadronic matter other than baryons should also be addressed. For example, light mesons have large couplings to the baryons, and should be significant components in the composition of both the PNS and the NSR. Incorporation of mesons should also significantly alter the nuclear matter EOS.

Finally, the interactions of non-nucleon hadronic species with neutrinos should be incorporated in the simulations of the neutrino flux. The details of the interactions of all the relevant hadrons with neutrinos have not yet been addressed, especially given the recently discovered neutrino flavor mixing effects, and the resulting changes in the flux of neutrinos from the PNS and NSR may have important effects on the evolution of a supernova.

Acknowledgments

This research was sponsored by the Division of Nuclear Physics, U.S.Department of Energy, under contract DE-AC05-00OR22725 with UT-Battelle, LLC.

References

1. J. Stone, these proceedings.
2. S. Reddy, these proceedings.
3. D. J. Gross and F. Wilczek, Phys. Rev. Lett. **30**, 1343 (1973).
4. H. D. Politzer, Phys. Rev. Lett. **30**, 1346 (1973).
5. G. S. Bali, K. Schilling and C. Schlichter, Phys. Rev. D **51**, 5165 (1995) [arXiv:hep-lat/9409005].
6. S. Eidelman *et al.* [Particle Data Group Collaboration], Phys. Lett. B **592**, 1 (2004).
7. D. B. Kaplan and A. E. Nelson, Phys. Lett. B **175**, 57 (1986).
8. K. Hicks, arXiv:hep-ex/0412048.
9. A. R. Dzierba, C. A. Meyer and A. P. Szczepaniak, arXiv:hep-ex/0412077.

10. C. J. Morningstar and M. J. Peardon, Phys. Rev. D **56**, 4043 (1997) [arXiv:hep-lat/9704011].
11. T. Barnes, F. E. Close and E. S. Swanson, Phys. Rev. D **52**, 5242 (1995) [arXiv:hep-ph/9501405].
12. H.Yukawa, Proc. Phys. Soc. Japan **17**, 48 (1935).
13. C.M.G.Lattes, G.P.S.Occhialini, and C.F.Powell, Nature **160**, 453, 486 (1947).
14. K. Maltman and N. Isgur, Phys. Rev. D **29**, 952 (1984).
15. K. Maltman and N. Isgur, Phys. Rev. Lett. **50**, 1827 (1983).
16. D. A. Liberman, Phys. Rev. D **16**, 1542 (1977).
17. T. Barnes, S. Capstick, M. D. Kovarik and E. S. Swanson, Phys. Rev. C **48**, 539 (1993) [arXiv:nucl-th/9302007].

Table 1. Baryons to 1.5 GeV. Above this mass there is a rapid increase in the density of excited baryon levels.

state	quark content (n=u,d)	mass (MeV)	isospin	spin-parity	strong width (MeV)
N	nnn	p(938.3); n(939.6)	1/2	$1/2^+$	-
Λ	uds	1115.7	0	$1/2^+$	-
Σ	nns	Σ^+(1189.4); Σ^0(1192.6); Σ^-(1197.4)	1	$1/2^+$	-
Δ(1232)	nnn	≈ 1232	3/2	$3/2^+$	≈ 120
Ξ	nss	Ξ^0(1314.8); Ξ^-(1321.3)	1/2	$3/2^+$	-
Σ(1385)	nns	Σ^+(1382.8); Σ^0(1383.7); Σ^-(1387.2)	1	$3/2^+$	≈ 36
Λ(1405)	uds	≈ 1407	0	$1/2^+$	50
N(1440)	nnn	≈ 1440	1/2	$1/2^+$	≈ 350

Table 2. Mesons to 1.02 GeV. (The neutral kaons and antikaons are actually mixed by the weak interaction.)

state	quark content (n=u,d)	mass (MeV)	isospin	spin-parity	strong width (MeV)
π	$n\bar{n}$	$\pi^\pm$(139.6); π^0(135.0)	1	0^-	-
K, $\bar{K}$	$n\bar{s}(K^+,K^0)$; $s\bar{n}(K^-,\bar{K}^0)$	$K^\pm$(493.7); K^0(497.7)	1/2	0^-	-
η	$n\bar{n}$, $s\bar{s}$	547.8	0	0^-	-
ρ	$n\bar{n}$	≈ 770	1	1^-	≈ 150
ω	$n\bar{n}$	782.6	0	1^-	8.5
K^*, $\bar{K}^*$	$n\bar{s}(K^{*+},K^{*0})$; $s\bar{n}(K^{*-},\bar{K}^{*0})$	$K^{*\pm}$(892.7); K^{*0}(896.1)	1/2	1^-	51
η'	$n\bar{n}$,$s\bar{s}$	957.8	0	0^-	-
ϕ	$s\bar{s}$	1019.5	0	1^-	4.3

Table 3. Diagram weights for S-wave baryon-baryon interactions.

channel	ω_1	ω_2	ω_3	ω_4
NN; (I,S) = (1,0)	$+\frac{31}{27}$	$+\frac{7}{27}$	$+\frac{7}{27}$	0
(0,1)	$+\frac{19}{27}$	$+\frac{7}{27}$	$+\frac{7}{27}$	$+\frac{2}{27}$
$\Lambda\Lambda$; (I,S) = (0,0)	$+\frac{2}{3}$	$+\frac{1}{6}$	$+\frac{1}{6}$	0
$\Sigma\Sigma$; (I,S) = (2,0)	$+\frac{31}{27}$	$+\frac{7}{27}$	$+\frac{7}{27}$	0
(1,1)	$+\frac{13}{81}$	$+\frac{10}{81}$	$+\frac{10}{81}$	$+\frac{8}{81}$
(0,0)	$-\frac{8}{27}$	$-\frac{1}{54}$	$-\frac{1}{54}$	0

Section 7
Nucleosynthesis and Light Curves

THE CHALLENGES OF COUPLING SUPERNOVA NUCLEOSYNTHESIS TO THE CENTRAL ENGINE

W. R. HIX, S. PARETE-KOON

Physics Division, Oak Ridge National Laboratory
Oak Ridge, TN 37831 USA and
Department of Physics & Astronomy, University of Tennessee
Knoxville, TN 37996 USA
E-mail: raph@ornl.gov

C. FRÖHLICH, F.-K. THIELEMANN

Departement für Physik & Astronomie, Universität Basel
CH-4056 Basel, Switzerland

G. MARTÍNEZ-PINEDO

ICREA and Institut d'Estudis Espacials de Catalunya, Universitat Autònoma de
Barcelona, E-08193 Bellaterra, Spain

Core collapse supernovae are the leading actor in the story of the cosmic origin of the chemical elements. The existing models, which assume spherical symmetry and parameterize the explosion, have been remarkably able to replicate the gross elemental pattern observed in core collapse supernovae. However, recent improvements in the modeling of core collapse supernovae, including detailed tracking of the neutrino distributions and better accounting for the multi-dimensional nature of the hydrodynamic flows, will have noticeable impact on the predicted composition and distribution of the ejecta. We will review recent explorations of these effects and discuss the means needed to achieve self-consistent models of the core collapse supernova mechanism together with the concomitant nucleosynthesis.

1. Supernovae and the Origins of the Elements

As "new stars" that have, throughout history, made their appearance in the sky, supernovae have long been a source of wonderment. With a visual display that can compete in brightness with its entire host galaxy, core collapse supernovae are among the most energetic astrophysical phenomena. The ejecta of a supernova delivers 10^{51} erg of kinetic energy and a rich mix of recently synthesized elements into the interstellar medium. It represents a major source of heat in the ISM as well as a potential trigger for star formation.[1] One of the central pursuits of modern astronomy is the search for our cosmic origins. Core collapse supernovae are a vital link in the chain of origins that links the Big Bang to the formation of galaxies, stars, planets, and ultimately living organisms. Among all astrophysical phenomena, core collapse supernovae are the richest producers of heavy elements, producing and/or disseminating most of the heavy elements between oxygen and iron, elements without which life as we know it could not exist. They are also thought to be the most likely site for the *r-process*, which is responsible for the production of half the elements heavier than iron.

A wide variety of observations at infrared, optical, ultraviolent and X-ray wavelengths of supernovae and young supernovae remnants, reveal the composition and distribution of the ejecta.[2–4] Observations of γ-ray lines reveal not just the elemental but isotopic production of supernovae and are therefore of particular interest, both to supernova theory in general, and to the nucleosynthesis we discuss here. Observations like the measurement of the ratio of ^{57}Ni to ^{56}Ni in SN1987A[5] or the detection of ^{44}Ti in the Vela SN remnant[6] probe the deepest layers of the supernova ejecta,[7] placing powerful constraints on conditions deep in the interior of supernovae and their progenitors. Only the neutrino and gravitational wave signals, which follow the evolution of the proto-neutron star, provide information from deeper within the explosion. Therefore the combination of realistic nucleosynthesis models and γ-ray line observations provide strong constraints on explosion models as well as shedding light on the supernova contribution to our origins.

2. Current Modeling of Core Collapse Supernova Nucleosynthesis

Complete understanding of core collapse supernovae and their nucleosynthesis begins with detailed knowledge of the life cycle of massive stars. Many of the lighter elements that supernova propel into the interstellar medium are formed during the late stages of stellar evolution and liberated from the star's deep potential well by the supernova shock wave. Even for the heavier elements that are formed in the explosion, the progenitor evolution leaves its fingerprint, primarily in the density structure of the star and the degree of neutronization. The physics of massive stellar evolution is discussed in numerous textbooks.[8–10] Woosley et al.[11] have recently provided an exhaustive review of the current state of modeling massive stellar evolution. Multi-dimensional studies of massive stellar evolution are in an embryonic stage,[12,13] thus current understanding rests on one dimensional models. For many years, the vital coupling between thermonuclear and hydrodynamic processes in these models was treated in an operator split fashion, with coupling occurring only after independent evolution of the individual processes over each timestep. Further only small networks or tabulated data were coupled,[14,15] though in some cases compositional changes were also followed using larger networks in a *post-processing* approach.[16,17] Post-processing calculations compute the isotopic evolution but neglect the feedback of the thermonuclear energy generation on the thermodynamic state. The fidelity of energy generation rates for small reaction networks is however limited, especially for later stellar burning stages.[18] As a result, in recent years the energy generation from increasingly large reaction networks have been fully coupled to the hydrodynamic evolution,[19,20] largely removing this source of uncertainty.

One of the most significant differences among current massive stellar evolution models is the treatment of convection. In spherical symmetry, convection is generally treated via mixing-length theory, but even with this common assumption, variations in the adoption of the Ledoux versus Schwarzschild stability condition and the degree of inertial overshooting significantly impact the star's evolution. To date, no such treatment of convection matches all the constraints available from observations of massive stars.[11] Until recently, computational limitations necessitated that the compositional evolution due to convective mixing be operator split from the thermonuclear evolution, in spite of the similarities of the convective timescales to the longer of the nuclear timescales during the late stages of stellar evolution. At present, efforts are underway to fully couple convective mixing and thermonuclear processes.[20,21] Mixing and convection are also influenced by rotation. While the assumption of spherical symmetry greatly limits the ability to examine the influence of rotation (and magnetic fields) on stellar evolution, recent years have also seen consideration of these effects. In the absence of magnetic fields, high angular momenta persists through late stellar evolution,[22,23] however magnetic braking seems to be very effective at transporting angular momentum out of the stellar core.[24]

The frequent failure of self-consistent models for core collapse supernovae to produce explosions has left modeling of core collapse supernova nucleosynthesis in the purview of stellar evolution modelers rather than modelers of the central engine. These simulations replace the central engine of the supernova with a parameterized kinetic energy *piston*[14,16,20] or a thermal energy *bomb*.[15,25] The energy of this blast wave, together with the placement of the *mass cut* that demarcates ejecta from matter that is assumed to fall back onto the neutron star, are tuned to recover the desired explosion energy and ejected ^{56}Ni mass. Aufderheide et al.[26] demonstrated that these two methods are largely compatible, with the largest differences coming in the inner regions of the ejecta. It is this inner region that is most affected by the details of the explosion mechanism. As we will discuss in the following sections, in the case of the neutrino reheating mechanism, these effects include interaction with the tremendous flux of neutrinos and the large scale convective behavior driven by the neutrino heating.

In the inner most regions of the ejecta, the passage of the shock heats matter to temperatures where nuclear statistical equilibrium (NSE) applies and is dominated by free nucleons and α particles.[27,28] As this matter expands outwards, it cools, allowing the light nuclei to recombine into iron, nickel and neighboring nuclei. How long this matter remains in NSE depends on whether the balanced reactions that maintain the equilibrium can keep up with the declining temperature and density. For conditions and expansion timescales typical of supernovae, NSE is only maintained while the temperature exceeds 6 GK,[29] whereupon the reactions linking ^{4}He to ^{12}C can no longer keep up. Much of the iron synthesized in core collapse supernovae results from *α-rich freezeout*,[30] which occurs if a significant fraction of the matter is still in the form of free nucleons and α particles when NSE breaks down. Lower densities favor a more α-rich freezeout. It is important to note that while NSE has broken down, there remains a large cluster of nuclei from carbon through iron and beyond that obey a quasi-equilibrium (QSE) relation.[31] As the matter cools below 4 GK, this QSE group divides into 2 groups near calcium,[32] before fragmenting completely as charged particle captures cease.

Spherically symmetric models find that above the innermost helium, iron and nickel dominated regions, the passage of the shock leaves a layer rich in the α isotopes; ^{40}Ca, ^{36}Ar, ^{32}S and ^{28}Si, the products of incomplete silicon burning. Above this is a layer of ^{16}O, in the outer portions of which, significant fractions of ^{20}Ne, ^{24}Mg and ^{12}C are found. Finally comes the helium layer and hydrogen envelope, assuming they were not driven off as part of the stellar wind. Though the passage of the shock does not grossly alter the composition of the outer layers, Woosley et al.[33] have shown that appreciable amounts of some rarer isotopes can be produced by the combination of the shock passage and the neutrino flux. This including several isotopes, like ^{19}F, ^{138}La, and ^{180}Ta, for which this *ν-process* could be the dominant production mechanism.

3. Limitations of Current Models

While the spherically symmetric bomb and piston models have been quite successful in matching the gross features of the composition and distribution of supernova ejecta, there remains a need to examine supernova nucleosynthesis without parameterizing the explosion. The tuning of explosion energy and mass cut are, at least, philosophically unsatisfying. With self-consistent models, the explosion energy and ejected mass of ^{56}Ni (and every other γ-ray emitter) will become diagnostics of the simulation's accuracy. Further, spherically symmetric models can not account for the grossly asymmetric elemental distributions seen in supernova remnants[4] and inferred from spectral line shapes[34,35] and polarization[36] in the earliest days after the explosion. Two issues which have significant impact on the nucleosynthesis, and are particularly relevant to any attempt to couple modeling of the nucleosynthesis to the central engine, are the influence of neutrinos and the multi-dimensional character of the central engine. In the following sections we will discuss recent models that seek to examine these issues and use their shortcomings to frame the discussion of the needed improvements.

3.1. *Open Issue: Convection and Asphericity*

Clearly convection and hydrodynamic instabilities greatly complicate our current spherical supernova picture, altering any compositional layering in the progenitor. Large scale convection occurs in the oxygen shell before the explosion,[13,37] breaking spherical symmetry even before the explosion occurs. Furthermore, it has been known for more than a decade that the passage of the shock through the Si/O and (C+O)/He boundaries initiates strong Rayleigh-Taylor instabilities,[25,38–41] however these instabilities alone do not seem to mix nickel to sufficiently high velocities to account for observations of SN 1987A[34] and other core collapse supernovae. The implication is that gross asymmetries must be present in the core and part of the mechanism itself. Furthermore the connection of some supernovae with γ-ray bursts[42] and a presumed jet-driven explosion introduces an additional source of asymmetry, for one subclass of supernovae (and perhaps for all[43]). Clearly the structure of the exploding star is quite different from the expanding spherical layer cake assumed in current models of supernova nucleosynthesis. However, only a handful of simulations to date directly considered the impact of this multi-dimensional behavior on the nucleosynthesis and even these have had to make significant computational compromises.

Nagataki et al.[44],[45] and Maeda et al.[46] have employed parameterized *thermal bomb* models that demonstrate that in aspherical models a larger fraction of the ejecta may experience α-rich freezeout. Unfortunately, these simulations tracked the composition only by post processing – no evolution of the composition was included within the hydrodynamic simulation. Post-processing has frequently been used in past supernova simulations, justified by the observation that nucleosynthesis contributes only $10 - 20\%$ of the explosion energy. While this is globally true, this energy release could be of local importance,[10] especially for matter at the cusp between ejection and accretion. For this innermost supernova ejecta, the nuclear energy released by the recombination of α particles into iron (and neighboring) nuclei ($1 - 2 \times 10^{18}$ erg/g) is comparable to the change in the thermal energy of the plasma due to expansion during the same time, thus there is significant feedback between the rate of this nuclear recombination and the temperature evolution that is driving the recombination. The α-richness of the matter, and thus the abundance of species like ^{44}Ti, ^{57}Fe, ^{58}Ni and ^{60}Zn,[14] therefore depends critically on this feedback, providing an excellent example of the need for a self-consistent treatment encompassing the radiation hydrodynamics and nucleosynthesis.

Multi-dimensional post processing nucleosynthesis calculations face an additional complication. Such calculations take as input the pre-computed temperature and density history of individual matter parcels, along with their initial nuclear abundances. Performing post-processing calculations based on one dimensional models is relatively straightforward, since most one dimensional simulations are Lagrangian. The needed temporal histories of temperature and density are simply those of the individual Lagrangian mass elements. However, in multi-dimensions non-smooth fluid motions result in highly tangled Lagrangian grids. As a result, Eulerian hydrodynamics, where the discretization occurs in space rather than mass, is used to perform most multi-dimensional stellar astrophysics simulations. Because Eulerian codes use spatial discretization, the thermodynamic histories that are a natural result in a Lagrangian code are unavailable. To overcome this problem, Nagataki et al. and Maeda et al. implemented a test (or tracer) particle approach within their Eulerian hydrodynamics models. These test particle methods are becoming popular in astrophysics but they are relatively untested against self-consistent calculations. The principle worry is the treatment of compositional mixing. Rarely is mixing considered in post-processing calculations, whereas Eulerian advection provides strong mixing. As a result, post-processing calculations may err not just in how the nuclear energy release alters the fluid flow but also how the fluid flow impacts the composition. Of course, the extent to which (or speed with which) compositional mixing at the macroscopic scale of the hydrodynamics simulation translates into mixing at the microscopic scale of the colliding nuclei, is also an open question.

Kifonidis et al.[47] examined the consequences of core hydrodynamic instabilities on the spatial and velocity distribution of the ejecta, finding significantly higher velocities for heavy elements than previous models.[38-40] These simulations represent a significant advance in that they employed a parameterized neutrino luminosity and spectrum[48] to drive the supernova explosion instead of a bomb or piston and followed the nuclear composition with a small network (though the thermonuclear energy release was ignored). Beyond the first .8 sec after bounce, the evolution was followed using an adaptive mesh, but hydrodynamics only, simulation for 20,000 sec. Unfortunately, reaching such long evolution times required ignoring the innermost regions of the supernova in the later simulation. This compromise prevented the examination of the impact of fallback and the neutrino-driven wind on the nucleosynthesis, factors that greatly impact the innermost ejecta. Removing this limitation will require running models of the central engine for several seconds longer than is done at present.

Kifonidis et al. also highlight the necessity of more detailed tracking of the composition. In most simulations of the central engine, the equation of state provides the nuclear composition for sufficiently high temperatures, 4 species (free protons and neutrons, α particles and a representative heavy nucleus) in NSE and the composition is largely ignored at lower temperatures. In some spherically symmetric simulations, an α-network, composed of the 13-14 α-like nuclei from ^{4}He to ^{56}Ni or ^{60}Zn, is used. While an α-network performs well for helium and carbon burning and adequately for oxygen burning, it does not include many of the reaction channels important for the production of iron and nickel.[18,49] Thus its predictions for the energetics of the α-rich freezeout are suspect. Another important limitation of an α-network is that can not include the effects that electron or neutrino capture have on the neutronization. Kifonidis et al. add a phantom neutron-rich iron isotope in place of ^{56}Ni as the endpoint of the network under neutron-rich conditions. While an improvement over a conventional α-network, this approach still does not provide the level of detail needed for comparison to Galactic chemical evolution. Not only is the nuclear composition sensitive to the degree of neutronization, but the structure of the QSE groups and the speed of the formation of heavy elements are also affected.[49,50] For these reasons, supernova simulations that follow the composition only as part of the equation of state or even with an α-network are limited in their ability to calculate realistic nucleosynthesis and these limitations can not be completely redressed by post-processing with a more complete nuclear network.

3.2. *Open Issue: Neutrino Interactions*

While the importance of neutrino interactions is manifest in the name of the ν-*process*, neutrinos potentially impact all stages of supernova nucleosynthesis. During explosive nucleosynthesis, in the inner layers of the ejecta, where iron group nuclei result from α-rich freezeout, interactions with neutrinos alter the neutronization, changing the ultimate composition. The r-process is thought by many to occur in the neutrino-driven wind that emanates from the proto-neutron star after the explosion is initiated. The neutrinos both drive the wind and interact with the nuclei in it.[51-53] Early phases of this wind have also been suggested as the source of light p-process nuclei.[54,55] Thus, neutrino-nucleus interactions are important to all phases of core collapse supernova nucleosynthesis.

Galactic chemical evolution calculations and the relative neutron-poverty of terrestrial iron and neighboring elements place strong limits on the amount of neutronized material that may be ejected into the interstellar medium by core collapse supernovae.[56] Hoffman et al.[54] placed a limit of $10^{-4} M_\odot$ on the typical amount of neutron-rich ($Y_e \lesssim 0.47$) ejecta allowed from a core collapse supernova. The simulations of Kifonidis et al., like all prior multi-dimensional simulations of the central engine that produce explosions,[48,57] predict the ejection of much larger quantities of neutron-rich iron group elements than this limit. In an effort to compensate, modelers have been forced to invoke the fallback of a considerable amount of matter onto the neutron star, occurring on a timescale longer than was simulated. One common property exhibited by recent multi-group Boltzmann simulations[58-61] is a decrease in the neutronization of the inner layers of the ejecta due to these neutrino interactions. This is a feature that current parameterized nucleosynthesis models can not replicate because they ignore the neutrino transport. As we discussed in the previous section, neither can models that employ α-networks fully investigate this effect, since their ability to track the neutronization is limited. While the decreased neutronization seen in Boltzmann transport models reduces the need to invoke fallback, it also makes any fallback scenario more complicated, since the most neutron-rich material may no longer be the innermost.

Because of the impact of the neutrinos on the nucleosynthesis, it is likely that the nucleosynthesis products from future explosion simulations (utilizing multi-group neutrino transport) will be qualitatively different, both in composition and spatial distribution, from either parameterized bomb or piston nucleosynthesis models. This was demonstrated by exploratory calculations by McLaughlin et al.[62]. The dominant processes are $\nu/\bar{\nu}$ and $e^{\pm}$ captures on shock dissociated free nucleons, though at later times (and in regions with cooler peak temperatures) the more poorly known $\nu/\bar{\nu}$ and $e^{\pm}$ captures on heavy nuclei may contribute significantly. In addition to their impact on the electron fraction, these interactions, as well as neutral current inelastic neutrino scattering off these nuclei,[63] are also important to the thermal balance, potentially affecting the α-richness of the ejecta, thereby altering the abundance of important nuclei like ^{44}Ti, ^{57}Fe, ^{58}Ni and ^{60}Zn.[14] Travaglio et al.[64] used a parameterized neutrino luminosity and spectrum[48] to drive the supernova explosion and a tracer particle approach with a large nuclear network to examine the nucleosynthesis. They found significant impact of nuclear electron capture since some ejected zones reached densities $> 10^8\,\mathrm{g\,cm^{-3}}$ as well as significant differences in the production of $45 < A < 55$ and $A > 65$ nuclei between one and two dimensional models. However, since these authors ignored neutrino captures, their simulations tell only half of the story.

Simulations by Fröhlich et al.[65] and Pruet et al.[66] have examined the effects of both electron and neutrino captures in the context of recent multi-group supernova simulations. The models of Fröhlich et al. are based on fully general relativistic, spherically symmetric simulations,[58] while Pruet et al. used tracer particles from two dimensional simulations.[61] In both cases, artificial adjustments to the simulations were needed to remedy the failure of the underlying models of central engine to produce explosions. Also in both cases, the simulations were mapped to simplified models as later times, because the neutrino tranport simulations could not be run to sufficiently late times. While both of these shortcomings need to be addressed, these simulations nonetheless reveal the significant impact that neutrino interactions have on the composition of the ejecta.

One observes three distinct phases in the evolution of the electron fraction of the matter that will become the innermost ejecta as it collapses,

378

passes through the stalled shock and is driven off by neutrino heating. During core collapse, the electron fraction in these lower density, silicon-rich, regions is little changed either by the electron capture that is deleptonizing denser regions or the relatively weak neutrino flux. However, the combination of the larger neutrino flux after core bounce and the burning of silicon to iron in the still infalling matter greatly enhances the neutrino capture rates. With most of the matter still tied up in relatively inert heavy nuclei, the greater abundance of free protons over free neutrons allows antineutrino captures to dominate, lowering Y_e. The passage of the matter through the stalled shock raises the temperature, dissociating the nuclei, but the concomitant increase in density prevents the lifting of the electron degeneracy. As a result of the high electron chemical potential, the balance of electron and positron captures strongly favors lower electron fraction. However, the combination of neutrino and antineutrino captures favors higher Y_e because of the slight dominance of neutrinos over antineutrinos as well as slightly higher abundance of neutrons compared to protons in the fully dissociated, mildly neutron-rich matter. As a result, the electron fraction undergoes only mild excursions in this phase. Eventually, continued neutrino heating (or perhaps some other mechanism) is sufficient to reenergize the shock, in the process lifting the electron degeneracy in this innermost ejecta. As a result, the rate of electron captures drops while the rate of positron captures increases causing Y_e to rise. While the dominance of neutrino captures over antineutrino captures drops as the matter becomes neutron-poor, their sum continues to favor higher Y_e. Eventually, the electron chemical potential drops below half the mass difference between the neutron and proton, allowing positron and neutrino captures to dominate electron and antineutrino captures.[67] With both neutrino emission and absorption processes favoring a higher electron fraction, Y_e rises markedly in this phase, reaching values as high as 0.55 in both simulations.

The global effect of this proton-rich ejecta is the replacement of previously documented overabundances of neutron rich iron peak nuclei (near the N=50 closed shell)[14,15] with a mix of ^{56}Ni and α-particles. Production of ^{58,62}Ni is suppressed while ^{45}Sc and ^{49}Ti are enhanced. The results are however sensitive to the details of the simulations. Pruet et al. found a significant sensitivity in the nuclear production to the expansion rate of the matter, which was a parameter in their late time extrapolation. In addition to the global effects on the neutronization and entropy of the matter, Fröhlich et al., who included neutrino and antineutrino capture rates on heavy nuclei,[68] find that these reactions have direct impact on the abun-

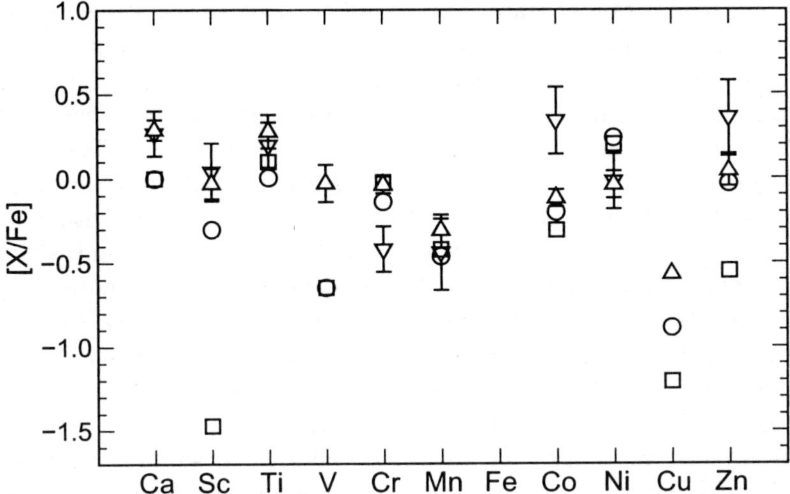

Figure 1.   Comparison of elemental abundances for Ca to Zn between models by Fröhlich et al.[65] (circles) and Thielemann et al.[15] (squares) and observation determinations for metal-poor[69] (upward pointing triangles) and extremely metal-poor[70] (downward pointing triangle) stars.

dances of species like ^{53,54}Fe, ^{55,56,57}Co, ^{59}Ni and ^{59}Cu. A mild rp-process is also seen in this proton-rich ejecta. Pruet et al., who neglected neutrino capture on heavy nuclei, find this effect limited to zinc (A=64) by the lifetimes of waiting-point nuclei that are longer than the expansion timescale. Fröhlich et al. found significant flow as high as A=78, producing ^{64,68}Zn, ^{72}Ge, ^{76}Se and ^{78}Kr. The result (Fig. 1) is the elemental abundances of scandium, cobalt, copper and zinc are significantly closer to those observed. These results clearly illustrate the need to include the full effect of the supernova neutrino flux on the nucleosynthesis if we are to accurately calculate the iron-peak nucleosynthesis from core collapse supernovae.

4. The Future

Ultimately, what will be needed to fully investigate the impact of the central engine on the nucleosynthesis are multi-dimensional supernova models with multi-group neutrino transport and self-consistent nuclear evolution throughout the ejecta. In a practical sense, this requires the marriage of two disciplines. The required elements that are currently missing from simulations of the central engine, detailed nuclear reaction networks and the appropriate equation of state, have long been part of the stellar modelers toolkit while few models of the core collapse supernova mechanism contain any means to follow the elemental or isotopic composition when it is not in nuclear statistical equilibrium. For this reason, the nucleosynthesis effects of the central engine have been discussed in only gross terms like the bulk neutronization or electron fraction. As the simulations discussed in prior sections illustrate, this is insufficient. For more refined comparisons, both with direct observations of supernovae and their remnants and with the constraints of galactic chemical evolution, detailed nucleosynthesis calculations that adequately include the effects of neutrino interactions, convection and the other important features of the mechanism are necessary.

Coupling the nucleosynthesis to the central engine requires several costly additions to the neutrino radiation-hydrodynamics code. The need to examine a larger portion of the stellar core and to do so for a longer time significantly expands the scope of the simulations. However, the primary obstacle is the addition of the computational cost of a full nuclear network to the already considerable cost of multigroup neutrino transport and multi-dimensional hydrodynamics. It is worth noting that only in recent years[21] have one dimensional parameterized models for supernovae nucleosynthesis included large networks self-consistently. To do this within two and three dimensional neutrino radiation hydrodynamics models, we will need to maximize computational efficiency. The computational cost of the neutrino transport is the dominant cost in current simulations of the central engine and sets the scale for any proposed additions. In current multi-group simulations, for each spatial zone the neutrino energy is binned into 12-40 groups. This is done for each of 3-4 neutrino species, the electron neutrino and antineutrino and either 1 or 2 species amalgamating the muon and tauon neutrinos and antineutrinos. The anisotropy is monitored by a few moments of the neutrino distribution in the case of Flux-limited diffusion or variable Eddington tensor methods or by 6 or more discrete angles with discrete ordinate Boltzmann solution methods. Thus for each zone there are a few hundred variables characterizing the neutrino distribution, a figure comparable to the number of species followed in detailed nucleosynthesis calculations. Also like the nuclear species, these neutrino quantities must be implicitly updated, for each of the $\sim 100,000$ timesteps in a typical supernova simulation. The vital difference is the nature of this implicit update and its frequency or the size of the numerical timestep. For the neutrino transport, a global solution is required and the physical processes involved, nuclear weak interactions and motion of the neutrinos, give typical timesteps of 10 microseconds. For the nuclear reactions, the timesteps are highly temperature dependent and can be orders of magnitude smaller. The feasibility of nucleosynthesis coupled to the central engine is largely a question of managing this timestep disparity.

In current multi-dimensional simulations, the implicit neutrino transport is coupled to explicit hydrodynamic evolution by operator splitting. While Jordan et al.[21] demonstrate the advantages of fully coupling compositional advection and the reaction network, operator splitting the nucleosynthesis affords several advantages. First, existing explicit hydrodynamics codes can be used. Second, the nuclear processes in each zone can be evolved according to their local timestep, restricting the smallest timesteps to a relatively few zones. A scheme that fully couples reactions and advection would use the smallest timestep for all zones. Third, local approximations can be utilized in place of the full reaction network for the shortest timescales. For temperatures less than roughly 4 GK, the typical reaction network timestep is comparable to or longer than those of the transport, offering the minimum additional computational cost – one network step per transport step. For temperatures in excess of 6 GK, the use of nuclear statistical equilibrium can greatly simplify the compositional evolution and achieve considerable cost saving, even while considering 100s of species rather than 4. Thus it is the intermediate region, 4-6 GK, with timesteps 100s of times smaller than the transport timesteps, which presents the greatest challenge to self-consistent coupling of the nucleosynthesis and the central engine.

4.1. *Tools: Nuclear Statistical Equilibrim*

The evolution of the nuclear abundances in most simulations of the central engine to date is predicated on the fact that for sufficiently high temperature and density, the strong and electromagnetic thermonuclear reaction rates are sufficiently rapid to achieve equilibrium within the timescale set by the hydrodynamics of the astrophysical setting. This permits considerable simplification of the calculation of the nuclear abundances. Since the weak reactions are not equilibrated, the resulting nuclear statistical equilibrium (NSE) requires monitoring of weak reaction activity. Even with this stricture, NSE offers several advantages, since hundreds of abundances are uniquely defined by the thermodynamic conditions and a single measure of the weak interaction history or the degree of neutronization. Computationally, this reduction in the number of independent variables greatly reduces the cost of nuclear abundance evolution. Because there are fewer variables to follow within a hydrodynamic model, the memory footprint of the nuclear abundances is also reduced, an issue of importance in modern multi-dimensional models. Finally, the equilibrium abundance calculations depend on binding energies and partition functions, quantities that are better known than many reaction rates. This is particularly true for unstable nuclei and for conditions where the mass density approaches that of the nucleus itself, resulting in exotic nuclear structures.

The expression for NSE is commonly derived using either chemical potentials or detailed balance.[8,10] For a nucleus $^A Z$, composed of Z protons and $N = (A - Z)$ neutrons, in equilibrium with these free nucleons, the chemical potential can be expressed in terms of the chemical potentials of the free nucleons

$$\mu_{Z,A} = Z\mu_p + N\mu_n \ . \tag{1}$$

Substituting the expression for the Boltzmann chemical potential (including rest mass) into Eq. 1 allows derivation of an expression for the abundance of every nuclear species in terms of the abundances of the free protons (Y_p) and neutrons (Y_n),

$$Y(^A Z) = \frac{G(^A Z)}{2^A}\left(\frac{\rho N_A}{\theta}\right)^{A-1} A^{\frac{3}{2}} \exp\left(\frac{B(^A Z)}{k_B T}\right) Y_n{}^N Y_p{}^Z, \tag{2}$$

where $G(^A Z)$ and $B(^A Z)$ are the partition function and binding energy of the nucleus $^A Z$ (including Coulomb/screening corrections[49,71]), N_A is Avagadro's number, k_B is Boltzmann's constant, ρ and T are the density and temperature of the plasma, and $\theta = (m_u k_B T/2\pi\hbar^2)^{3/2}$.

Abundances of all nuclear species can therefore be expressed as functions of two quantities. Nucleon number conservation ($\sum AY = 1$) provides one constraint. The second constraint is the amount of weak reaction activity, often expressed in terms of the total proton abundance, $\sum ZY$, which charge conservation requires equal the electron abundance, Y_e. Thus the nuclear abundances are uniquely determined for a given (T, ρ, Y_e). Alternately, the weak interaction history is sometimes expressed in terms of the neutron excess $\eta = \sum(N - Z)Y$.

As with any equilibrium distribution, there are limitations on the applicability of NSE. For NSE to provide a good estimate of the nuclear abundances the temperature must be sufficient for the endoergic reaction of each reaction pair to occur. Since for all particle-stable nuclei between the proton and neutron drip lines (with the exception of nuclei unstable against alpha decay), the photodisintegrations are endoergic, with typical Q-values among (β) stable nuclei of 8-12 MeV, $T > 3\,\mathrm{GK}$ is required for sufficient high energy photons to be present. While this requirement is necessary, it is not sufficient. In the case of hydrostatic silicon burning, even when this condition is met, appreciable time is required to convert Si to Fe-peak elements. As discussed in §2, in the case of expanding matter, the adiabatic cooling on timescales of seconds can cause conditions to change more rapidly than NSE can follow, breaking down NSE first between ^{4}He and ^{12}C, at $T \sim 6\,\mathrm{GK}$[29] and later between the species near silicon and the Fe-peak nuclei, at $T \sim 4\,\mathrm{GK}$.[30,32] Thus it is clear that in the face of sufficiently rapid thermodynamic variations, NSE provides a problematic estimate of abundances. However, in spite of the breakdown of the global NSE, many nuclei under these conditions do obey local equilibrium relations with their neighbors, and this fact can be used to construct hybrid reaction networks.

4.2. *Tools: Merging Equilibria with Reaction Networks*

In spite of the limitations on the applicability of NSE, the reduced computational cost provides a strong motivation to maximize the use of equilibria. The use of partial equilibrium expressions for single abundances is, in fact, common in nuclear reaction networks, typically to track the abundances of short-lived unstable intermediates in "three-particle" processes. The most common example of this is the triple α process, $^4\mathrm{He} + {}^4\mathrm{He} \rightleftharpoons {}^8\mathrm{Be} + {}^4\mathrm{He} \rightleftharpoons {}^{12}\mathrm{C}^* \rightarrow {}^{12}\mathrm{C}$, by which Helium burning occurs. Only rarely does a ^{8}Be nucleus survive long enough for a second α to capture. Likewise for temperatures in excess of .1 GK, the most likely result following the second α capture to form an excited state of ^{12}C is a decay back to ^{8}Be. Thus the effective triple α reaction rate is simply that of the equilibrium population of $^{12}\mathrm{C}^*$, with respect to ^{4}He, multiplied by the rate of decay from the excited state to the ground state.[72] This use of local equilibrium within a rate equation shares many characteristics with the more elaborate schemes we will discuss next. The number of species tracked by the network is reduced since $Y(^8\mathrm{Be})$ need not be directly evolved. Problematically small timescales like $\tau(^8\mathrm{Be})(\sim 10^{-16}\,\mathrm{sec})$ are removed, replaced by larger time scales ($\tau_{3\alpha} \sim 10^5 - 10^7$ years during core helium burning). The non-linearity of network time derivatives is increased ($\dot{Y}(^{12}\mathrm{C}) \propto Y_\alpha^3$) under this scheme and this approximation also breaks down at low temperature.[73]

As we noted in § 4.1, while global NSE may not always apply for temperatures in the range 3-6 GK, many nuclei are in local equilibrium with their neighbors. Beginning with Bodansky et al.[31], a number of attempts have been made to take advantage of these partial equilibria (termed quasi-equilibria or QSE) to reduce the number of independent variables evolved via rate equations and thereby reduce the computational cost of modeling these burning stages. To evolve the abundances of every member of a QSE group, it is sufficient to evolve the abundance of any single group member along with the abundances of the free nucleons. One can thereby reduce the number of abundances that are evolved, while still calculating, from QSE relations, the abundances of all members of a QSE group and the resultant reaction rates, including the electron and neutrino capture reactions responsible for changes in the neutronization. The result is a more computationally efficient method that retains the accuracy of the full network and yields abundances for all nuclei found in the full network.

As a example, we will briefly discuss the QSE-reduced α-network. Its mission is to evolve the same 14 abundances evolved by a conventional α-network (α— ^{60}Zn, which we'll term $\vec{Y}^{\mathcal{F}}$), and calculate the resulting energy generation, in a more efficient way. Under conditions where QSE applies, the existence of the silicon and iron peak QSE groups (which are separated by the nuclear shell closures Z=N=20 and the resulting small Q-values and reaction rates) allows calculation of these 14 abundances from 7. For the members of the silicon group (^{28}Si, ^{32}S, ^{36}Ar, ^{40}Ca, ^{44}Ti) and the iron peak group (^{48}Cr, ^{52}Fe, ^{56}Ni, ^{60}Zn) the individual abundances can be calculated by expressions similar to Eq. 2,

$$Y_{QSE,\mathrm{Si}}(^{A}Z) = \frac{C(^{A}Z)}{C(^{28}\mathrm{Si})} Y(^{28}\mathrm{Si}) Y_{\alpha}^{\frac{A-28}{4}}$$

$$Y_{QSE,\mathrm{Ni}}(^{A}Z) = \frac{C(^{A}Z)}{C(^{56}\mathrm{Ni})} Y(^{56}\mathrm{Ni}) Y_{\alpha}^{\frac{A-56}{4}}, \tag{3}$$

where $C(^{A}Z)$ is defined in Eq. 2 and $(A-28)/4$ and $(A-56)/4$ are the number of α-particles needed to construct ^{A}Z from ^{28}Si and ^{56}Ni, respectively. Where QSE applies, $\vec{Y}^{\mathcal{F}}$ is a function of the abundances of a reduced nuclear set, $\mathcal{R}$, defined as α, ^{12}C, ^{16}O, ^{20}Ne, ^{24}Mg, ^{28}Si, ^{56}Ni, and we need only evolve $\vec{Y}^{\mathcal{R}}$. It should be noted that ^{24}Mg is ordinarily a member of the silicon QSE group,[10,30,49] but for easier integration of prior burning stages with a conventional nuclear network, we will evolve ^{24}Mg independently. The main task when applying such hybrid schemes is finding the boundaries of QSE groups and where individual nuclei have to be used instead. Treating marginal group members as part of a group increases the efficiency of the calculation, but may decease the accuracy.

While $\vec{Y}^{\mathcal{R}}$ is a convenient set of abundances for calculating $\vec{Y}^{\mathcal{F}}$, it is not the most efficient set to evolve, primarily because of the non-linear dependence on Y_α. Instead we define a set of group abundances, $\vec{Y}^{\mathcal{G}}=$ $[Y_{\alpha G}, Y(^{12}C), Y(^{16}O), Y(^{20}Ne), Y(^{24}Mg), Y_{SiG}, Y_{FeG}]$ where

$$Y_{\alpha G} = Y_\alpha + \sum_{i \in Si\ group} \frac{A_i - 28}{4} Y_i + \sum_{i \in Fe\ group} \frac{A_i - 56}{4} Y_i ,$$

$$Y_{SiG} = \sum_{i \in Si\ group} Y_i , \tag{4}$$

$$Y_{FeG} = \sum_{i \in Fe\ group} Y_i .$$

Physically, $Y_{\alpha G}$ represents the sum of the abundances of free α-particles and those α-particles required to build the members of the QSE groups from ^{28}Si or ^{56}Ni, while Y_{SiG} and Y_{FeG} represent the total abundances of the silicon and iron peak QSE groups. This method, which here is applied only to the chain of α-nuclei can also be generalized to arbitrary networks.[74] For larger networks that contain nuclei with $N \neq Z$, one must be able to follow the abundances of free neutrons and protons, particularly since weak interactions will change the global ratio of neutrons to protons. In place of $Y_{\alpha G}$ in Eq. 4, one constructs $Y_{NG} = \sum_{i,light} N_i Y_i + \sum_{i,Si}(N_i - 14)Y_i + \sum_{i,Fe}(N_i - 28)Y_i$ and $Y_{ZG} = \sum_{i,light} Z_i Y_i + \sum_{i,Si}(Z_i - 14)Y_i + \sum_{i,Fe}(Z_i - 28)Y_i$, if ^{28}Si and ^{56}Ni are chosen as the focal nuclei for the Si and Fe groups.

Corresponding to this reduced set of abundances $\mathcal{G}$ is a reduced set of reactions, with quasi-equilibrium allowing one to ignore the reactions among the members of the QSE groups. Unfortunately, the rates of these remaining reactions are functions of the full abundance set, $\vec{Y}^{\mathcal{F}}$, and are not easily expressed in terms of the group abundances, $\vec{Y}^{\mathcal{G}}$. Thus, for each $\vec{Y}^{\mathcal{G}}$, one must solve for $\vec{Y}^{\mathcal{R}}$ and, by Eq. 3, $\vec{Y}^{\mathcal{F}}$, in order to calculate $\dot{\vec{Y}}^{\mathcal{G}}$ that is needed to evolve $\vec{Y}^{\mathcal{G}}$. Furthermore, the calculation of the Jacobian, $\partial \dot{\vec{Y}}^{\mathcal{G}} / \partial \vec{Y}^{\mathcal{G}}$, which can not be calculated directly since $\dot{\vec{Y}}^{\mathcal{G}}$ cannot be expressed in terms of $\vec{Y}^{\mathcal{G}}$, is required. Instead it is sufficient to use the chain rule,

$$\frac{\partial \dot{\vec{Y}}^{\mathcal{G}}}{\partial \vec{Y}^{\mathcal{G}}} = \frac{\partial \dot{\vec{Y}}^{\mathcal{G}}}{\partial \vec{Y}^{\mathcal{R}}} \frac{\partial \vec{Y}^{\mathcal{R}}}{\partial \vec{Y}^{\mathcal{G}}} \tag{5}$$

to calculate the Jacobian. Analytically, the first term of the chain rule product is easily calculated from sums of reaction terms, while the second term requires implicit differentiation using Eq. 4. Additional details and comparisons with the full α-network are demonstrated by Hix et al.[75] (see also Timmes et al.[18]).

The result is a network that is twice as fast as the minimal α-chain network, without significantly affecting the nuclear evolution. For small networks, the calculation of the reaction rates is the computationally limiting phase of the evolution, hence for small sizes, the QSE-reduction of the network achieves a near linear improvement. For larger networks, the limitation is the speed of the matrix solution, hence much larger speed-ups are possible. Reductions of an order of magnitude in computational cost have been achieved[74,76] for QSE-reduced networks of the size necessary to capture the essential features of supernova nucleosynthesis and we believe greater savings are possible with further refinement. In Figure 2 we demonstrate the accuracy of a QSE-reduced network for 300 species for a test problem in which a parcel of matter undergoes adiabatic expansion at the freefall timescale, with initial temperature of 8 GK and initial density of $10^8 \, \mathrm{g \, cm^{-3}}$. While the smallest abundances show considerable spread, for the most abundant species $Y > 10^{-6}$, those which determine the energetics, differences $< 5\%$ are seen even as the temperature approaches 3.5 GK. For intermediate abundances ($10^{-6} < Y < 10^{-12}$), typical errors are less than 10%. At least some of the outliers in this group may be improved by adjusting QSE group boundaries, while others, for example ^{57}Cu, can be remedied by finding improperly balanced reaction rates.

In addition to silicon burning and cooling from NSE, there are a number of astrophysically important situations where temperatures are large enough relative to the Q values for at least some of the relevant reactions, and so equilibrium groups exist, but NSE is not globally valid. This includes the r-process[77] and the rp-process in X-ray bursts,[78] where neutron or proton separation energies (Q_n or Q_p) of 2 MeVand less are often encountered.

5. Conclusion

The challenges in coupling simulations of core collapse supernova nucleosynthesis to multi-dimensional models of the central engine are great, however this effort will yield rich rewards. Not only will we better understand the contribution of supernovae to our cosmic origins, but by comparing nucleosynthesis estimates with observations we will also improve our understanding of the central engine. Fully coupled, multi-dimensional models

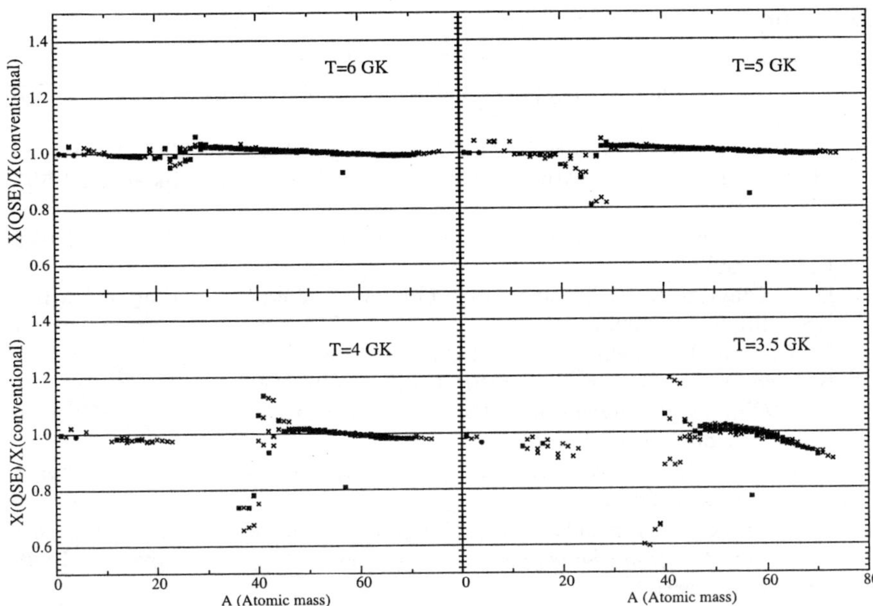

Figure 2. Comparison between QSE-reduced and conventional reaction networks for adiabatic expanding gas with an initial temperature of 8 GK and density of $10^8 \, \mathrm{g\,cm^{-3}}$. Abundances from $1 - 10^{-6}$, $10^{-6} - 10^{-12}$ and $10^{-12} - 10^{-20}$ are denoted by circles, squares and x's respectively,

will remain untenable for quite some time. However, by keeping the shorter nuclear reaction timescales localized and taking advantage of equilibria that Nature provides, the next few years should see the first simulations coupling (in the operator split limit) neutrino transport, hydrodynamics and non-equilibrium thermonuclear evolution with large networks.

Acknowledgements The authors wish to thank M. Liebendörfer, H.-Th. Janka, A. Mezzacappa for fruitful discussions as well as E.J. Lentz for helpful comments on the manuscript. The work has been partly supported by the U.S. National Science Foundation under contract PHY-0244783, by the U.S. Department of Energy, through the Scientic Discovery through Advanced Computing Program, by the Swiss NSF grant 2000-061031.02 and by the Spanish MCyT and European Union ERDF under contracts AYA2002-04094-C03-02 and AYA200306128. Oak Ridge National Laboratory is managed by UT-Battelle, LLC, for the U.S. Department of Energy under contract DE-AC05-00OR22725.

References

1. T. Preibisch and H. Zinnecker, AJ **117**, 2381 (1999).
2. E. Oliva, D. Lutz, S. Drapatz, and A. F. M. Moorwood, A&A **341**, L75 (1999).
3. W. P. Blair, J. A. Morse, J. C. Raymond, R. P. Kirshner, J. P. Hughes, M. A. Dopita, R. S. Sutherland, K. S. Long, and P. F. Winkler, ApJ **537**, 667 (2000).
4. J. P. Hughes, C. E. Rakowski, D. N. Burrows, and P. O. Slane, ApJ **528**, L109 (2000).
5. D. D. Clayton, M. D. Leising, L.-S. The, W. N. Johnson, and J. D. Kurfess, ApJ **399**, L141 (1992).
6. W. Chen and N. Gehrels, ApJ **514**, L103 (1999).
7. R. Diehl and F. X. Timmes, PASP **110**, 637 (1998).
8. D. D. Clayton, *Principles of Stellar Evolution and Nucleosynthesis* (Chicago: University of Chicago Press, 1983).
9. R. Kippenhahn and A. Weigert, *Stellar Structure and Evolution* (Springer-Verlag, Berlin, 1990).
10. W. D. Arnett, *Supernovae and nucleosynthesis : an investigation of the history of matter, from the big bang to the present* (Princeton University Press, Princeton, 1996).
11. S. E. Woosley, A. Heger, and T. A. Weaver, Rev. Mod. Phys. **74**, 1015 (2002).
12. P. P. Eggleton, G. Bazán, R. M. Cavallo, D. S. P. Dearborn, D. D. Dossa, S. C. Keller, A. G. Taylor, and S. Turcotte, in *ASP Conf. Ser. 293: 3D Stellar Evolution* (2003), p. 15.
13. G. Bazan and D. Arnett, ApJ **496**, 316 (1998).
14. S. E. Woosley and T. A. Weaver, ApJS **101**, 181 (1995).
15. F.-K. Thielemann, K. Nomoto, and M. Hashimoto, ApJ **460**, 408 (1996).
16. T. Rauscher, A. Heger, R. D. Hoffman, and S. E. Woosley, ApJ **576**, 323 (2002).
17. M. F. El Eid, B. S. Meyer, and L.-S. The, ApJ **611**, 452 (2004).
18. F. X. Timmes, R. D. Hoffman, and S. E. Woosley, ApJS **129**, 377 (2000).
19. H. Umeda and K. Nomoto, ApJ **565**, 385 (2002).
20. M. Limongi and A. Chieffi, ApJ **592**, 404 (2003).
21. G. C. Jordan IV, B. S. Meyer, and E. D'Azevedo, in *Open Issues in Core Collapse Supernovae*, eds. G. Fuller and A. Mezzacappa, (World Scientific, Singapore, 2005).
22. G. Meynet and A. Maeder, A&A **361**, 101 (2000).
23. A. Heger, N. Langer, and S. E. Woosley, ApJ **528**, 368 (2000).
24. A. Heger, S. E. Woosley, N. Langer, and H. C. Spruit, in *IAU Symposium* (2004), p. 591.
25. S. Nagataki, T. M. Shimizu, and K. Sato, ApJ **495**, 413 (1998).
26. M. B. Aufderheide, E. Baron, and F. K. Thielemann, ApJ **370**, 630 (1991).
27. F. E. Clifford and R. F. Tayler, MNRAS **129**, 104 (1965).
28. D. Hartmann, S. E. Woosley, and M. F. El Eid, ApJ **297**, 837 (1985).
29. B. S. Meyer, T. D. Krishnan, and D. D. Clayton, ApJ **498**, 808 (1998).

30. S. E. Woosley, W. D. Arnett, and D. D. Clayton, ApJS **26**, 231 (1973).

31. D. Bodansky, D. Clayton, and W. Fowler, ApJS **16**, 299 (1968).

32. W. R. Hix and F.-K. Thielemann, ApJ **511**, 862 (1999).

33. S. E. Woosley, D. H. Hartmann, R. D. Hoffman, and W. C. Haxton, ApJ **356**, 272 (1990).

34. R. McCray, ARA&A **31**, 175 (1993).

35. J. Spyromilio, MNRAS **266**, L61 (1994).

36. L. Wang, D. A. Howell, P. Höflich, and J. C. Wheeler, ApJ **550**, 1030 (2001).

37. S. M. Asida and D. Arnett, ApJ **545**, 435 (2000).

38. I. Hachisu, T. Matsuda, K. Nomoto, and T. Shigeyama, ApJ **358**, L57 (1990).

39. E. Müller, B. Fryxell, and D. Arnett, A&A **251**, 505 (1991).

40. M. Herant and W. Benz, ApJ **387**, 294 (1992).

41. J. Kane, D. Arnett, B. A. Remington, S. G. Glendinning, G. Bazán, E. Müller, B. A. Fryxell, and R. Teyssier, ApJ **528**, 989 (2000).

42. K. Z. Stanek, T. Matheson, P. M. Garnavich, P. Martini, P. Berlind, N. Caldwell, P. Challis, W. R. Brown, R. Schild, K. Krisciunas, et al., ApJ **591**, L17 (2003).

43. J. C. Wheeler and S. Akiyama, in *Open Issues in Core Collapse Supernovae*, eds. G. Fuller and A. Mezzacappa, (World Scientific, Singapore, 2005).

44. S. Nagataki, M. Hashimoto, K. Sato, and S. Yamada, ApJ **486**, 1026 (1997).

45. S. Nagataki, M. Hashimoto, K. Sato, S. Yamada, and Y. S. Mochizuki, ApJ **492**, L45 (1998).

46. K. Maeda, T. Nakamura, K. Nomoto, P. A. Mazzali, F. Patat, and I. Hachisu, ApJ **565**, 405 (2002).

47. K. Kifonidis, T. Plewa, H.-T. Janka, and E. Müller, A&A (2003).

48. H.-T. Janka and E. Müller, A&A **306**, 167 (1996).

49. W. R. Hix and F.-K. Thielemann, ApJ **460**, 869 (1996).

50. L.-S. The, D. D. Clayton, L. Jin, and B. S. Meyer, ApJ **504**, 500 (1998).

51. Y.-Z. Qian, W. C. Haxton, K. Langanke, and P. Vogel, Phys. Rev. C **55**, 1532 (1997).

52. W. C. Haxton, K. Langanke, Y.-Z. Qian, and P. Vogel, Phys. Rev. Lett. **78**, 2694 (1997).

53. B. S. Meyer, G. C. McLaughlin, and G. M. Fuller, Phys. Rev. C **58**, 3696 (1998).

54. R. D. Hoffman, S. E. Woosley, G. M. Fuller, and B. S. Meyer, ApJ **460**, 478 (1996).

55. B. S. Meyer, Nucl. Phys. A **719**, 13 (2003).

56. V. Trimble, A&AR **3**, 1 (1991).

57. M. Herant, W. Benz, W. R. Hix, C. L. Fryer, and S. A. Colgate, ApJ **435**, 339 (1994).

58. M. Liebendörfer, A. Mezzacappa, F.-K. Thielemann, O. E. B. Messer, W. R. Hix, and S. W. Bruenn, Phys. Rev. D **63**, 103004 (2001).

59. M. Rampp and H.-T. Janka, A&A **396**, 361 (2002).

60. T. A. Thompson, A. Burrows, and P. A. Pinto, ApJ **592**, 434 (2003).

61. R. Buras, M. Rampp, H.-T. Janka, and K. Kifonidis, Phys. Rev. Lett. **90**, 241101 (2003).

390

62. G. C. McLaughlin, G. M. Fuller, and J. R. Wilson, ApJ **472**, 440 (1996).

63. S. W. Bruenn and W. C. Haxton, ApJ **376**, 678 (1991).

64. C. Travaglio, K. Kifonidis, and E. Müller, in *Carnegie Observatories Astrophysics Series, Vol. 4: Origin and Evolution of the Elements*, eds. A. McWilliam and M. Rauch (Carnegie Observatories, Pasadena, 2003), `http://www.ociw.edu/ociw/symposia/series/symposium4/proceedings.html`.

65. C. Fröhlich, P. Hauser, M. Liebendörfer, G. Martinez-Pinedo, F. . Thielemann, E. Bravo, N. T. Zinner, W. R. Hix, K. Langanke, A. Mezzacappa, et al., ApJ (2005), submitted, `astro-ph/0410208`.

66. J. Pruet, S. E. Woosley, R. Buras, H.-T. Janka, and R. D. Hoffman, ApJ **623**, 325 (2005).

67. A. M. Beloborodov, ApJ **588**, 931 (2003).

68. N. T. Zinner and K. Langanke, private communication.

69. R. G. Gratton and C. Sneden, A&A **241**, 501 (1991).

70. R. Cayrel, E. Depagne, M. Spite, V. Hill, F. Spite, P. François, B. Plez, T. Beers, F. Primas, J. Andersen, et al., A&A **416**, 1117 (2004).

71. E. Bravo and D. García-Senz, MNRAS **307**, 984 (1999).

72. H. O. U. Fynbo, C. A. Diget, U. C. Bergmann, M. J. G. Borge, J. Cederkäll, P. Dendooven, L. M. Fraile, S. Franchoo, V. N. Fedosseev, B. R. Fulton, et al., Nature **433**, 136 (2005).

73. K. Nomoto, F.-K. Thielemann, and S. Miyaji, A&A **149**, 239 (1985).

74. W. R. Hix, C. Freiburghaus, S. Parete-Koon, and F.-K. Thielemann, ApJS (2005), in preparation.

75. W. R. Hix, A. M. Khokhlov, J. C. Wheeler, and F.-K. Thielemann, ApJ **503**, 332 (1998).

76. C. Freiburghaus, W. R. Hix, and F.-K. Thielemann, in *Proceeding of Nuclei in the Cosmos V*, edited by N. Prantzos (Paris:Editions Frontiere, 1999), p. 140.

77. J. J. Cowan, F. Thielemann, and J. W. Truran, ARA&A **29**, 447 (1991).

78. F. Rembges, C. Freiburghaus, T. Rauscher, F.-K. Thielemann, H. Schatz, and M. Wiescher, ApJ **484**, 412 (1997).

TOWARD *IN SITU* CALCULATION OF NUCLEOSYNTHESIS IN SUPERNOVA MODELS

G. C. JORDAN, IV AND B. S. MEYER

Department of Physics and Astronomy,
Clemson University,
Clemson, SC 29634-0978, USA
E-mail: mbradle@clemson.edu

ED D'AZEVEDO

Computer Science and Mathematics Division,
Oak Ridge National Laboratory,
P.O.Box 2008, Bldg 6012,
Oak Ridge, TN 37831-6367

We have constructed a matrix solver for large-scale sparse matrices that arise from the treatment of nuclear burning in convective or advective environments in one dimension. We use this solver to compute nuclear abundances in a parameterized problem relevant for nucleosynthesis in supernova explosions. We discuss some details of our solutions, possible future extensions to higher dimensionality, and possible inclusion in supernova codes themselves.

1. Introduction

Solution of single-zone nuclear reaction networks to understand formation of the isotopes by nuclear reactions is by now a well-established art[1]. The next frontier in the field is the study of element formation *in situ* in models of astrophysical fluid flows. For example, future calculations will seek to study as accurately as possible nuclear reactions in the highly turbulent region between the neutrinosphere and the stalled shock in core-collapse supernova calculations. This will require calculations to follow the change in nuclear abundances not only from nuclear reactions but also from convection or advection between spatial zones.

To understand the computational aspects of this problem, consider the types of matrices that arise in the situations under consideration. Figure 1 shows the non-zero elements of the (634×634) matrix that results when we

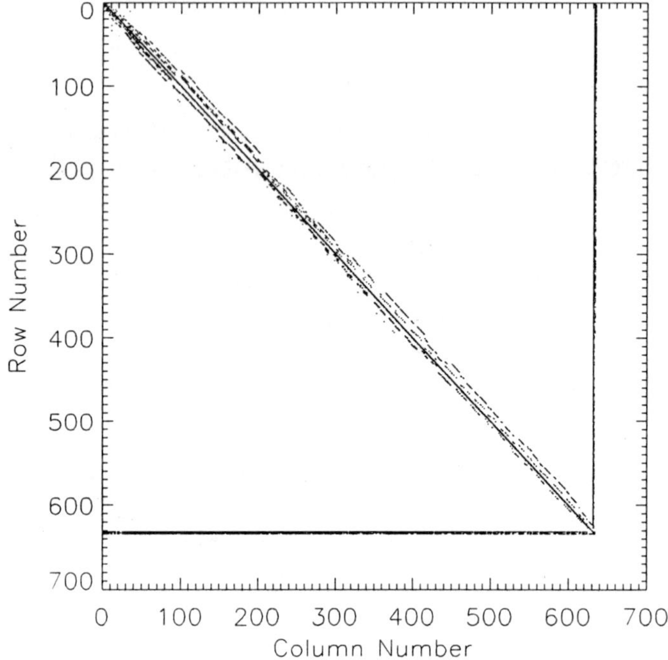

Figure 1. Matrix for a single-zone network with 634 nuclear species.

use an implicit scheme to finite difference in time the coupled differential
equations describing the rate of change of abundances in a particular single-
zone nuclear reaction network with 634 nuclear species. In this network,
the neutrons, protons, and alpha particles are elements 632, 633, and 634
in the network. Since these nuclei react will all other species, they form the
wings of the sparse arrow pattern present. By constrast, most other species
interact with only a handful of neighbors in the network, so the other non-
zero elements lie within a narrow band of the diagonal. Efficient routines
allow us to solve such networks easily on a normal desktop computer.

When the problem is extended to multiple zones, however, the matrix
grows in size and complexity. Figure 2 shows the matrix that results from
the 634 species network of Figure 1 in nine concentric shells that mix only
with their nearest neighbor. This matrix is thus a block diagonal matrix
with nine single-zone submatrices making up the blocks. The blocks are
then linked with mixing bands connecting each zone with the one above and

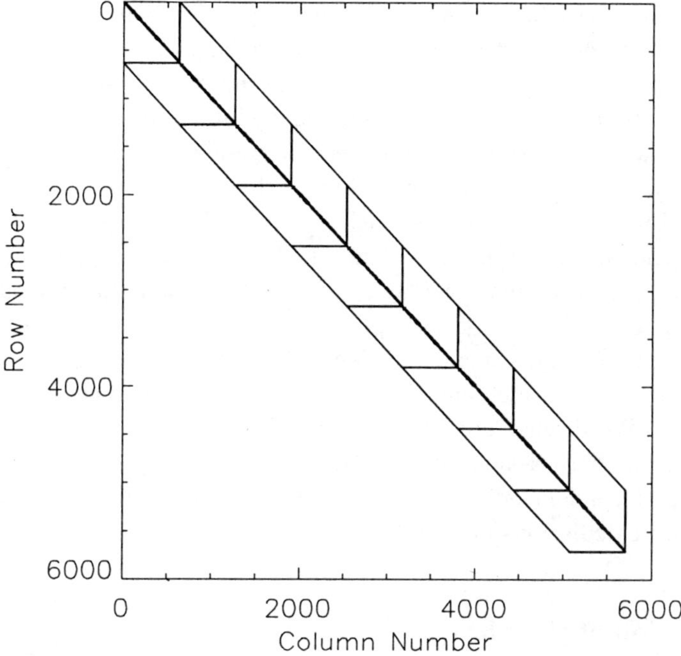

Figure 2.   Matrix for a network with nine concentric zones and 634 nuclear species in each zone.

the one below, except for the first and last zones, which mix with only the zone above or below, respectively. If there is more complex mixing among the zones such as non-nearest neighbor mixing or higher dimensionality, the mixing bands will be more numerous and not as regular as shown in Figure 2.

While the matrix in Figure 2 is sparse, it is easy to see that it will become quite large with extension to many zones. This can challenge the storage capacity of most computers. A frequent solution is to "operator split", that is, to split the mixing and the nuclear burning into separate steps. This is frequently done in stellar evolution where one computes the nuclear burning in each zone and then mixes the material among the zones according to an appropriate convection scheme[2]. Operator splitting does not require as much memory as solving the full matrix because one needs to store and solve only one nuclear sub-block at a time. The technique works well when the timescale for convection dominates that for nuclear burning, as it does in the early stages of stellar evolution. In later stages, burning and mixing timescales become comparable, and a more careful treatment becomes necessary. A popular technique in this latter case is to assume that the nuclear reactions become sufficiently fast that the nuclear populations are in a local equilibrium so that, instead of keeping track of all the abundances, one keeps track of only a limited set of nuclei[3]. This reduces the size of the matrix required. The difficulty is that one imposes an equilibrium that may or may not be present, and this imposition may mask interesting and important physics.

It is therefore desirable to consider solving the fully coupled mixing and burning problem, which entails solving large-scale matrices of the type in Figure 2. We discuss one class of such solvers and present results for a particular model relevant to supernova nucleosynthesis. We discuss some details and challenges of our technique and the prospects for using it in supernova calculations directly.

2. Solution of the Matrix

The large size and sparse nature of the coupled mixing and burning matrices under consideration lead us to employ iterative type methods for their solution[4]. We restricted ourselves to the two most popular iterative methods: the stabilized biconjugate gradient method (BICGSTAB) and the generalized minimum residual method (GMRES). Our implementation of these methods uses routines from SparseKit, which is a package available from the URL

http://www-users.cs.umn.edu/ saad/software/SPARSKIT/sparskit.html.

We use a two-step preconditioner on matrices of the type shown in Figure 2. We first solve the tridiagonal system formed by the bands (the diagonal and the two mixing bands). We then perform an incomplete-LU decomposition on the resulting matrix. Our experience is that this preconditioner works particularly well for the coupled mixing and burning matrices of the type in Figure 2 and a solution is obtained after only a few iterations for all but the most ill conditioned matricies.

To evolve our solutions, we typically begin with a small initial timestep of $dt = 1.0 \times 10^{-15}$ seconds. The coupled mixing and burning code computes the matrix from the available abundances and reaction and mixing rates. The code then solves the matrix equation for the new abundances by applying our iterative routine. To determine when to stop the iterations, we compute the residual, defined as $\| Ax - b \|_2$, where A is the original matrix, x is the solution vector, b is the original right-hand side vector, and the symbol $\| \ \|_2$ indicates the Euclidean norm. We consider that the solution has converged when the residual is less than a prescribed tolerance.

Due to inaccuracies that can be introduced by the stiffness of the problem, the possibility arises that the new abundances are negative or that the sum of the mass fractions in a given zone differ from unity by too large an amount. If this happens, we reduce the timestep and solve the matrix equation again. Note that this does not require rebuilding the matrix; it only requires changing the diagonal elements to accomodate the new timestep. Once a satisfactory solution is found, the time is incremented based on the requirement that no abundance above a certain threshold changes by more than 15%. This allows the timestep to grow as the abundances tend toward a steady state.

3. A Model of Nucleosynthesis Behind the Supernova Shock

In order to test our matrix solver for a scenario relevant to the supernova problem, we consider nucleosynthesis in the region between a nascent neutron star and a stalled supernova shock, an environment that is known to be highly convective[5,6]. An exact treatment of the fluid dynamics in this region lies at the forefront of supernova modeling and is, therefore, well outside the scope of this paper. Nevertheless, we present a simple model to understand the issues and explore the consequences of treating simultaneous mixing and nuclear burning that might arise in the future when this problem is treated in a more sophisticated manner.

Our model treats the region between the surface of a proto-neutron star and the stalled supernova shock in one dimensional and with spherical symmetry. The surface of the proto-neutron star lies at Lagrangian mass ("Interior Mass") coordinate 1.4 $M_\odot$, that is, 1.4 $M_\odot$ of matter lies inside the proto-neutron star surface. The stalled shock lies at Lagrangian mass coordinate 2.0 $M_\odot$; thus, the region under consideration comprises a total of 0.6 $M_\odot$ of material. We divided this region into 200 equal mass concentric spherical shells, each consisting of $0.6/200 = 3 \times 10^{-3}$ $M_\odot$ of material. We took the temperature T_9 in billions of Kelvins to fall off with increasing radial coordinate in the star as shown in Figure 3. We took the photon-to-

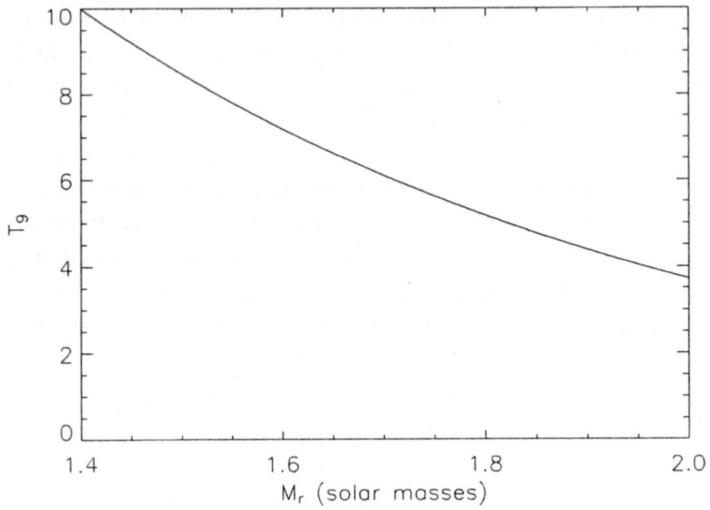

Figure 3. Temperature (in 10^9 K) as a function of Lagrangian mass coordinate in the model studied in the paper.

nucleon ratio ϕ in the modeled region to be unity, which, by the equation[7]

$$\phi = 3.4 \times 10^4 \frac{T_9^3}{\rho},\tag{1}$$

gives the mass density ρ shown in Figure 4. The run of temperature and of density were taken to be fixed in time through the calculation.

For each mass shell, we followed the nuclear reactions among 236 nuclear species ranging from neutrons and protons up to bromine ($Z = 35$). The nuclear reaction rates used are those described in Jordan and Meyer[8]. The initial composition was taken to be pure ^{28}Si, and the calculations followed one second of evolution.

The mixing of material between the shells was taken to occur by diffusion, which means the rate of change of the mass fraction X_i of any species i in any mass shell due solely to the mixing followed Fick's Law:

$$\frac{\partial X_i}{\partial t} = \frac{\partial}{\partial M_r}\left(D\frac{\partial X_i}{\partial M_r}\right)\tag{2}$$

where D is the appropriate diffusion coefficient. For simplicity, we took D to be a constant throughout the region modeled. When we finite differenced the coupled diffusion/nuclear network differential equations for this

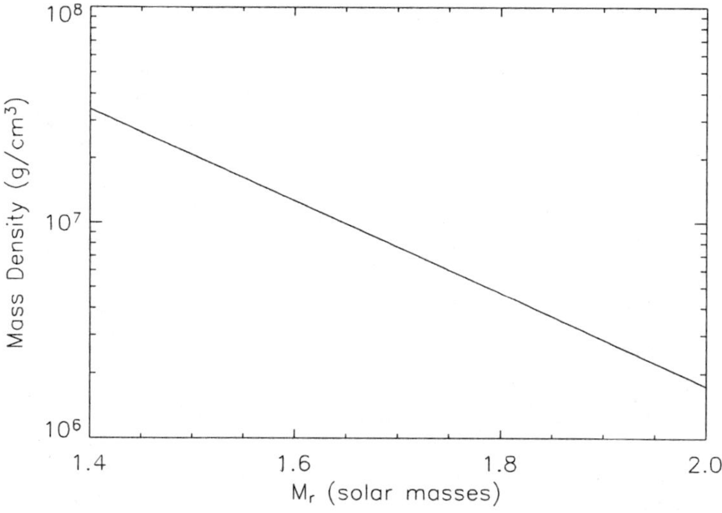

Figure 4.   Mass density as a function of Lagrangian mass coordinate in the model studied in the paper.

problem, we found a matrix of the form shown in Figure 2.

Mixing in the supernova is certainly not simple diffusion, as treated in this model. In general, we expect blobs or plumes of matter will maintain their integrity over a period of time, which violates the diffusion approximation's assumption that species mix independently of each other. Furthermore, blobs of matter are likely to "punch" through shells without directly mixing with the surroundings, which violates the diffusion approximations assumption of mixing only with neighboring shells. Nevertheless, despite these limitations of our model, we expect more realistic treatments to be elaborations on our model, and the features of the solutions we present below to carry over to the more sophisticated treatments.

As a first test of the model and the underlying matrix solver, we performed a calculation with $D = 0$. As is evident from Eq. (2), this corresponds to no mixing between the mass shells, only nuclear burning within each zone. Figure 5 shows the evolution of the mass fraction of ^{4}He during the calculation. For comparison, we also present in the figure, the mass fraction of ^{4}He in nuclear statistical equilibrium (NSE). Given sufficient time, strong and electromagnetic reactions will drive the nuclear composition into NSE, and it is clear this happens throughout almost the entire

398

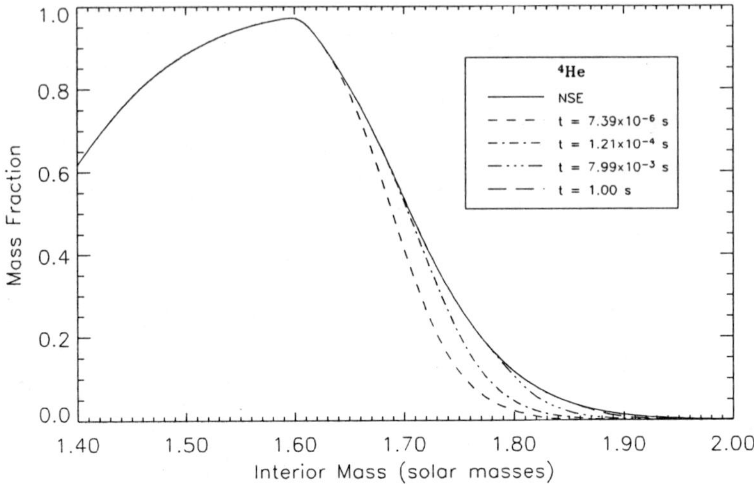

Figure 5.   Mass fraction of ^{4}He as a function of Lagrangian mass coordinate in the $D = 0$ model at four different times. The NSE mass fraction is also shown for comparison.

0.6 $M_\odot$ in the model within one second. Only the outer layers, where the density and temperature are lower and, hence, where the nuclear reactions are slower, do not reach NSE within the allotted time. That the matrix solver does drive the composition to NSE shows that it is working correctly. It is worth noting that weak reactions are so slow that the bulk number of neutrons equals that of the protons in each mass shell throughout the calculation.

For the next calculation we used a diffusion coefficient $D = 1$. From Eq. (2), we see that a typical mixing timescale τ is given by

$$\tau = \frac{(\Delta M_r)^2}{D}, \tag{3}$$

where ΔM_r is the mass range mixed. From Eq. (3), we see that the full mass range 0.6 $M_\odot$ will mix in $\tau = 0.36$ seconds with $D = 1$. Figure 6 shows the time evolution of the ^{4}He mass fraction in this calculation. By the end of the calculation, the ^{4}He mass fraction matches that of NSE in the inner 0.3 $M_\odot$ of material. In these zones, the nuclear burning occurs much faster than the mixing. Outside of $M_r = 1.7\ M_\odot$, however, the mixing dominates the nuclear burining. This gives rise to a ^{4}He mass fraction greater than in NSE in these shells because the rapid mixing is able to

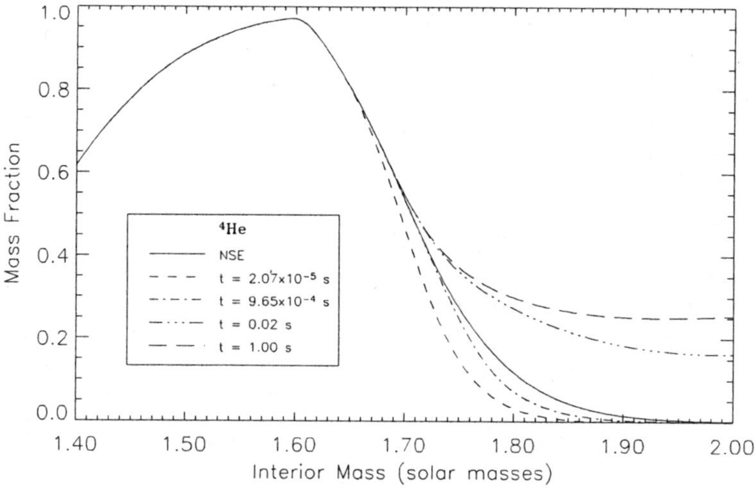

Figure 6. Mass fraction of ^{4}He as a function of Lagrangian mass coordinate in the $D = 1$ model at four different times. The NSE mass fraction is also shown for comparison.

bring large concentrations of ^{4}He out from the inner zones.

For our third calculation, we chose $D = 100$. In this case, the mixing is 100$\times$ faster than in the previous calculation. Figure 7 shows the ^{4}He mass fraction in this case. As expected, the larger D allows the mixing to compete more effectively with the nuclear burning. The result is that the ^{4}He mass fraction in the outer zones is even higher than in the $D = 1$ case as material from even deeper inside is mixed out before burning. It is also worth noting that the $t = 3.81 \times 10^{-3}$ seconds and $t = 1$ second curves lie on top of each other in this figure. This shows that the system has established a steady state by 3.81 milliseconds and changes very little after this time since nuclear production and destruction and mixing rates have come into balance.

The success of our calculations strongly encourages us that iterative methods of the type we employ can be used in larger-scale calculations and, even, *in situ* in supernova codes, given sufficient memory. For those seeking more details on our results, movies of the time evolution of the abundances of key species in the $D = 0$, $D = 1$, and $D = 100$ calculations are available on the world-wide web via the movies link at the Clemson Nuclear Astrophysics web site:

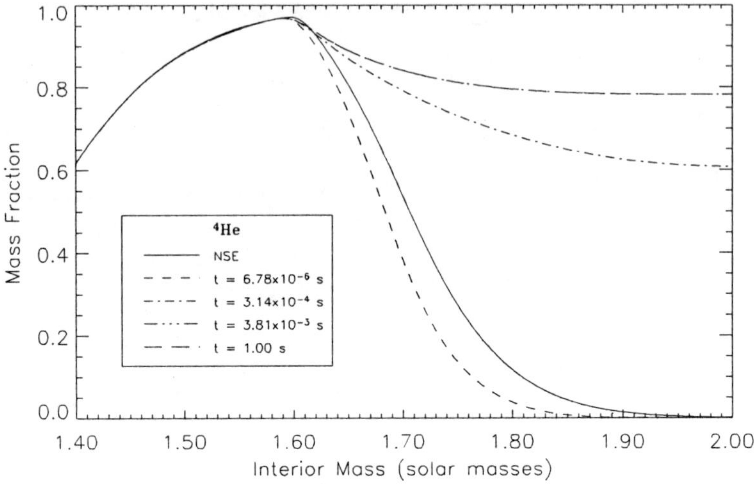

Figure 7.   Mass fraction of ^{4}He as a function of Lagrangian mass coordinate in the $D = 100$ model at four different times. The NSE mass fraction is also shown for comparison.

http://nucleo.ces.clemson.edu/

4. Some Details of the Matrix Solutions

In this section we compare the performance of two iterative solvers, GMRES and BICGSTAB, in terms of their overall speed and accuracy. To achieve this, we use some gross performance indicators to evaluate which solver evolves a test calculation to the end the quickest and with the least amount of cumulative error over several values of the convergence tolerance. The test calculation we used to test the performance of the solvers was the aforementioned supernova calculation with $D = 1$. The calculation was set to run until it reached a simulation time of 1 second. For reference, all calculations discussed in this section and the previous one were performed on a SUN V480 computer.

For the first performance indicator, we examind the total number of iterations by the solver over the course of the calculation. Though we realize the amount of work per iteration is different for each method, this information does allow a straightforward comparison between the methods and also a direct analysis of a single method with different convergence tolerances. Figures 8 and 9 show the cumulative number of iterations taken by BICGSTAB[a]. First notice that in the beginning of the calculation, only a few iterations are needed (on the order of 1 to 2 per timestep), but as the system evolves (between $1.0 \cdot 10^{-10}$ and $1.0 \cdot 10^{-6}$ seconds) the number of iterations increases to about 10 per timestep. Note also that at a simulation time of about $1.0 \cdot 10^{-6}$ seconds the total number of iterations begins to diverge for different values of the convergence tolerance due to the extra work required to meet the stricter convergence criteria. As the calculation nears completion, the total number of iterations greatly increases along with a marked divergence between the various values of convergence tolerances. This is due to the fact that as the simulation time approaches that of the diffusion timescale, the system begins to enter into its steady state configuration. In this case, the matrix becomes quite ill conditioned and each additional iteration by the solver produces little or no improvement. Thus, the solver iterates until it reaches the preset limit (for this test problem, the limit is 500 iterations), at which point it stops iterating and checks the isotopic mass fractions produced in the particular timestep for accuracy. If the mass fractions are acceptable, then the simulation is allowed to go to the next timestep, if the mass fractions are not acceptable, then the current timestep is reduced. Unfortunately, a large portion of time can be used on unnecessary iterations at these late times and we plan to address this issue in the future. Finally, if one compares the Tol=$1.0 \cdot 10^{-9}$ line in Figure 8 with the others, it is apparent that this line levels off at the very end of the calculation. This is interesting because it reveals that once the system achieves a steady state, the initial solution has a residual less than 10^{-9} but greater than 10^{-10}. Thus, a stricter convergance tolerance poses problems at late times. The similar comparison for GMRES is presented in Figures 10 and 11. The behavior of this iterative solver is virtually identical except that this solver requires fewer numbers of iterations for a given tolerance.

Figure 12 shows the total amount of time taken for each calculation to complete. The one data point at Tol=$1.0 \cdot 10^{-9}$ seems to indicate that when fewer iterations are needed, GMRES outperforms BICGSTAB when solving the matrix equations that occur in our test calculation. For smaller values

---

[a]The Tol=10^{-8} is not included in Figure 8 because this relatively large tolerance setting introduced too much error into the mass fractions from which the calculation could not recover.

of Tol, BICGSTAB finishes faster by an average of about 10 minutes. This effect may be due to the large number of iterations taken when the system gets into a near steady state (as mentioned above). Since GMRES is more costly per iteration than BICGSTAB, the large number of iterations taken at the end of the calculation may be adding more time to the calculations using GMRES than to those using BICGSTAB. More work is needed to confirm this hypothesis so that one could truly tell which method would optimize the calculation for time. As is shown in Figure 12, both methods seem to scale similarly with decreasing values of tolerance.

An important issue when solving any problem is the amount of error introduced by the particular method of solution. Figures 13 and 14 show an estimate of the error for various tolerances with both BICGSTAB and GMRES. To conserve mass, the sum of the mass fractions in each zone is required to equal 1. Given this, one can estimate the error in the solution in zone i with equation $error_i = 1 - X_{sum,i}$ where the quantity $X_{sum,i}$ is the sum of mass fraction of all of the isotopes in zone i. With this, one can get an estimate of the total error by summing $error_i$ over all of the zones. This sum is plotted in Figures 13 and 14 against simulation time. In these plots, one can see that as the tolerance gets tighter the amount of estimated error is reduced. BICGSTAB gives lower values of the estimated error over the course of the calculation than GMRES, until the very end of the calculation. At very late times, the solution obtained by GMRES has a slightly lower value of the estimated error than does BICGSTAB. A second interesting feature at late times for both methods is that the solutions obtained using a smaller value for the convergance tolerance actually introduces more error than those with a larger value for the tolerance. Upon examination of Figures 8 and 10 during these late times, it is possible that the large number of extra iterations added to the estimated error. It would be interesting to reexamine this question having addressed issues that occur when the system approaches a steady state.

Finally, Figure 15 shows the scaling of the test calculation if one varies the number of spatial zones. GMRES was used in each of these calculations with a tolerance of 10^{-10}. In each calculation, the nucleosynthesis network was kept the same. As expected, the time to complete a calculation increases as the number of spatial zones is increased. We plan to carry out a full comparison between BICGSTAB and GMRES over a range of zone numbers and tolerances in the future.

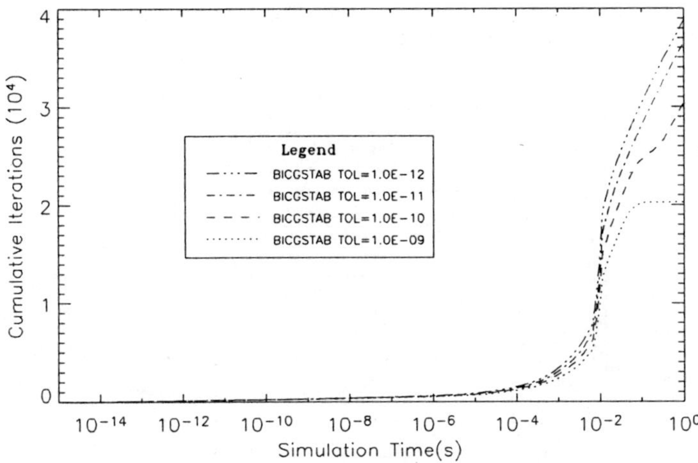

Figure 8. Cumulative number of iterations for BICGSTAB calculations with different values of Tol.

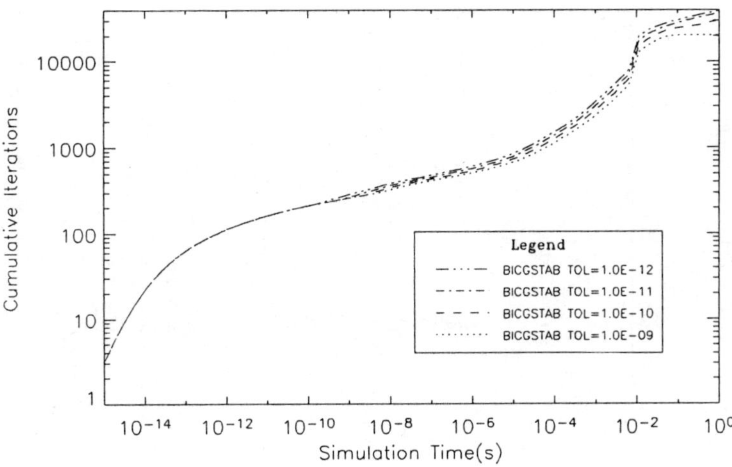

Figure 9. Cumulative number of iterations for BICGSTAB calculations with different values of Tol.

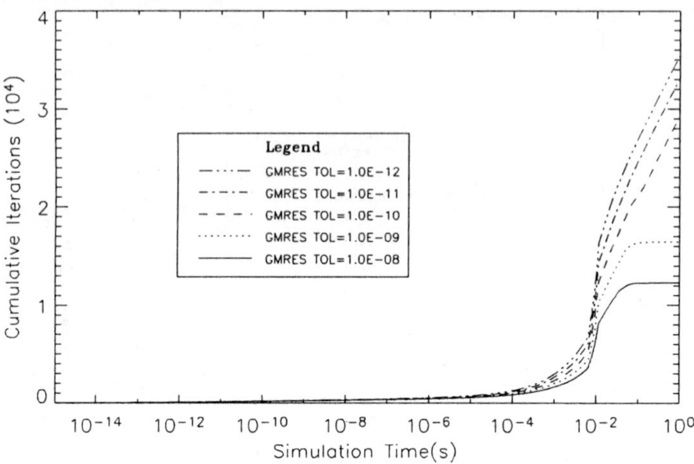

Figure 10. Cumulative number of iterations for GMRES calculations with different values of Tol.

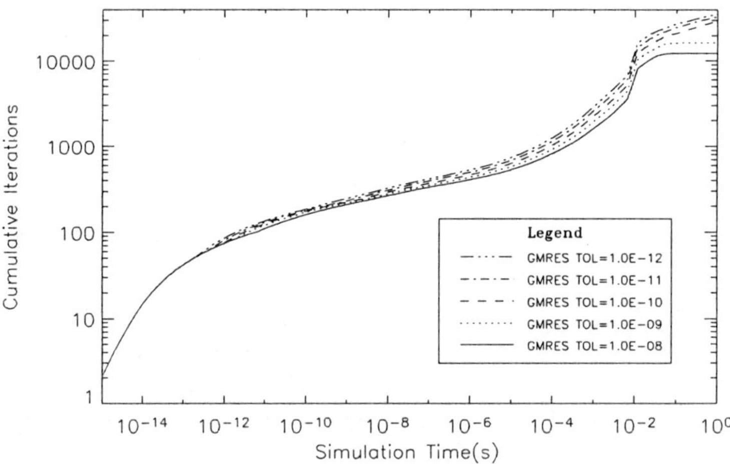

Figure 11. Cumulative number of iterations for GMRES calculations with different values of Tol.

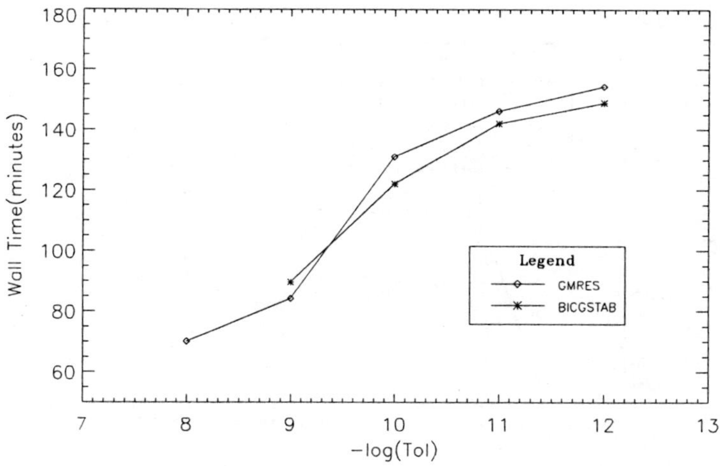

Figure 12.   Scaling of Wall Time with Tol.

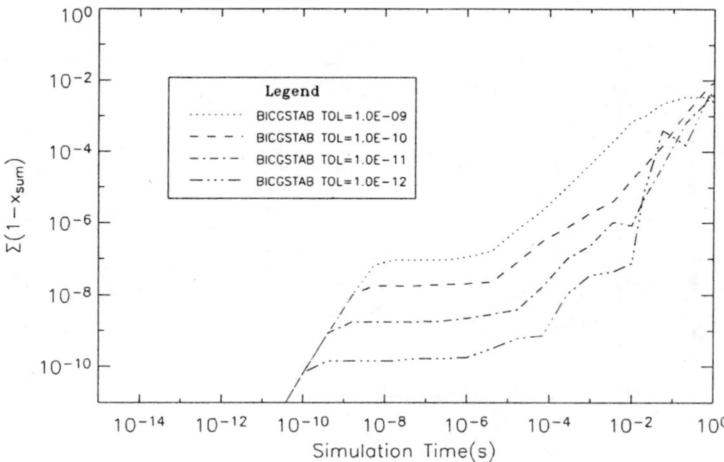

Figure 13.   Estimated errors of calculations using BICGSTAB over different values of Tol.

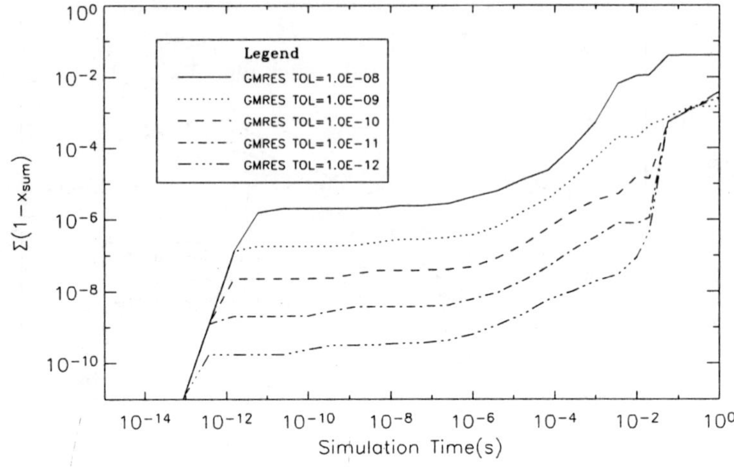

Figure 14. Estimated errors of calculations using GMRES over different values of Tol.

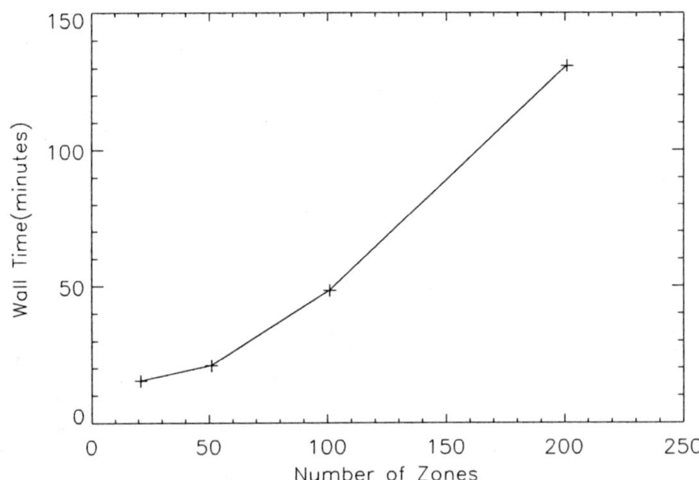

Figure 15. Scaling of GMRES calculation with Tol$= 1.0 \cdot 10^{-10}$ for different numbers of spatial zones.

5. A Comparison with Operator Splitting

In this section we compare our $D = 1$ calculations with some results from operator splitting calculations. For our operator splitting calculations, we first mixed the compositions of each zone according to our diffusion prescription in Eq. (2) and then solved the nuclear burning in each mass shell with a one-zone nuclear network. Each time the composition is mixed, it disturbs the steady state that is established in the system. This means we cannot use the fractional change in any nuclear mass fraction to determine the next timestep in the overall problem; hence, there is no prefered timestep for the mixing. We chose to divide the one second into an equal number of mixing timesteps denoted by dt. Thus, for our $dt = 0.05$ second calculation, the composition was mixed on a timestep of 0.05 seconds. Then the one zone nuclear calculations were run over a period of 0.05 seconds, and the process was repeated 20 times to give the full evolution. We repeated this with $dt = 0.005$ seconds (200 total mixing steps) and $dt = 0.001$ seconds (1000 total mixing steps).

Figure 16 shows the ^{4}He mass fraction in each of the operator splitting calculations. For comparison, the fully coupled calculation and the NSE ^{4}He mass fractions are shown. All calculations work well in the inner regions where the nuclear burning dominates the mixing; however, for the outer regions where mixing becomes important, the operator splitting calculations deviate from the (correct) coupled calculation. Clearly, the smaller the dt, the better the fully coupled calculation is approximated, which makes sense because the two approaches are equivalent when $dt \rightarrow 0$. It is worth pointing out, however, that, because the operator splitting calculations had to re-establish a steady state after each mixing, they take much longer than the fully coupled calculation. For comparison the fully coupled calculation took only two hours while the $dt = 0.001$ second operator splitting calculation took almost thirty seven hours!

Where such differences in the treatment of mixing and burning may be particularly important is in the prediction of the production of radioactivities like ^{44}Ti. The abundance of this isotope may be measured in supernova remnants[9] and is therefore a sensitive indicator of the location of the supernova mass cut[10]. Figure 17 shows the ^{44}Ti mass fraction in the fully-coupled and the operator splitting calculations. Also shown is the NSE mass fraction for comparison. It is evident that the fully-coupled calculation produces the least amount of ^{44}Ti. Indeed, the total yield of ^{44}Ti in the fully-coupled calculation is 1.96×10^{-6} $M_\odot$. This is in comparison to 1.27×10^{-5} $M_\odot$,

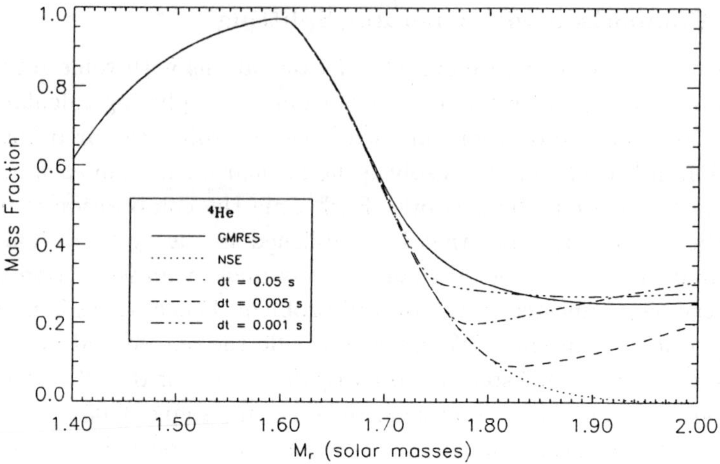

Figure 16.   Comparison of the ^{4}He mass fraction at 1 second in the model as a function of Lagrangian mass coordinate for $D = 1$ for the fully coupled diffusion and nuclear network solution and for three different operator-splitting calculations. The NSE solution is also shown for comparison.

6.19×10^{-6} $M_\odot$, and 3.39×10^{-6} $M_\odot$ for the $dt = 0.05$, $dt = 0.005$, and $dt = 0.001$ second runs, respectively, and 2.81×10^{-5} $M_\odot$ for NSE. Important caveats are that our model is not a full supernova model and the mixing is treated in a diffusion approximation. Nevertheless, we expect a careful treatment of coupled hydrodynamics and nuclear burning will be important in reliable predictions of the yields of supernova observables in the coming years.

6. Conclusions

The success of our calculations shows that coupling of large-scale nuclear reaction networks with mixing schemes in a fully implicit manner is feasible. It is clearly necessary to employ iterative methods, but a two-step preconditioner works well, and, in our experience our solutions tend to break down only when the system reaches a global steady state such that the matrix becomes close to ill conditioned. In practice, this will not generally be a problem since, unlike in our parameterized studies, in the full supernova problem, the temperature and density will in fact be changing on timescales comparable to the mixing or burning timescales. This will

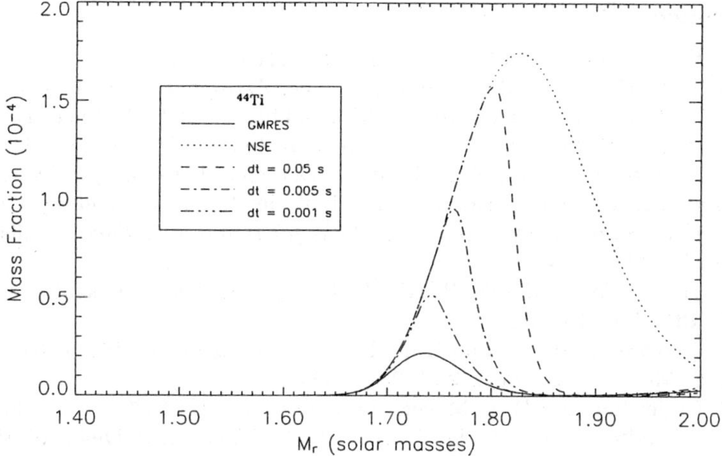

Figure 17.   Comparison of the ^{4}4Ti mass fraction at 1 second in the model as a function of Lagrangian mass coordinate for $D = 1$ for the fully coupled diffusion and nuclear network solution and for three different operator-splitting calculations. The NSE solution is also shown for comparison.

prevent the full system from reaching a steady state and the matrix from becoming singular.

While our calculations are restricted to one-dimensional simulations in the diffusion approximation, extension to three dimensions with more complicated mixing should be straightforward. The underlying matrices will be similar to that in Figure 2 but with more mixing bands that are not all aligned. This will require a rewrite of the first step of the preconditioner, but this should not be difficult. The more onerous task may be implementing a parallel version of the code, but we again do not anticipate any fundamental difficulties. Indeed, the only limitations we now foresee on directly incorporating a large-scale nuclear network into a supernova code are hardware ones, and, given the recent and impressive increase in the speed and size of computers, even these may be removed sooner than one might suppose.

Acknowledgments

This work was supported by DOE grant DE-FC02-01ER41189 and by NASA grant NAG5-10454.

References

1. F. X. Timmes, *Astrophys. J. Suppl* **124**, 241 (1999).
2. M. F. El Eid, B. S. Meyer, and L.-S. The, *Astrophys. J.*, **611**, 396 (2004).
3. W. R. Hix, F.-K. Thielemann, A. M. Khokhlov, and J.-C. Wheeler, in *Stellar Evolution, Stellar Explosions and Galactic Chemical Evolution*, 599 (1998).
4. R. Barrett, M. Berry, T. F. Chan, J. Demmel, J. Donato, J. Dongarra, V. Eijkhout, R. Pozo, C. Romine, and H. V. der Vorst, *Templates for the Solution of Linear Systems: Building Blocks for Iterative Methods*, (Philadelphia, PA: SIAM) (1994).
5. M. Herant, W. Benz, W. R. Hix, C. L. Fryer, and S. A. Colgate, *Astrophys. J.*, **435**, 339 (1994).
6. A. Burrows, J. Hayes, and B. A. Fryxell, *Astrophys. J.*, **450**, 830 (1995).
7. B. S. Meyer, *Phys. Rep.*, **227**, 257 (1993).
8. G. C. Jordan, and B. S. Meyer, *Astrophys. J. Lett.*, **617**, L131 (2004).
9. A. F. Iyudin, V. Schönfelder, K. Bennett, H. Bloemen, R. Diehl, W. Hermsen, G. C. Lichti, R. D. van der Meulen, J. Ryan, and C. Winkler, *Nature*, **396**, 142 (1998).
10. F. X. Timmes, S. E. Woosley, D. H. Hartmann, and R. D. Hoffman, *Astrophys. J.*, **464**, 332 (1996).

NUCLEOSYNTHESIS FROM EXOTIC SUPERNOVAE

G. C. MCLAUGHLIN AND J. P. KNELLER*

*Department of Physics
North Carolina State University
Raleigh, NC, 27608 USA
E-mail: Gail_McLaughlin@ncsu.edu, Jim_Kneller@ncsu.edu*

R. A. SURMAN

*Department of Physics
Union College
Schenectady, NY 12308
E-mail: surmanr@union.edu*

We discuss the neutrinos and nucleosynthesis in exotic supernovae. We consider the case where the collapse of a massive star forms a black hole surrounded by an accretion disk instead of the usual proto-neutron star. We examine accretion rates of $0.1 M_\odot/s - 10 M_\odot/s$, which are sufficiently large to produce strong fluxes of neutrinos. In many cases significant regions of trapped neutrinos are created in the accretion disk. We discuss the implications for the nucleosynthesis that occurs in these objects as well as for the neutrino-antineutrino annihilation above the disk.

1. Introduction

While gamma ray bursts were first observed more than thirty years ago, it is only recently that great strides have been made observationally. The current evidence points to explosions of massive stars, i.e. some sort of exotic supernova as the origin of these burst.

Gamma ray bursts represent a new frontier in the study of neutrino astrophysics and element synthesis. There is recent observational evidence that at least some long duration gamma ray bursts are "exotic supernova" [1,2,3,4,5,6,7,8]. These bursts have afterglow spectra that are similar to Type Ib and Ic supernovae. Type Ib/c are core collapse supernovae that have lost their hydrogen and helium envelopes, respectively. According to the collap-

---

*Work partially supported by grant de-fg-02er4166 of the Department of Energy sar model [9] a gamma ray burst is produced when a massive star undergoes core collapse, but is rotating too rapidly for the "normal" collapse and bounce associated with supernovae. Instead an accretion disk is produced that surrounds a black hole. A similar configuration of an accretion disk surrounding a black hole is produced in neutron star-neutron star mergers. Neutron star mergers are a candidate site for the short duration bursts [10,11]. Both the massive star core collapse and the neutron star merger events are expected to occur at rates on the order of $\sim 10^{-5} - 10^{-6}$ per year for the Galaxy.

As in the case of supernovae, the neutrinos from accretion disks may be responsible for powering a gamma ray burst. We explore the energy deposition that comes from neutrino-antineutrino annihilation above the gamma ray burst. Another similarity with supernovae is that the outflow will produce a variety of elements that will contribute significantly to the abundance of certain isotopes observed in our solar system today. We explore the elements that are produced in the course of the outflow from the accretion disk. We begin by discussing the neutrinos in the disk.

2. Neutrinos from Accretion Disks

We calculate the nuclear composition of matter in accretion disks surrounding stellar mass black holes which are thought to accompany GRBs. We use accretion disk parameters from [12,13]. As in reference [14] we follow mass elements in the accretion disk starting at the point of nuclear dissociation and calculate the evolution of the electron fraction. We find that the neutronization of the disk material by electron capture can be reversed by neutrino interactions in the inner regions of disks with accretion rates of 1 $M_\odot/s$ and higher. For these cases the inner disk regions are optically thick to neutrinos. In order to estimate the emitted neutrino fluxes we calculate the surface of last scattering for the neutrinos (the equivalent of the proto-neutron star neutrinosphere) for each optically thick disk model. Fig 1 shows the surfaces of last scattering for the neutrinos in two different accretion disks. We estimate the influence of neutrino interactions on the neutron-to-proton ratio in outflows from GRB accretion disks, and find that it can be significant even when the disk is optically thin to neutrinos. Optical depths of neutrinos in accretion disks have also been considered in [15].

In order to determine what sort of nucleosynthesis may occur we must examine the composition of matter as it flows away from gamma ray burst

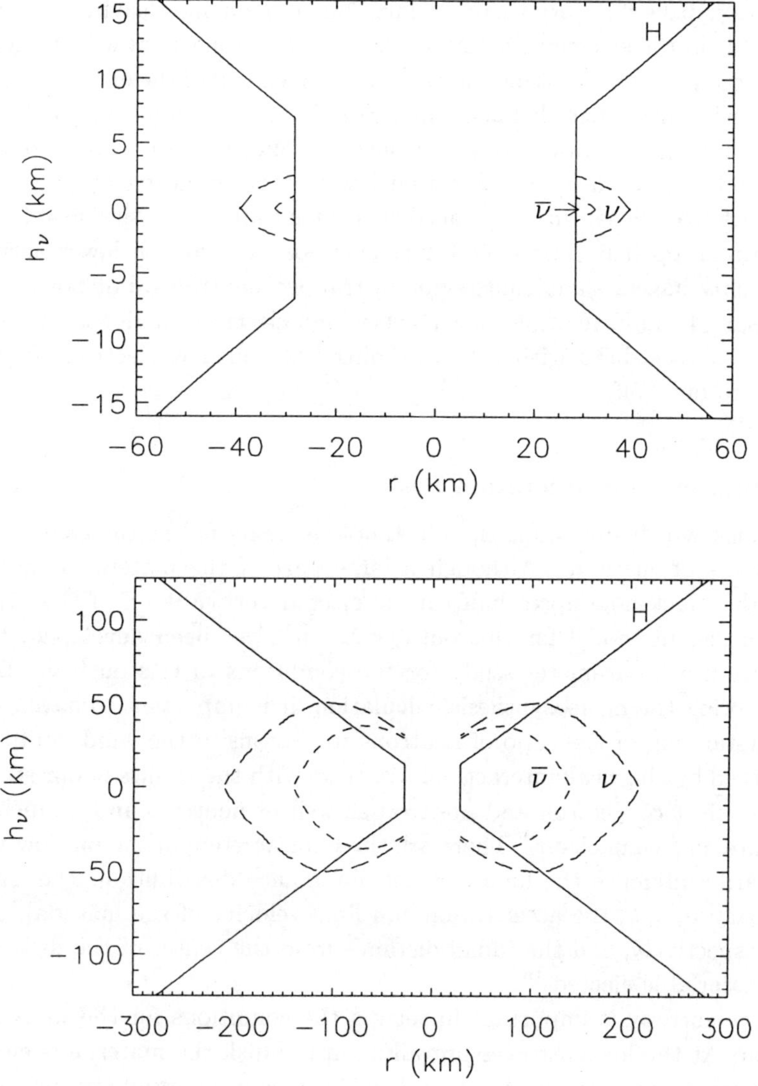

Figure 1.   Neutrino and antineutrino surfaces in a $\dot{M} = 1\,\mathrm{M_\odot/s}$, $a = 0$ (a, on the top) and a $\dot{M} = 10\mathrm{M_\odot/s}$, $a = 0$ (b, on the bottom) accretion disk. The solid line shows the density scale height of the disk.

accretion disks [16]. Since there is a large flux of neutrinos leaving the surface of the disk, the electron fraction of the outflowing material will change due to charged current neutrino interactions. We calculate the electron fraction in the wind using detailed neutrino fluxes from every point on the disk and study a range of trajectories and outflow conditions for several different accretion disk models. We find that low electron fractions only appear in outflows from disks with high accretion rates that have a significant region of both trapped neutrinos and antineutrinos. Disks with lower accretion rates only have a significant region of trapped neutrinos, but not antineutrinos, and can have outflows with very high electron fractions. The lowest accretion rate disks with little trapping have outflow electrons fractions closer to one half.

3. Outflow from Accretion Disks

The disk which surrounds the black hole is being fed at the edges by the remnants of the star. Although a large part of the material is accreted into the black hole up to half can be ejected vertically off of the disk. A self-consistent model for this outflow has not yet been developed, therefore we use a parameter study for the conditions in this outflow. Before attempting the nucleosynthesis calculation, it is important to have a good understanding of the ratio of neutrons to protons in the wind, and this is governed by the weak interaction. We start with the results of our study [14] of the effect of electron and positron as well as neutrino and antineutrino capture on free nucleons. There are three parameters in the outflow which primarily influence the final element abundance distribution: the entropy per baryon, s/k, the acceleration and final velocity of the material, β and v_∞ respectively, and the radial distance from the center of the disk where the material is ejected [16].

The entropy is important in setting the conditions for the weak interaction. At the low entropies prevailing in the disk the material is electron degenerate and therefore has very low Y_e. This was initially proposed as a mechanism for making the r-process in low accretion rate disks [17,18]. However, it was subsequently noted that for estimated entropies of 40, positron capture in the outflow drives the electron fraction up to values that are unacceptable for the r-process [19]. In the highest accretion rate cases the entropy is not quite as crucial as it is in the lower accretion rate cases, since the determining factor in the former is the neutrino and antineutrino capture reactions. These will force the electron fraction to be quite low.

Similarly in the medium accretion rate cases, neutrino capture dominates the determination of the relative numbers of neutrons and protons. However, in the lower accretion rate cases, electron and positron capture plays a significant role and therefore the entropy is a crucial indicator of the final element synthesis. Fig. 2 shows the final electron fractions of material ejected from an accretion disk. In the case of lower accretion rate disks, the neutron-to-proton ratio is strongly influenced by the value of the entropy.

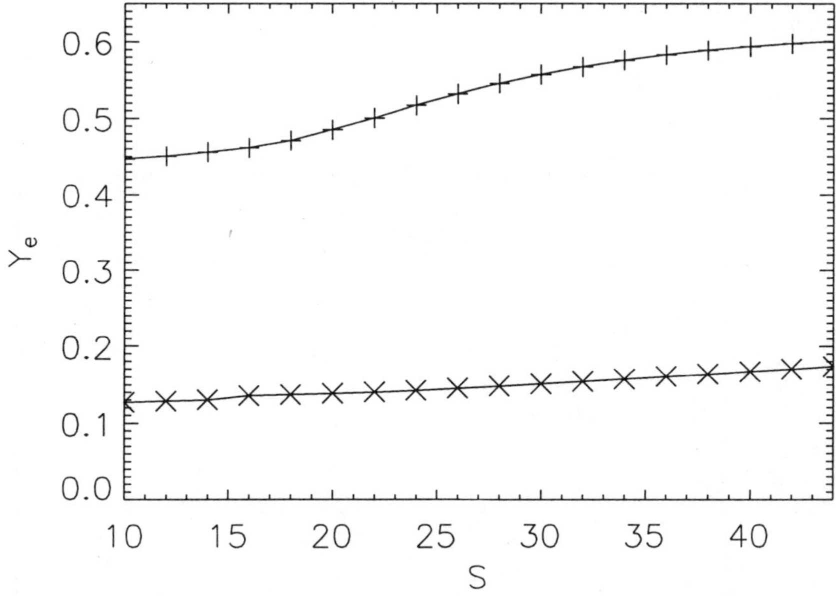

Figure 2. Shows the value of the final electron fraction in the outflow from a gamma ray burst accretion disk (with $\beta = 2.4$ and $v_\infty = 3 \times 10^4$ km/s) for the $\dot{M} = 10\,M_\odot/s$ disk model trajectories (lower line, starting disk radius $r_0 = 250$ km) and the $\dot{M} = 0.1\,M_\odot/s$ disk model trajectories (upper line, $r_0 = 100$ km).

Outflow timescale (β and v_∞) will have a very similar role to the outflow timescale in the neutrino driven wind of the core collapse supernova. Fast outflow will mean less time for alpha particles to combine into heavy nuclei and therefore a smaller number of seed nuclei, whereas slow outflow will mean more time in which a greater number of heavy nuclei can form.

Disk position is a new variable that is not considered in the core collapse supernova case, since there the proto-neutron star geometry is spherically symmetric. Here, due to the geometry, this new parameter must be taken into account. Material ejected close to the center of the accretion disk will have lower initial electron fraction, but will also be subjected to more neutrino and antineutrino capture. If the ejection occurs farther out on the disk, the initial electron fraction will be higher, but there will be less neutrino capture on the way out of the disk. Fig. 2 shows the results of a calculation of the electron fraction in the outflow for two different ejection points on the disks. Material which is ejected from the inner regions of the disk will be more strongly influenced by the neutrinos than that ejected farther out.

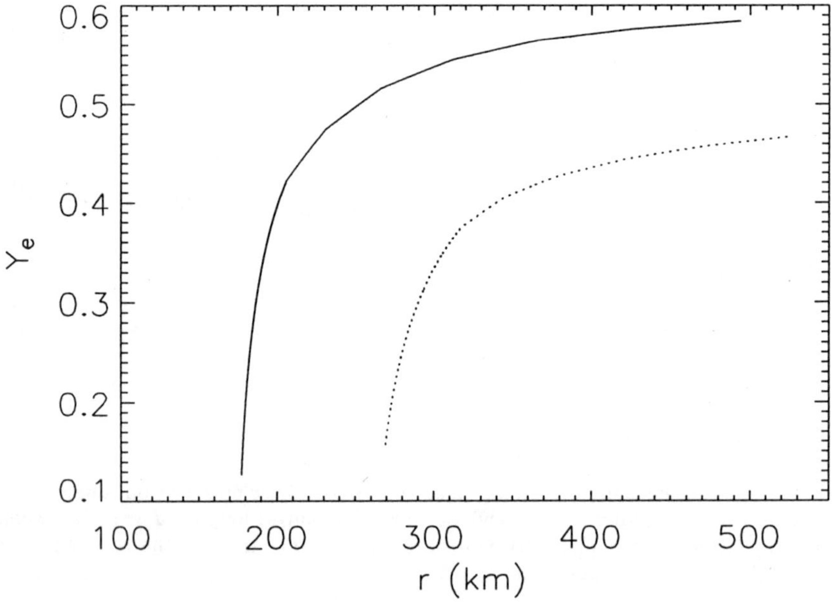

Figure 3. This figure shows two trajectories from the DPN $\dot{M} = 1\,M_\odot/s$ disk model, with the same outflow parameters ($\beta = 0.8$, $s/k = 10$, $v_\infty = 10^4$ km/s) but with different starting disk positions. The solid line is for a trajectory with starting disk radius $r_0 = 167$ km and the dashed line is for a trajectory with $r_0 = 250$ km.

4. Nucleosynthesis in the Outflow of Accretion Disks

We study three different types of disk models, relatively high accretion rates, $\dot{M} = 10M_\odot/s$, medium accretion rates $\dot{M} = 1M_\odot/s$, and low accretion rates $\dot{M} = 0.1M_\odot/s$. For the highest and medium accretion rate disks, the electron neutrino and antineutrino interactions play the dominant role in determining the nucleosynthesis that takes place in the material which is ejected from the disk. These accretion rates reflect those that would occur in the neutron star - neutron star merger scenario. In the highest accretion rate disks the electron fraction is driven quite low in the ejecta, so the material is very neutron rich. Our preliminary calculations show that this is a potentially viable environment for the r-process [20]. Fig. 4 shows two nucleosynthesis calculations using the high accretion disk model.

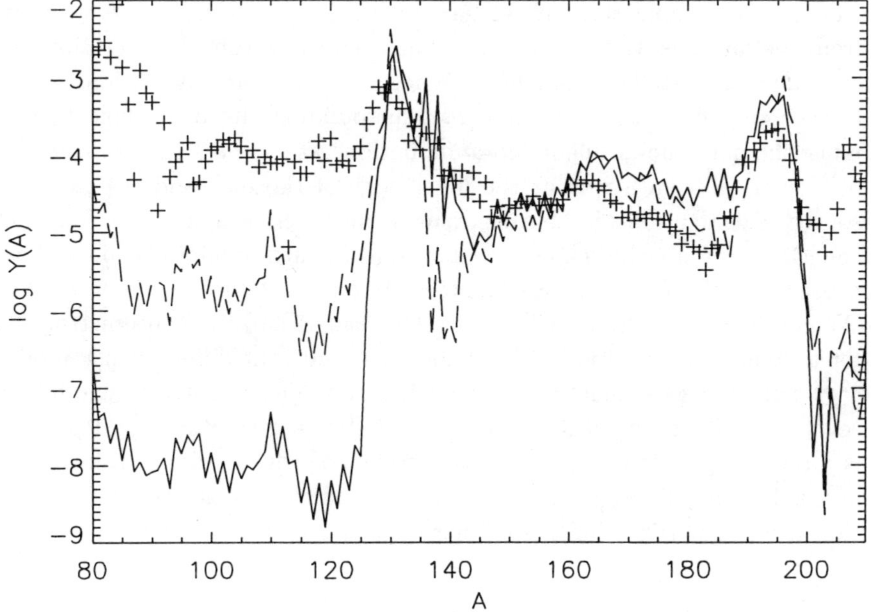

Figure 4.   R-process nucleosynthesis abundances resulting from a reaction network calculation using a $\dot{M} = 10M_\odot/s$ model and outflow conditions of $\beta = 2.0$, $s/k = 48$ (solid line) as well as $\beta = 0.8$ and $s/k = 12$ (dashed line). The starting disk radii for both trajectories is $r_0 = 250$ km. The crosses show the solar system abundances.

For the medium accretion rate disks the nucleosynthesis products come from very high electron fraction environments. These occur when there is a significant region of trapped neutrinos, but not of trapped antineutrinos. This is an unusual environment since it is very proton rich, relatively low entropy ($s \sim 10-40$) *and* it will produce a primary nucleosynthesis process, i.e. the material begins in free nucleons and as it cools nuclei are formed. It is similar to the environment in the big bang, except that the entropy is many orders of magnitude lower.

Lower accretion rate disks are currently predicted by the collapsar model [9,21]. For low accretion rates and entropies per baryon around $s/k \sim 40$ the electron fraction is closer to 1/2, i.e. equal numbers of neutrons and protons. In this case rare nuclei such as Scandium and p-process nuclei may be produced [22,23].

5. Neutrino - Antineutrino Annihilation from Accretion Disk Neutrinos

The central engine that drives gamma ray burst explosions may derive from the large neutrino luminosity that is emitted from the accretion disk surrounding a stellar mass black hole. We use our calculations of the flux coming from the accretion disk surrounding the black hole to estimate the neutrino-antineutrino annihilation rate. The strongest neutrino flux emerges from a ring at the inner edge of the disk and the extended source leads to a considerable departure from spherical symmetry. The non-spherical geometry increases the efficiency of neutrino annihilation and can cause a collimation of scattered material into a jet along the axis of the disk. We consider annihilation efficiencies (see Fig.5) and momentum transfers for a number of accretion disk models. We find that different disk models [12,13] not only give neutrino-antineutrino annihilation rates that differ by several orders of magnitude, but also that the pattern of energy deposition is altered[24]. As more sophisticated models of the accretion disk become available this will also affect the total rate and the deposition pattern, particularly in the region near the inner edge of the accretion disk.

6. Conclusions

Accretion disks that form in the collapse and subsequent explosion of massive stars represent a rich new area in which to study neutrinos and nucleosynthesis. A wide variety of nucleosynthesis is possible in the outflow from the disk. This depends primarily on the accretion rate, but also on the

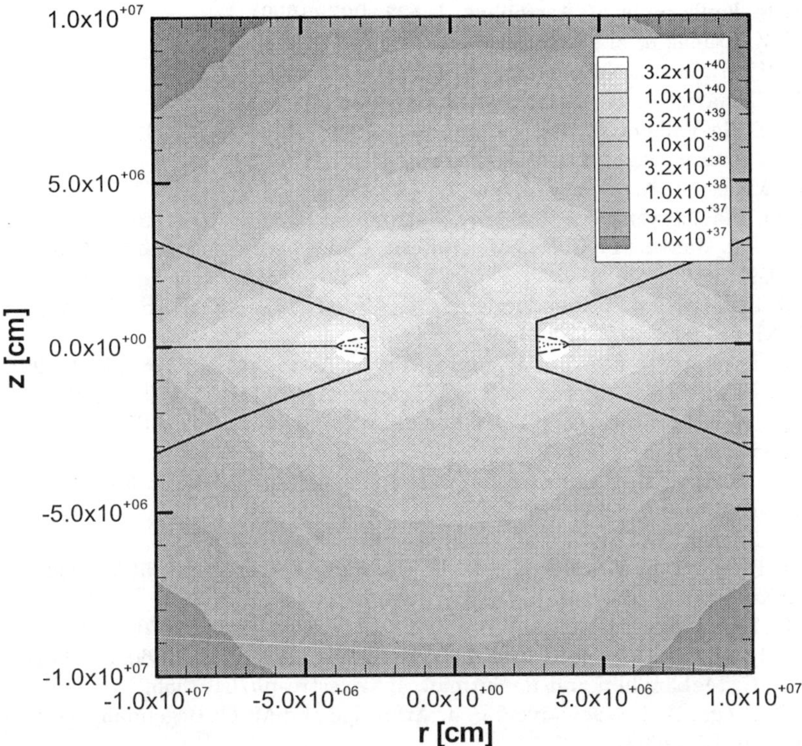

Figure 5. Shows the rate of neutrino-antineutrino annihilation, in units of eV/cm^{-3}/s above a GRB accretion disk disk. The black hole is in the center and the solid lines show the density scale height of the disk.

entropy and outflow timescale of the ejected material. For a rapid accretion rate and rapid outflow, the conditions are viable for producing an r-process. In the future, more sophisticated models will determine definitively what sort of nucleosynthesis comes from these 'exotic supernovae'.

In addition to the disk outflow, these accretion disks may produce a jet which in turn produces a gamma ray burst. The neutrino flux emitted from the disk may be instrumental in driving the jet. Again however, the flux depends greatly on the accretion rate of the of the disk. Future models will tell us whether neutrino-antineutrino annihilation is responsible for providing the energy which drives gamma ray bursts.

References

1. T. J. Galama et al., Nature 395, 670 (1998).

2. S. R. Kulkarni et al. Astrophys. J. 522, L97 (1999).

3. K. Z. Stanek et al., Astrophys. J., 591, L17 (2003).

4. J. Hjorth et al., Nature 424, 751 (2003).

5. P. Garnavich et al., Astrophys. J. 582, 924 (2003).

6. M. Della Valle et al., Astron. and Astrophys. 406, L33 (2003).

7. B. Thomsen, Astron. and Astrophys. 419, L21 (2004).

8. D. Malesani, Astrophsy. J. 609, L5 (2004).

9. A. I. MacFadyen, S. E. Woosley, Astrophys. J., 524, 262 (1999).

10. H.-Th. Janka, T. Eberl, M. Ruffert, C. L. Fryer, Astrophys. J. 527, L39 (1999).

11. S. Rosswog, M. Liebendoerfer, MNRAS, 342, 673 (2003), S. Rosswog et al. MNRAS, 345, 1077 (2003).

12. R. Popham, S. E. Woosley, and C. Fryer, Astrophys. J., 518, 356 (1999).

13. T. DiMatteo, R., Perna, and R. Narayan, Astrophys. J. 579, 706 (2002).

14. R. Surman and G. C. McLaughlin, Astrophys. J. **603**, 611 (2004) [arXiv:astro-ph/0308004].

15. W. H. Lee, E. Ramirez-Ruiz and D. Page, Astrophys. J. **608**, L5 (2004) [arXiv:astro-ph/0404566].

16. R. Surman and G. C. McLaughlin, arXiv:astro-ph/0407206.

17. J. Pruet, S. E. Woosley and R. D. Hoffman, Astrophys. J. **586**, 1254 (2003) [arXiv:astro-ph/0209412].

18. K. Kohri, R. Narayan and T. Piran, arXiv:astro-ph/0502470.

19. J. Pruet, T. Thompson, and R. D. Hoffman, Astrophys. J., 606, 1006 (2004).

20. G. C. McLaughlin and R. Surman, arXiv:astro-ph/0407555.

21. D. Proga, A. I. MacFadyen, P. J. Armitage and M. C. Begelman, Astrophys. J. **599**, L5 (2003)

22. J. Pruet, R. Surman and G. C. McLaughlin, Astrophys. J. **602**, L101 (2004) [arXiv:astro-ph/0309673].

23. S. Fujimoto, M. Hashimoto, K. Arai and R. Matsuba, "Nucleosynthesis inside Gamma-Ray Burst Accretion Disks," Astrophys. J. **614**, 847 (2004) [arXiv:astro-ph/0405510].

24. J. P. Kneller, G. C. McLaughlin and R. Surman, submitted to MNRAS, arXiv:astro-ph/0410397.

TOWARD THREE-DIMENSIONAL MODELS OF CORE-COLLAPSE SUPERNOVA SPECTRA AND LIGHT CURVES: MOTIVATIONS AND CHALLENGES

R. C. THOMAS

Lawrence Berkeley National Laboratory,
1 Cyclotron Road MS 50R5008,
Berkeley, CA 94720-8158, USA
E-mail: rcthomas@lbl.gov

Multiple lines of evidence indicate that some fraction of core-collapse supernovae are nonspherical. The observational evidence includes spectropolarimetry, neutron star kicks, gamma-ray burst/supernova coincidences, speckle interferometry and direct imaging, and the morphologies of Galactic supernova remnants. Otherwise spherically symmetric theoretical models of observed supernovae often invoke multidimensionality to explain certain phenomena. Understanding spectra and light curves arising from multidimensional explosions requires the solution of the multidimensional model supernova atmosphere problem. This contribution presents the problem and its context, and outlines some current efforts at a solution.

1. Introduction

Though optical radiation constitutes but one percent of one percent of the total energy-release budget of core-collapse events, it transmits a plethora of facts to be decoded in light curves and spectra. Core-collapse supernova (SN) light curves constrain the energy imparted to progenitor envelopes by the collapse mechanism, the configuration of the progenitor and its environment prior to explosion, and the amount of energy deposited by freshly synthesized radioactive nuclei into the ejecta.[44,86] Spectroscopic time-sequences provide further details, including ejecta composition, density, temperature, and velocity structure.[7,21] Such findings inform a variety of astrophysical problems, including stellar evolution and cosmic nucleosynthesis, and leverage the utility of core-collapse SNe as cosmic distance indicators.

Detections of the youngest core-collapse SNe occur a few days to a week after outburst (usually they are a few weeks old or much older at discovery). Hence, by the time of earliest practical observations, the dynamics of the outer layers of a SN are described by the velocity law $v = r/t$ (homologous expansion). Characteristic velocities are on the scale of $10,000$ km s^{-1}. The diffusion and expansion timescales in the outer layers are still comparable, and emergent spectra evolve on timescales of days and weeks. A SN gradually becomes more optically thin to continuum in its outermost layers, below which a photosphere forms and recedes as the ejecta expand, cool and the density drops as t^{-3}.

This configuration is characterized in line spectra by the P Cygni profile, consisting of a blueshifted absorption feature and a broad emission peak centered at the rest wavelength of the transition. Ejected material in front of the photosphere scatters photospheric radiation away from the line of sight to the observer, creating the blueshifted absorption feature. Material external to the region in front of the photosphere produces a broad emission feature, since it arises from ejecta moving both toward and away from the observer. At later times (exactly when depends on circumstances described below), the photosphere disappears and the SN spectrum becomes "nebular," typified by strong emission features from forbidden ions and a lack of an appreciable continuum.

In the overall picture, some core-collapse mechanism ejects the envelope; energy stored in the ejecta (shock-deposited and from radioactivity) as well as the distribution of elemental abundances (manifested as opacity) result in the light show. Differences among core-collapse spectra and light curves reflect variations in initial conditions such as progenitor envelope mass, the amount and distribution of radioactives produced, and the density of the circumstellar environment. Hence, various classes of core-collapse SNe arise from corresponding regions in the parameter space of these variables.

SNe are classified by the appearance of their spectra (see examples in Figs. 1 and 2)[73] and sometimes the appearance of their light curves (see examples in Fig 3).[18,16,77,72] The basic dichotomy is the hydrogen Balmer line litmus test: Type II SNe have conspicuous hydrogen features at early time while Type I SNe do not, though there are exceptions. The most notorious exception is the extremely peculiar SN 2002ic,[27] spectra of which possessed Hα in emission: Other factors favored a classification of Type Ia, with the most likely scenario being the explosion of a white dwarf in a dense circumstellar medium. Neglecting this case and the other SNe Ia leaves us with the core-collapse types.

The two most familiar classes of SNe II are based on light curve shape.[3] Light curves of SNe IIP possess a "plateau" phase of nearly constant luminosity following maximum light persisting for three or four months,[26]

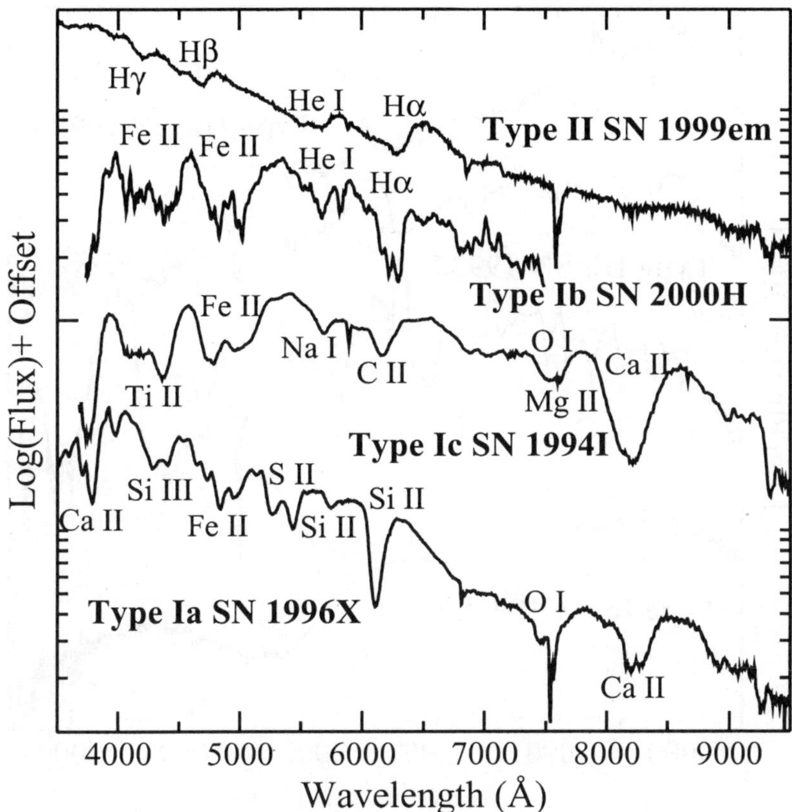

Figure 1. Example photospheric phase SN spectra. The top three examples are of "typical" core-collapse events, the bottom one thermonuclear (included for comparison). The sharp absorptions around 6800 Å and 7600 Å are telluric absorption, not intrinsic to SN spectra.

while those of SNe IIL diminish "linearly" after maximum (compare the top two curves Fig. 3). Initially, objects from both subclasses have nearly featureless blackbody spectra which are quickly populated with strong hydrogen Balmer and He I features. As the spectra evolve, they accumulate lines from Fe II and Ca II. In the nebular phase, both light curves become exponential tails.

The SN IIP plateau is driven by steady hydrogen recombination in the shock-heated ejected envelope as it cools through the recombination temperature.[17] Hence, the plateau phase duration is correlated to the thickness of the ejected hydrogen envelope, and models suggest red supergiant

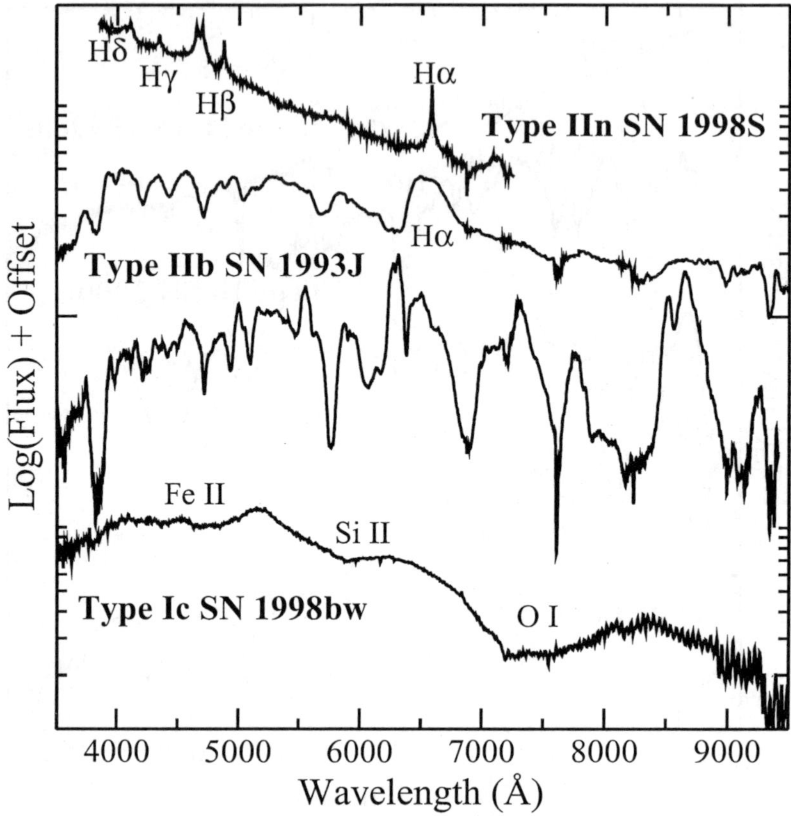

Figure 2. More example SN spectra. The spectra of SNe IIn (like SN 1998S) are characterized by sharp emission features at early times. SNe IIb form an intermediate class of object, characterized by a short-lived hydrogen signature. A small fraction of core-collapse SN spectra are characterized by extreme Doppler shifting and blending, the so-called "hypernovae," and have been observed in coincidence with gamma-ray bursts.

progenitors with initial masses above 8 $M_\odot$. Recent Hubble Space Telescope and ground-based observations[79,67] lend support to these models, and future observations will increase our understanding of the progenitors of type II SNe. The thickness of the envelope is believed to mask nonsphericity in the collapse mechanism, so SNe IIP are thought to be outwardly spherically symmetric during the plateau. Hence, several distance measurement techniques capitalizing on this symmetry have been developed: The expanding photosphere method,[42] the spectral-fitting expanding atmosphere method,[5] and the standard candle method.[25] Progenitors of SNe IIL presumably have

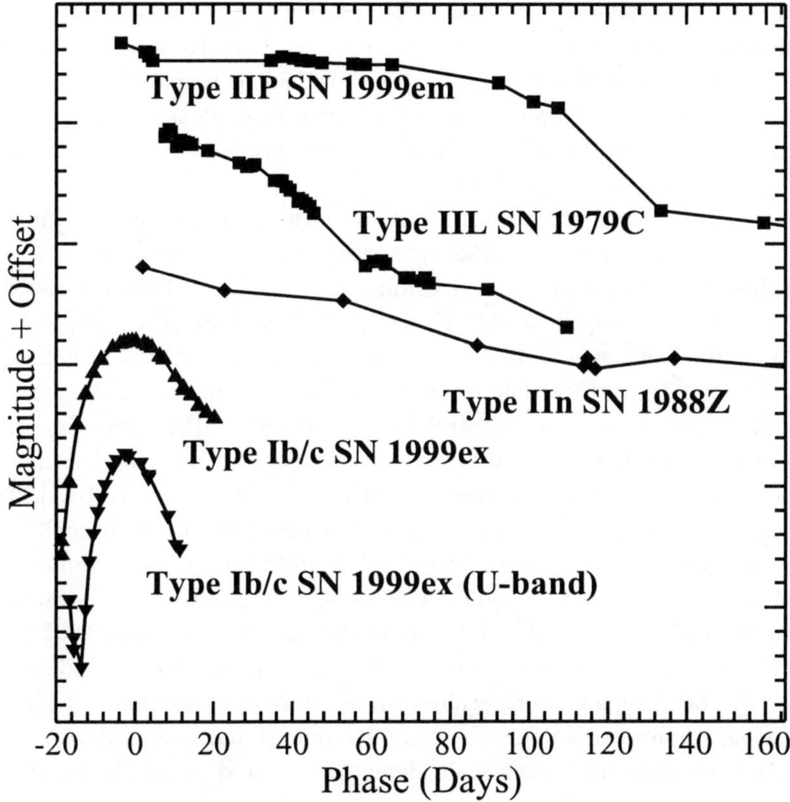

Figure 3. A sample of V-band (except where noted) light curves of core-collapse SNe. The U-band light curve of SN 1999ex is included, as it exhibits a clear signature of shock breakout at very early times.

thinner envelopes than those of SNe IIP at explosion, possibly due to late-stage stellar mass loss or interaction with a companion star. The long-term radio light curve of SN IIL 1979C reveals a complex phase of heavy mass loss prior to explosion (10^{-4} $M_\odot$ yr^{-1} over $t > 10^4$ yr).[83,84]

The spectra of SNe IIn[65] (such as SN 1998S, at the top of Fig. 2) exhibit narrow (hence the "n") emission lines, occasionally superposed on top of broader features. The narrow feature arises from slow-moving circumstellar material from a pre-SN stellar wind ionized by shock breakout. The broader components arise from interaction of the SN ejecta with the circumstellar material.[14]

SNe IIb are an "intermediate type," the most clear example being SN

1993J. At early times, the spectra of this event presented a typical Hα feature, but gradually this feature was replaced while He I grew strong (see the middle two spectra in Fig. 2). The progenitor star of SN 1993J had likely lost almost all of its outer envelope prior to explosion. The hydrogen features were evident during a brief recombination phase, but after that hydrogen became optically thin.

Spectra of SNe Ib/c are deficient in both hydrogen and helium features, indicating that they arise from the collapse of massive stars that shed their hydrogen and perhaps helium layers before collapse. Lines from low-excitation ions such as Ca II, O I, Fe II and Si II are typical (see Fig. 1, middle two spectra). An absorption feature near 10,000 Å in some SNe Ic might be attributable to He I λ10,830 but this remains somewhat unclear.[60] Examples of published SN Ib spectra are rather rare. Spectra of these events are characterized by the presence of He I lines and sometimes weak, detached Hα.[50] An interesting subset of SNe Ic, such as SN 1998bw (Fig. 2, at bottom) are dominated by heavily blended line profiles with very large Doppler shifts and hence, very high kinetic energy measurements.

Most of the above variations in core-collapse behavior are largely the products of the state of the stellar envelope and environment at the time of collapse, and not so much the collapse mechanism itself. Understanding of the mechanism, and perhaps robust explosion scenarios, may only arise from numerical simulations of multidimensional core collapse. Some models have exhibited large-scale deviation from spherical symmetry — emphasizing the need for finding solutions to the multidimensional model SN atmospheres problem. In this contribution, I review the observational and theoretical motivations for computing multidimensional light curves and spectra, describe the current situation, and outline some initial steps toward the answer.

2. Motivations

A litany of observational and theoretical evidence suggests that at least some core-collapse SNe are asymmetric in some way. Direct arguments, including spectropolarimetry, the SN/GRB (gamma-ray burst) connection, imaging, and SN remnant morphology are bolstered to some extent by more indirect arguments (the invocation of macroscopic mixing, neutron star birth-kicks, insufficiencies in spherically symmetric models). Such evidence provides motivation for and constraints on the multidimensional solutions to the model SN atmosphere problem.

2.1. *Spectropolarimetry*

First proposed as a useful SN diagnostic by Shapiro & Sutherland in 1982,[66] polarization measurements have arguably become the most important means of characterizing ejecta asymmetry on large scales. The envelopes of young SNe are dominated by electron scattering and line opacity, the former process more or less governing the continuum polarization and the latter modulating it. An unpolarized beam of light incident on an electron and scattered to an angle θ with respect to incidence departs with polarization

$$P = \frac{1 - \cos^2 \theta}{1 + \cos^2 \theta}.$$

Forward scattering results in zero polarization, orthogonal scattering results in complete linear polarization. This direction-dependence of scattering results in higher polarization from the limb than from the center of a source. Hence, to a nearby observer, individual patches of a SN atmosphere would appear polarized to some extent.

However, since all SNe observed in the modern era have been too distant to have been resolved directly, only net polarizations are measured. Thus the measured quantity is a superposition, where one linear polarization vector cancels another of equal magnitude but orthogonal phase. An unresolved, spherically symmetric SN will exhibit zero net polarization since each polarization vector originating at one point is cancelled by another from somewhere else in the envelope. On the other hand, an ellipsoidal geometry results in incomplete cancellation, and a nonzero net polarization. Both situations are compared in the cartoon Fig. 4.

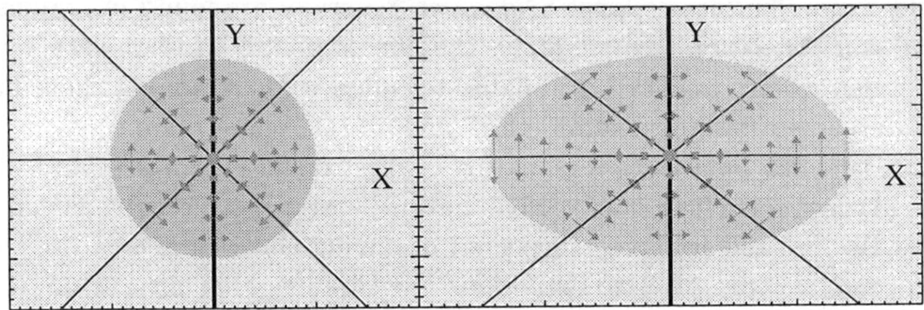

Figure 4. Spherical symmetry of the atmosphere on the left results in zero net polarization when observed as an unresolved source, since each polarization vector cancels another. A net polarization signature arises from the nonspherical geometry on the right, due to incomplete cancellation. (The line of sight is perpendicular to the page.)

Still more complicated cases may arise. An otherwise spherically symmetric atmosphere may be partially blocked at some wavelengths by line opacity distributed in a manner deviating from spherical symmetry. Such a local overdensity of opacity can block some portions of the envelope, leaving some polarization vectors uncancelled. Fig. 5 illustrates this case plus a still more complicated situation where both local and global asymmetries conspire to result in a nonzero net polarization measurement.

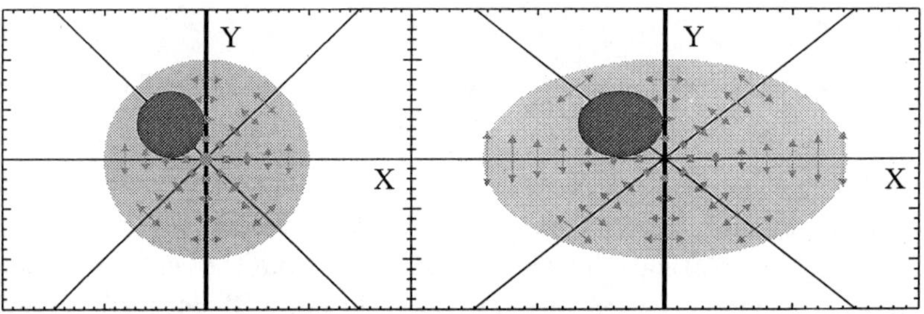

Figure 5. If localized "depolarizing" opacity blocks a spherically symmetric photosphere as on the left, incomplete cancellation can arise. On the right is a more complicated example, combining both localized and global asymmetries that would be more difficult to constrain.

Spectropolarimetric observations are difficult to obtain and interpret. Even using such instruments such as the Very Large Telescope, long exposure times and complicated set-ups are required to achieve adequate signal-to-noise. Furthermore, the SN polarization signature itself is typically at most a few percent. The interstellar medium itself contributes a polarization signature (ISP) of about the same magnitude, a particularly important consideration for core-collapse SNe which generally are found in regions of active star formation. Fortunately, the effects of ISP are not very wavelength dependent, and rather reliable techniques for removing them have been developed (see, for example Ref. 45 and references therein).

Overall, net polarization signatures are stronger in core-collapse SNe than in thermonuclear ones. Polarization is strongest among events of Types IIb, Ib, and Ic, those suspected of having shed some or all of their hydrogen envelopes prior to explosion.[81,85] Circumstellar interactors (SNe IIn) are also characterized by some degree of polarization, possibly imparted by asymmetric density variations in the circumstellar region.[45] Core-collapse events enshrouded by thick hydrogen envelopes (SNe IIP) exhibit little polarization at early times, but develop stronger signatures as their photospheres recede to reveal deep-down asymmetry.[46] A most interesting case is that of SN IIP 2004dj, which exhibits a marked increase in its continuum polarization at the end of its plateau phase, when the photosphere is dissipated.[47] This signature may indicate intrinsic asymmetry in the collapse mechanism, but radiative transfer simulations are necessary to confirm this hypothesis.

Spectropolarimetry results broadcast a clear message to core-collapse modellers. The collapse mechanism itself is likely to be markedly nonspherical, and the magnitude of its observational signature is controlled by the mass of the surrounding envelope. To determine the nature of the asymmetry, three-dimensional radiation transfer models including spectropolarimetry are necessary.

2.2. SN/GRB Connection

The detection of the Type Ic SN 2003dh loitering in the afterglow of the GRB 030329[29,71] has vaporized lingering doubts of a connection between SNe and GRBs. Prior to this discovery, the SN Ic 1998bw was associated with GRB 980425,[24] but because of the poor spatial localization of the GRB and the several day uncertainty in the SN's explosion date, skepticism persisted. The fact that the early-phase spectra of SN 2003dh resemble those of SN 1998bw supports the 1998bw/980425 association. Likewise, the transition of SN 2003dh spectra at later times to resemble those of SN 1997ef may retroactively support the tenuous association between that SN and GRB 971115 (for a listing of possible SN/GRB associations, consult Ref. 80).

More circumstantial evidence for positing a common evolutionary channel to core-collapse SNe and GRBs includes their shared association with star-forming regions[32] and the presence of bumps in GRB afterglow light curves[6] that are consistent with the shape of SN light curves. Taken together, the evidence suggests that some fraction of core-collapse SN channels result in the production of GRBs.[63]

Explosion energies inferred from these SNe appear to be up to ten times higher than is typically measured in other core-collapse events. Some or all of this excess may be due to beaming effects in asymmetric core-collapse events.[30,56] Axisymmetry in core-collapse events may be generic, and in some cases could result in a GRB as well.

Unfortunately, ISP effects could be masking a polarization signature from SN 2003dh, but there have been difficulties in modelling its rather narrow light curve and spectra assuming spherical symmetry.[87,58] Decreased densities along certain directions might permit gamma-rays to leak out faster, resulting in a narrower light curve than in the spherically symmetric case. In order to evaluate these claims, light curves and spectra computed without the assumption of spherical symmetry are necessary.

2.3. *Imaging*

SN 1987A was sufficiently close to us that its shape and size could be almost directly determined. The diffraction limit of a circular telescope at the wavelength of Hα, with the aperture D measured in meters is $165/D$ mas. At the distance of the Large Magellanic Cloud (50 kpc), a typical SN expansion velocity of 5000 km s^{-1} after one year results in an object subtending about 40 mas, at about the resolution limit of a 4m telescope. Some speckle interferometry observations suggested asymmetry in the explosion.[59]

Thirteen years after SN 1987A, Hubble Space Telescope observations of the expanding remnant revealed an interesting geometry confirming early suspicions of asymmetry in the ejecta from speckle interferometry and polarization. The resolved images clearly show an axially symmetric geometry in the distribution of ^{56}Fe, and account for the particular asymmetries in the speckle and polarization data.[82] Unfortunately, imaging (direct or otherwise) is applicable only to the most nearby events. Technological improvements in the next century may enable larger detectors to image at greater distances, however. Of course, we *are* overdue for a Galactic SN.

2.4. *Observations within the Milky Way: SN Remnants and Pulsars*

Possibly the most conspicuous indication of macroscopic mixing in SN envelopes comes from the interesting morphologies of Galactic SN remnants. Obviously, the problem is complicated by hydrodynamical interaction between SN remnants and the interstellar medium. Still, the presence of fast-moving "shrapnel" in remnants like Cas A^{20,34} and Vela^{76} have been interpreted in the context of hydrodynamical SN explosion models. Such models predict the growth of initial instabilities, resulting in the ejection of a "clumpy" envelope.[12,9,41] The resulting clumps evolve into the observed knots and filaments as the remnant ages.[1]

The space velocity distribution of pulsars as inferred from radio observations is significantly different from that of normal stars. Statistical analysis and population synthesis imply some sort of phenomenon accelerating pulsars to velocities that can eject them from the Galaxy.[15,2] The most spectacular example is the pulsar B2224+65, moving at around 1000 km s^{-1}, creating a bow shock nebula as it interacts with its environment.[11] The acceleration phenomenon may be an impulse imparted to a young neutron star during asymmetric collapse.

2.5. *Modelling Requirements*

In the interpretation of some flux spectra, a clumpy or macroscopically mixed ejecta distribution (or at least an ionization structure deviating from spherical symmetry) has been invoked to explain certain features. Most notably, the "Bochum event" in the Hα feature of SN 1987A (starting around 20 days after explosion)[62] has been attributed to asymmetry in the distribution of ^{56}Ni, resulting in an asymmetric overexcitation of hydrogen in the ejecta.[78] In other SNe, spectroscopic features from nebular phases (a few months to years after explosion) have been used to obtain envelope filling factors for some ions that are much less than unity, and possibly indicative of clumping[68,69,48,70] (but see also Ref. 33). Some amount of freshly synthesized radioactive nickel must be mixed up from the core to higher velocity in order for spectrum synthesis calculations to match spectra of core-collapse SNe 1987A[61] and 1995V.[19] In the case of SN 1987A, the mixing appears to be macroscopic, and the clumps must evolve in size.[49]

3. Approaches to the Problem

The problem is to interpret the spectra and light curves of SNe in order to place constraints on, identify, or rule out hydrodynamical explosion models. Empirical techniques include light curve fitting and parameterized spectrum synthesis. The theorist's approach is to begin with a given hydrodynamical explosion model, compute the time-evolution of its model atmosphere self-consistently, and compare its observables to those found in nature. Here I describe three generic approaches that all overlap to some extent, and how they have been or can be extended to multidimensional SN modelling.

3.1. *Prospects for Spectroscopic Inversion*

Within certain limits, SN line profiles may be inverted to place constraints on ejecta composition and geometry. The details of the technique used at early times differ from those at late times, but both capitalize on a useful consequence of the homologous velocity structure. That is, a locus of points moving with common velocity (Doppler shift) with respect to an observer at $z = \infty$ is a plane perpendicular to the line of sight ($z = const$). Hence, the effects of emission or scattering on a given plane combine to manifest as the flux at a common wavelength in a given line profile.

In nebular phase, the young remnant is optically thin to the continuum, densities are low and the role of scattering in the radiation transfer may be more safely neglected. Emission from forbidden lines populates the spectrum. A single line (rest wavelength λ_0 with multidimensional emissivity distribution $\eta(\vec{r}; \lambda_0)$ has a line profile given approximately by

$$F(\lambda) = \iint \eta(\vec{r}; \lambda_0)\delta(z - z_0)dy\,dx.$$

where $z_0 = c(\lambda - \lambda_0)/\lambda_0$. Hence, variations in flux across the line profile reflect variations in the emissivity distribution, up to a point. Integration over the plane at $z = z_0$ introduces a degeneracy that may only be removed by symmetry arguments. Assuming spherical symmetry can allow constraints to be placed on the energy emitted in some transitions as a function of radius, allowing the envelope structure and composition to be discerned to some extent.[23]

At the photospheric phases, the problem is complicated by the inclusion of scattering. Avoiding blended line profiles, or assuming that line blending effects are small, individual line profiles may be inverted analytically under the Sobolev approximation, assuming spherical symmetry in the envelope (though the technique is less robust if the data are noisy).[35,37] Again, since the flux at a given wavelength is determined by integration over a plane, the exact multidimensional distribution of opacities and source function cannot be reconstructed with certainty. Still, the inversion exercise yields important heuristics for understanding SN line profiles — phenomena such as shifting emission peaks on otherwise unblended lines are impossible under spherical symmetry. Ripples in a line profile between the absorption trough and emission peak indicate clumpiness in the ejecta near the photosphere.

3.2. *Direct Analysis*

Photospheric inverse analysis is currently limited by both data quality and the fact that the technique only applies to instances of isolated, unblended lines (a comparative rarity in SN spectra). Another approach that can be applied to the entire spectrum from the ultraviolet through the infrared, including line blending explicitly, is direct analysis. Using a simplified radiation transfer model, and approaching the problem again in an empirical spirit, constraints on explosion models are extracted from heavily line-blended SN spectra. These constraints come in the form of line identifications, and specification of ejection velocity intervals (in the multidimensional case, regions) of the parent ions. Such constraints have proved useful to hydrodynamical explosion modellers as a sort of input.

Direct analysis codes are basically Schuster-Schwarzschild simulators, and employ a sharply-defined, blackbody-emitting photosphere as a lower boundary placed at some radius *ad hoc*. Above the photosphere, line transfer is computed using the Sobolev approximation.[10,64,36] By neglecting some self-consistency in the model atmosphere, or selectively reducing the opacity load (discarding weaker lines), performance is increased to the point where direct analysis codes may be used iteratively to interpret SN spectra.

The classic direct analysis code is SYNOW,[22] and has been hacked in various ways to include effects such as circumstellar interaction,[8] and deviation from spherical composition symmetry.[74] SYNOW does not solve the model SN atmospheres problem. Instead, Sobolev optical depth profiles are parameterized in velocity space. The optical depth in the lines of a given ion are scaled relative to that of some "reference" line according to Boltzmann excitation of the levels. To compute a radiation field, a pure resonance scattering source function is used. SYNOW includes the full line list of R. L. Kurucz.[43] Example SYNOW output is compared to an observed spectrum of the Type IIP SN 1999em in Fig. 6.[4]

The Monte Carlo SN spectrum synthesis code of Mazzali and Lucy[57] uses a reduced line list, but uses an innovative technique to construct a self-consistent temperature structure, includes more transfer physics than pure resonance scattering, and electron scattering opacity. The code is still a Sobolev-based Schuster-Schwarzschild simulator, but the optical depths are not parameterized, they are derived from a solution to the model SN atmospheres problem. Convergence is achieved by virtue of the equal-energy packet Monte Carlo technique. The radiation field is always explicitly divergenceless, so given enough Monte Carlo packet trajectories, a simple Λ-iteration scheme converges quickly.

The compactness of these codes (only a few hundred lines of source each), their simplicity and small memory footprint result in rapid computations that run in just minutes on a desktop machine. Less than an hour of wall clock time is needed to produce a good fit to an observed spectrum. However, both codes depend on the simplifying assumption of spherical symmetry. Direct analysis codes constructed without this simplifying assumption have a far greater memory footprint just to store all the opacities

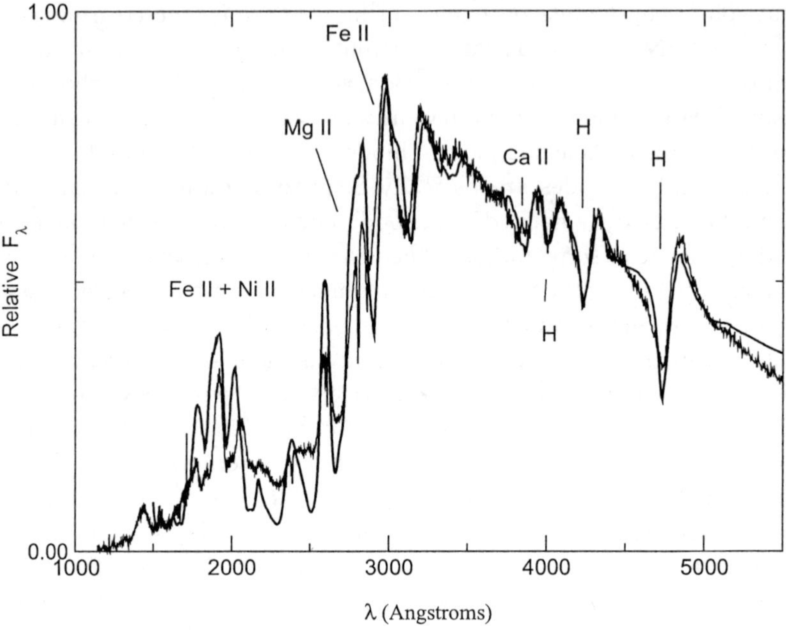

Figure 6. Example SYNOW fit to a spectrum of the Type IIP SN 1999em SYNOW is used for rapid line identifications, and exploring the velocity/composition structure of SN envelopes.

and radiation field estimates. Hence, the first attempts at direct analysis with multidimensional codes (some include polarization at minimal cost) were restricted to only one or a few lines.[13,38,75]

3.3. *Detailed Analysis*

Approaching the problem from a more theoretical side, detailed analysis is intended to match a given (constructed or computed) hydrodynamical explosion model to observational phenomena. A successful model would reproduce spectroscopic and broadband behavior at every epoch for which data were available. Every physical phenomenon relevant to radiative transfer is included to produce a faithful representation of the explosion model, otherwise the results are ambiguous. Accurate, up-to-date, and complete atomic data must be included.

One code used for this purpose is the generalized stellar atmosphere code PHOENIX.[28] PHOENIX solves the relativistic comoving radiation transfer equation, including full NLTE rate equations for a large number of ionic species, energy deposition and nonthermal excitation. This one-dimensional code is massively parallelized, and in addition to solving SN atmospheres has been applied to a variety of other astrophysical problems. An example fit to SN IIP 1999em appears in Fig. 7.[4] Like other codes of its kind, PHOENIX depends on vast amounts of atomic data, which are not always well-determined. PHOENIX is also not explicitly time-dependent but development continues.

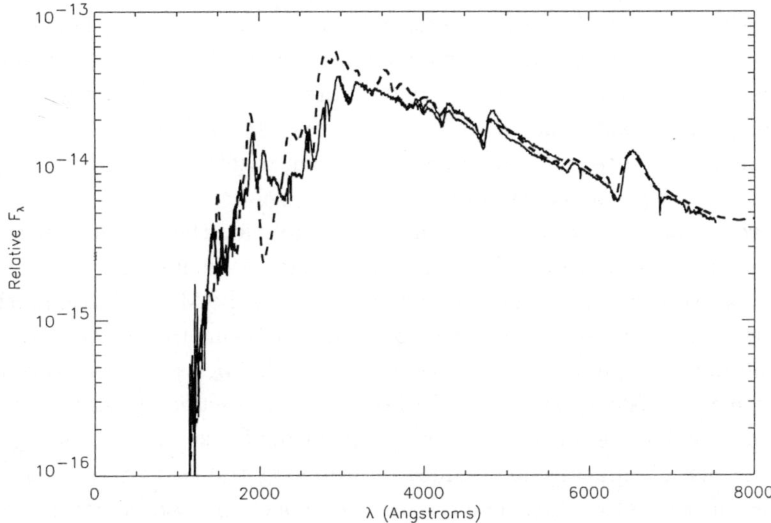

Figure 7.   Example PHOENIX fit to the Type IIP SN 1999em. Given a hydrodynamic model composition, a prescription for energy deposition, and a target luminosity, PHOENIX generates synthetic spectra for comparison to observations.

HYDRA is a HYDrodynamical RAdiation transfer code specifically used for the analysis of spectra and light curves of SNe of all types.[31] It is highly modular and parallelized, includes nuclear reaction networks, statistical physics solvers, as well as hydrodynamics and radiation transfer modellers. Various levels of dimensionality can be achieved through configuration of the code, with some simplification of the transfer model.

4. Multidimensional Models

The ultimate goal is to compute spectra and light curves for any observer's line-of-sight, given a multidimensional hydrodynamical explosion model.

Such a model would consist of the three-dimensional specification of composition, density, and energy deposition at the time homologous expansion sets in. However, for illustrative purposes, we consider the various complications encountered as one approaches the problem by first implementing the direct analysis solution in three dimensions, moving gradually toward the more detailed one.

As alluded to earlier, the most obvious problem in multidimensional transfer modelling is simply computing and storing the opacity grid. Pathological nonlocality in the transfer problem generates a demand for large-memory nodes, but a strong preference by a more powerful market favoring heavily distributed systems results in a slide toward domain decomposition. One solution is to accept the accuracy limitations imposed by the Sobolev method and dispense with a well-sampled opacity grid. Further reduction in the memory footprint is achieved by binning together collections of weak lines to form a pseudocontinuum.

Given some distribution of opacities on a grid, the problem becomes solving for the radiation field and computing the emergent spectrum. A convenient solution that is geometrically generic is the Monte Carlo method. Packets of photons or energy may be launched from the lower boundary and followed through random interactions with lines and electrons until they are re-absorbed at the lower boundary or escape toward a virtual detector. Polarization is included by propagating Stokes vectors along with the photons, computing new values by applying the scattering matrix on interaction with electrons. The polarization spectrum can be computed by combining polarization from packets emerging in the same direction.

The practical benefits of Monte Carlo are obvious. Accuracy can be achieved faster by employing more nodes and generating more realizations of the same atmosphere, combining the results at the end. The actual Monte Carlo engine is easy to maintain and extend when needed. The main caveats are that enough particles are used in the calculation, and that if the required number for the desired accuracy is very high that the random number generator employed be robust and perhaps explicitly for parallelized use.

Another difficulty is the impact of randomness on communication in the face of domain decomposition. If the strategy employed breaks up the envelope in physical space, internode communication caused by packets moving among nodes will be unpredictable and may bottleneck. An alternative strategy might capitalize on the fact that photons propagating through a homologously expanding atmosphere redshift as they move through comoving space. Chunks of adjacent wavelength bins may be stored on the same processor, and when a packet reaches the end of the reddest bin it can move to the next processor. This greatly reduces randomness in communication, especially in scattering-dominated cases where packets simply move from blue to red, from one node to the next.

Self-consistency in the model is achieved by introducing radiative equilibrium. A promising technique is that explored by Lucy in a series of interesting papers.[51,52] Instead of propagating packets of constant photon number, the scheme follows packets of constant total *energy*. Given a large number of packets moving through a transition, the photon-number rates are recovered. The key benefit of the equal-energy packet formalism is that the radiation field is always guaranteed to be divergenceless for a given opacity realization. The absence of divergence eliminates one of the problems of Λ-iteration, and permits convergence to a self-consistent temperature structure in just a few iterations. The radiation field is estimated by tallying the amount of energy passing through each cell during the simulation using the total luminosity (or total emitted luminosity) as a normalizer. A simplified local thermodynamic equilibrium calculation of a spherically symmetric hydrogen atmosphere test case through a few iterations to convergence appears in Fig. 8.[39]

Further improvements emphasize physical accuracy. The lower boundary is replaced with a realistic energy deposition prescription. Transfer can be extended beyond as simple equivalent two-level atom formalism to include branching or fluorescence. Monte Carlo treatment of the transition rates has already been proposed, but not yet included in a production code.[53,54] Very recent work on a proof-of-concept code for time-dependent[55] Monte Carlo models looks very promising, provided the domain decomposition and communication problems can be overcome.

5. Conclusion

We have reviewed some of the observational and theoretical motivations for pursuing multidimensional models of SN light curves and spectra. Development of spherically symmetric transfer codes provides us with a rich legacy of techniques and expertise for application to the multidimensional regime. A particularly promising technique for multidimensional transfer is the Monte Carlo technique, which is both powerful and simple to implement. To a large degree, the problems encountered in the coming years as we reach toward fully self-consistent models with as much physics transfer

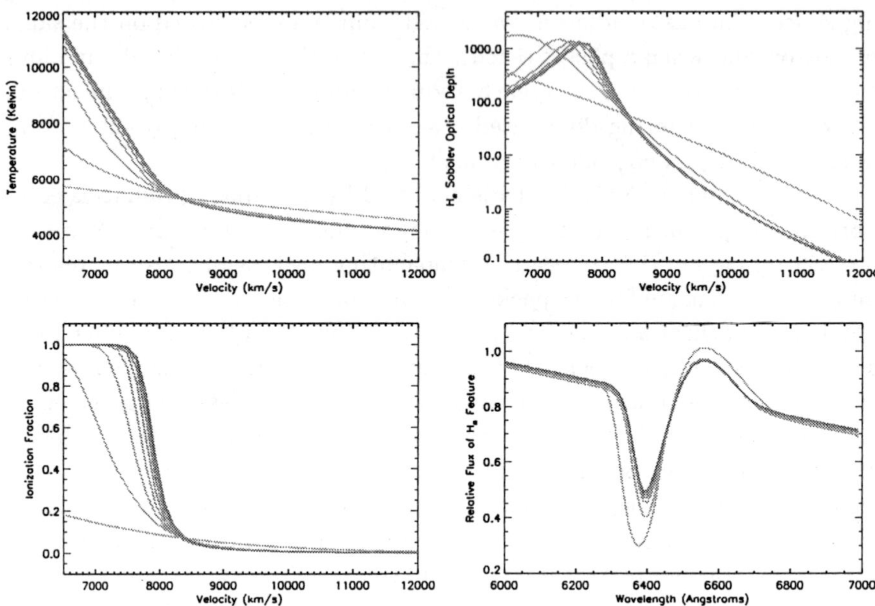

Figure 8.   Simplified LTE calculation of a hydrogen-rich SN envelope over 10 iterations (darker lines are more converged). Using the equal-energy packet Monte Carlo techniques results in relatively rapid convergence even in scattering-dominated atmospheres.

as possible will be limited by computational technology. As these barriers are overcome, a complete picture of core-collapse SNe from the end of the collapse process to the production of light curves and spectra will emerge.

References

1. M. Anderson, *et al.*, *ApJ*, **421**, L31 (1994).
2. Z. Arzoumanian, D. F. Chernoff, & J. M. Cordes, *ApJ*, **568**, 289 (2002).
3. R. Barbon, F. Ciatti & L. Rosino, *A&A*, **72**, 287 (1979).
4. E. Baron, *et al.*, *ApJ*, **545**, 444 (2000).
5. E. Baron, *et al.*, *ApJ*, **616**, 91L (2004).
6. J. S. Bloom, *et al.*, *Nature*, **401**, 453 (1999).
7. D. Branch, E. Baron & D. J. Jeffery: 'Optical Spectra of Supernovae.' In: *Supernovae and Gamma-Ray Bursters*, ed. by K. Weiler (Springer: Berlin 2003), 47
8. D. Branch, *et al.*, *PASP*, **112**, 217 (2000).
9. A. Burrows, J. Hayes & B. Fryxell, *ApJ*, **450**, 830 (1995).
10. J. I. Castor, *MNRAS*, **149**, 111 (1970).
11. S. Chatterjee & J. M. Cordes, *ApJ*, **600** , L51 (2004).
12. R. Chevalier & R. Klein, *ApJ*, **219**, 994 (1978).

13. N. N. Chugai, *SvAL*, **18**, 168 (1992).

14. N. N. Chugai & I. J. Danziger, *MNRAS*, **268**, 173 (1994).

15. J. M. Cordes & D. F. Chernoff, *ApJ*, **505**, 315 (1998).

16. G. de Vaucouleurs, *et al.*, *PASP*, **93**, 36 (1981).

17. R. G. Eastman, *et al.*, *ApJ*, **430**, 300 (1994).

18. A. Elmhamdi, *et al.* *MNRAS*, **338**, 939 (2003).

19. A. Fassia, *et al.* *MNRAS*, **299**, 150 (1998).

20. R. Fesen & K. Gunderson, *ApJ*, **470**, 967 (1996).

21. A. V. Filippenko, *ARA&A*, **35**, 309 (1997).

22. A. K. Fisher, PhD Thesis, University of Oklahoma, Norman (2000).

23. C. Fransson & R. Chevalier, *ApJ*, **343**, 323 (1989).

24. T. J. Galama, *et al.*, *Nature*, **395**, 670 (1998).

25. M. Hamuy & P. A. Pinto, *ApJ*, **566**, L63 (2002).

26. M. Hamuy, *ApJ*, **582**, 905 (2003a).

27. M. Hamuy, *et al.*, *Nature*, **424**, 651 (2003b)

28. P. Hauschildt & E. Baron, *J. Comp. Appl. Math.*, **109**, (1999).

29. J. Hjorth, *et al.*, *Nature*, **423**, 847 (2003).

30. P. Höflich, J. C. Wheeler & L. Wang, *ApJ*, **521**, 179 (1999).

31. P. Höflich, 'Parallelization of Stellar Atmosphere Codes.' In: *Stellar Atmosphere Modelling*, ed. by I. Hubeny, D. Mihalas, & K. Werner (ASP, San Francisco 2003), **288**, 371.

32. S. T. Holland & J. Hjorth, *A&A*, **344**, L67 (1999).

33. J. Houck & C. Fransson, *ApJ*, **456**, 811 (1996).

34. U. Hwang, S. Holt & R. Petre, *ApJ*, **537**, L119 (2000).

35. R. Ignace & M. A. Hendry, *ApJ*, **537**, L131 (2000).

36. D. J. Jeffery & D. Branch, 'Analysis of Supernova Spectra.' In: *Supernovae, 6th Jerusalem Winter School for Theoretical Physics*, ed. by J. C. Wheeler, T. Piran, S. Weinberg (World Scientific, Singapore 1990), 149.

37. D. Kasen, *et al.*, *ApJ*, **565**, 380 (2002).

38. D. Kasen, *et al.*, *ApJ*, **593**, 788 (2003).

39. D. Kasen, PhD Thesis, University of California, Berkeley (2004).

40. K. S. Kawabata, *et al.*, *ApJ*, **593**, L19 (2003).

41. K. Kifonidis, *et al.*, *ApJ*, **532**, L123 (2000).

42. R. P. Kirshner & J. Kwan, *ApJ*, **193**, 27 (1974).

43. R. L. Kurucz, CD-ROM 1, Atomic Data for Opacity Calculations, (Cambridge, SAO 1993).

44. B. Leibundgut & N. B. Suntzeff: 'Optical Light Curves of Supernovae.' In: *Supernovae and Gamma-Ray Bursters*, ed. by K. Weiler (Springer: Berlin 2003), 77

45. D. C. Leonard, *et al.*, *ApJ*, **536**, 239 (2000).

46. D. C. Leonard, *et al.*, *ApJ*, **553**, 861 (2001).

47. D. C. Leonard, *et al.*, *BAAS*, *205*, 7108 (2004).

48. H. Li & R. McCray, *ApJ*, **387**, 309 (1992).

49. H. Li, R. McCray & R. Sunyaev, *ApJ*, **419**, 824 (1993).

50. L. B. Lucy, *A&A*, **409**, 737 (1991).

51. L. B. Lucy, *A&A*, **344**, 282 (1999a).

52. L. B. Lucy, *A&A*, **345**, 211 (1999b).
53. L. B. Lucy, *A&A*, **384**, 725 (2002).
54. L. B. Lucy, *A&A*, **403**, 261 (2003).
55. L. B. Lucy, *A&A*, **429**, 19 (2005).
56. A. I. MacFadyen & S. E. Woosley, *ApJ*, **254**, 262 (1999).
57. P. A. Mazzali & L. B. Lucy, *A&A*, **279**, 447 (1993).
58. P. A. Mazzali, *et al.*, *ApJ*, **599**, L95 (2003).
59. W. P. S. Meikle, *PASAu*, **7**, 47 (1988).
60. J. Millard, D. Branch, E. Baron, K. Hatano, *et al.*, *ApJ*, **527**, 746 (1999).
61. R. C. Mitchell, *et al.*, *ApJ*, **556**, 979 (2001).
62. M. Phillips & S. Heathcote, *PASP*, **101**, 137 (1989).
63. P. Podsiadlowski, *et al.*, *ApJ*, **607**, L17 (2004).
64. G. B. Rybicki & D. G. Hummer, *ApJ*, **219**, 654 (1978).
65. E. M. Schlegel, *MNRAS*, **244**, 269 (1990).
66. P. Shapiro & P. Sutherland, *ApJ*, **263**, 902 (1982).
67. S. J. Smartt, *et al.*, *Science*, **303**, 499 (2004).
68. J. Spyromilio & P. A. Pinto, 'Oxygen in Supernovae.' In: *Supernova 1987A and Other Supernovae, ESO Conference and Workshop Proceedings*, ed. by I. Danziger and K. Kjär (ESO: Garching 1991), 423.
69. J. Spyromilio, *MNRAS*, **253**, 25P (1991).
70. J. Spyromilio, *MNRAS*, **266**, L61 (1994).
71. K. Z. Stanek, *et al.*, *ApJ*, **591**, L17 (2003).
72. M. Stritzinger, *et al.*, *AJ*, **124**, 2100 (2002).
73. SUSPECT Online SN Spectrum Archive, http://suspect.nhn.ou.edu.
74. R. C. Thomas, *et al.*, *ApJ*, **567**, 1037 (2002).
75. R. C. Thomas, *et al.*, *ApJ*, **601**, 1019 (2004).
76. H. Tsunemi, E. Miyata, B. Aschenbach, *PASJ*, **51**, 711 (1999).
77. M. Turatto, *et al.*, *MNRAS*, **262**, 128 (1993).
78. V. Utrobin, N. Chugai & A. Andronova, *A&A*, **295**, 129 (1995).
79. S. D. Van Dyk, W. Li & A. V. Filippenko, *PASP*, **115**, 1289 (2003).
80. L. Wang & J. C. Wheeler, *ApJ*, **504**, 87L (1998).
81. L. Wang, *et al.*, *ApJ*, **550**, 1030 (2001).
82. L. Wang, *et al.*, *ApJ*, **579**, 671 (2002).
83. K. W. Weiler, *et al.*, *ApJ*, **380**, 161 (1991).
84. K. W. Weiler, *et al.*, *ApJ*, **399**, 672 (1992).
85. J. C. Wheeler, 'Cosmic Explosions: Rapporteur Summary of the 10th Maryland Astrophysics Conference.' In: *Cosmic Explosions*, ed. by S. Holt and W. Zhang (AIP, New York 2000), 445.
86. S. E. Woosley & T. A. Weaver, *ARA&A*, **24**, 1986 (1986).
87. S. E. Woosley & A. Heger, astro-ph/0309165 (2003).

OPEN ISSUES IN SUPERNOVA LIGHTCURVES: ASYMMETRIES AND NICKEL/COBALT DECAY

A. L. HUNGERFORD

CCS-4, D409
Los Alamos National Laboratory
Los Alamos, NM 87545
E-mail: aimee@lanl.gov

C. L. FRYER *

T-6, MS B227
Los Alamos National Laboratory
Los Alamos, NM 87545
E-mail: fryer@lanl.gov

The optical emission from supernovae is primarily driven by the high-energy photons emitted in the decay of ^{56}Ni and ^{56}Co. The most direct means to study this decay process is to observe the hard X-ray and gamma-ray flux. We discuss the interplay between supernova asymmetries and both the emergence time and spectral distribution of this high energy flux. We conclude with a discussion of how this asymmetry may affect calculations of supernova lightcurves.

1. The Source of Supernova Lightcurves

In a supernova (SN) explosion, the supernova shock causes the inner $\sim 0.1 M_\odot$ of the ejecta to burn into ^{56}Ni. This can only occur when the supernova shock is extremely hot, and hence is limited only to this inner material. Although this ^{56}Ni represents only a small amount of the ejecta, it dominates the supernova lightcurve, which is powered by the γ-ray photons emitted as ^{56}Ni decays to ^{56}Co which, in turn, decays to ^{56}Fe. These high-energy photons are downscattered to lower energies, heating the matter, and ultimately producing optical photons that are observed in over a hundred supernovae every year. To construct a supernova lightcurve, theo-

---

*C.L.F. is also affiliated with the Physics Department, University of Arizona, Tucson, AZ 87521

rists must take the explosions produced in collapse, calculate the deposition of the high-energy photons from this decay, convert this energy into optical photons, and follow the transport of these photons out of the star. This multi-step process, equally as difficult as the supernova engine itself, produces theoretical fluxes and spectra for which we have an abundance of observations to compare against.

Unfortunately, this process is filled with a number of steps, each of which is fraught with uncertainty. One of the least studied uncertainties is the effects of asymmetries on this problem. In stellar collapse, we have the potential for direct observations of the asymmetry in the collapse by studying the neutrino and gravitational wave emission (Fryer, Holz, & Hughes 2004). But these signals are so difficult to measure that such observations are limited to nearby supernovae (we have no gravitational wave observations and only a limited neutrino observation from SN 1987A). In a similar manner, we have a direct probe of the asymmetry in the supernova explosion by studying the high-energy emission (e.g. Hungerford, Fryer & Warren 2003; Hungerford, Fryer & Rockefeller 2005). Unfortunately, this probe is also limited to nearby supernovae, and we only have really good data of SN 1987A.

2. High Energy Emission from Asymmetric Supernovae

In principal, high-energy line calculations are much simpler than the optical transport. The source of photons arises from the decay of ^{56}Ni and ^{56}Co and is a function only of the ^{56}Ni abundance. It doesn't depend (or depends weakly) on the thermodynamic conditions in the matter. In addition, above $\sim 0.1 - 0.2$MeV, the dominant opacity arises from Compton scattering which is also independent of the thermodynamic conditions in the matter (to an MeV photon, a free electron appears identical to an electron bound to an atom). Time independence can essentially be assumed, and most calculations of the gamma-ray flux use Monte-Carlo techniques postprocessing explosion models (see Milne et al. 2004 for a review). Milne et al. (2004) made a detailed review of all of the leading codes used to calculate high energy flux and found that, even with this first step in the lightcurve calculations, some codes obtained discrepant answers. In this code comparison, most of the discrepant answers were found as "coding bugs" in the calculations. Comparison projects are an often neglected aspect of "verification and validation" and can, as it did in the Milne et al. (2004) code comparison project, demonstrably improve our understanding (and accuracy) of the calculations.

In this proceedings, we use results from the Maverick Monte-Carlo transport code (Hungerford et al. 2003,2005) and focus on studying the observed features of SN 1987A. Three specific features of SN 1987A made scientists realize that their simple 1-dimensional models were not sufficient to explain the explosion: the high-energy emission appeared roughly 150 days earlier than expected if the ^{56}Ni were indeed limited to the inner region where it was synthesized, the observed high energy lines were broader than the expected velocities for this inner material, and the lines were redshifted (implying the observed ^{56}Ni is traveling away from us). Such results require asymmetries.

2.1. *Asymmetries in 1987A: The Need for Single Lobe Explosions*

Hungerford et al. (2005) focused on trying to find a qualitative effect to explain the redshifted lines in 1987A. To make the redshifted line, there must be either more ^{56}Ni material (or at least more ^{56}Ni at low optical depths to dominate the observations) traveling away from us than toward us in the explosion. A natural way to explain this effect is to assume that the explosion is stronger in one direction. This single-lobe explosion is becoming more justified by recent results of core-collapse modelers (Scheck et al. 2004). Figure 1 shows the line-of-sight velocity distribution of the 56-weight elements (initially ^{56}Ni) for an explosion where the velocity in a 40° cone was 3 times higher than the explosion in the rest of the star. Viewed from behind, this single-lobe explosion produces a great deal of nickel moving away from us.

But the velocity distribution of the nickel can be misleading. This is the velocity distribution of the actual nickel, not the velocity distribution of the nickel that is observed through high-energy emission. Figure 2 shows the ^{56}Co 847 keV line profiles for the explosion model from Figure 1 for the same set of 5 lines of sight. Observing the single-lobe explosion exactly from behind leads to a fairly weak line that is not tremendously redshifted. However, if the observer is located 126° off of the single-lobe explosion (instead of the 180° corresponding to orbserving the explosion from behind), redshifted lines are measureable. Such a vantage point may be the easiest explanation of the redshifted line profiles for SN 1987A. The inferred inclination angle for SN 1987A ($\theta = 44°$ based on the observed ring geometry) may then suggest a loose correlation between low-mode explosion

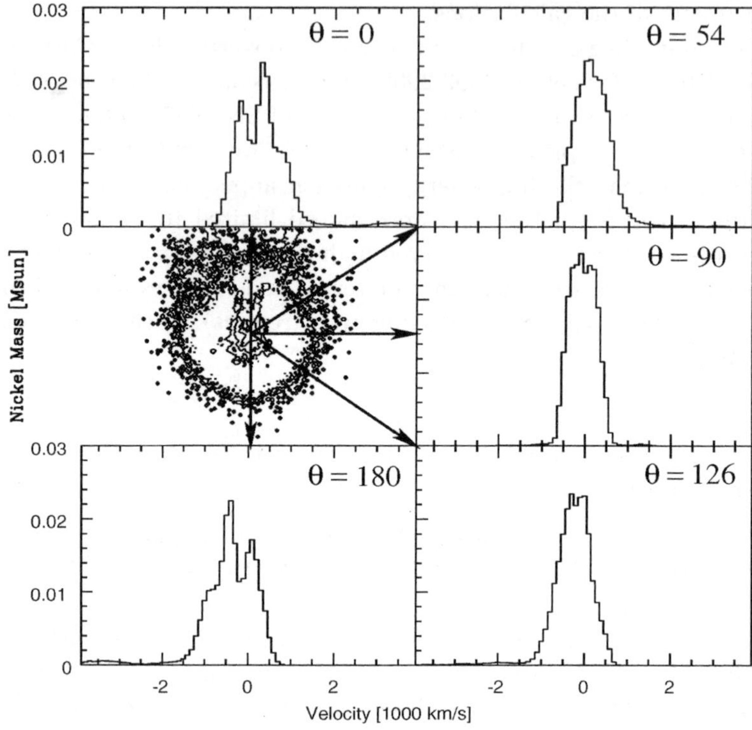

Figure 1. Mass of 56-weight elements (e.g. initial ^{56}Ni mass) versus line of sight velocity for a number of viewing angles at t = 365 days. Central panel on the left shows a contour plot of density (outer contour) and cobalt number density (inner contours). Due to the homologous nature of the expansion, these distributions represent the line shapes one would expect from nickel, cobalt or iron emission in the absence of significant ionization or opacity effects. From Hungerford et al. (2005).

asymmetry and rotation axis.

A range of explosion asymmetries can produce such redshifted lines. Figure 3 shows the same observed flux of the 847 keV line, but for a single-lobe explosion where the velocities were increased by a factor of 5 in a cone with a 20° opening angle. In addition, we have assumed in these explosions that there is a step function describing the velocity profile in and outside of the cone. Smoother profiles may produce different net results, but still explain the qualitative features of the observed lines. It may be difficult to determine exactly the explosion profile with the coarsely resolved spectral

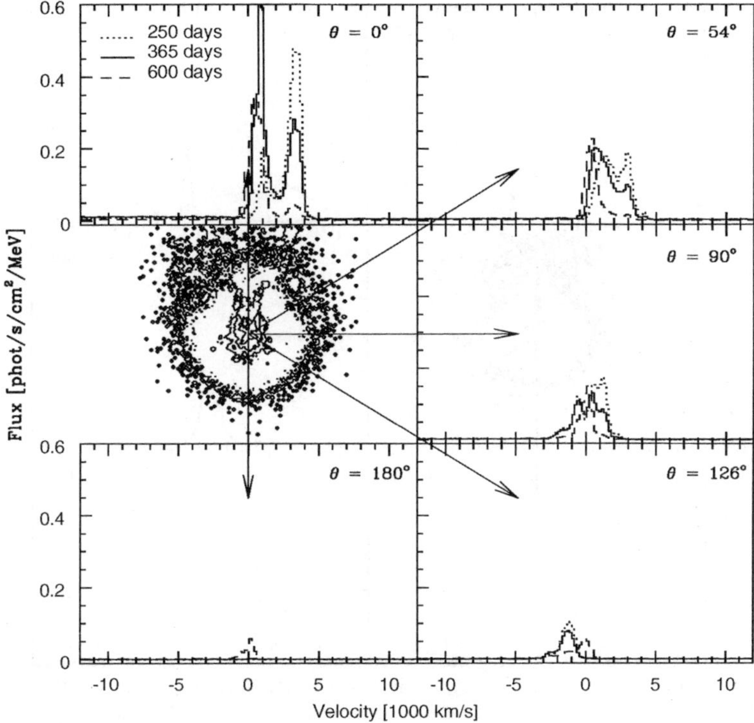

Figure 2. Line profiles of the ^{56}Co 847 keV decay line for explosion model in Figure 1 at t = 365 days. Central panel on the left shows a contour plot of density (dark) and cobalt number density (light). Surrounding panels represent line profiles for the set of viewing angles depicted by the black vectors overplotted on the density contours. The emission in the line profiles arises predominantly from the cobalt ejected along the enhanced explosion lobe. Due to the homologous nature of the ejecta, the structure in the lines can be understood by summing this extended cobalt material along lines perpendicular to the viewing angle vectors. From Hungerford et al. (2005).

features in the observations. One feature that may help us to constrain the level of the explosion asymmetry is the time dependence of the line profile. In Figure 2, we see that the line profile shifts with time as we see deeper into the explosion. A time profile of the high-energy emission will allow us to extract much more information about the supernova asymmetry.

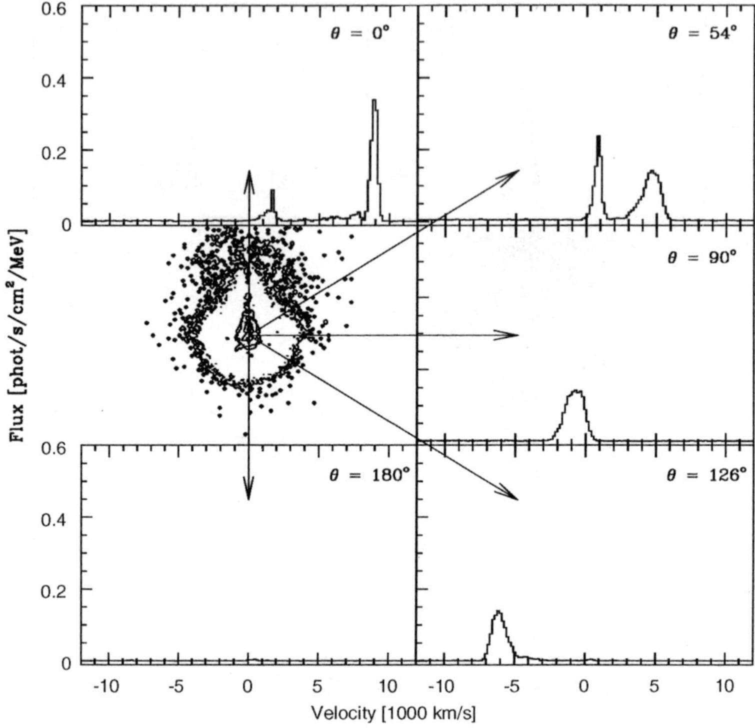

Figure 3.   Same as Figure 2, but for a single-lobe explosion where the cone was only 20° large, but the velocity was increased by a factor of 5 in this cone over the surrounding material. From Hungerford et al. (2005).

2.2. *Asymmetries in 1987A: Open Issues for the High Energy Continuum*

Such extreme single-lobe explosions seem capable of explaining the red-shifted high-energy lines but, alone, may not be able to explain the broadened line profiles or early emergence of the hard X-ray continuum. Figures 1 and 4 show the underlying nickel distributions for the suite of models presented here. A first order test of a model's ability to match the broadened line observations is to compare the theoretical nickel distribution with the observed infrared iron line profiles. Haas et al. (1990) argued that the [FeII] line profile reflects the entirety of the spatially distributed radioactive nickel, as any optically thick component of iron is likely distributed

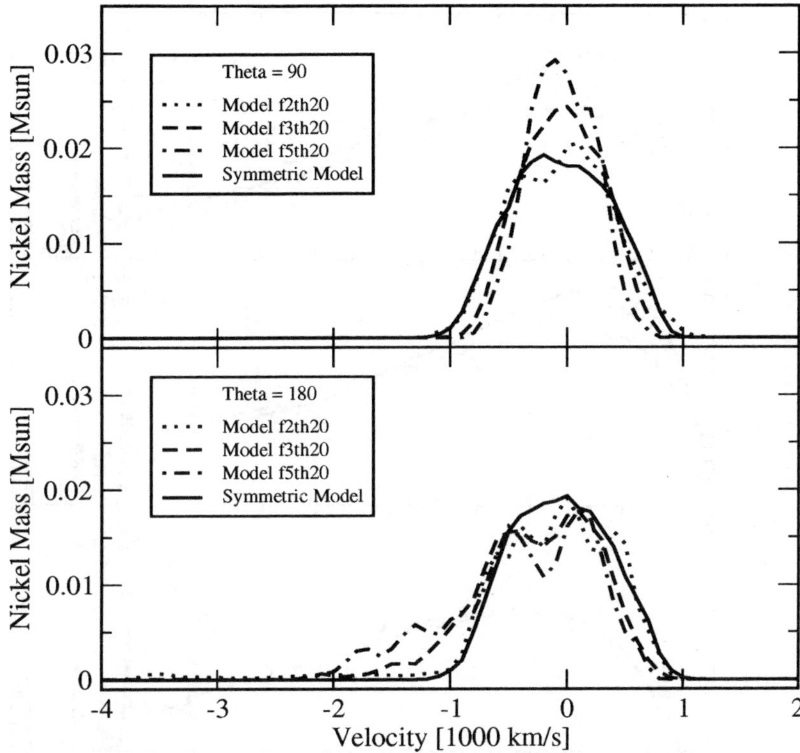

Figure 4. Underlying nickel distributions for the fXth20 models, with symmetric model distribution plotted for comparison. Here "X" corresponds to the magnitude the velocity is greater in the cone than out of the cone, "th20" corresponds to cone opening-angles of 20°.

throughout the ejecta in clumps. Since the stellar progenitor and explosion properties used in these simulations were not specified to match SN 1987A, it is difficult to draw quantitative conclusions. That said, trends in the line profiles relative to the globally symmetric explosion suggest that broadened lines *can* be achieved (as in models f5th20, f3th20), though are not always the case (models f2th20, f3th40).

It is even more difficult to draw meaningful conclusions regarding the early emergence of the high energy continuum. Keep in mind that the "earlier emergence" time was relative to theoretical predictions based on spherically symmetric (i.e. unmixed explosion models.) Although multi-dimensional explosions (capable of modeling the mixing instabilities) have been simulated using a stellar progenitor appropriate for SN 1987A (Ki-

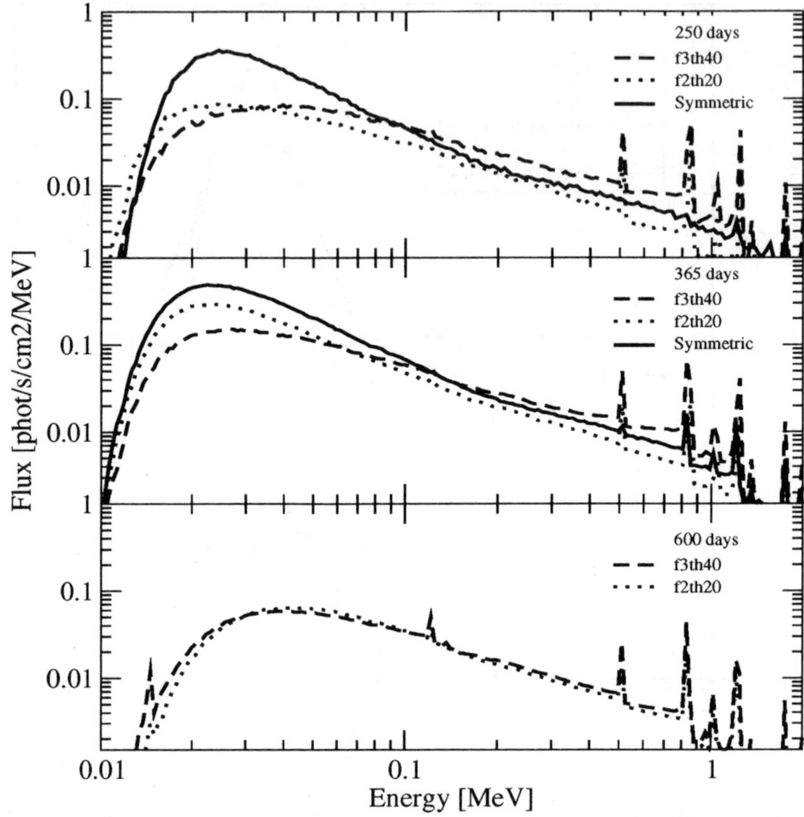

Figure 5.   Logarithmic plot of total hard X- and γ-ray spectrum at t = 365 d for models f3th40 (light line), f2th20 (dark line) and Symmetric (black line) for 3 different times (t = 250 d, 365 d and 600 d.) The flux was calculated assuming a distance of 60 kpc. The f3th40 model evolves more slowly in time, and the Symmetric model has a brighter low energy continuum at all epochs. The nomenclature of these models is described in Figure 4.

fonidis et al. 2003; Nagataki 2000; Herant & Benz 1992), the nickel distributions from these globally symmetric explosion models were not broad enough for the first order comparison against IR line profiles. Having already fallen short on the line width, further modeling of the high energy spectrum was not performed. So these globally symmetric explosion models told us that additional mixing (possibly arising from a global asymmetry) is needed to match the broad lines, but they told us nothing regarding the effect a global asymmetry should have on the continuum. The high energy spectra from a subset of our explosion simulations (using a generic red supergiant SN progenitor) are shown in Figure 5. It's clear that the asymmetric explosions produce a lower continuum relative to the symmetric explosion model, and that the time evolution of models with different input asymmetries is qualitatively different. Previous work modelling these explosions (even the ones employing global asymmetries) has judged its mixing successes and failures based on the spatial distribution of nickel alone, without regard to the early high energy continuum observations. However, these results suggest that the hard X-ray emission can vary widely with asymmetry type. Clearly, high energy calculations are necessary for determining the ultimate adequacy of any SN 1987A mixing calculation.

3. From Gamma-Rays to Optical Light Curves

Why is it so important to determine the level of asymmetry? Remember that our final goal is to determine the optical outburst of these supernovae. We must know the explosion asymmetry to determine just exactly where the ^{56}Ni gets mixed to during the explosion. The decay products of ^{56}Ni will provide the energy of the light curve. Hopefully, using our high-energy probe, we will eventually be able to determine this asymmetry. But this marks only the beginning of the work required to calculate a supernova light curve.

The next step requires knowing where this decay energy is deposited. Figure 6 shows the energy deposition from the gamma-rays for a single-lobe explosion where the velocities are a factor of 5 higher in a cone with a 20° opening angle 30 days after the launch of the explosion. The deposition is neither spherically symmetric, nor does it closely trace the ^{56}Ni abundance.

To produce a lightcurve, we take this deposited energy as a source of thermal photons which are then transported out of the exploding star to the observer. The current state-of-the-art takes time snapshots and uses the deposition profiles to calculate the spectra from the supernova explosion. Such a calculation does not include effects of time evolution in the model, but probably gives reasonable first approximations of the supernova light curve. However, considerable work is needed before we can use the optical outbursts from supernovae to constrain our explosion models.

Acknowledgments

These simulations were run on LANL's Space Simulator and a workstation provided by Beomax. This work is supported by a DOE SciDAC grant num-

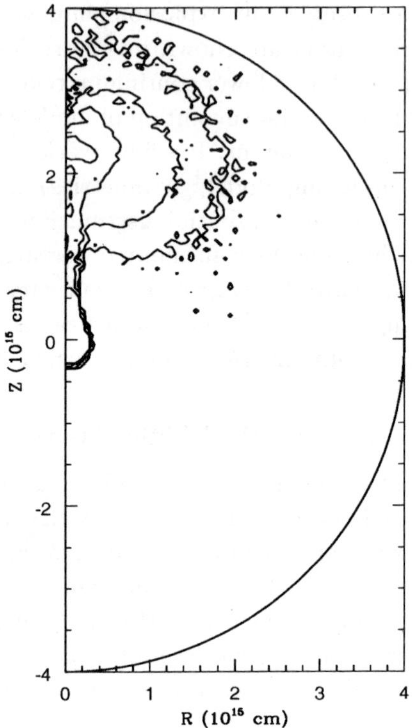

Figure 6. Energy deposition profile for model f5th20 at 30 days. Outer circle represents outer edge of computational domain and contours show energy deposition in ergs s^{-1} from gamma-ray absorption and down scattering processes. Contour levels are $10^{47}, 10^{48}, 10^{49}$ and 10^{50} erg s^{-1}.

ber DE-FC02-01ER41176 and by NASA Grant SWIF03-0047-0037 under the auspices of the U.S. Dept. of Energy, contract number W-7405-ENG-36.

References

1. L. Scheck, T. Plewa, H.-Th. Janka, K. Kifonidis, and E. Müller *PhRvL*, **92**, 011103 (2004).
2. C. L. Fryer, D. E. Holz, and S. A. Hughes *ApJ*, **609**, 288 (2004).
3. A. L. Hungerford, C. L. Fryer, M. S. Warren *ApJ*, **594**, 390 (2003).
4. A. L. Hungerford, C. L. Fryer, G. Rockefeller *submitted to ApJ*.
5. P. A. Milne et al. *ApJ*, **613**, 1101 (2004).
6. C. L. Fryer and M. S. Warren *ApJ*, **601**, 391 (2004).

PROBING NEUTRINO PROPERTIES WITH SUPERNOVAE

C. LUNARDINI*

Institute for Advanced Study
Einstein Drive
Princeton, 08540, New Jersey, USA
E-mail: lunardi@ias.edu

I review the effects of flavor conversion of neutrinos from stellar collapse due to masses and mixing, and discuss the motivations for their study. I consider the sensitivity of a supernova neutrino signal to the 13-mixing ($\sin^2 \theta_{13}$) and to the type of mass hierarchy/ordering ($\mathrm{sign}[\Delta m^2_{13}]$). In particular, I illustrate the possibility to get model-independent information from regeneration effects on ν_e and $\bar{\nu}_e$ in the matter of the Earth.

1. Introduction and motivations

The mechanism of neutrino flavor conversion due to masses and flavor mixing has been recently established by the combination of the results of solar neutrino detectors and those of the KamLand experiment [1]. Results from the detection of atmospheric neutrinos and the preliminary data from the K2K experiment [2] strongly support the existence of this phenomenon.

From the analysis of all the available data, we get a partial reconstruction of the neutrino masses m_i (the label $i = 1, 2, 3$ denotes the neutrino mass eigenstates) and of the mixing matrix U, defined by $\nu_\alpha = \sum_i U_{\alpha i} \nu_i$, where ν_α ($\alpha = e, \mu, \tau$) are the flavor eigenstates. Using the standard parameterization of the mixing matrix in terms of three angles, $\theta_{12}, \theta_{13}, \theta_{23}$, we have:

$$m_2^2 - m_1^2 \equiv \Delta m_{21}^2 = (4 - 30) \cdot 10^{-5} \mathrm{eV}^2, \qquad \tan^2 \theta_{12} = 0.25 - 0.85 , \quad (1)$$

from solar neutrinos and KamLand, and

$$m_3^2 - m_2^2 \equiv \Delta m_{32}^2 = \pm(1.5 - 4) \cdot 10^{-3} \mathrm{eV}^2, \qquad \tan^2 \theta_{23} = 0.48 - 2.1 \quad (2)$$

---

*Work supported by the Keck fellowship program and by the grant PHY-0070928 of the National Science Foundation.

from atmospheric neutrinos. The sign of Δm^2_{32} is unknown. The two possibilities, $\Delta m^2_{32} \approx \Delta m^2_{31} > 0$ and $\Delta m^2_{32} \approx \Delta m^2_{31} < 0$, are referred to as *normal* and *inverted* mass hierarchies/ordering respectively (abbreviated as n.h. and i.h. in the text).

The mixing angle θ_{13}, which describes the ν_e content of the third mass eigenstate, ν_3, is still unmeasured. We have an upper bound from the CHOOZ and Palo Verde experiments [3,4]: $\sin^2 \theta_{13} \lesssim 0.02$. The identification of the neutrino mass hierarchy and the determination of θ_{13} have become the main issues of further studies.

To achieve these, and other important goals, the study of neutrinos from core collapse supernovae is particularly interesting. Indeed, these neutrinos are produced and propagate in *unique* physical conditions of high density and high temperature, and therefore can manifest effects otherwise unaccessible. As will be discussed in the following, due to the very large interval of matter densities realized there, the interior of a collapsing star is the only environment where a neutrino of a given energy undergoes two MSW resonances, associated to the two mass squared splittings of the neutrino spectrum. This implies a richer phenomenology of flavor conversion, and therefore wider possibilities to probe the relevant parameters, with respect to the case of neutrinos in solar system, where only one resonance, i.e. one mass splitting, is relevant at a time.

Since observables depend both on the features of the original fluxes and on the flavor conversion effects, it is clear that the extraction of information on the neutrino mixing and on the neutrino mass spectrum requires a careful consideration of astrophysical uncertainties.

2. Properties of supernova neutrino fluxes and density profile of the star

Neutrinos and antineutrinos of all the three flavors are produced in a supernova and emitted in a burst of ~ 10 seconds duration. At a given time t from the core collapse the original flux of the neutrinos of a given flavor, ν_α, can be described by a "pinched" Fermi-Dirac (F-D) spectrum[a],

$$F^0_\alpha(E, T_\alpha, \eta_\alpha, L_\alpha, D) = \frac{L_\alpha}{4\pi D^2 T^4_\alpha F_3(\eta_\alpha)} \frac{E^2}{e^{E/T_\alpha - \eta_\alpha} + 1}, \tag{3}$$

where D is the distance to the supernova (typically $D \sim 10$ kpc for a galactic supernova), E is the energy of the neutrinos, L_α is the luminosity in the

---

[a]An alternative parameterization has been suggested recently[5].

flavor ν_α, and T_α represents an effective temperature. The normalization factor equals: $F_3(\eta_\alpha) \equiv \int_0^\infty dx\ x^3/(e^{x-\eta_\alpha}+1)$. Supernova simulations provide the indicative values of the average energies [5]:

$$\langle E_{\bar{e}} \rangle = (14-22)\ \mathrm{MeV}, \quad \langle E_x \rangle/\langle E_{\bar{e}} \rangle = (1.1-1.6), \quad \langle E_e \rangle/\langle E_{\bar{e}} \rangle = (0.5-0.8), \tag{4}$$

and the typical value of the (time integrated) luminosity in each flavor: $L_\alpha \sim (1-5)\cdot 10^{52}$ ergs. The luminosities of all neutrino species are expected to be approximately equal, within a factor of two or so [5]. The ν_μ and ν_τ ($\bar{\nu}_\mu$ and $\bar{\nu}_\tau$) spectra are equal with good approximation, and therefore the two species can be treated as a single one, ν_x ($\bar{\nu}_x$). The pinching parameter η_α can vary between 0 and ~ 3 for ν_e and $\bar{\nu}_e$, while smaller pinching is expected for ν_μ, ν_τ: $\eta_\mu = \eta_\tau \sim 0-2$.

The matter density profile met by the neutrinos can be approximated, at least in the first few seconds of their emission, by that of the progenitor star [6]. The latter is well described by the radial power law [6]:

$$\rho(r) = 10^{13}\ C \left(\frac{10\ \mathrm{km}}{r}\right)^3\ \mathrm{g}\cdot\mathrm{cm}^{-3}, \tag{5}$$

with $C \simeq 1-15$.

3. Conversion in the star, jump probability and θ_{13}

Let us consider the conversion of neutrinos as they propagate from the production region outward in the star, for the case of normal mass hierarchy ($\Delta m_{32}^2 > 0$). As shown in Fig. 1 (positive density semi-plane), the eigenvalues of the Hamiltonian in matter and the flavor composition of its eigenstates change with the variation of the matter density along the neutrino trajectory.

At production, the mixing is suppressed due to the very large density ($\rho \sim 10^{11}\ \mathrm{g}\cdot\mathrm{cm}^{-3}$), therefore the eigenstates of the Hamiltonian coincide with the flavor states. At lower densities, the neutrinos undergo two MSW resonances (level-crossings). The inner resonance (H) is governed by the parameters Δm_{32}^2 and θ_{13} and is realized at $\rho \sim 10^3\ \mathrm{g}\cdot\mathrm{cm}^{-3}(10\mathrm{MeV}/E)$. The probability of transition between the eigenstates of the Hamiltonian (jump probability) in this resonance, P_H, strongly depends on θ_{13} as discussed later in this section. The second resonance, (L) is determined by Δm_{21}^2 and θ_{12} and happens at lower density, $\rho \sim (30-140)(10\mathrm{MeV}/E)\ \mathrm{g}\cdot\mathrm{cm}^{-3}$. For the values of parameters in Eq. (1) the jump probability in this resonance is negligibly small (adiabatic propagation). The neutrinos leave the

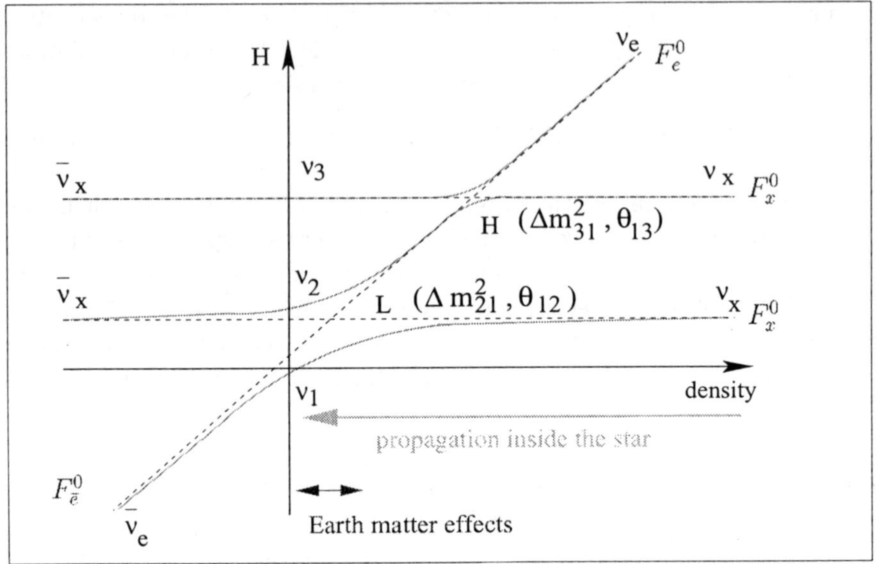

Figure 1. The level-crossing diagram for normal mass hierarchy. The solid curves represent the eigenvalues of the Hamiltonian in matter.

star as mass eigenstates and therefore do not oscillate on the way from the star to the Earth. If they cross the Earth before detection, oscillations are restarted due to Earth matter effects[7].

Since they have opposite sign of the matter potential, antineutrinos do not undergo any resonance in the matter of the star (negative density semi-plane in Fig. 1).

As an effect of conversion, the ν_e and $\bar\nu_e$ fluxes in the detector, F_e and $F_{\bar e}$, are combinations of the original ν_e and ν_x ($\bar\nu_e$ and $\bar\nu_x$) fluxes. Considering for simplicity the case of no Earth crossing, one gets:

$$F_e = P_H \sin^2 \theta_{12} F_e^0 + (1 - P_H \sin^2 \theta_{12}) F_x^0 \ ,$$
$$F_{\bar e} = \cos^2 \theta_{12} F_{\bar e}^0 + \sin^2 \theta_{12} F_{\bar x}^0 \ . \tag{6}$$

For inverted hierarchy ($\Delta m_{32}^2 < 0$), the H resonance is in the antineutrino channel, while the L resonance is unaffected. In this case the fluxes in the detector equal:

$$F_e = \sin^2 \theta_{12} F_e^0 + \cos^2 \theta_{12} F_x^0 \ ,$$
$$F_{\bar e} = P_H \cos^2 \theta_{12} F_{\bar e}^0 + (1 - P_H \cos^2 \theta_{12}) F_{\bar x}^0 \ . \tag{7}$$

As expected, here the jump probability P_H appears in the expression of the

$\bar{\nu}_e$ flux, in contrast with Eqs. (6).

In summary, the supernova neutrino signal is sensitive to the mass hierarchy and to θ_{13} for the following reasons: (i) depending on the hierarchy, the H resonance affects either neutrinos or antineutrinos; (ii) the observed ν_e or $\bar{\nu}_e$ fluxes depend on the value of θ_{13} via the jump probability P_H. The latter can be calculated using the Landau-Zener formula and the profile (5). The result is:

$$P_H = \exp\left[-\left(\frac{1.08 \cdot 10^7 \text{ MeV}}{E}\right)^{2/3}\left(\frac{\Delta m_{32}^2}{10^{-3} \text{ eV}^2}\right)^{2/3} C^{1/3} \sin^2\theta_{13}\right] . (8)$$

It follows that three regions exist:
(i) Adiabaticity breaking region: $\sin^2\theta_{13} \lesssim 10^{-6} (E/10\text{MeV})^{2/3}$, where $P_H \simeq 1$;
(ii) Transition region: $\sin^2\theta_{13} \sim (10^{-6} - 10^{-4}) \cdot (E/10\text{MeV})^{2/3}$, where $0 \lesssim P_H \lesssim 1$;
(iii) Adiabatic region: $\sin^2\theta_{13} \gtrsim 10^{-4} (E/10\text{MeV})^{2/3}$, where $P_H \simeq 0$.

Notice that if $P_H = 1$ (adiabaticity breaking region) Eqs. (6) and (7) coincide. Thus, we get equal predictions for normal and inverted hierarchy and any sensitivity to the mass hierarchy is lost. Furthermore, from Eqs. (6) and (7) it is easy to see that, in the extreme case in which the original fluxes in the different flavors are equal ($F_{\bar{e}}^0 = F_{\bar{x}}^0$, $F_e^0 = F_x^0$), conversion effects cancel and one has $F_e = F_e^0$, $F_{\bar{e}} = F_{\bar{e}}^0$.

4. Earth matter effects

If the neutrinos cross the Earth before detection, they undergo regeneration effects due to the interaction with the matter of the Earth. Indeed, the matter density in the Earth, $\rho \sim (1 - 13) \text{ g} \cdot \text{cm}^{-3}$, is close to the density at the L resonance in the star (see Sec. 3), for which the mixing angle θ_{12} is resonantly enhanced. This implies that the amplitude of neutrino oscillations in the Earth can be significant and lead to observable effects.

The results (6) and (7) can be immediately generalized by the replacements:

$$\sin^2\theta_{12} \to P_{2e} \qquad\qquad \cos^2\theta_{12} \to P_{1\bar{e}} , \qquad\qquad (9)$$

where P_{2e} ($P_{1\bar{e}}$) is the probability that a neutrino (antineutrino) arriving at Earth in the state ν_2 ($\bar{\nu}_1$) is detected as ν_e ($\bar{\nu}_e$) in the detector. It depends on the oscillation parameters θ_{12} and Δm_{21}^2, on the Earth density profile, on the neutrino arrival direction and on the neutrino energy. The dependence on the energy has an oscillatory character, resulting in characteristic distortions of the energy spectra of observed events. The distortions are different for detectors at different locations on the planet.

As an example, consider two detectors, D1 and D2, with D2 shielded by the Earth and D1 unshielded. The differences of the ν_e and $\bar{\nu}_e$ fluxes in the two detectors follow from Eqs. (6), (7) and (9). For n.h. they are:

$$F_e^{D2} - F_e^{D1} = P_H(P_{2e} - \sin^2\theta_{12})(F_e^0 - F_x^0) \ ,$$
$$F_{\bar{e}}^{D2} - F_{\bar{e}}^{D1} = (P_{1\bar{e}} - \cos^2\theta_{12})(F_{\bar{e}}^0 - F_{\bar{x}}^0) \ , \tag{10}$$

while for i.h. one gets:

$$F_e^{D2} - F_e^{D1} = (P_{2e} - \sin^2\theta_{12})(F_e^0 - F_x^0) \ ,$$
$$F_{\bar{e}}^{D2} - F_{\bar{e}}^{D1} = P_H(P_{1\bar{e}} - \cos^2\theta_{12})(F_{\bar{e}}^0 - F_{\bar{x}}^0) \ . \tag{11}$$

The dependence of these differences on the neutrino mass hierarchy and on θ_{13} has the same origin as that of conversion effects in the star and can be described in analogous terms (Sec. 3).

Figure 2 gives an illustration of the energy spectra of events predicted from Eq. (10) for two identical water Cerenkov detectors.

5. Probing θ_{13} and the mass hierarchy

There are several approaches to probe the neutrino oscillation parameters and at the same time take into account the uncertainties on the features of the original fluxes:

1. to perform a global fit of the data, determining both the oscillation parameters and the parameters of the original fluxes simultaneously [9,10]. With this method a completely general analysis is not possible due to the large number of parameters involved.

2. to single out and study (numerically and analytically) specific observables which (i) have maximal sensitivity to the oscillation parameters of interest and (ii) whose dependence on the astrophysical uncertainties is minimal or well understood [11].

3. to study Earth matter effects [7,12,16].

4. to study the effects of shock-wave propagation on the neutrino signal. The shockwave driving the supernova explosion modifies the density profile of the star at the resonance points, thus changing the conversion pattern inside the star and the observed neutrino energy spectra [17,18,11,19].

Here I summarize some aspects of the method 3. [16].

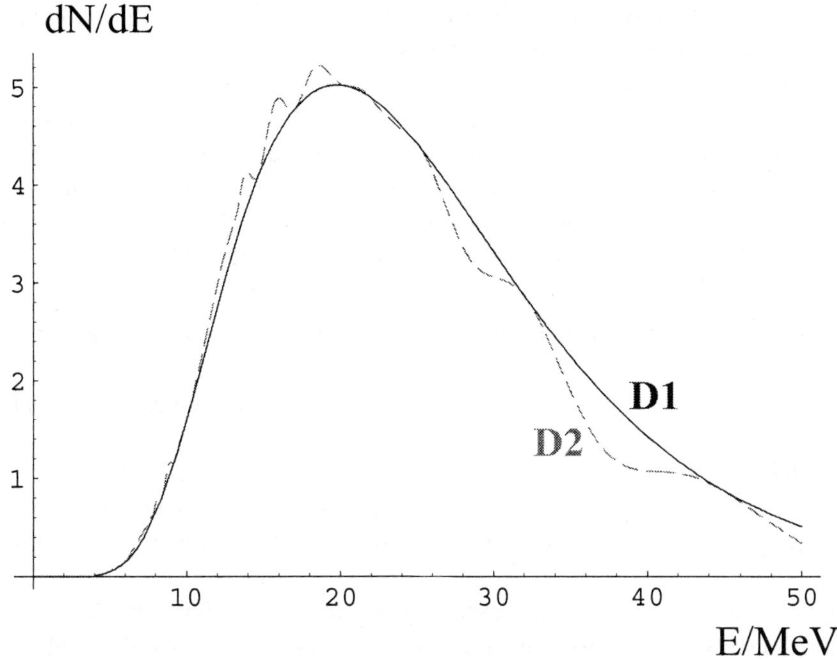

Figure 2. Predicted energy spectra of positrons (arbitrary units on the vertical axis) from the $\bar{\nu}_e + p \to n + e^+$ reaction at two identical water Cerenkov detectors, one of which (D2) is shielded by the Earth with nadir angle $\theta_n = 70°$. I used the detection efficiency of the SuperKamiokande detector and the parameters: $T_{\bar{e}} = 5$ MeV, $T_{\bar{x}} = 7$ MeV, normal hierarchy (or i.h. with $P_H = 1$) and $\Delta m_{21}^2 = 7 \cdot 10^{-5}$ eV^2. Equal luminosities have been taken in $\bar{\nu}_e$ and $\bar{\nu}_x$ and the detailed Earth density profile[8] has been used.

5.1. *Analyzing the Earth matter effects*

The study of Earth regeneration effects is particularly promising. The reason is that the information on θ_{13} and on the mass hierarchy which can be obtained by this method is largely independent of astrophysical uncertainties.

This can be understood considering that:

- The main signature of Earth matter effects – consisting in oscillatory modulations of the observed energy spectra (see Sec. 4) – is unambiguous since it can not be mimicked by any astrophysical phenomenon. Moreover the pattern of oscillation minima and maxima depends only on Δm_{21}^2 and θ_{12}, which probably will be known precisely from solar neutrino experiments and KamLand before the next galactic supernova event. This will improve the possibility of identification of Earth matter effects.

- To obtain unambiguous conclusions on θ_{13} and on the mass hierarchy it is enough that experiments exclude or establish the Earth matter effect, without measuring its size precisely, especially if the effect is probed in both the ν_e and $\bar{\nu}_e$ channels. The fact that precision is not necessary is very important in this specific problem, where large astrophysical uncertainties are present.

To illustrate the latter point in more detail consider the following scenario.

(i) At the time of arrival of the neutrino burst, at least one running detector is shielded by the Earth.

(ii) The shielded detectors record data due to both ν_e and $\bar{\nu}_e$. The two (ν_e and $\bar{\nu}_e$) sets of events can be efficiently separated and for each of them the energy spectrum of the incoming neutrinos can be reconstructed with good resolution.

(iii) The statistics of both the data sets are sufficiently large so that (oscillatory) spectral distortions as large as $\sim 10 - 20\%$ can be established with high statistical significance.

If these conditions are fulfilled, we have four possible experimental results. The corresponding conclusions on θ_{13} and on the neutrino mass hierarchy are discussed below. They easily follow from Eqs. (10) and (11). For simplicity two possibilities are considered in the discussion. The first is the case in which substantial differences in the fluxes of different flavors are assumed, on the basis of (future) precise theoretical predictions. In the second case astrophysical uncertainties are large and allow equality of the original fluxes. The generalization of the discussion to intermediate cases is straightforward.

1. <u>The Earth effects are established in both ν_e and $\bar{\nu}_e$ channel.</u> In this case *unique* conclusions are obtained on the oscillation parameters and on the original fluxes at the same time. A first result is the difference of the original fluxes in the different flavors: $F_e^0 \neq F_x^0$ and $F_{\bar{e}}^0 \neq F_{\bar{x}}^0$. This would be an important test of calculations of neutrino spectra formation inside the star. Secondly, we get that P_H is significantly different from zero, $P_H \sim 1$. This gives the upper bound $\sin^2 \theta_{13} \lesssim 10^{-6}$ (see Sec. 3). The mass hierarchy remains undetermined.

2. <u>The Earth effect is seen in the ν_e channel only</u>, and excluded in the

$\bar{\nu}_e$ channel. This tells us that $F_e^0 \neq F_x^0$. If $F_{\bar{e}}^0 \neq F_{\bar{x}}^0$ is assumed, again the conclusion is unambiguous: the mass hierarchy is inverted and $P_H \sim 0$, corresponding to $\sin^2 \theta_{13} \gtrsim 10^{-4}$. In absence of *a priori* assumptions on the fluxes, we must consider that the equality $F_{\bar{e}}^0 \simeq F_{\bar{x}}^0$ could suppress the Earth matter effect on antineutrinos, allowing other scenarios of hierarchy and θ_{13}. Nevertheless, the case of normal hierarchy with $P_H \sim 0$ remains excluded.

3. <u>The Earth effect is established in the $\bar{\nu}_e$ channel only</u>, while excluded in the ν_e channel. Similarly to the previous case, here we conclude that $F_{\bar{e}}^0 \neq F_{\bar{x}}^0$. Assuming that the neutrino original fluxes are different, we have that the normal hierarchy is singled out and $P_H \sim 0$ ($\sin^2 \theta_{13} \gtrsim 10^{-4}$). The possibility that $F_e^0 \simeq F_x^0$ appears exotic, since both numerical calculations and simple physical considerations predict that the difference $(F_x^0 - F_e^0)$ should be larger than $(F_{\bar{x}}^0 - F_{\bar{e}}^0)$. If the case $F_e^0 \simeq F_x^0$ is allowed, still the scenario of inverted hierarchy with $P_H \sim 0$ is excluded.

4. <u>No Earth effect is seen in both ν_e and $\bar{\nu}_e$ channel</u>. As can be easily realized, this result requires that at least in one channel (ν_e or $\bar{\nu}_e$) oscillations are suppressed by equality of original fluxes: $F_e^0 \simeq F_x^0$ or $F_{\bar{e}}^0 \simeq F_{\bar{x}}^0$. This would be interesting as a test of predictions of the neutrino fluxes and energy spectra. No definite conclusions on oscillation parameters are possible.

The present discussion serves as illustration; it can be generalized to different experimental setups, for which we refer to the literature [16].

Acknowledgments

I would like to thank the organizers of the workshop "Open Issues in Understanding Core Collapse Supernovae" for their invitation and for financial support. I am grateful to them and to the other participants for the stimulating atmosphere at the workshop.

References

1. See e.g. the review by C. Giunti, hep-ph/0305139, for a summary and references.
2. See e.g. the analysis and references in G. L. Fogli, E. Lisi, A. Marrone and D. Montanino, Phys. Rev. D **67**, 093006 (2003), hep-ph/0303064.
3. **CHOOZ** Collaboration, M. Apollonio *et. al.*, Phys. Lett. **B466** (1999) 415–430.
4. F. Boehm *et. al.*, Phys. Rev. **D62** (2000) 072002.

5. See e.g. M. T. Keil, G. G. Raffelt, and H.-T. Janka, astro-ph/0208035, and references therein.

6. G. E. Brown, H. A. Bethe and G. Baym, Nucl. Phys. A **375** (1982) 481.

7. A. S. Dighe and A. Y. Smirnov, Phys. Rev. **D62** (2000) 033007.

8. A. M. Dzewonski and D. L. Anderson, Phys. Earth. Planet. Inter. **25** (1981) 297.

9. V. Barger, D. Marfatia, and B. P. Wood, Phys. Lett. **B547** (2002) 37–42.

10. H. Minakata, H. Nunokawa, R. Tomas, and J. W. F. Valle, Phys. Lett. **B542** (2002) 239–244.

11. C. Lunardini and A. Y. Smirnov, hep-ph/0302033, to appear in JCAP.

12. C. Lunardini and A. Y. Smirnov, Nucl. Phys. B **616**, 307 (2001), hep-ph/0106149.

13. K. Takahashi and K. Sato, Phys. Rev. D **66**, 033006 (2002), hep-ph/0110105.

14. A. S. Dighe, M. T. Keil and G. G. Raffelt, hep-ph/0303210.

15. A. S. Dighe, M. T. Keil and G. G. Raffelt, hep-ph/0304150.

16. The present discussion is based on the paper by C. Lunardini and A. Y. Smirnov, in preparation.

17. R. C. Schirato, G. M. Fuller, astro-ph/0205390.

18. K. Takahashi, K. Sato, H. E. Dalhed and J. R. Wilson, astro-ph/0212195.

19. G. L. Fogli, E. Lisi, D. Montanino and A. Mirizzi, hep-ph/0304056.

20. The expected supernova neutrino signal is discussed in O. L. Peres and A. Y. Smirnov, Nucl. Phys. B **599**, 3 (2001), hep-ph/0011054.

21. See e.g. C. Waltham, proceedings from the International Conference on Cosmic Rays (ICRC) 2001, available at www.copernicus.org/icrc/papers/ici7144_p.pdf .

22. D. B. Cline, F. Sergiampietri, J. G. Learned and K. McDonald, Nucl. Instrum. Meth. A **503**, 136 (2003), astro-ph/0105442.

23. A. Bueno, I. Gil-Botella and A. Rubbia, hep-ph/0307222.

24. J. F. Beacom and M. R. Vagins, in preparation; see e.g. the talk by M. R. Vagins at the NOON 2003 conference, Kanazawa, Japan, February 4th -10th, available at http://www-sk.icrr.u-tokyo.ac.jp/noon2003/.

25. S. R. Elliott, Phys. Rev. C **62**, 065802 (2000), astro-ph/0006041.

26. J. Engel, G. C. McLaughlin and C. Volpe, Phys. Rev. D **67**, 013005 (2003), hep-ph/0209267.

27. C. J. Horowitz, K. J. Coakley and D. N. McKinsey, astro-ph/0302071.